Plasma Physics by Laser and Applications (PPLA 2017)

EPJ Web of Conferences Volume 167 (2018)

Messina, Italy
5 - 7 July 2017

Editors:

L. Torrisi
M. Cutroneo

ISBN: 978-1-5108-5573-1

Printed from e-media with permission by:

Curran Associates, Inc.
57 Morehouse Lane
Red Hook, NY 12571

Some format issues inherent in the e-media version may also appear in this print version.

Printed by Curran Associates, Inc. (2018)

For additional information, please contact EDP Sciences – Web of Conferences at the address below.

EDP Sciences – Web of Conferences
17, Avenue du Hoggar
Parc d'Activité de Courtabœuf
BP 112
F-91944 Les Ulis Cedex A
France

Phone: +33 (0) 1 69 18 75 75
Fax: +33 (0) 1 69 28 84 91

contact-edps@webofconferences.org

Additional copies of this publication are available from:

Curran Associates, Inc.
57 Morehouse Lane
Red Hook, NY 12571 USA
Phone: 845-758-0400
Fax: 845-758-2633
Email: curran@proceedings.com
Web: www.proceedings.com

TABLE OF CONTENTS

PPLA is a biannual appointment for the international research community interested in the field of the laser generated plasmas. This meeting is focused on the fundamental interaction between long/short pulsed lasers and solids, liquids, gases and the plenty of their induced effects and applications as well as target preparation and characterization, diagnostics technologies.

PPLA has a long tradition, this year it reached the 8th edition and it took place exactly where it was born in 2003: Messina. The past successful editions were in Giardini Naxos (2005), Scilla (2007), Messina (2009), Catania (2011), Lecce (2013); Frascati (2016).

In 2013 in Lecce was established the first edition of the Leos Laska prize for a young researcher, dedicated to the memory of our colleague Leos Laska, prematurely passed away in the same year. This year, several successful side-events have been introduced: the tutorial session held by prof. D. Neely (UK) prof. I. Foldes (HU), prof. D. Giulietti (IT) one day before the start of the conference; the participation of Prof. L. Roso, director of the Center for Pulsed Lasers, Salamanca, Spain; the participation of Prof. S. D. Dabagov, representing Channeling conference to sow the seeds for the synergy between PPLA and Channeling international conferences.

PREFACE

Plasma Physics by Laser and Applications (PPLA) is an international conference taking place every two years and involving the scientific community interested in plasmas generated by laser. After 7 editions, it was an honor for Messina to be host of the last prestigious event held in 2017. The organization of the PPLA 2017 was carried out by the organizing committee (OC) consisting of : Prof. L. Torrisi,-chairman, Dr. R. De Angelis, Prof. D. Giulietti, Prof, V. Nassisi and Prof. F. Neri. The conference was co-financed by the University of Messina, Center of Accelerators and Nuclear Analytical Methods (CANAM), Agenzia nazionale per le nuove tecnologie, l'energia e lo sviluppo economico sostenibile (ENEA), Italian National Institute for Nuclear Physics (INFN), the Physics Department of the Pisa University and the Mathematics and Physics Department of the Lecce University.

University of Messina has supported the conference by the local Project "Research and Mobility" coordinated by Prof. L. Torrisi. The Sicily Region has supported the conference ceremony expenses.

The scientific PPLA program covers topics concerning the laser-matter interaction, plasma physics, theoretical models relative to laser plasmas, laser induced photons, electron and ion sources, plasma diagnostics and detectors, applications and technologies inherent the plasma use.

The program was developed in three effective days and consisted of 7 invited speakers, 22 oral and 50 poster contributions. Detailed information about the conference program and venue has been available on the webpage:

http://www.ppla2017.it/ or contacting by email the secretary: secretary@ppla2017.it

All invited speakers presented the frontiers of laser matter interaction research. Their important contributions were : "Escaping Electrons from Intense Laser-Solid Interactions as a Function of Laser Spot Size", by Prof. D. Neeley [a)], "Nanoimaging using soft X-ray and EUV laser-plasma sources", by Dr. P. Wachulak [b)], "High repetition rate Petawatt lasers" by Prof. L. Roso[c)], "Toward ultra-intense plasma based soft X-ray lasers" by Prof. S. Sebban[d)], "Plasma Mirrors for Cleaning Laser Pulses from the Infrared to the Ultraviolet" by Prof. I. B. Foldes[e)], "Aneutronic p-B Reaction in Magnetized Inertial Confined Plasma of Laser-Accelerated Particles", by Prof. S. Guskov[f)] , "High intensity laser interactions with optically levitated microtargets" by Prof. R. A. Smith[g)].

A **tutorial session** on various topics related to laser processing, plasma parameters, ions acceleration was organized. It consisted of three lectures given by three of the most important experts in Europe in this field. Prof. David Neely presented a lecture titled: "Emerging laser driven application source developments"; Prof. István B. Foldes: "High-harmonics and attosecond pulse generation from plasmas on solid surfaces"; Prof. Danilo Giulietti: "Laser Absorption and Electromagnetic Radiation in Laser Produced Plasmas".

Prof. D. Neely[a)], Prof. I.B. Foldes[b)], Prof. D. Giulietti[c)]

The tutorial session, was especially addressed to the numerous young participants, undergraduates and PhD students, who participated enthusiastically to the conference, to introduce them to the basis of the laser-matter/plasma interaction phenomena.

Prof. S. B. Dabagov, the chair of one of the most successful conferences on "Charged & Neutral Particles Channeling Phenomena" co-financed by the INFN and the European Organization for Nuclear Research (CERN), presented a report on the channeling radiation, X-rays polycapillary optics and channeling of neutral and charged particles. Prof. Dabagov emphasized the synergy between CHANNELING and PPLA conferences hoping that it can be further increased by organizing together training seminars for students and scientists on the common interest themes. The first appointment will be scheduled at the next CHANNELING-2018 conference.

Prof. S. B. Dabagov

PPLA has been characterized by an increasing international participation of scientists. Roughly 100 researchers coming mostly from Europe, but also from other continents.

Some participating to PPLA 2017: Prof. S. Cuzzocrea on behalf of Rector of the University of Messina [a)], Dr. M. Okamura[b)] (Brookhaven National Lab., USA), Prof. , A. Kozyrev[c)] (Institute of High Current Energy, RU) Prof. S. Gocheva-Ilieva [d)](Technical Univ., BG), Prof. Aix Feng[e)](Wenzhou Univ.,CN), Dr. M. Rosinski[f)] (IPPLM, PL), Prof. W. Nazarov[g)] (Dundee University, UK) Prof. P. Ossi[h)](Milan Univ., IT).
The quality and variety of the contributions presented, ranging from particle acceleration to nuclear fusion, material science, biology and medicine demonstrated once again the central role played by laser plasma physics in modern science.

The **Leos Laska prize** was awarded to the best contribution by a young researcher. The prize is dedicated to the memory of Dr. Leos Laska from Academy of Science of the Czech Republic, prematurely passed away in 2013. The prize was delivered by Prof. L. Torrisi to Mr. G. Ceccio, PhD at the University of Messina (IT).

Prof. L. Torrisi, chairman PPLA 2017 and Mr. G. Ceccio winner of the Leos Laska prize 2017

The Local Organizing Committee (LOC)

The conference offered the participants several opportunities to interact friendly while enjoying the visit of Messina town, Taormina and Forza D'Agrò, among the most celebrated attractions of the Sicily Region .

Participants to the PPLA conference 2017

A special thanks to the sponsors: ARS, OSA, Sear, centro studi Diodoro and citta' Metropolitana Messina as well as to the PhD students in Physics and to the students of Physics Master's degree, and to all the technicians and administrative staff of the MIFT Physical Sciences Department of the University of Messina.

The next PPLA edition will take place in Pisa (Italy), hosted by the Physics Department of the University.

A large part of the presented contributions will be included in a dedicated issue of the European Physical Journal, after peer review.

Guest Editors

Prof. L. Torrisi

Dr. M. Cutroneo

edp sciences

EPJ Web of Conferences

Statement of Peer review

In submitting conference proceedings to EPJ Web of Conferences, I certify to the Publisher that I adhere to the **Policy on Publishing Integrity** of the journal in order to safeguard good scientific practice in publishing.

1. All articles have been subjected to peer review administered by the proceedings editors.
2. Reviews have been conducted by expert referees, who have been requested to provide unbiased and constructive comments aimed, whenever possible, at improving the work.
3. Proceedings editors have taken all reasonable steps to ensure the quality of the materials they publish and their decision to accept or reject a paper for publication has been based only on the merits of the work and the relevance to the journal.

Title, date and place of the conference

PLASMA PHYSICS BY LASER AND APPLICATIONS (PPLA 2017)

5-7 JULY 2017, MESSINA (ITALY)

Proceedings editor(s):

PROF. LORENZO TORRISI

DR. MARIAPOMPEA CUTRONEO

Date and editor's signature

20.12.2017

20.12.2017

17, avenue du Hoggar - PA de Courtabœuf - BP 112 - 91944 Les Ulis Cedex A (France)
Tél. : 33 (0)1 69 18 75 75 - Fax : 33(0)1 69 07 45 17 – www.edpsciences.org

EPJ Web of Conferences **167**, 01001 (2018) https://doi.org/10.1051/epjconf/201816701001
PPLA 2017

High repetition rate Petawatt lasers

Luis Roso*

Center for Pulsed Lasers, CLPU, Villamayor, Salamanca, Spain.

Abstract. Petawatt lasers are now available in a number of facilities around the world and are becoming a very useful tool in physics and engineering. Some of such lasers are able -or will be able soon- to fire at high repetition rates (one shot per second or more). Experiments at such repetition rates have certain peculiarities that are to be briefly exposed here, based on the author's experience with the Salamanca VEGA-3 laser. VEGA-3 is a 30 fs PW laser, firing one shot per second.

1 Introduction

Laser acceleration of electrons and protons is now a reality at low repetition rate. However there are many lasers troughout the world beyond 100 TW and with repetition rates around one shot per second, or more. High repetition rate lasers are now under consideration, kHz at the 10 TW level and 100 Hz at the 100 TW level and 10 Hz at the PW level. Those systems represent a major technological revolution so far. A lot of problems have been solved or are under solution regarding those systems, as cryo-cooling, large crystal growing, large diffraction gratings construction, transverse lasing, big femtosecond broadband mirrors, pumping lasers, and many more. However one point is underdeveloped so far: targetry (with associated shot to shot diagnostic). This is going to be analyzed in the present contribution.

Table 1. Basic parameters of the VEGA laser system installed at CLPU. It is a Ti:Sapphire laser emitting around 800 nm. Pulse duration is 30 fs in all three lines.

	Energy per shot	Peak Power	Repetition rate
VEGA 1	0.6 J	20 TW	10 Hz
VEGA 2	6 J	200 TW	10 Hz
VEGA 3	30 J	1 PW	1 Hz

2 Present day technology

Recent advances in Chirped Pulse Amplification (CPA) allowed dramatic improvements of the laser peak power. CPA was developed in mid eighties at the University of Rochester. After the initial paper of Donna Strickland and Gerard Mourou [1], the technique was rapidly developed [2]. CPA consists in lengthening the duration of the pulse being amplified in a reversible fashion. The reversibility, or ability to recompress the pulse after amplification relies on the technique of optical pulse compression developed by Treacy [3] and Martínez [4]. By lengthening the pulse in time, energy can be efficiently extracted from the laser gain medium, avoiding at the same time damage to the laser amplifier components (typically Ti:Sapphire or other doped crystals). CPA is particularly useful for efficient utilization of solid-state laser media with high stored energy density (1 J/cm2), where full energy extraction in a short pulse would lead to intensities above the damage threshold of the amplifier materials.

During more than three decades CPA has evolved and is now a developed technology with many different directions of evolution. The frontier land now is the 1 to 10 PW peak power range. It is well know that there are several possibilities to arrive to the PW peak power. One Megajoule shots with duration of one nanosecond correspond to one PW, however this limit has not been achieved so far. There are two stadium-size lasers in the world, the National Ignition Facility [5] at Livermore, California, USA, and the Laser MegaJoule [6] at Bordeaux, France. Both arrive to the Megajoule per shot but combing incoherently many beams.

One kilojoule shot with duration of one picosecond corresponds to one PW. This limit has been achieved so far. One very representative example is the PETAL laser (adjacent to the French Laser MegaJoule). PETAL was the first kilojoule-class laser to arrive to 1.2 PW peak power. Of course these lasers work at single shot regime, and they need hours from shot to shot.

The intensity limit corresponds to PW's of few femtoseconds. With conventional CPA technology the limit -at those amplification limits is 23-30 fs. 30 J /30 fs correspond to one PW. With OPCPA (Optical Parametric CPA) it seems feasible to arrive to 15 fs outpout, and 15 J/15 fs correspond to one PW also. In this contribution we will refer to those systems as intensity PW (we call them high intensity PW's because the wavefront quality is much better that for the kilojule

* Corresponding author: roso@clpu.es

or Megajoule lasers and then it is possible to focus them to extreme intensities. The possibility to go to much shorter durations (5 J/ 5 fs) at the PW peak power level is under consideration. Sub fs PW are not feasible for the moment.

Today, arriving to the PW level is more or less the state of the art of the reliable technology. Several projects to arrive to the next step -10 PW- are well advanced particularly for intensity lasers. At the time of writing the two leading systems are at the Shanghai Superintense-Ultrafast Lasers Facility, SULF, Shanghai [7] and at the Center for Relativistic Laser Science, Institute for Basic Science, Gwangju, South Korea [8] both close to 5 PW, while 10 PW systems are under construction in the Extreme Laser Infrastructure ELI,

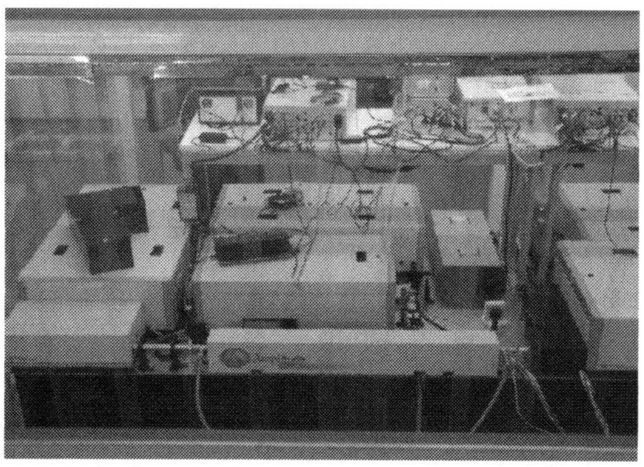

Fig. 1. General view of the VEGA laser front end, with the XPW system in front. XPW helps to get a high contrast of the pulse, something very necessary for solid targets.

Fig. 2. VEGA-3 (the PW laser) compressor installed at Salamanca. The four gratings are at vacuum (10^{-6} mbar) and their vibrational stability is guaranteed by an optical table bound directly to the ground trough legs and bellows that are decoupled from the main vacuum vessel.

3 High repetition rates

One of the advantages of the development of the CPA techniques is the possibility of high repetition rates. The first Petawatt lasers were single shot systems with a long time for cooling between shot and shot. Today if we want a laser accelerator able to be compared to conventional accelerators, we need to work to increase the repetition rate.

Repetition rate at Megajoule or kilojoule levels is not under consideration because the huge pumping lasers they have, that require very long cooling times (from several minutes to hours). For the intensity PW's the situation is much different.

Fig. 3. Beam transport from the PW compressor to the experimental station. The beam diameter of VEGA-3 is 24 cm flat-top. The tubes are almost 50 cm in diameter

High repetition rate implies certain technological problems in the cooling of the different components. Among those problems, we must mention laser pumping, amplifiers cooling and damage of the optical elements. Price is also a major problem, but we are not going to comment on that.

The most relevant problem is the pumping system. For example we can comment on the Ti:Sapphire laser, the most common one in those lasers. Ti:Sapphire is a vibronic laser that has to be pumped by a blue-green laser with pulses around 10 ns (Q-switched). The most common pump is Nd:glass or Nd:solid. Nd:glass is the best and cheapest option for low repetition rate. But the duty cycle of the glass cooling prevents going beyond a few shots per minute. Nd:solid, typically Nd:YAG, is a good option for systems in the 0.1-100 Hz regime. However the problem in turn is the system to provide energy to the laser used to pump Ti:Sapphire. For high energies (beyond Joule) Nd:YAG is nowadays pumped by flash lamps and those lamps have a practical limit at 10 Hz. The future is diode lasers. Diode pumped Nd:YAG are the best solution now below 10 TW and will become the main future option as the diode price drops. Diode pump opens the door to Nd:YAG lasers at 100 Hz or even at kHz (although in this case the best option is Nd:YLF) and soon will allow a generation of extreme high repetition rate lasers [9].

Another possibility that is becoming real is the direct diode pumping of Ti:Sapphire without the need to use Nd:YAG doubled in frequency. In this case blue diodes are necessary (this is not a problem now) but also 10 ns pulse durations (and this is the main problem so far) [10].

Another very relevant problem is the cooling of the amplifiers, particularly the last amplifiers in the chain. The quantum efficiency of the Ti:Sapphire laser is not good. We pump it with 532 nm light and we get laser light around 800 nm, this means that approximately one third of the pumping energy is lost. But not all pumping photons are used to generate laser photons, a percent of them is also lost. As a result more than 40 percent of the energy ends heating the amplifier crystal. Sapphire is a crystal very resistant mechanically and thermally, so the heating does not represent a structural problem. However a small change in the temperature ends in a local change of the refractive index, and therefore the energy lost heats locally the crystal and changes the light propagation through it. This is a major problem that can ruin the lasing capacity of any multi-pass amplifier. The way to solve this is to cool down the amplifier crystal. The cooling helps to the temperature homogeneity and moreover helps to distribute the dissipated heat throughout all the crystal because the thermal conductivity of materials as Ti:Sapphire increases as the temperature goes down.

A third important problem relevant enough to be mentioned here is the heat dissipation at the gratings and third quick damage in those cases. Damage threshold of today's gratings requires fluences not greater that 100 mJ/cm2 (for 30 fs pulse durations). For example this implies that the beam size of the VEGA PW is 24 cm flat top (that corresponds to 450 cm2, with an uncompressed beam around 45 J and a compressed beam around 30 J). The diameter of the beam transport can be seen in Fig. 3. Reducing the beams size would require gratings more resistant to damage. Of course this is more relevant as the repetition rate increases. If gratings were to close to the damage thresholds they could be specified for a few ten thousand shots. At one shot per hour this can be reasonable. With a shot per second this imply an impossible replacement of the gratings every day!

With all this in mind, a multi-Hz multi-PW system seems quite possible now. At the moment of writing there are only three lasers at the PW peak power level and at the Hz repetition rate operative in the World. The first in being operative was the BELLA laser system [11] at Lawrence Berkeley National Laboratory in California USA. The second one is the DRACO laser system [12] laser system at Dresden, Germany, and the third one is the VEGA laser system [13] at Salamanca, Spain. There are 10 Hz PW and multi PW lasers under construction or under design in several places. Among them, particularly advanced are the three pillars of ELI [14].

The characteristics of the Salamanca VEGA laser can be found in Table 1. It is a Ti:Sapphire laser working at a central wavelength close to 800 nm and with pulse durations of 30 fs. VEGA has a common front end (see Fig. 1) and three lines, VEGA-1 arriving to 20 TW at 10 Hz, VEGA-2 arriving at 200 TW at 10 Hz, and the PW, VEGA-3, at only one Hz, i.e. at one shot per second. The compressor for VEGA-3 is shown in Fig. 2, it is a typical four grating compressor arranged in two levels.

At low powers (GW) kHz repetition rates are easy to obtain and moreover one can work in air in several cases.

For example Fig. 4 shows a set-up developed at Salamanca to generate soft X-rays (tens of keV) with a 120 fs Ti:Sapphire laser delivering pulses of only a few milijoules per shot. In that case we focused the laser on a thick copper plate, as indicated in Fig. 4, using a microscope objective (and therefore a very wide numerical aperture). Ionization releases electrons that move back and forth driven by the laser electric field oscillation and that when colliding back to the copper plate generate Bremsstrahlung X-rays.

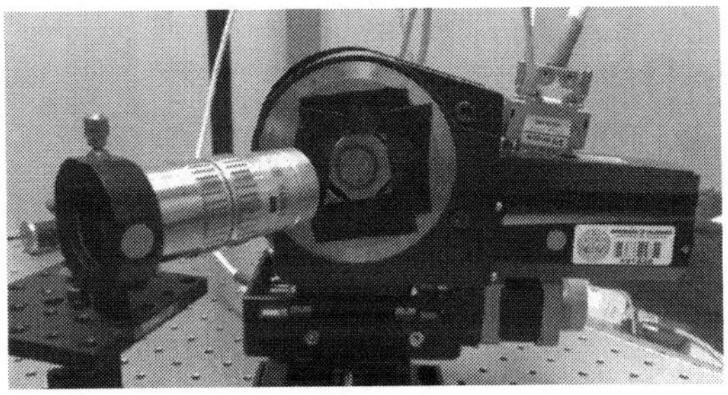

Fig. 4. CLPU soft X-ray system using a multi-GW laser and working in air. The system runs at kHz repetition rate. Unfortunately this easy geometry is not efficient at higher intensities.

4 Targetry

Assuming that there are a few 1 PW / 1 Hz lasers ready to fire for long times, then is the problem to find targets for them. And this is a major bottleneck. Of course, the targets depend on the experiments we plan to perform with the laser. The most common uses of the high intensity lasers are acceleration (electrons, protons or ions), warm dense plasmas, and betatron radiation. In some case the accelerated beams or the generated radiation are used in turn as secondary sources (for example, neutrons obtained from accelerated protons).

For electron acceleration studies, the problem is a bit easier. Underdense (underdense relative to the 1.5 eV photons of the near infrared lasers as Ti:Sapphire) gas targets are possible and so the replacement of the gas is straightforward. However for Laser Wakefield Acceleration (LWFA) and related techniques it is necessary to reach certain density of the gas. A pulsed valve is a good choice at low repetition rate but problematic beyond 0.1 Hz. Moreover shooting so quickly implies a large amount of gas introduced in the chamber and this has to be carefully engineered. Fortunately, typically electron experiments use long focal lengths, long Rayleigh lengths and so are not so exigent with the longitudinal position of the target. Some acceleration approaches beyond the GeV require hollow fibers and the transverse alignment is critical.

Proton and ion acceleration have slightly different requirements. For TNSA (Target Normal Sheath

Acceleration) studies thin targets with overcritical or at least near critical densities are necessary. Moreover TNSA is not possible with thick targets. The mechanism requires at least targets of 1 to 10 micrometer thick and works even better with submicron thick targets. Preparing one thin target is not complicated, but this represents one shot. The next approach is to prepare a set of targets (a plate with a number of targets), a carousel or other techniques. Each laser shot hits (and destroys) one of the targets. Between shot and shot the target holder moves and positions the next one (and this implies typically recheck of the focus position). All that is easy at single shot, possible at few shots per minute, but impossible at Hz or beyond. So it is necessary to find other strategies.

As a result of this targetry is a very active field now inside the high intensity laser community. Some approaches are very interesting, as the idea developed by CEA Grenoble of continuous production of a thin ribbon of solid hydrogen [15], obviously at very high pressure and extremely low temperature. However several laser teams would like to study this interaction with pure targets. In this context, the low temperature laboratory, at CEA-Grenoble has developed a cryostat able to continuously produce a thin hydrogen ribbon (50 and 100 micrometer thick). A new extrusion concept, without any moving part has been carried out, using only the thermodynamic properties of the fluid. This implies state of the art techniques in cryogenic cooling. Besides the use of pure hydrogen, that guarantees the existence of protons in the plasma, the use of pure hydrogen has other advantages and peculiarities: it is a pure atomic species, so there is no influence of other ions; the plasma densities are due to the electronic densities (electrons per cubic centimeter) and in this case the density of this solid hydrogen plus the fact that each atom has only one electron constitute an experimental scenario a bit different than the one with Al foils (where protons come from adsorbed hydrogen atoms and water molecules) and hydrocarbon plastic sheet products made from the resin Polyethylene Terephthalate (Mylar) where besides water the same molecule has hydrogen.

The scenario for such targets is more complicated with recent studies that show that proton/ion acceleration is more efficient with structured targets [16]. Mass production of those targets is possible if they provide to be efficient enough, but it is going to be expensive and time consuming.

Solid targets are not the only possibility to work with plasma densities critical or near-critical. One alternative is to work with high-pressure gases and with specifically designed nozzles that allow a high gas density a few millimeters away from the nozzle [17]. We stress on the need to have the nozzle away from the laser focal spot in order to get a reasonable lifetime of the nozzle. Gas clusters generated by the expansion of a gas jet plume into vacuum are also used [18].

Another interesting approach to continuous targets is the possibility of specifically designed liquid targets. The idea of using a liquid jet is not new in the laser community. For example remember the old times of the dye lasers. However in this case there are several extra considerations. The first one is the need of very small sources. The second one is the contamination generated by the liquids. Regarding the size of the liquid jet, nozzle technology has advanced dramatically during the last decades and now liquid columns comparable to the focal spot are possible (see Fig 5). Also the column can split in micro-droplets with the help of a piezoelectric attached at the nozzle tip. This also allows synchronization between the droplets and the laser pulses. The most remarkable evolution of the micro-jets is the possibility to make a controlled collision of two micro-jets to get a flat liquid curtain conveniently thin for proton acceleration (see Fig 6).

Fig. 5. CLPU liquid column set-up. General view (right) and zoom of the detail of the liquid column (left). The liquid column itself is too narrow (well bellow 100 microns) to bee seen in the picture and has been highlighted (it is indicated by the arrow at the left).

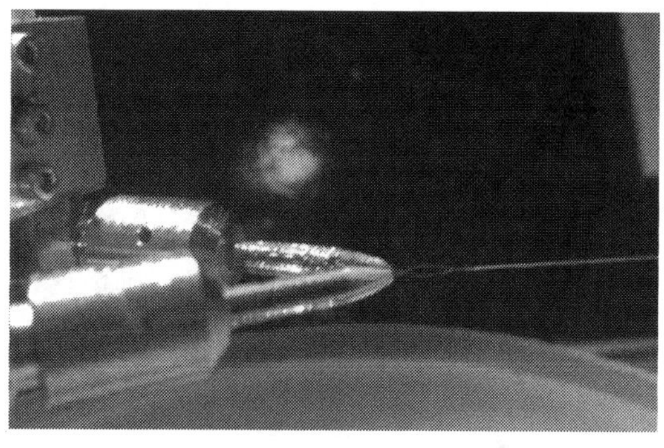

Fig. 6. CLPU flat liquid curtain made by the precise collision of two jets. Observe that all that has to be operative at vacuum and compatible with the low pressure (10^{-6} mbar) and therefore a catcher (not shown in the figure) to collect the final jet and cold finger to collect the vaporized liquid are fundamental.

5 Additional problems

Working at high repetition rates, debris from the target is a fundamental problem that needs to be analyzed carefully before any experiment. The situation is quite different with liquids than with solids.

Micro liquid jets are a promising technology for continuous reliable targets. However there are several

problems related to the targets that are far from being solved, the most relevant one and specific for liquids is the extraction of the unused liquid. For this purpose it is necessary to prepare a catcher that gets the remainder of the liquid column. The catcher can be seen in Figure 6, in the zoom at right. The catcher extracts the liquid column. Moreover liquids can evaporate quite quickly in vacuum at room temperature. So it is necessary to place a cold finger close to the jet to trap most of the vaporized part.
When considering solid targets, nano-particle or micro-particle debris can not move too far and probably are not a big deal with the gratings or other optical components provided that the ventilation of the vacuum lines is planned in the correct way (ventilation procedures are not always considered carefully enough). Such solid targets are a problem only for the focusing parabolas (for that reason in many labs the cleaning or replacement of the main parabola has to be done quite often). The damage of the parabola can be minimized placing a pellicle in front of it.
Also need to comment on another problem specifically relevant at high repetition rates: shock waves. At one Hz, or more, the shock waves that one shot generates can affect the next one. Shock waves are very relevant at PW or multi-PW laser shots. They generate shock waves that can damage nozzles or other key components of the continuous flux systems. Moreover, in high repetition rate, one shock can feel the target still modified by the previous pulse.
Target composition is not regarded in general as a hazard for the laser longevity at high repetition rates, but it must be carefully considered. Most experimental setups work in high vacuum (typically in the 10-6 millibar region) and -what is more restrictive- free of hydrocarbons. Released hydrocarbons can move across the vacuum space that is generally connected to the gratings. The high fluence of the laser beam allows polymerization at the surface of the optical components, and this is particularly dangerous for the gratings. Obviously this is a noticeable at high repetition rates. The effect is the same that appears on some femtosecond oscillators working at high fluences, as the case of certain CEP (Carrier Envelope Phase) oscillators.
A last point to comment is radiation shielding. Radiation is generally proportional to the number of shots, so multiplying the repletion rate multiplies the radiation shielding problems, and the experiments need quite heavy bunkers to shield the ionizing radiation generated. Fortunately not all acceleration experiments generate the same amount of radiation. And a balance between different experiments has to be carefully planned.
In principle, it could be possible to work in air also. The fluences involved are much bigger that the critical power for the onset of filamentation. So a PW in air will filament and because of the intensity clamping, the beam is going to break in multiple filaments.

6 Vacuum polarization

Let's conclude with an optimistic message. Besides all those experiments where targets represent a bottleneck, there is a family of very relevant experiments that do not require a target: they just require vacuum. Therefore those experiments can take full advantage of the multi-PW multi-Hz lasers on construction now.
There are two big families of experiments in the optical domain (infrared, visible or ultraviolet). One is the nonlinear propagation in vacuum. Oversimplifying, the experiment consists in focusing such a big laser arriving to a high intensity at vacuum and then studying the propagation of the beam after the focus to detect small deviations from the linear regime.
The other family of experiments is the collision of two laser beams. Photon-photon cross section in the optical regime is extraordinarily small (as can be seen in Fig. 7). Therefore the way to get a measurable signal is to prepare the collision of two big laser pulses [19]. There is no target so the experiment can run using the laser full repetition rate. With multi PW we are still far, too far, from the Schwinger limit, as a consequence the possibilities to get a collision are very small and require long measuring times (i.e. many laser shots). On the contrary, this can provide a direct measure on the heart of QED (Quantum Electro-Dynamics). Direct information on the existence or not of axions can be obtained from such experiments.

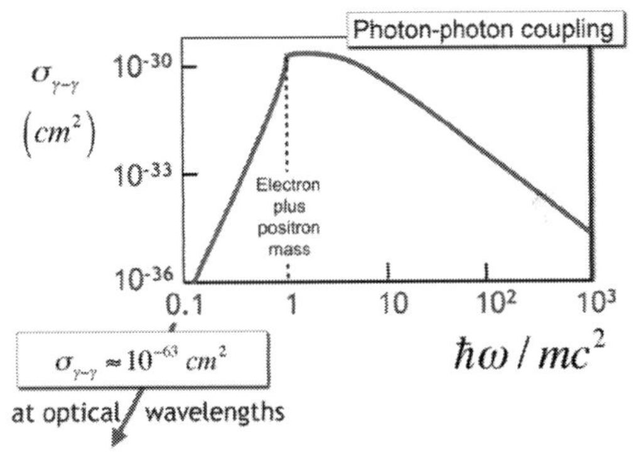

Fig. 7. Photon-photon cross section versus photon energy (for two photons of the same energy) versus the photon energy (in units of the electron mass, mc^2). The photon-photon coupling is extraordinarily small when one considers infrared photons. However, the direct study of such coupling making "collisions" of two big lasers can be of paramount relevance in theoretical physics.

7 Conclusions

In conclusion, with today's available, or almost ready, high repetition rate lasers we can -we must- plan a new generation of experiments able to take profit of those relevant laser tools. Among those experiments, particle acceleration, warm dense matter, betatron radiation, etc., are relevant. But also quantum vacuum experiments have to be considered in order to get a fundamental step forward on the laser implications on technology and fundamental physics

Support from Spanish Ministerio de Economía, Industria y Competitividad (MINECO) through the PALMA Grant No. FIS2016-81056-R, from LaserLab Europe IV Grant No. 654148, and from Junta de Castilla y León Grant No. CLP087U16 is acknowledged. We particularly acknowledge an equipment grant PUU15-EE-3784 from MINECO that funded the CLPU liquid target system.

References

1. D. Strickland and G. Mourou, Optics Communications 56, 219 (1985)
2. P. Maine, D. Strickland, P. Bado, M. Pessot and G Mourou, IEEE Journal on Quantum Electronics 24 398 (1988)
3. E. B. Treacy, IEEE J. Quant Electron 5 (1969) 454
4. O. E. Martinez, J. P. Gordon and R. L. Fork, Journal Optical Society of America A 1, 1003 (1984)
5. https://lasers.llnl.gov/
6. http://www-lmj.cea.fr/
7. X. Liang, et al, in Laser Congress 2017 (ASSL, LAC), OSA Technical Digest (online) (Optical Society of America, 2017), paper AW4A.1.
8. Jae Hee Sung, et al., Optics Lett. 42, 2058 (2017)
9. R Lera, PhD Thesis, University of Salamanca, 2018
10. Ch. G. Durfee, Charles et al., Optics Express 20 13677 (2012)
11. http://bella.lbl.gov/
12. https://www.hzdr.de
13. https://www.clpu.es
14. https://eli-laser.eu
15. S. Garcia, D. Chatain, J. Perin, Laser and Particle Beams, 32, 569 (2014)
16. A. Macchi, M. Borghesi, M. Passoni, Reviews of Modern Physics 85, 751 (2013)
17. F. Sylla, et al, Phys. Rev. Lett. 110, 085001 (2013)
18. T. Ditmire, et al, Nature 398, 489 (1999)
19. D. Tommasini, D. Novoa, L. Roso, in K. Yamanouchi et al eds, Progress in Ultrafast Intense Laser Science, Springer Series in Chemical Physics, vol 106 (2014), Springer

Advanced Channeling Technologies in Plasma and Laser Fields

Sultan B. Dabagov*
[1] *INFN Laboratori Nazionali di Frascati, Frascati (RM), Italy*
[2] *NR Nuclear University MEPhI, Moscow, Russia*
[3] *PN Lebedev Physical Institute, Moscow, Russia*

Abstract. Channeling is the phenomenon well known in the world mostly related to the motion of the beams of charged particles in aligned crystals. However, recent studies have shown the feasibility of channeling phenomenology application for description of other various mechanisms of interaction of charged as well as neutral particle beams in solids, plasmas and electromagnetic fields covering the research fields from crystal based undulators, collimators and accelerators to capillary based X-ray and neutron optical elements. This brief review is devoted to the status of channeling-based researches at different centers within international and national collaborations. Present and future possible developments in channeling tools applied to electron interactions in strong plasma and laser fields will be analyzed.

1 Introduction

As known the beam channeling in crystals takes place during relativistic charged particles motion in periodic structures close to the main crystallographic directions, which define the crystal axes or planes. For a relativistic particle traveling almost parallel to that directions the potential of interaction between the electron and a set of the lattice atoms (ions) could be averaged along the propagation direction. The interaction potential formed at these conditions might limit transverse motion of the projectile within well defined channels, i.e. relativistic particle becomes undulating in transverse plane at relativistic longitudinal motion down to the long narrow channel [1, 2].

For more than 50 years of intense studies the basics of crystal channeling for charged beams have been in details defined and described in many scientific papers and books, discussed in a number of conferences and workshops (see, for instance in [3–14]). Presently crystal channeling is known as useful technique for beam steering [15, 16], while the related to crystal channeling phenomena are promising candidates for coherent radiation sources [17]. Moreover, the phenomenology of beams channeling becomes very useful to describe neutral beams handling via various beam guiding structures [18, 19]. Besides, channeling conditions could be realized for particles not only in medium (crystals [20], capillaries [21, 22] and nanotubes [23–25], plasma [26–29]) but also in high intensity electromagnetic fields of specific configurations [30].

*e-mail: sultan.dabagov@lnf.infn.it

2 Charged and Neutral Beams Channeling

2.1 Crystal channeling

Channeling phenomenon was discovered in the beginning of 1960th at studying the passage of ion beams through crystals. As a result of computer simulations of charged particles penetration into crystals, Robinson, Oen and Beeler, Besco [31, 32] explained the effect of anomalous large penetration for fast ions moving with initial velocities close to principal axial directions. Almost simultaneously the newly predicted phenomenon[1] was revealed experimentally [33, 34]. In 1965 Lindhard explained the existence of new phenomenon [1]; that work, presently well known as a classical one, has stimulated the great interest of physicists to the effect of channeling.

The channeling effect is observed when fast particles penetrate into a crystal under small angles (less than so-called Lindhard angle, the critical angle of channeling) to crystallographic planes (planar channeling) or axes (axial channeling). Channeled projectile penetrates along planes or axes into a crystal without significant energy loss at the distance, which is much larger than the corresponding distance in the case of amorphous solids. To describe this effect the model of continuous interaction potential, when fast particle interacts with the averaged field of atoms placed on a plane or axis instead of the field of separate atoms of a crystal was suggested. The corresponding averaged fields were called "continuous potentials" of the planes or axes that form well defined planar or axial channels, respectively, while the motion in that potentials - "channeling" motion.

[1] In reality for many years before this phenomenon was known as unusual deep penetration of ions at defined crystal orientations taking place at ion doping processes

The classical channeling theory for charged heavy particles as well as the quantum theory was developed in successive works. Later on Andersen with coauthors have shown that channeling effect takes also place for electrons and positrons traversing crystals [35]. Further development of channeling theory has been mainly provoked by Kumakhov's work [17], in which powerful spontaneous γ-radiation by relativistic channeled particles has been predicted. It appears as a result of a classical charged particle undulation in the averaged channeling potential. Later on the classical approach was extended to quantum one, which explains the radiation process via transitions between quantum states in the potentials of atomic rows and planes (for Refs. see in the reviews [36–38]).

2.2 Beam steering for collimation and extraction

One of the rapidly developing application of channeling phenomenon relates to the problem of the heavy particles beam shaping (deflection, steering, splitting) in the modern accelerator facilities. The use of optimally bent crystals to fulfill the channeling condition has been first suggested in [15, 16] to deflect the charge particle beams over the angle much exceeding the critical angle of channeling. Indeed, when a crystal is bent along crystallographic planes, the channeled particle moves along bent channels and it can be deflected from initial direction at large angles (much exceeding the critical channeling angle).

At present the use of planar channeling in bent crystals to control a charge particle beam is of great interest as one of the accelerator and storage ring techniques for beam collimation. For the last decade several technologies for crystal collimation have been proposed and developed (single bent crystals and strip crystals in various configurations). During technological test measurements new phenomena for beam deflection have been observed (channeling, volume capture, volume reflection, etc) [39–44]. While for a long period the experiments have been dealing with the proton beams of 400 GeV/c at SPS CERN revealing the fine features of proton channeling, like beam mirroring, focusing, collimation, the last measurements were performed for LHC proton beams of 6.5 TeV/c that prove relativistic proton channeling [45]. Channeling-based parasitic beam extraction for high energy particles at present accelerator facilities has been recently proposed within the EU Crysbeam project [46] that aims in developing an alternative technique, which will be able to substitute massive magnetic kickers.

Recent studies have also shown rather efficient beam deflection for light particles. Indeed, the tests performed at MAMI and SLAC as well allowed observing very efficient sub-GeV (855 MeV MAMI) and multi-GeV (3.5 and 6.3 GeV) electron beam steering [47–49]. Simultaneously with the technique application, theoretical background for the beam propagation in bent crystal planar channels has been developed. Within collaboration with Tomsk Polytechnic University a new computer code to simulate the processes of relativistic proton/ion beams propagation into various crystals have been proposed and developed in order to characterize the beam deflection inside bent crystals [50–53]. This activity takes place within international CERN project UA9 and aims in LHC beams collimation based on crystal channeling phenomenon. UA9 LNF team has proposed to use parametric X-ray radiation (PXR) for diagnostics of both beam-crystal orientation and crystal quality; first PXR measurements for SPS CERN protons in a bent crystal have been performed and successfully proved the predictions for PXR making evident the peak structure in the radiation both spectral and angular distributions [54–56]. A new phenomenon on effective deflection of relativistic hadron and lepton beams by crystal miscut surface has been predicted [57, 58]. Due to extremely fine structure of the miscut surface the phenomenon might be applied for future accelerators dealing with nanoscale emittance.

Another very important application of charged beam channeling relates, based on undulatory motion at channeling, to the possibility of creating powerful radiation sources in X-ray and γ-ray frequencies. This topic has been strongly investigated since the middle of 1970s, and till now remains of great interest for the researchers resulting in a number of new brilliant ideas. It would be difficult to review this subject in a short paper, hence, I do not discuss it in this work.

2.3 X-ray and neutron channeling in micro- and nanoguides

Some types of processes based on the self-organization of materials that have recently attracted considerable interest because of the possibility of preparing fine patterns of nanometer dimensions over larger areas, can be used for the fabrication of X-ray waveguides [59, 60]. Among them the formation of highly-ordered aligned carbon nanotubes and ordered arrays of uniform-sized porous in alumina is of great interest. X-ray propagation in nanochannels [59] is important due to potential applications in X-ray optics. A special feature of these structures is a long, hollow, inner cavity, which could act as a channel for selective radiation penetration, similar to channeling of charged particles in crystals (see [18] and Refs. therein); nanochannels can be considered as capillaries (the base of capillary/polycapillary optical elements [61–63]). Research in X-ray propagation in capillary structures shows that diminishing the capillary internal radius from microns to nanometers results in a change of the character of radiation propagation, from the surface channeling in microcapillaries down to bulk channeling in nanocapillaries. Numerical simulations [64, 65] have shown that carbon nanotubes will act as soft X-ray waveguides and support modes of propagation when coated with various materials [66]. Other researchers have already demonstrated experimentally the coherent propagation of X-rays in a planar waveguide with a tunable air gap [67]. The angular dependence of the intensity of characteristic K_α radiation in carbon vs the aligned carbon nanotubes orientation suggests the possibility of X-ray channeling as well as radiation diffraction on nanotube multiwalls [68].

The modes of radiation propagation in a waveguide are revealed at interference between the incident and reflected waves forming a standing wave pattern [59]. However, it becomes constructive just for specific angles. This phenomenon, valid for reflection from a flat surface, takes place just in the vicinity of the surface. Similar phenomena could be observed at radiation reflection from a curved surface (so called 'whispering modes') [69]. Strong radiation redistribution also takes place behind capillary systems (which is actually a simple example of the curved surface system); some structural features in the distribution are due to the geometry of the system (typically, hexagon type in the transverse cross section). However, some fine features could not be interpreted by ray optics, and require solution of the wave equation of radiation propagation.

The interference phenomena were observed for X-rays propagating in capillary structures practically 25 years ago. A first theoretical note regarding such a possibility for capillary optical systems, with X-ray interference behind capillaries, was published as an internal note [70]. After that, during 1993-94 at the S-60 LPI synchrotron[2] the phenomenon was observed and successfully studied in a set of experiments on propagation of synchrotron radiation through transversely-ordered polycapillary optical elements (channels sizes of ~100 μm, radiation wavelength ~1÷10 Å). Then, the first paper appeared [71] with the wave interpretation of the features recorded. Phenomenology of the phenomenon given in [72] has shown that the fine features of X-ray propagation in micron-size channels, which have been observed behind capillary structures, can be explained in view of the radiation interference due to the various channels curvatures. Later the trapped radiation propagation in the very vicinity of a surface was carefully studied in a number of papers [73–81] where the wave theory of X radiation propagation along a curved surface was developed (for more complete list of citations, see Refs. in [18]).

3 Channeling in a Plasma-Ion Cavity

Recent studies have shown that interaction of intense ultrashort laser pulses in a medium can be used as a source of compact electron beams with successful beams acceleration over short distances. Studying the propagation of nonlinear plasma electromagnetic waves being of great practical interest due to the progress in laser technologies [84, 85] has shown that powerful ultrashort laser pulses propagating in a plasma can accelerate the plasma electrons to ultra-relativistic energies (so called "laser wakefield acceleration")[86–89]. Considering the problem of powerful ultrashort laser pulse interaction with a plasma based on numerical simulations, a model for a strongly nonlinear regime of interaction of intense laser pulses with plasmas has been proposed. The model has demonstrated that in the process of laser interaction with the plasma the electrons, due to the ponderomotive forces, are displaced from the region at the higher laser intensity. This process has been described within the phenomenology of "moving bubble" [90] (Fig. 1). Indeed, the laser pulse propagating in the direction Oz pushes out quasi-free plasma electrons that results in creating a nearly spherical moving cavity, "bubble". This bubble, as said, is free of electrons and, hence, represents a rather big positive charge moving with the velocity of a laser pulse. Thus, this bubble forms laser-pulse-oriented field gradient, or, in other words, an plasma-ion channel, which is a distinct potential well for electrons. This potential can trap the plasma electrons consistently accelerating them.

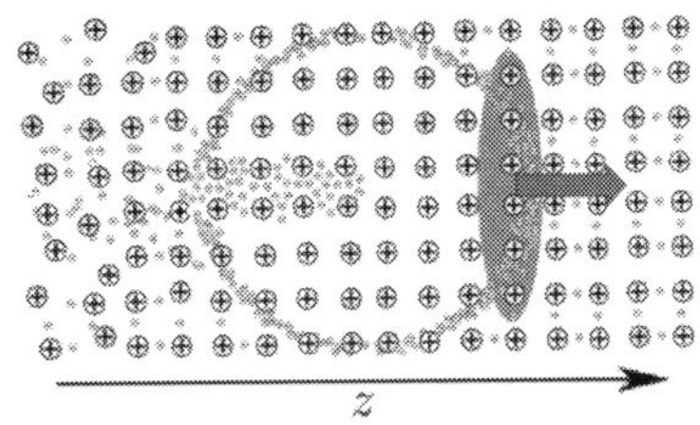

Figure 1. *Schematic drawing of a strongly nonlinear regime of interaction of an ultrashort laser pulse with a plasma. Laser pulse propagating down the Oz axis with the velocity close to the light speed in vacuum produces an ion cavity, which involves plasma electrons into corresponding plasma-ion channel*

The equations of electron motion in above described field at the condition that longitudinal momentum much exceeds transverse one, $p_z >> |\mathbf{p}_\perp|$, can be written in the following way

$$\frac{dp_z}{dt} \approx -2\pi e^2 n_0(z - v_l t), \; \frac{d\mathbf{p}_\perp}{dt} \approx -2\pi e^2 n_0 \mathbf{r}_\perp, \quad (1)$$

where e and n_0 are the electron charge and the charge concentration in a medium, correspondingly, v_l is the propagation velocity of a laser pulse. The force responsible for the electron acceleration is proportional to ~ $(z - v_l t)$. Hence, for electrons moving at a speed greater than the ion bubble speed the force exerted by the latter slows down the electrons. On the contrary, electrons moving at a speed less than the cavity speed become accelerated by the same force (see in [82, 83] and Refs. therein). Neglecting electrons acceleration down the channel at a slight difference between the velocities of electrons and ion chamber, the moving bubble can be presented by an infinite cylindrical plasma-ion cavity characterized by the following field potentials (for an infinite cylindrical channel containing "frozen" ions)

$$\begin{cases} \mathbf{A} \approx 0, \\ \varphi = -\pi n_0 e r_\perp^2, \end{cases} \quad (2)$$

Obviously, we deal with a long ion cavity, plasma-ion channel, characterized by quasi axially-symmetric potential well that binds plasma electrons. The bound projectile can be considered as a channeled one. Assuming relativistic electron motion down to the ion channel with transverse

[2]The cyclic electron accelerator S-60 LPI is used as a source of synchrotron radiation (SR) in vacuum ultraviolet (VUV) and soft X-ray spectral regions. The main characteristics are the maximum electron energy of 630 MeV, the electron orbit radius of 2 m, the average current in the orbit of 130 mA, the maximum of the SR spectrum is 0.66 keV. The synchrotron has been designed in the beginning of 1950s by Vladimir Veksler.

to the Oz axis oscillations, we can write the Hamiltonian in the form

$$H = E_z + \frac{c^2 p_\perp^2}{2E_z} - e\varphi, \tag{3}$$

where $E_z = c\sqrt{p_z^2 + m^2c^2}$ is the longitudinal energy and the potential φ is given by Eq.(2). As it results from the equation of motion, a channeled electron moving fast down to the channel becomes undulating in the transverse plane with the frequency of oscillations $\omega_0^2 = 2\pi e^2 n_0 c^2/E_z$ (for typical electron longitudinal energies $\sim$ 1 GeV and plasma concentrations $\sim 10^{19}$ cm^{-3} this frequency is of the order of $\omega_0 \approx 4\cdot 10^{12}$ s^{-1}), characteristic trajectories of such motion are shown in Fig. 2.

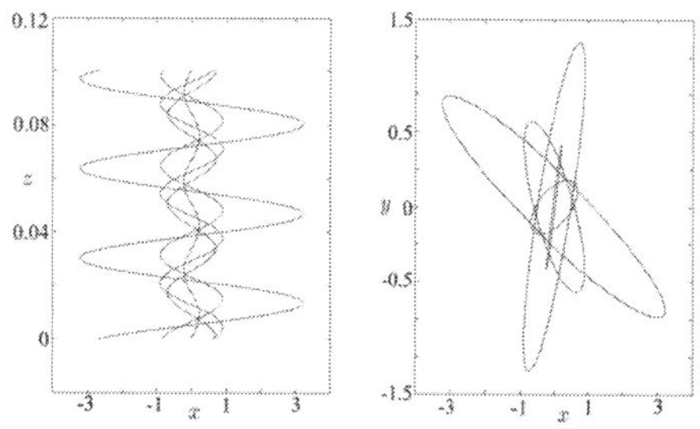

Figure 2. *The electron trajectory in two transverse planes at different initial conditions. The graphs show that initial conditions determine the projectile trajectory. The longitudinal coordinate z is given in cm, while the vertical coordinates x, y - in 10^{-5} cm.*

Due to transverse undulations, the electron emits electromagnetic radiation mostly of X-ray frequencies, and we can define this type of radiation as channeling radiation of electrons in a plasma-ion cavity (radiation by electrons channeled in a plasma-ion channel) that simplifies essentially calculation of radiation functions. Utilizing channeling phenomenology for describing the radiation features the components of both vector $\mathbf{A}_n$ and scalar φ_n potential expansions can be expressed in known sum of the spherical monochromatic waves as functions of the radiation frequency $\omega = \omega_0/(1-\beta_z n_z)$, where $\xi = n\omega(\mathbf{n}_\perp, \mathbf{v}_0)/\omega_0 c$, and $\mathbf{n}$ is the unit vector in the direction of radiation. The angular distribution of electromagnetic radiation intensity is also defined by the potential parameters

$$\frac{dI_n}{d\Omega} = \frac{\omega^2 n^2}{2\pi c}\left(|\mathbf{A}_n|^2 - |\varphi_n|^2\right) \tag{4}$$

The electron becomes trapped into the ion cavity under the following condition $\gamma_z mc >> p_\perp >> mc$ that results in main contribution to the radiation yield by high harmonics components and in a quasi-continuous spectrum of radiation as well. Simplified spectral intensity distribution in forward direction (relativistic projectiles emit in a narrow ahead directed cone) with respect to the longitudinal channel axis can be written as

$$\left(\frac{dI}{d\omega}\right)_{\theta=0} = \frac{e^2\gamma_z^4\omega'^2}{3\pi^3 c\omega_0}\left(\frac{3\omega_0}{\omega_m}\right)^{2/3} f\left(\frac{\omega}{\omega_m}\right), \tag{5}$$

where the function $f(\xi)$ is graphically presented in Fig. 3, and the maximum of radiation distribution function is at ω_m ($\sim 10^{19}$ s^{-1}).

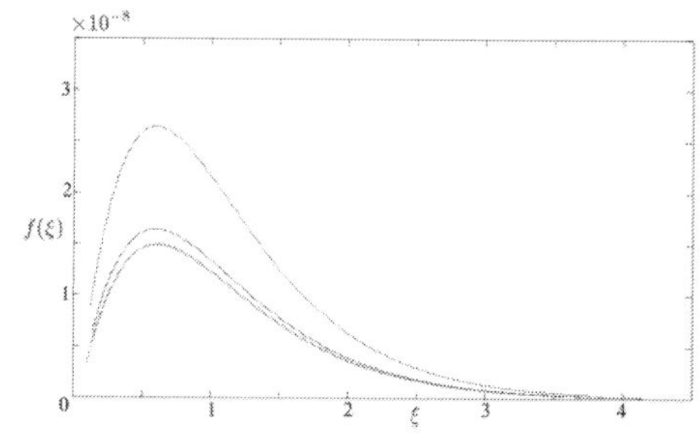

Figure 3. *The function of spectral distribution of electromagnetic radiation by plasma-ion-channeled electron in forward direction at different initial parameters. The graph shows the shape of the spectrum to be almost independent of the initial conditions. Various curves correspond to various electron trajectories in the transverse plane xOy: 1 - a circular trajectory of the radius $r_0 = 10^{-4}$ cm; 2 - an elliptical trajectory with the semi-major axis of $a = r_0/2$, which coincides with the Ox axis, and the semi-minor axis of $b = r_0/4$, which coincides with the Oy axis; 3 - an elliptical trajectory with the semi-major axis of $a = r_0/3$, which coincides with the Oy axis, and the semi-minor axis of $b = r_0/4$, which coincides with the Ox axis.*

Despite the fact that the shape of the spectrum does not depend on the initial conditions, as seen from the graphs of Fig. 3 in the description of the radiation by channeled electrons the spatial distribution function of electrons depends strongly on initial conditions. This fact has a significant impact on coherent and incoherent radiation by the electron beam [91].

Described radiation is rather similar to channeling radiation in crystals [9]. However, on the contrary to channeling radiation in crystals when the electron is considered to be free along the channel direction (the continuous crystal potentials are transverse coordinate dependent), in our case, the electron is accelerated down the channel and known in literature as betatron radiation at laser wakefield acceleration[3] [27, 92, 93]. The latter results in shifting the projectile's frequency for transverse oscillations during the electron channeling in a plasma-ion channel, and thus, the radiation spectrum becomes changed [94]. However, for X-ray frequencies this change is not significant due to the fact that only electrons already being relativistic might be trapped in a channeling mode by the plasma-ion potential. Hence, for these frequencies the spectral radiation distribution by electrons in a plasma-ion channel, previously calculated and observed successfully, is close to that described in this work as channeling radiation.

[3]Obviously, the definition "betatron radiation" is far from the origin of radiation at the electron wakefield acceleration and can be accepted only because of the similarity of the trajectories for betatron electrons and wakefield accelerated ones.

4 Crossed Laser Field to Channel Charged Particles

Nowadays the topic of electrons dynamics in crossed laser beams is gaining growing attention. Usually the case of standing electromagnetic wave being the result of two counter-propagating laser beams is considered. Kapitza and Dirac were the first who referred to the electron dynamics in such an interference field in well known paper [95] evaluating for the first time the electron beam diffraction in periodical field of an optical lattice. Since that many papers have been published proposing a tool for monitoring the electron beam size [96], a new-type free electron laser with optical undulator [97–101], a novel channeling radiation source [102–105] as well as several works clarifying processes taking place at interaction of charged beams in various periodical optical systems [106–108]. Practically all those papers have been dedicated to the particle dynamics in the field of two laser beams.

There are much in common between particles dynamics in crossed laser beams and processes found in FEL undulators, betatron oscillations in plasma channels and crystal channeling. We should give a short remark on why the channeling point of view could be the most appropriate for considering the phenomenon. First of all, interference of two crossed laser beams creates electromagnetic field peaks and nodes, i.e. optical lattice, which is similar to the crystal lattice. This forms semblance of crystal lattice in absence of actual medium that becomes important due to the possibility of beam handling without inelastic scattering of the beam by the medium subsystems. Furthermore, averaging interaction of a particle with both crystal and optical lattices one derives effective potential responsible for particle channeling in these systems. And this descriptive similarity is not the only reason for treating the considered process as channeling.

The typically used ponderomotive force to describe interaction of a charged particle in a laser field is defined by the laser frequency and the gradient of the laser electrical field amplitude [109, 110]

$$\mathbf{F}_p = \frac{e^2}{4\omega_0^2 m}\vec{\nabla}\,|\mathbf{E}(x)|^2\,, \qquad (6)$$

where m is the mass of electron, ω_0 is the laser frequency and $\mathbf{E}(x)$ is the electrical field amplitude distribution over x-coordinate. However, this force depends not only on laser intensity but also on its polarization as well as on the energy of particle [106, 107]. Situation becomes even more complex when we deal with the field formed by two interfering laser waves [108].

The electron motion in optical lattice formed by crossed linearly polarized laser beams, in plane electromagnetic waves approximation, can be described in terms of particles channeling. In the region of two laser beams overlapping the ponderomotive force characterized by averaged effective potential affects charged particles. The ponderomotive potential forms planar (one-dimensional) potential wells (Fig. 4). This structure is similar in some way to the crystal lattice, seen by a high-velocity projectile, that is a set of crystal nodes, forming crystallographic planes and axes. The standing waves can be represented as the effective potential wells, the channels, periodically situated in a transverse plane that can trap the electron at specific conditions.

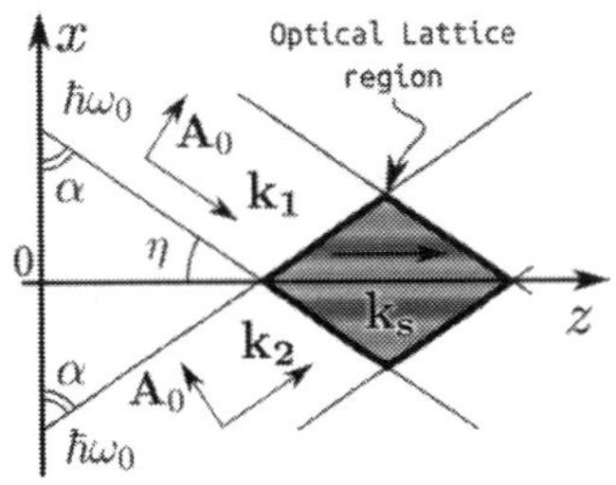

Figure 4. The general scheme of the considered system. Two crossed laser beams of the ω_0 frequency with the wave vectors $\mathbf{k}_1$ and $\mathbf{k}_2$, which lie in the xOz plane at the acute angles $\eta = \pi/2 - \alpha$ with respect to the Oz direction. The effective potential channels (shown in color) are parallel to $0z$-axis.

To examine the bound motion of charged particles in optical lattice, Kapitza method [111] is rather effective at non-relativistic electron energies when classical approximation might be applied to the solution of the equation of electron motion. Within this approach, due to the presence of high frequency electromagnetic field, the electron trajectory at defined conditions looks like frequent oscillations δx_i along a smooth curve $\bar{x}_i$: $x_i = \bar{x}_i + \delta x_i$. Neglecting the oscillations in the equation of motion (small by the def-

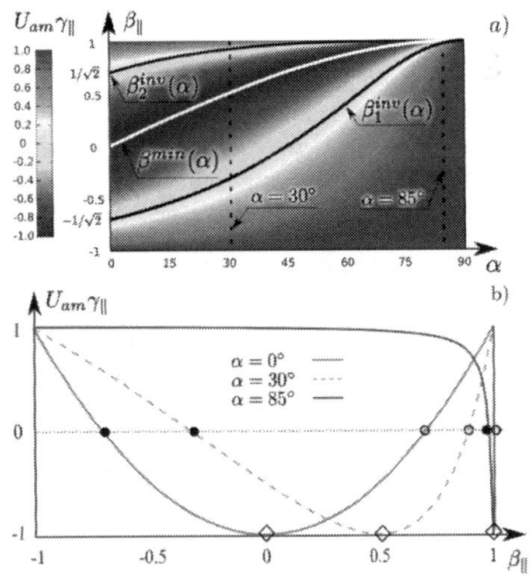

Figure 5. a) The averaged potential amplitudes normalized by $(e^2A_0^2)/(2\gamma_\parallel mc^2)$ as a function of $\beta_\parallel$ and α is shown in color. Inversion speed values $\beta_{1,2}^{inv}$ as a functions of α are shown in black. The electron speed value for which at chosen α the value of $U_{am}\gamma_\parallel$ becomes minimal, $\beta^{min}(\alpha)$, is shown in white. b) Normalized potential amplitudes for two p-polarized laser beams, crossed at different angles. The inversion speed values β_1^{inv} are marked with •, values of β_2^{inv} – with ∘, and β^{min} – with ⋄ for each α.

inition, $\delta x_i \sim eE_0/m\omega_0^2$), an effective potential energy of electron interaction in combined field can be defined. Applying some modification to Kapitza method the problem can be solved also for relativistic velocities [108] when after averaging by rapid oscillations one derives the effective potential expression

$$U_{eff} = -U_{am}\cos(2kx\cos\alpha),$$
$$U_{am} = \frac{e^2A_0^2}{2\gamma_{\|}mc^2}\frac{(1+\cos^2\alpha)\beta_{\|}^2-\cos(2\alpha)-2\beta_{\|}\sin\alpha}{(1-\beta_{\|}\sin\alpha)^2}, \quad (7)$$

where $\beta_{\|} = v_0/c$ and positive $\beta_{\|}$ means that the electron moves in the direction of $\mathbf{k}_s = \mathbf{k}_1 + \mathbf{k}_2$, while negative $\beta_{\|}$ – in opposite direction; A_0 is the longitudinal vector potential amplitude. The above derived expression describes the effective potential in the region of two p-polarized laser beams crossed at arbitrary angle (Fig. 5). The motion equation could be linearized allowing its description via known elliptic integral trajectory characterized by both frequencies

$$\Omega_1 = 2k\cos\alpha\sqrt{U_{am}/\gamma_{\|}m}, \quad \Omega_2 = \omega_0(1-\beta_{\|}\sin\alpha), \quad (8)$$

where $\Omega_{1,2}$ are the frequencies for channeling oscillations (within standing wave structure) and for rapid oscillations due to the particle interaction with original laser field, respectively.

At $\alpha = 0$ the beams are counter-propagating and the potential amplitude in this case becomes positive for $|\beta_{\|}| > 1/\sqrt{2}$ and negative otherwise. This means that for an electron moving at the speed $|\beta_{\|}| = 1/\sqrt{2} = -\beta_1^{inv} = \beta_2^{inv}$ (so called "*inversion speed*") *no* periodic potential is formed. When the electron speed is $|\beta_{\|}| > 1/\sqrt{2}$, the regions of interference electric field peaks are scattering ones for it. And for $|\beta_{\|}| < 1/\sqrt{2}$ these regions are characterized by the attractive potentials. This potential inversion is observed only for p-polarized crossed laser beams.

The phase space trajectories projected onto the plane of transverse momentum and coordinate can be presented for both channeled and over-barrier electrons in the field of laser beams crossed as shown in Fig. 6, for instance, at the angle of $2\eta = 30°$ ($\alpha = 75°$). The inner closed curves (in blue) correspond to channeled electrons with transverse energies less than potential well height. Over-barrier electrons (in red) are characterized by transverse energies greater than the barrier height and not limited within the channels. For electron with $\beta_{\|} = \beta^{min}$ the channel borders are situated at $x/d_{ch} = 0.5 + n$ and, correspondingly, the channel centers – at $x/d_{ch} = n$ (see Fig. 6a). On the contrary, electron with longitudinal velocity $\beta_{\|} \to 1 > \beta_2^{inv}$ could be trapped by the channels, central axes of which are placed at $x/d_{ch} = 0.5 + n$ (see Fig. 6b), where $n = 0, \pm1, \pm2, \ldots$

Obviously, a particle oscillating in external field should radiate at frequencies related to the undulatory motion. The radiation spectrum of a charged particle emitting in a crossed laser field is characterized by two peak frequencies $\omega_{1,2} = \Omega_{1,2}/(1-\beta_{\|}\cos\theta)$ [105], where $\Omega_{1,2}$ were defined before by the expressions (8). Both of them being measured in the forward direction ($\theta = 0$) are shifted by

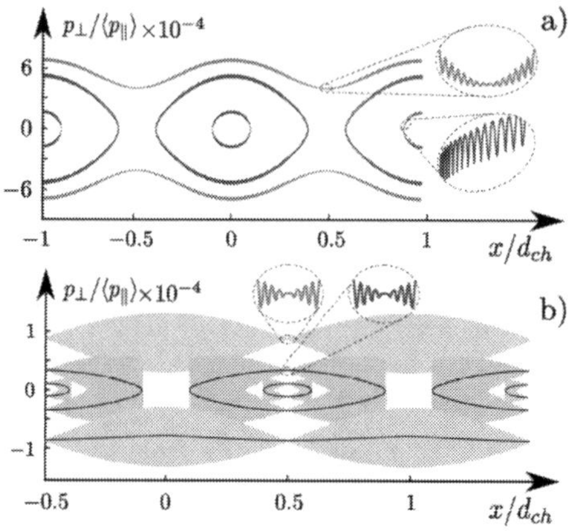

Figure 6. Simulated transverse phase-space trajectories for two extreme electron energies - $\beta_{\|} = \beta^{min}$ (a) and $\beta_{\|} = 0.99986$ (b). The channel has formed by two p-polarized lasers of $\lambda = 800$ *nm* wavelength and 5 TW/cm^2 intensity for each one. The electron transverse momentum is normalized by the averaged longitudinal momentum, while the transverse coordinates – by the channel width $d_{ch} \equiv \lambda_0/(2\sin\eta)$. Channeled (under-barrier) trajectories are shown in blue, while quasichanneled (over-barrier) – in red. The rapid oscillations could not be properly visualized and are outlined near the plots. The averaged phase space trajectories are also demonstrated in solid black lines.

the factor of $\sim 2\gamma_{\|}^2$ due to the Doppler effect. The first one (ω_1) corresponds to slow channeling electron oscillations in the potential well of a system and does not depend on the laser frequency ω_0. The ω_2 radiation frequency is caused by the electron interaction with the original laser beams field.

Taking into account that relativistic particle emits radiation in a narrow forward directed solid cone ($\sim \gamma^{-1}$), one could define the analytical expression for radiation spectrum of relativistic channeled electron moving near the bottom (center) of the cross-laser channel

$$\frac{dP}{d\omega} = \sum_{i=1}^{2}\frac{e^2\Omega_i^3a_i^2\gamma_{\|}^2}{c^3}\zeta_i\left(1-2\zeta_i+2\zeta_i^2\right)\Theta\left(\pi N_i(1-\zeta_i)\right), \quad (9)$$

where ω is the radiation frequency, a_1 is the amplitude of channeling oscillations, and a_2 is the amplitude of rapid oscillations, N_i is the number of Ω_i-frequency particle oscillations, $\zeta_i = \omega/(2\gamma_{\|}^2\Omega_i)$, and $\Theta(x) = 0.5 + \mathrm{Si}(2x)/\pi - \sin^2(x)/\pi x$. Total low-frequency power emitted by the channeled electron can be then evaluated by

$$P[W] = \frac{2e^2\Omega_1^4a_1^2\gamma_{\|}^4}{3c^3} \approx 10^{-42}\left(I\left[\frac{W}{cm^2}\right]\right)^2 \quad (10)$$

for the channeling oscillations amplitude $a_1 \approx 0.1d_{ch}$. The channel width for the external laser wavelength of $\lambda_0 = 1\ \mu m$ crossed at $\eta = 15°$ would be $d_{ch} \equiv \lambda_0/(2\sin(\eta)) \approx 2\lambda_0 \approx 2\ \mu m$.

One should note that, first, both electrons and lasers parameters could be chosen so that the radiation wavelength is equal to the external laser one. Hence, this could be interesting as amplification mechanism for increasing external laser intensity. Moreover, due to rather wide tunable channel width it becomes possible to use a large bunch. In case of, for instance, 10^{10} electrons (monoenergetic, for simplicity), forming a bunch of several μm transverse width, low-frequency *coherent* radiation power is

$$P[W] \approx 10^{-22}\left(I\left[\frac{W}{cm^2}\right]\right)^2 \qquad (11)$$

As seen, in the case of initial laser intensity of $10^{17}\ W\,cm^{-2}$ the total low-frequency radiation power is estimated to be equal to $10^{12}\ W$.

The radiation spectrum classically calculated is characterized by two major frequencies: the first one is due to external laser field scattering on relativistic electron, while the second corresponds to the radiation of electron trapped by the effective potential well. The maximum radiation intensity falls on the frequency $\omega_m^{rad} \sim \gamma_\parallel \sqrt{I}$ defined by both electron energy and external laser intensity, while the integral radiated power depends on squared external laser intensity $P \sim I^2$. Combined with high radiation coherence for a channeled electron bunch it can result in a high intensity gain. The main issue here would be not only electron bunch low-frequency radiation coherency but also electrons cooling due to high-frequency radiation which would be considerable in strong fields.

The expressions for the effective potential and radiation yield were derived mostly for one-electron case. And the most interesting questions concerning beam dynamics, cooling and coherent radiation in crossed lasers are the subject of our current research that will help to determine the feasibility and conditions for creating tools based on the considered phenomena. However, in order to analyze the interaction of charged particles beam in laser optical lattices we have to use computer simulations that allows taking into account space charge interaction which involves all the beam particles. It should be underlined that in a beam we deal simultaneously with practically all possible solutions for a single particle motion that has been previously analyzed (Fig. 7). Moreover, only in multiple particle dynamics we can predict the peculiarities such as beam shaping, deflection, focusing.

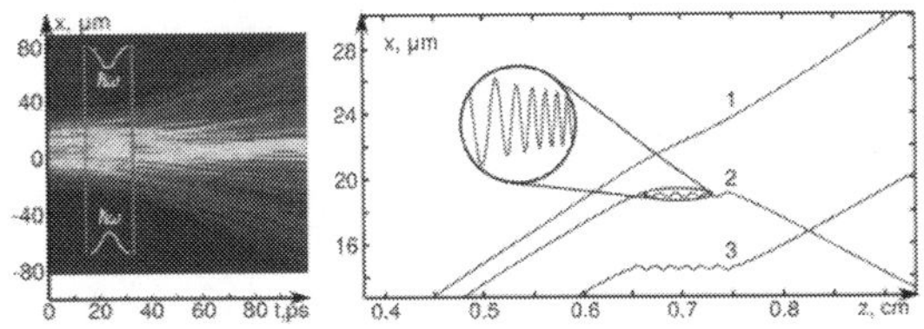

Figure 7. Numerical simulations for a Gaussian profile ~ 1 pC nondivergent electron beam of ~ 2 MeV interacting with a laser standing field of the Gaussian transverse profile with σ_z = 1 mm. The laser fan has positioned at 23 ps from the initial electron beam position. The density color graph of electron beam passing through optical lattice depicts the particle time distribution. Curve 1 - non channeled particles, curve 2, 3 - channeled particles.

5 Conclusions

For more than 50 years of its history the channeling phenomenon having started from trapped motion in aligned crystals has been observed at propagation of X-ray radiation and thermal neutrons in capillary structures, and presently becomes very useful to examine fine features of charged beam dynamics in laser and plasma based guides.

As shown the similarity of possible applications and effects may take place in optical lattice and in crystal channeling. The effective potential of both crystal and optical lattices could be very similar and both are capable of trapping electrons. So that channeled electron beams can be transported, focused and reflected by potential of both lattices.

Bending a crystal one obtains a tool for charged particles beams steering and bent laser channels might be also applied for this. Such channels can be formed by illuminating a curved reflecting surface with a laser at some angle that creates interference region with potential channels near the surface. Besides, recently researches have shown different regimes of charged particles dynamics in presence of intense laser fields that are analogous to crystal volume reflection, to both planar and axial crystal channeling and optical undulating.

Electrons deflection by curved laser channels, formed by laser grazing incidence on curved reflecting surface is a new research inspired by analogy of optical lattice channeling and crystal channeling, which could be done in bent crystals for charged particle beam steering. The reported derivation of curved laser channel effective potential is done for the region close to the reflecting surface and the resulting critical radius values for ultra-relativistic electrons are extremely high, but the idea may find application for moderate energies. In any case additional investigation is needed.

Channeling phenomenology applied to the description of wake-filed acceleration becomes very powerful tool that enables analyzing the dynamics of beam acceleration as well as the radiation by accelerated particles. Application of channeling technique can simplify prediction of various new effects.

Besides of beam shaping studies based on channeling technique, it is worthy to underline that particle channeling is always accompanied by rather strong radiation. Channeling motion provides high-frequency and low amplitude undulation of trapped particles that results in high flux electromagnetic emission of wide frequency range. The frequency is defined by the parameters of the beam and external field. In major cases this radiation, additionally to the interest of creating powerful tunable radiation source, might be of the reason for efficient beam cooling that is under our present investigations.

First of all, I would like to emphasize that I am extremely grateful to the organizers of the PPLA 2017 conference for the invitation to present the results of our recent studies on channeling of charged and neutral particles in strong external fields of various origin.

I would like to acknowledge the work done by my young colleagues Alexey Dik and Evgeny Frolov. I have included in this short review the results of our common research for several last years.

This work was supported by the Competitiveness Program of National Research Nuclear University "MEPhI".

References

[1] J. Lindhard, Kgl. Dan. Vid. Selsk. Mat.-Fys. Medd., 34 (14) (1965) 1.

[2] D.S. Gemmell, Rev. Mod. Phys., 46 (1974) 129.

[3] M.A. Kumakhov, and G. Shirmer, Atomic Collisions in Crystals (G. & B. Sci. Pub. Ltd, London) 1979.

[4] V.G. Baryshevsky, Channeling, Radiation and Reactions in Crystals at High Energies (BSU, Minsk) 1982 (in Russian).

[5] Y.-H. Ohtsuki, Charged Beam Interaction with Solids (Taylor & Francis Ltd, London-New York) 1983.

[6] A.W. Saenz, and H. Uberall, Eds, Coherent Radiation Sources (Springer) 1985.

[7] V.A. Bazylev, and N.K. Zhevago, Radiation of Fast Particles in Matter and External Fields (Nauka, Moscow) 1987 (in Russian).

[8] N.P. Kalashnikov, Coherent Interactions of Charged Particles in Single Crystals. Scattering and Radiative Processes in Single Crystals. Harwood Acad. Pub. London and New York. 1988.

[9] M.A. Kumakhov, and F.F. Komarov, Radiation from Charged Particles in Solids (AIP, New York) 1989.

[10] A.I. Akhiezer, and N.F. Shuloga, High-Energy Electrodynamics in Matter (G. & B. Sci. Pub., Amsterdam) 1996.

[11] V.N. Baier, V.M. Katkov, and V.M. Strakhovenko, High Energy Electromagnetic Processes in Oriented Single Crystals (World Scientific, Singapore) 1998.

[12] P. Rullhusen, X. Artru, and P. Dhez, Novel Radiation Sources Using Relativistic Electrons (World Scientific, Singapore) 1998.

[13] Proceedings of the "Channeling" conferences for the period of 2004-2016, Dabagov S.B., Ed.: Proc. of SPIE 5974 (2005); Proc. of SPIE 6634 (2007); Proc. of the 51st Workshop of the INFN Eloisatron Project, World Scientific, (2010); Nuovo Cimento C 34 (4) (2011); Nucl. Instr. Meth. B309 (2013); Nucl. Instr. Meth. B355 (2015); Nucl. Instr. Meth. B402 (2017).

[14] S.B. Dabagov, and M.N. Strikhanov, Eds., the books "Charged & Neutral Particles Channeling Phenomena", MEPhI Pub., Moscow, ISBN 978-5-7262-1864-9 (2013); ISBN-978-5-7262-2232-5 (2016).

[15] E.N. Tsyganov, Fermilab TM-682, Batavia (1976).

[16] R. Carrigan, Phys. Rev. Lett.,35 (1975) 206.

[17] M.A. Kumakhov, Physics Letters A57 (1976) 17.

[18] S.B. Dabagov, Phys. Usp., 46 (2003) 1053.

[19] I. Bukreeva et al., Phys. Rev. Lett. 97 (2006) 184801.

[20] V.M. Biryukov, Y.A. Chesnokov, and V.I. Kotov, Crystal Channeling andIts Application at High-Energy Accelerators (Springer, Berlin) 1997.

[21] M.A. Kumakhov and F.F. Komarov, Phys. Rep. 191 (5) (1990) 289.

[22] C. MacDonald, X-Ray Opt. Instr. 867049-17 (2010).

[23] N.K. Zhevago and V.I. Glebov, Phys. Lett. A310 (2003) 301.

[24] V.V. Klimov and V.S. Letokhov, Phys. Lett. A222 (1996) 424.

[25] A. Karabarbounis, S. Sarros, and C. Trikalinos, Nucl. Instr. Meth. B316 (2013) 160.

[26] E. Esarey et al., Phys. Rev. E 65, 056505 (2002).

[27] I. Kostyukov, S. Kiselev, and A. Pukhov, Phys. Plasmas 10 (2003) 4818.

[28] J. Faure et al., Phys. Plasmas 17, 083107 (2010).

[29] A.V. Dik, A.Z. Ligidov, and S.B. Dabagov, Nucl. Instr. Meth. B309 (2013) 210.

[30] E.N. Frolov, A.V. Dik, and S.B. Dabagov, Nucl. Instr. Meth. B309 (2013) 157.

[31] M.T. Robinson, and O.S. Oen, Phys. Rev. 132 (1963) 2385.

[32] J.R. Beeler, and D.S. Besco, J. Appl. Phys. 34 (1963) 2873.

[33] G.R. Piercy et al., Phys. Rev. Lett. 10 (1963) 399.

[34] H. Lutz, and R. Sizmann, Physics Letters 5 (1963) 113.

[35] J.U. Andersen, S.K. Andersen, and W.M. Augustyniak, Kgl. Dan. Vid. Selsk. Mat.-Fys. Medd., 39 (10) (1977) 1.

[36] V.V. Beloshitsky, and F.F. Komarov, Phys. Rep. 93 (1982) 117.

[37] R.A. Carrigan, Jr., and J.A. Ellison, Eds., Relativistic Channeling (Plenum Press, New York) 1987.

[38] S.B. Dabagov, and N.K. Zhevago, La Rivista del Nuovo Cimento 31 (9) (2008) 491.

[39] W. Scandale et al, Physics Letters B703 (2011) 547.

[40] W. Scandale et al, Physics Letters B726 (2013) 182.

[41] W. Scandale et al, Physics Letters B733 (2014) 366.

[42] W. Scandale et al, Physics Letters B734 (2014) 1.

[43] W. Scandale et al, Nucl. Instr. Meth. B338 (2014) 108.

[44] W. Scandale et al, Physics Letters B748 (2015) 451.

[45] W. Scandale et al, Physics Letters B758 (2016) 129.

[46] http://crysbeam.roma1.infn.it/index.html

[47] A. Mazzolari et al., Phys. Rev. Lett. 112 (2014) 135503.

[48] U. Wienands et al., Phys. Rev. Lett. 114 (2015) 074801.

[49] U. Wienands et al., Nucl. Instr. Meth. B402 (2017) 11. doi: 10.1016/j.nimb.2017.03.097

[50] A. Babaev, G. Cavoto, and S.B. Dabagov, Nucl. Instr. Meth. B309 (2013) 120.

[51] A. Babaev, and S.B. Dabagov, Journ. of Physics: Conf. Ser. 357 (2012) 012032.

[52] A. Babaev and S.B. Dabagov, Eur. Phys. J. Plus 127 (2012) 62.

[53] A. Babaev, and S. Dabagov, Il Nuovo Cimento C34 (4) (2011) 417.

[54] W. Scandale et al, Physics Letters B701 (2011) 180.

[55] S.B. Dabagov, et. al., INFN-16-05/LNF, 17th March 2016.
[56] K.B. Korotchenko, Yu.L. Eykhorn, and S.B. Dabagov, Phys. Lett B774 (2017) 47-475.
[57] A. Babaev, G. Cavoto, and S.B. Dabagov, Nucl. Instr. Meth. B355 (2015) 356.
[58] A. Babaev, G. Cavoto, and S.B. Dabagov, JETP Letters 100 (9) (2014) 550.
[59] E. Spiller, and A. Segmuller, Appl. Phys. Lett. 24 (1974) 60.
[60] S.P. Pogossian, Opt. Comm. 114 (1995) 235.
[61] M.A. Kumakhov, and F.F. Komarov, Phys. Rep. 191 (1990) 289.
[62] P. Engstrøm, S. Larsson, and A. Rindby, Nucl. Instr. Meth. A302 (1991) 547.
[63] D.J. Thiel et al., Nucl. Instr. Meth. A317 (1992) 597.
[64] N.K. Zhevago, and V.I. Glebov, Phys. Lett. A250 (1998) 360.
[65] G.V. Dedkov, Nucl. Instr. Meth. B143 (1998) 584.
[66] P.A. Childs, and A.G. O'Neill, Physica E19 (2003) 153.
[67] M.J. Zwanenburg et al., Phys. Rev. Lett. 82 (1999) 1696.
[68] A.V. Okotrub et al., JETP Lett. 81 (2005) 34.
[69] A. Vinogradov et al., Sov. Phys. - Tech. Phys. 30 (1985) 335.
[70] S.B. Dabagov, Redistribution of X-rays Trapped in Bound States by Capillary Systems (Research Report of FIROS: Nalchik-Moscow, 1992).
[71] S.B. Dabagov et al., J. Synchrotron Rad. 2 (1995) 132.
[72] S.B. Dabagov, M.A. Kumakhov, and S.V. Nikitina, Phys. Lett. A203 (1995) 279.
[73] S.B. Dabagov, and M.A. Kumakhov, Proc. SPIE 2515 (1995) 124.
[74] Yu.M. Alexandrov et al., Nucl. Instr. Meth. B134 (1998) 174.
[75] N. Artemiev et al., Phys. Scripta 57 (1998) 228.
[76] S.B. Dabagov et al., Appl. Opt. 39 (2000) 3338.
[77] S.B. Dabagov et al., Proc. SPIE. 3444 (1998) 486.
[78] C. Liu, and J.A. Golovchenko, Phys. Rev. Lett. 79 (1997) 788.
[79] S.V. Kukhlevsky et al., Nucl. Instr. Meth. B168 (2000) 276.
[80] G. Cappuccio, et al., Appl. Phys. Lett. 78 (2001) 2822.
[81] S.B. Dabagov, X-ray Spectrom. 32 (2003) 179.
[82] D. Giulietti, J. Phys.: Conf. Ser. 508 (2014) 012001.
[83] L. Torrisi, et al., J. Applied Phys. 107 (2010) 123303.
[84] G. Mourou, C. Barty, and M. Perry, Phys. Today 51 (1) (1998) 22.
[85] S.P.D. Mangles et al., Phys. Rev. Lett. 94 (2005) 24500.
[86] T. Tajima, and J.M. Dawson, Phys. Rev. Lett. 43 (1979) 267.
[87] S. Bulanov, V. Kirsanov, and A. Sakharov, Plasma Phys. 16 (1990) 543.
[88] A. Pukhov, S. Kiselyov, and I. Kostyukov, Appl. Phys. 6 (2006) 35 (in Russian).
[89] J. Faure et al., Nature 431 (2004) 541.
[90] E. Esarey, C. Schroeder, and P. Leemans, Rev. Mod. Phys. 81 (2009) 1229.
[91] A. Dik, and S.B. Dabagov, Izv VUZov 12 (2012) 77 (in Russian).
[92] A. Rousse, et al., Phys. Rev. Lett. 93 (13) (2004) 135005.
[93] A. Curcio, and D. Giulietti, NIMB 355 (2015) 214.
[94] S. Corde, et al., Rev. Mod. Phys. 85 (2013) 045004.
[95] P.L. Kapitza, and P.A.M. Dirac, Math. Proc. Cambridge Philos. Soc. 29 (1933) 297.
[96] T. Shintake, Nucl. Instr. Meth. A311 (1992) 453.
[97] M.V. Fedorov, K.B. Oganesyan, and A.M. Prokhorov, Appl. Phys. Lett. 53 (1988) 353.
[98] P. Balcou, Eur. Phys. J. D 59 (2010) 525.
[99] I.A. Andriyash, P. Balcou, and V.T. Tikhonchuk, Eur. Phys. J. D 65 (2011) 533.
[100] I.A. Andriyash et al., Phys. Rev. Lett. 109 (2012) 244802.
[101] I.A. Andriyash et al., J. Phys. Conf. Ser. 414 (2013) 012008.
[102] M. Bertolotti, C. Sibilia, and L. Fuli, in Trends in Quantum Electronics, A.M. Prokhorov, I. Ursu, Eds. (Springer, Berlin Heidelberg, 1986).
[103] A. V. Andreev, and S. A. Akhmanov, JETP Lett. 53 (1991) 18.
[104] E.N. Frolov, A.V. Dik, and S.B. Dabagov, Nucl. Instr. Meth. B309 (2013) 157.
[105] E.N. Frolov, A.V. Dik, and S.B. Dabagov, J. Phys. Conf. Ser. 517 (2014) 012002.
[106] A.E. Kaplan, and A.L. Pokrovsky, Phys. Rev. Lett. 95 (2005) 053601.
[107] A.L. Pokrovsky, and A.E. Kaplan, Phys. Rev. A72 (2005) 043401.
[108] S.B. Dabagov, A.V. Dik, and E.N. Frolov, Phys. Rev. ST 18 (2015) 064002.
[109] A.V. Gaponov, and M.A. Miller, Zh. Eksp. Teor. Fiz. 34 (1958) 242 (in Russian).
[110] B.M. Bolotovskii, and A.V. Serov, Usp. Fiz. Nauk. 37 (1994) 515.
[111] L.D. Landau and E.M. Lifshitz, Mechanics, Course of Theoretical Physics, Vol. 1, 3rd ed. (Elesevier, Amsterdam etc.) 1976.

Ray optics hamiltonian approach to relativistic self focusing of ultraintense lasers in underdense plasmas

Alessandro Curcio[1,2,⋆], *Maria Pia* Anania[1], *Fabrizio Giuseppe* Bisesto[1], *Massimo* Ferrario[1], *Francesco* Filippi[1], *Danilo* Giulietti[3], and *Massimo* Petrarca[4]

[1] *INFN-LNF, via Enrico Fermi 40, 00044 Frascati, Rome, Italy*
[2] *Department of Physics, "Sapienza" University of Rome, Piazzale A. Moro 2, I-00185 Rome, Italy*
[3] *Physics Department of the University and INFN, Largo Bruno Pontecorvo 3, 56127 Pisa, Italy*
[4] *Department of Basic and Applied Sciences for Engineering (SBAI) and INFN-Roma1, "Sapienza" University of Rome, Via A. Scarpa 14, 00161 Rome, Italy*

Abstract. The relativistic self focusing of an ultraintense laser propagating through an underdense plasma is analyzed from a geometrical optics point of view, exploiting the classical hamiltonian formalism. The distribution of the laser intensity along the self-generated plasma channel is studied and compared to measurements.

1 Introduction

The relativistic self focusing of ultraintense lasers in underdense plasmas is a well-known phenomenon, of great interest in last decades after the advent of ultrashort lasers of the femtosecond-class. Self focusing effects have been observed since the first era of pulsed laser systems of nanosecond-class, but only with new compression techniques (Chirped Pulse Amplification [1]) and consequently with the possibility to exploit ultrashort laser pulses it became possible to experimentally explore the variety of optical phenomena which relate to relativistic optics [2]. The relativistic optics is the study of the optical response of the matter when the electron quiver velocity $v_q = a_0 c/\sqrt{1+a_0^2}$ under the action of an ultraintense electromagnetic field reaches values close to the velocity of light in vacuum c. The relativistic parameter $a_0 = 8.5 \times 10^{-10} \lambda_0[\mu m]\sqrt{I_0[W/cm^2]}$ related to the laser wavelength λ_0 and intensity I_0, allows to define the threshold above which an optical phenomenon can be classified as relativistic, i.e. $a_0 \gtrsim 1$. The relativistic self focusing is actually a relativistic phenomenon typical for ultraintense laser pulses propagating through underdense plasmas [3–6]. In fact the Lorentz factor of a free electron quivering under the action of a linearly polarized laser pulse is given by $\gamma = \sqrt{1 + a^2/2}$ [7]. In reality the laser beam has a finite and well-defined radial profile, therefore, in a fluid picture of the plasma electrons, the factor $\gamma(r)$ is a function of the radial coordinate r via the laser intensity profile, contained in a. For sake of simplicity we are limiting our considerations to a radial-symmetric physical system. In most of practical and realistic cases the laser intensity profile is peaked in $r = 0$ and decreases for $r > 0$. It means that the on-axis plasma electrons are "heavier" than the off-axis electrons if looking at the Lorentz factor $\gamma(r)$. The index of refraction for a plasma medium is generally $n = \sqrt{1 - \omega_p^2/\gamma(r)\omega_0^2}$, where $\omega_p = \sqrt{n_e e^2/m\varepsilon_0}$ is the plasma frequency and $\omega_0 = 2\pi c/\lambda_0$ is the laser pulsation. The plasma frequency is defined through the elementary charge e, the electron plasma density n_e, the dielectric constant of vacuum ε_0 and the electron mass m. The r-dependence of the Lorentz factor γ in the expression of the plasma index of refraction is the cause for relativistic self focusing. In fact the plasma electron profile at the laser front acts like a focusing lens. This focusing effect can be effective as far as the laser intensity is maintained so high that relativistic quiver velocities are provided to the plasma electrons at the pulse front. The relativistic self focusing is responsible in guiding laser beams giving rise to typical laser envelope oscillations, manifesting the counterbalance of focusing/defocusing effects. We can find an optimum condition for the guiding of the laser in the linear and slightly non-linear regime of laser-plasma interaction, by equating the laser electric field to the limit field E_{wb} (also called wavebreaking field [7]), then supposing to have a diffraction-limited beam with cross-section $\sim \lambda_0^2$. The power associated to this beam is defined as critical power for relativistic self focusing P_{cr}:

$$P_{cr} \sim I_{wb}\lambda_0^2 = c\frac{\varepsilon_0}{2}E_{wb}^2\lambda_0^2 \sim 14\left(\frac{\omega_0^2}{\omega_{p0}^2}\right)GW \qquad (1)$$

where the wavebreaking intensity I_{wb} is the intensity of an electromagetic wave with electric field E_{wb}, i.e. $I_{wb} = c\varepsilon_0 E_{wb}^2/2$. A more precise and more complex calculation with an undulatory description of the laser through a non-linear Schroedinger equation would give a factor 17.5 in place of 14 [8], nevertheless our naive calculation clearly

⋆e-mail: alessandro.curcio@uniroma1.it

enlightens the physics of the laser guiding process, as a counterbalance between the laser and the plasma electric fields. In particular for power exceeding P_{cr} self focusing can occur, otherwise diffraction dominates. This is due to the fact that the laser field is not effective in shaping the electron density into such a profile useful for the guiding until the laser electric field reaches the maximum plasma electric field E_{wb}. In fact below this threshold the electrons move under the action of the plasma electric field, being the laser action only a small correction to their dynamics. When the pump is depleted, i.e. when the laser energy has been exhausted in the vastness of interaction processes which couple the laser energy to the plasma electrons (in particular to Langmuir waves), a net defocusing overcomes. Efficient relativistic self focusing has been experimentally demonstrated, so far that the laser propagation at high-intensity has been observed over paths greater than tens of Rayleigh lengths. In the present letter we introduce a ray optics hamiltonian approach [9] to the propagation of a ultraintese laser through an underdense plasma, with the main aim to study the effects of relativistic self focusing. Other fundamental works have been done in the past about the relativistic self focusing, all based on non linear wave-optics, eventually recalling a hamiltonian formalism related to non-linear Schroedinger equations. Starting from the consideration that the ratio $\lambda_0/\lambda_p << 1$ in typical cases of laser-plasma interaction, where $\lambda_p = 2\pi c/\omega_p$ is the plasma wavelength, we develop a different approach to the physical problem under study, based on ray-optics instead than wave-optics, in particular we are going to make use of hamiltonian optics. In fact the parameter λ_0/λ_p can set the border between a undulatory description with respect to a geometrical description of the laser light propagating through an underdense plasma, indeed when also the condition $\sigma >> \lambda_0$ is satisfied, where σ is the beam size parameter. The typical length scale to compare in this case to the laser wavelength in order to determine if a wave optics approach is needed is the plasma wavelength. For $\lambda_0/\lambda_p << 1$ a geometrical approach is justified. We show that, even with some limitations in the description of laser dynamics which nevertheless don't compete to the Relativistic Self Focusing (RSF) effect, a simple model can be built in order to study the RSF based on geometric, hamiltonian optics. We are going to study the trajectory of the light rays throughout an underdense plasma in the condition for RSF to occur. In the end a beam envelope trend is obtained, after the reconstruction of the laser beam starting from the single ray components. Computationally speaking, this approach turns out to be very fast, suitable for quick physical estimations which PIC codes, even if more complete and exhaustive in the physical description of phenomena, are not able to provide.

2 Theoretical background

The hamiltonian for a light ray of a light beam propagating in the z direction through a medium characterized by an index of refraction $n(r,z)$ is given by:

$$H = -\sqrt{n^2 - p_r^2} \tag{2}$$

where p_r is the light ray radial momentum. The Hamilton equation for the divergence $\theta = dr/dz$ is:

$$\frac{\partial H}{\partial p_r} = \theta = \frac{p_r}{\sqrt{n^2 - p_r^2}} \tag{3}$$

from which the useful relation is obtained:

$$p_r = \frac{n\theta}{\sqrt{1+\theta^2}} \tag{4}$$

where we used equations 2 and 3 to get the result 4. The second Hamilton equation gives:

$$\frac{dp_r}{dz} = \frac{n}{\sqrt{n^2 - p_r^2}}\frac{dn}{dr} = \frac{n\theta}{p_r}\frac{dn}{dr} = \sqrt{1+\theta^2}\frac{dn}{dr} \tag{5}$$

By deriving equation 3 by z (under the assumption $d\theta/dz >> dn/dz$, corresponding to the slowly varying envelope approximation for the laser) and equating to equation 4, an equation for the only variable θ is obtained:

$$\frac{d\theta}{dz} \approx (1+\theta^2)^2\frac{d\log n}{dr} \tag{6}$$

The radial coordinate of the light ray as function of z comes directly from the integration of θ after solving the equation 5, namely:

$$r(z) = r_0 + \int_0^z \theta dz \tag{7}$$

where r_0 is the initial radial position of the light ray. Notice that equation 5 is analytically integrable, yielding the following result for θ:

$$\frac{1}{2}\Big(\frac{\theta}{1+\theta^2} + arctan\theta\Big) = \Theta + \int_0^z \frac{d\log n}{dr}dz \tag{8}$$

where Θ is an integration constant. When considering for example a laser beam it is possible to define, once the radial distribution R(r) of the laser energy is known, the beam radius:

$$\sigma(z) = \sqrt{< r(z,r_0)^2 > - < r(z,r_0) >^2} \tag{9}$$

where we defined the average on the radial distribution of the laser beam energy as:

$$< f(r,r_0) > = \frac{\int_0^\infty dr_0 R(r_0) f(r,r_0)}{\int_0^\infty dr_0 R(r_0)} \tag{10}$$

For what concerns the index of refraction of the underdense plasma we consider here, we make the following assumptions: the regime of the acceleration is highly nonlinear, in particular we refer to the bubble regime, the electron plasma density at the front of the laser pulse is assumed unmodified during the propagation. We neglect longitudinal non-linear effects like steepening of the front pulse, only retaining in an empiric way the effect of the pump depletion. These choices are motivated by the fact that the main concern here is to study the transverse dynamics of the laser beam under consideration which undergoes relativistic self-focusing. The expression of the laser normalized potential we take into account is:

$$a = \frac{a_0}{\sqrt{2}} \exp\Big(-\frac{r^2}{\sigma_0^2} - \frac{z}{2L_{pd}}\Big) \tag{11}$$

where the pump depletion length is $L_{pd} = c\tau\omega_p^2/\omega_0^2$ and $R(r_0) = \exp(-2r^2/\sigma_0^2)$ [10]. The laser pulse duration is τ and the initial laser beam size parameter is σ_0.

3 Comparison with current experiments

In the following we develop a calculation referring to an experiment of Laser Wakefield Acceleration (LWFA) performed at the SPARC-LAB test facility through the interaction of the ultrashort ultraintense $Ti : Sa$ laser FLAME with a He gas-jet target [11, 12]. The initial normalized vector potential relative to the experiment was $a_0 \sim 4$, corresponding to a laser pulse of energy $\lesssim 1J$, duration $\tau = 30\ fs$ and focal spot radius $6 \pm 1\ \mu m$, equivalent to $\sigma_0 = 8.5\ \mu m$ in Eq. 10. A single-shot set of diagnostic were adopted to determine the extension of the plasma generated by the laser pulse and the relative plasma electron density. In particular a Mach-Zender interferometer was installed in order to retrieve the electron plasma density value. The result of the measurement was $n_e \sim 8 \times 10^{18} cm^{-3}$. In Fig. 1 the interferogram of the laser-produced plasma channel due to the propagation of the ultraintense laser FLAME through the supersonic He gas-jet, from left to right, is reported. The resolution of the image was $5\ \mu m$ per pixel. Once the laser and plasma

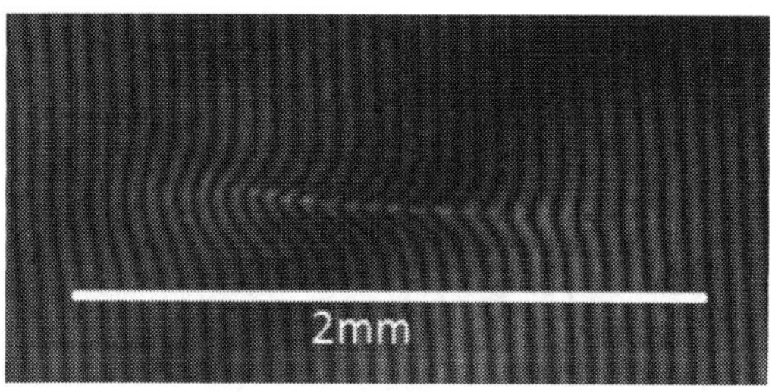

Figure 1. The interferogram of the laser-produced plasma channel due to the propagation of the ultraintense laser FLAME through the supersonic He gas-jet, from left to right.

parameters are known it becomes possible to exploit the model developed above for evaluations regarding relativistic self focusing, in order to verify the validity of the model itself in comparison to the experimental data. The first calculation we show is reported in Fig. 2 where the expected trend of the laser beam radius σ in vacuum is compared to the one obtained by virtue of the relativistic self focusing, when the ultraintense laser pulse FLAME propagates through the underdense He plasma inducing relativistic electron quivering. The beam radius in presence of the plasma fluctuates while it is guided by RSF. After a depletion length $L_{pd} \sim 1.9\ mm$ the laser radius starts to increases because of the upper hand of the defocusing. The region of high laser intensity is slightly more than one millimeter which is almost ten times the Rayleigh length corresponding to the considered laser beam. The situation is more evident if looking at Fig.3 where the normalized laser intensity a^2 is represented. The laser intensity is reconstructed by summing up the contributions of each light ray: the local density of the light rays is proportional to the laser local intensity. The imposition of the initial condition 10 ensures a self-consistent way to quantitatively trace the laser intensity along the plasma channel. The calculated extension of the laser high-intensity region is reasonably comparable to the extension of the high electron plasma density region of Fig. 1, recognizable as the region with more-pronounced curvature of the interferometric lines. In effect a laser high-intensity region must correspond to high electron plasma density by virtue of the local ionization efficiency, which increases with the local photon number.

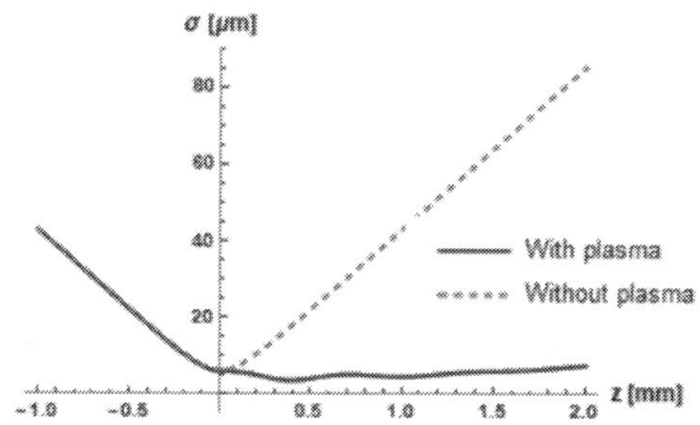

Figure 2. The comparison between the free propagation beam radius of the FLAME laser in vacuum (red) and the beam radius determined by the relativistic self focusing of FLAME propagating inside the He underdense plasma (blue). The zero point corresponds to the beginning of the plasma region ($z > 0$).

In fact the optical field ionization (OFI) is a multiphotonic process, for which the probability of tunneling ionization of a bound electron increases with the local laser intensity. Fig. 3 contains even another important information regarding the distribution of the laser energy along the plasma channel, very high in the beginning of the interaction and then slowly decreasing. This trend which we can observe in our calculation is in effect a physical feature we can observe even in Fig. 4. In fact a second diagnostics for the laser propagation inside the He gas-jet was adopted, based on optical Thomson scattering [13]. In this process the laser light is scattered by the plasma electrons escaping from the bound atomic states with a residual kinetic energy, and consequently it is Doppler-shifted in a way correlated to the electron velocities. Fig. 4 shows a RGB picture of the plasma channel generated by the laser propagating in the He gas. The colors are related to the local electron plasma temperature and also to broadening of the laser spectrum due to Self-Phase Modulation [14], which in turn is proportional to the local laser intensity. The plasma channel length is comparable to the length cal-

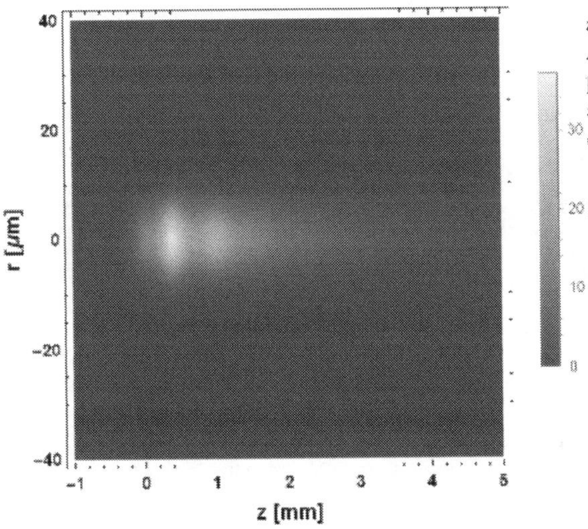

Figure 3. Normalized laser intensity a^2 of the laser FLAME propagating from left to the right from vacuum to the He gas-jet. The zero point corresponds to the beginning of the plasma region ($z > 0$).

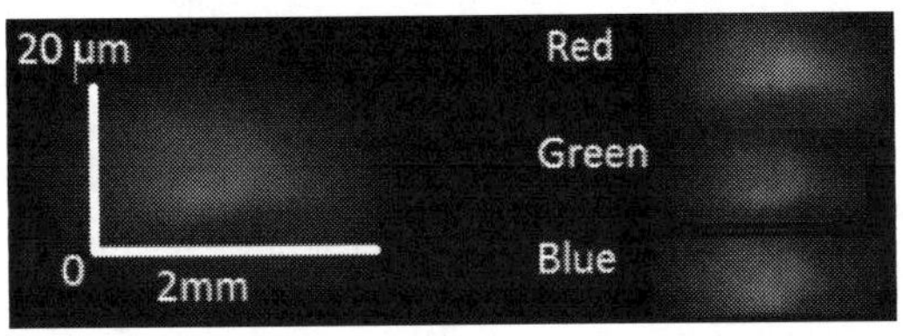

Figure 4. Picture of the plasma channel in RGB colors: red (570-750 nm), green (495-570 nm), blue (380-495 nm).

culated from Fig. 3, in particular the length corresponding to the blue light in Fig. 4 is comparable to the length of maximum intensity of the laser calculated from Fig. 3, which is about 1.2 *mm*. The ray optics model for relativistic self focusing can reproduce the extension of the region of the maximum laser intensity, which is an useful benchmark for laser plasma acceleration experiments where the knowledge of the laser maximum intensity region extension allows the determination of the most suitable plasma density for the acceleration. In fact the dephasing length for the accelerating electrons, i.e. the length after which the electrons start to experience a decelerating field, it is strongly determined by the plasma density and it is of fundamental importance to know to what extent the laser can propagate at high intensity producing wakefields that accelerate electrons with the maximum efficiency.

4 Conclusions

We have presented a model based on hamiltonian ray optics capable of reproducing experimental results related to relativistic self focusing phenomena. The ratio λ_0/λ_p determines the limit between the geometrical and wave optics. An underdense plasma, namely a plasma in which a laser pulse can propagate, always satisfies the condition $\lambda_p >> \lambda_0$, therefore a geometrical optics description of phenomena like the relativistic self focusing is possible and justifiable. The comparison with experimental data confirm this assumption.

References

[1] D. Strickland and G. Mourou, "Compression of amplified chirped optical pulses", Opt. Commun. 56, 219 (1985).

[2] Borovsky, Andrew V., et al. Laser physics at relativistic intensities. Vol. 34. Springer Science Business Media, 2013.

[3] Askar'yan, G. A. (1962). "Cerenkov Radiation and Transition Radiation from Electromagnetic Waves". Journal of Experimental and Theoretical Physics. 15 (5): 943–6.

[4] Max, Claire Ellen, Jonathan Arons, and A. Bruce Langdon. "Self-modulation and self-focusing of electromagnetic waves in plasmas." Physical Review Letters 33.4 (1974): 209.

[5] Hora, Heinrich. "Theory of relativistic self-focusing of laser radiation in plasmas." JOSA 65.8 (1975): 882-886.

[6] Feit, M. D., A. M. Komashko, and A. M. Rubenchik. "Relativistic self-focusing in underdense plasma." Physica D: Nonlinear Phenomena 152 (2001): 705-713.

[7] Esarey, E., C. B. Schroeder, and W. P. Leemans. "Physics of laser-driven plasma-based electron accelerators." Reviews of Modern Physics 81.3 (2009): 1229.

[8] Mourou, Gerard A., Toshiki Tajima, and Sergei V. Bulanov. "Optics in the relativistic regime." Reviews of modern physics 78.2 (2006): 309.

[9] Buchdahl, Hans Adolph. An introduction to Hamiltonian optics. Courier Corporation, 1993.

[10] Lu, Wei, et al. "Generating multi-GeV electron bunches using single stage laser wakefield acceleration in a 3D nonlinear regime." Physical Review Special Topics-Accelerators and Beams 10.6 (2007): 061301.

[11] Curcio, A., et al. "Trace-space reconstruction of low-emittance electron beams through betatron radiation in laser-plasma accelerators." Physical Review Accelerators and Beams 20.1 (2017): 012801.

[12] Curcio, A., et al. "Single-shot non-intercepting profile monitor of plasma-accelerated electron beams with nanometric resolution." Applied Physics Letters 111.13 (2017): 133105.

[13] Hutchinson, I.H. (2005). Principles of Plasma Diagnostics. Cambridge University Press.

[14] Giulietti, D. and Curcio, A. Self-Phase Modulation Eects as Laser Produced Plasma Diagnostics. Journal of Instrumentation 11(08):C08011-C08011 August 2016 DOI: 10.1088/1748-0221/11/08/C08011

Numerical studies of acceleration of thorium ions by a laser pulse of ultra-relativistic intensity

Jaroslaw Domanski*, and *Jan* Badziak

Institute of Plasma Physics and Laser Mirofusion, 01-497 Warsaw, Poland

Abstract. One of the key scientific projects of ELI-Nuclear Physics is to study the production of extremely neutron-rich nuclides by a new reaction mechanism called fission-fusion using laser-accelerated thorium (^{232}Th) ions. This research is of crucial importance for understanding the nature of the creation of heavy elements in the Universe; however, they require Th ion beams of very high beam fluencies and intensities which are inaccessible in conventional accelerators. This contribution is a first attempt to investigate the possibility of the generation of intense Th ion beams by a fs laser pulse of ultra-relativistic intensity. The investigation was performed with the use of fully electromagnetic relativistic particle-in-cell code. A sub-μm thorium target was irradiated by a circularly polarized 20-fs laser pulse of intensity up to 10^{23} W/cm^2, predicted to be attainable at ELI-NP. At the laser intensity ~ 10^{23} W/cm^2 and an optimum target thickness, the maximum energies of Th ions approach 9.3 GeV, the ion beam intensity is > 10^{20} W/cm^2 and the total ion fluence reaches values ~ 10^{19} ions/cm^2. The last two values are much higher than attainable in conventional accelerators and are fairly promising for the planned ELI-NP experiment.

1 Introduction

There are several different ways that heavy elements are produced in the Universe. Elements up to iron are produced via the thermonuclear fusion process in stars. Heavier nuclei could be produced in the slow neutron capture process (s-process) along the valley of stability. This process occurs inside the asymptotic giant branch (AGB) stars and lead to about half of the heavier nuclei [1,2]. The rest of the heavier nuclei are produced via a rapid neutron capture process (r-process). The astrophysical site of the r-process nucleosynthesis is still under debate: it may be cataclysmic core collapse supernovae (II) explosions with neutrino winds [2,3] or mergers of neutron-star binaries [2,4]. While the lower-mass path of the r-process for the production of heavy elements is well explored, the nuclei around the N = 126 (number of neutrons) waiting point critically determine this element production mechanism. At present, basically nothing is known about these nuclei [2]. For this reason an experiment was proposed for ELI-Nuclear Physics (ELI-NP) laser facility in which extremely neutron rich nuclei from this region will be produced via the fission-fusion reaction mechanism.

ELI-NP is a currently build large European research facility that will produce multi-PW femtosecond laser pulses of ultra-relativistic intensity (~10^{23} W/cm^2) to perform research in laser-based nuclear physics and related applications [5]. In the basic concept of the fission-fusion reaction mechanism [2,6], fissile isotopes (like ^{232}Th) accelerated by an intense laser pulse will enable the interaction of a dense beam of fission fragments with a second target, also consisting of fissile isotopes. So finally in a second step of the reaction process, fusion between (neutron-rich) beam-like and target-like (light) fission products will become possible, generating extremely neutron-rich ion species. A fission-fusion reaction scenario was proposed for the experiment, wherein the accelerated thorium ions will be fissioned in the CH_2 layer of the reaction target. Furthermore, the accelerated carbon ions and deuterons from the production target generate thorium fragments in the thick thorium layer of the reaction target. This scenario is more efficient than the one where fission would be induced by the thorium ions only. The required parameters for Th ion beams in this scenario are: energy –E_i/A=7 MeV/nucleon, fluence – $F_i=10^{18}$ ion/cm^2 and number of ions - $N_i=10^{11}$ [2,6].

This paper is a first attempt to investigate the possibility of the generation of intense Th ion beams by a fs laser pulse of ultra-relativistic intensity. The investigation was performed with the use of fully electromagnetic relativistic multi-dimensional (2D3V) particle-in-cell code. The code was elaborated at the Institute of Plasma Physics and Laser Microfusion and verified for example by a comparison of the results obtained using the code with the ones from measurements presented in [7,8] as well as with the results of 3D simulations of carbon ion acceleration carried out by [9]. The numerical simulations were performed for flat or spherical thorium targets of thickness ranging from 25 nm to 700 nm (which corresponds to the areal mass density from 0.07 mg/cm^2 to 0.9 mg/cm^2) and the transverse dimension equal to 12

* Corresponding author: jaroslaw.domanski@ifpilm.pl

 https://doi.org/10.1051/epjconf/201816701004

μm. Molecular density of the target corresponded to the solid-state density and was equal to $3.04*10^{22}$ molecules/cm^3 and the ionization level of Th was assumed to 10. A pre-plasma layer of 0.25 μm thickness and the density shape described by an exponential function was placed in front of the target. The target was irradiated by a circularly polarized 20 fs laser pulse of intensity up to 10^{23} W/cm^2 and energy equal to 150 J predicted to be attainable at ELI-NP. The laser beam shape in time and space (along the y-axis) was described by a super-Gaussian function of the power index equal to 6. The laser beam width (FWHM-Full Width at Half Maximum) was ranging from 3μm to 9 μm and the laser wavelength was equal to 800nm.

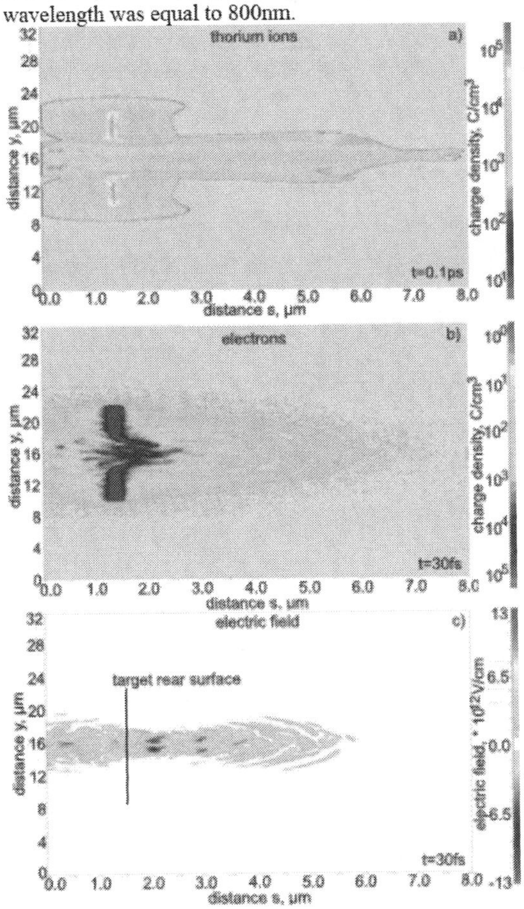

Fig. 1. 2D spatial distribution of charge density of thorium ions (a), electrons (b) and component of electric field parallel to the y axis (c). The 300 nm (0.4mg/cm^2) thick flat target was irradiated by the laser beam of 3μm in diameter and intensity equal to 10^{23} W/cm^2. The simulation time t=0.1ps for (a) and t=30fs for (b,c)

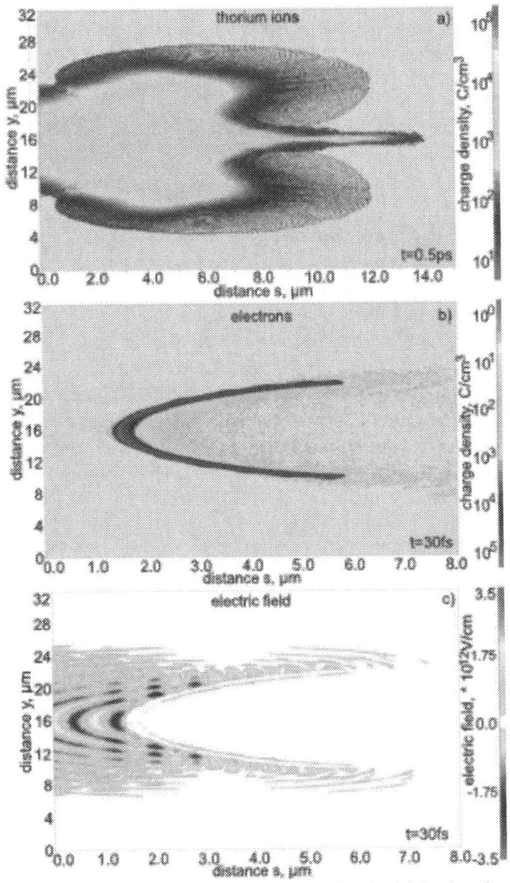

Fig. 2. 2D spatial distribution of charge density of thorium ions (a), electrons (b) and component of electric field parallel to the y axis (c). The 400 nm (0.5mg/cm^2) thick hemispherical target was irradiated by the laser beam of 9μm in diameter and intensity equal to $1.1*10^{22}$W/cm^2. The simulation time t=0.5ps for (a) and t=30fs for (b,c)

2 Results and discussion

Figures 1 and 2 present a 2D spatial distributions of charge density of thorium ions (a), electrons (b) and the component of electric field parallel to the y axis (c). The 300 nm (0.4mg/cm^2) thick flat target (Fig.1) and the 400nm (0.5mg/cm^2) thick hemispherical target (Fig.2) were irradiated by the laser beams of 3 μm (Fig.1) and 9 μm (Fig.2) diameters. The beam intensity was equal to $1.0*10^{23}$ W/cm^2 for the flat target and $1.1*10^{22}$ W/cm^2 for the hemispherical one. For both considered cases the laser pulse parameters correspond to the radiation pressure acceleration (RPA) mechanism domination range [10,11,12,13]. However, for the flat target the relativistic transparency [14,15] of the target is observed (Fig.1c) since the electron density in the interaction area decreases below the relativistic critical density ($e^* n_{cr\text{-}rel}=3.0*10^4$ C/cm^3). The relativistic transparency is the main reason here for the broad energy (velocity) spectrum of accelerated ions (Fig.1a). In case of the

* Corresponding author: jaroslaw.domanski@ifpilm.pl

hemispherical target, the electron density in the target remain higher than the relativistic critical density ($e^* n_{cr\text{-}rel}=9.8*10^3$ C/cm^3) and the target is not transparent for the light (Fig, 2c). However, the target is also penetrated by the laser pulse (the ions are not compressed by the laser pulse as effective as electrons) and the energy spectrum of accelerated ions is also broad.

The maximum ion energies as a function of target thickness (a), target focal length and laser beam diameter (b) is shown in Fig. 3. The results in Fig.3a correspond to the flat targets irradiated by the 3μm laser beam. The thickness of the targets for data in Fig.3b is equal to 400nm. It is visible that relativistic transparency of the targets hinders the achievement of high acceleration efficiency for thin targets (< 300nm) (Fig. 3a). Furthermore, the weak influence of the target shape on

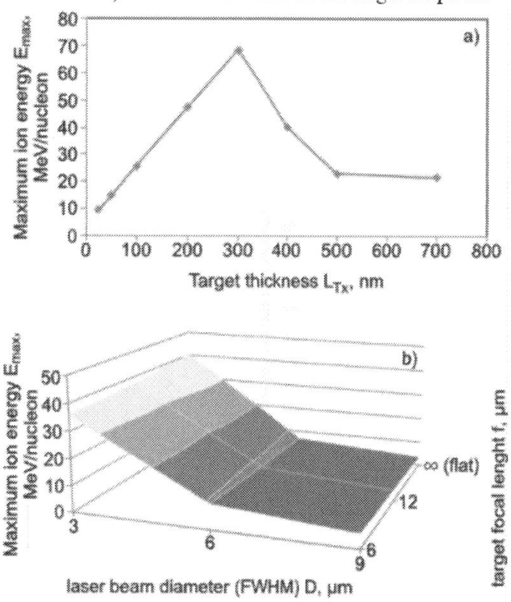

Fig. 3. The maximum ion energy as a function of the flat target thickness (a), and as a function of the hemispherical target focal length and the laser beam diameter (b). For all cases in figure (b) the targets are 400 nm thick. The laser beams energy is equal to 150 J.

the maximum ion energies is observed. Finally, it is worth mentioning that for all investigated cases the ion energy spectra are broad and the standard energy deviations expressed in the mean energy units are higher than one (Fig.4). Despite these unfavorable factors, the maximum thorium ion energies are much higher than required in experiment [2,6] and reach 70 MeV/nucleon.

From the point of view of the proposed nuclear experiment [2,6], the characteristics of ion beam such as the beam intensity, diameter and fluence as well as the total number of accelerated ions are also important. Figure 5 presents the maximum ion beam intensity (a) and the ion beam diameter (b), recorded 10 μm behind the rear surface of the target, as a function of the flat target thickness. For all investigated cases the laser beam diameter on the target is equal to 3 μm. The intensity of ion beam is higher for thicker targets (≥300nm) and the best results is achieved for the 400nm, the thinnest

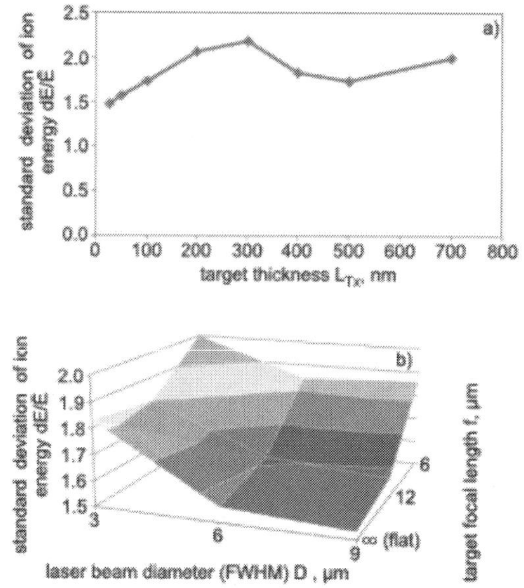

Fig. 4. The standard deviation of ion energy as a function of the flat target thickness (a), and as a function of the hemispherical target focal length and the laser beam diameter (b). For all cases in figure (b) the targets are 400nm thick. The laser beams energy is equal to 150 J.

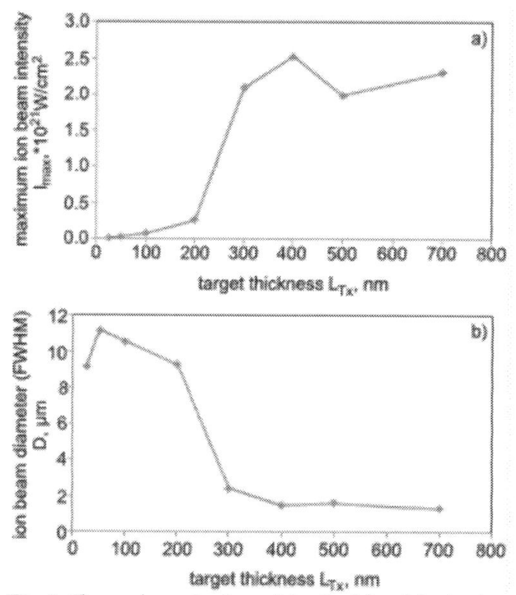

Fig. 5. The maximum ion beam intensity (a) and the ion beam diameter (b), recorded 10 μm behind the target rear surface, as a function of the flat target thickness. For all investigated cases the laser beam diameter is equal to 3 μm.

opaque, target. In this case the ion beam of intensity of $2.5*10^{21}$ W/cm^2 and diameter of 1.5 μm is produced.

The maximum ion beam intensity (a) and the ion beam diameter (b) as a function of the hemispherical target focal length and the laser beam diameter is shown

* Corresponding author: jaroslaw.domanski@ifpilm.pl

in figure 6. The target thickness is fixed at 400 nm, the value which ensures the highest ion beam intensity for the flat targets. For the 3 μm laser beam and the target of the focal length equal to 6 μm the ion beam of the highest intensity is produced. In this case the intensity is equal to $8*10^{21}$ W/cm^2, the beam width is ~1μm, the maximum thorium ion energy is ~9,3 GeV (40MeV/nucleon) and the ion pulse duration is ~70 fs.

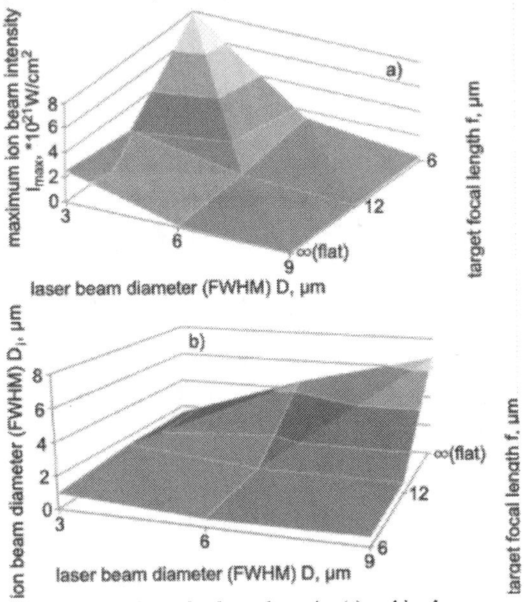

Fig. 6. The maximum ion beam intensity (a) and ion beam diameter (b) as a function of target focal length and laser beam diameter. The target thickness is fixed on 400 nm.

The estimated ion fluence is equal to $7.0*10^{18}$ ion/cm^2 and is even higher than required in the experiment. Furthermore, it is worth mentioning that curving the target leads to focusing of the ion beam and increasing of the beam intensity which is clearly visible in figure 6 and also in figure 2a.

The electron distribution in plasma does not cover exactly the thorium ion distribution and locally the net electric charge in the plasma is different from zero. The inhomogeneous distribution of the charge in the plasma leads to the generation of electric current with the total net value $J=J_{Th}-J_e$ of which can be fairly high (J_e and J_{Th} are the electron current and the thorium ion current, respectively). The total current J changes in time and is inhomogeneously distributed in space as demonstrated in figures 7 and 8. Fig. 7 presents the temporal run of the total current (integrated over the whole transverse size of the simulation box) at the distance of 10 μm from the target rear surface while Fig.8 shows the transverse distribution of the current density at this distance at t=325fs. The comparison of the results for flat 400 nm thick and hemispherical (D_i=6μm) targets irradiated by laser pulse of intensity equal to 10^{23} W/cm^2 and 3μm diameter is presented. It can be seen that the structure of the total current distribution is similar for both types of investigated targets and is also similar to that obtained for CH target [16]. It is also visible that at the time longer than 200fs (an order of magnitude longer than the laser pulse) the total current is dominated by fast electrons (the current is negative) and reaches the value of several MA both for flat and hemispherical targets. However, the maximum value of the current is clearly higher for the hemispherical target. This strong flux of

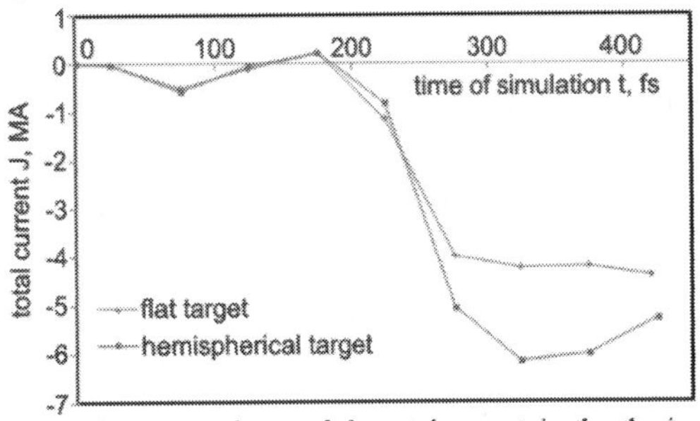

Fig. 7. The temporal run of the total current in the thorium plasma (integrated over the whole transverse size of the simulation box) at the distance of 10 μm from the rear surface of the target. The comparison of the results for the 400-nm flat and hemispherical (D_i=6μm) targets irradiated by the laser beam of intensity of 10^{23} W/cm^2 and diameter of 3μm.

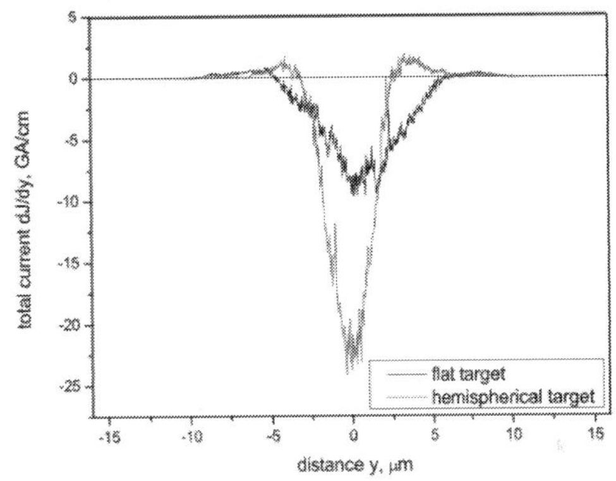

Fig. 8. The transverse distribution of the current density at the distance of 10 μm from the rear surface of the target at t=325fs. The comparison of the results for the 400-nm flat and hemispherical (D_i=6μm) targets irradiated by the laser beam of intensity of 10^{23} W/cm^2 and diameter of 3μm.

fast electron escaping from the target is the generator of an intense electromagnetic emission (usually in THz domain of frequency) and moreover, it is the cause of the creation of net positive charge of the target (an electric neutrality of the target is disturbed by loss of negative charge carried by the fast electron flux). The neutralization current flowing through the target and the target holder can be the source of a strong electromagnetic pulse (EMP), usually in the (0.1 – 10) GHz domain which propagates into the interaction chamber and may be very detrimental for diagnostic equipment and other electronic devices [17,18]. The strength of EMP is proportional to the power of the laser pulse interacting with the target. For this reason the problem of EMP generated in the laser-target interaction is currently a hot topic intensely studied and is especially important for multi-PW laser facilities like ELI. The results presented in this paper show briefly the way the

* Corresponding author: jaroslaw.domanski@ifpilm.pl

fundamental source of EMP is created in the initial stage of the process when the ion beam is formed and accelerated due to the laser-target interaction. More detailed analysis of this issue will be presented in our other paper.

3 Conclusions

In conclusion, the generation of intense Th ion beams by a fs laser pulse of ultra-relativistic intensity has been investigated with the use of relativistic 2D3V particle-in-cell code. A sub-μm thorium target was irradiated by a circularly polarized 20-fs laser pulse of intensity up to 10^{23} W/cm^2 predicted to be attainable at ELI-NP. It was found that a key effect that hinders the achievement of high efficiency of ion acceleration in the considered cases and leads to the broad energy spectrum of accelerated ions is relativistic transparency of the target plasma. Despite this, for the 3 μm laser beam of intensity of 10^{23} W/cm^2 and the hemispherical target of the focal length of 6 μm a thorium ion beam of the maximum ion energy ~9,3 GeV (40MeV/nucleon), the peak beam intensity of $8*10^{21}$ W/cm^2 and the beam fluence of $7.0*10^{18}$ ion/cm^2 is produced. Such ion beam parameters are much higher than attainable in conventional accelerators and are fairly promising for the planned ELI-NP experiment. The key problem which needs to be solved is the broad energy spectrum of the accelerated ions that will be the subject of our further studies.

This work was supported in part by the National Centre for Science (NCN), Poland under the Grant No. 2014/14/M/ST7/00024. The simulations were carried out with the support of the Interdisciplinary Center for Mathematical and Computational Modelling (ICM), University of Warsaw under grant no. G57-20.

References

1. R. Gallino, C. Arlandini, M. Busso, M. Lugaro, C. Travaglio, O. Straniero, A. Chieffi, M. Limongi, The Astrophysical Journal **497**,388-403 (1998)
2. F. Negoita, M. Roth, P. G. Thirolf, S. Tudisco, F. Hannachi, S. Moustaizis, I. Pomerantz, P. McKenna, J. Fuchs, K. Sphor, G. Acbas, A. Anzalone, P. Audebert, S. Balascuta, F. Cappuzzello, M. O. Cernaianu, S. Chen, I. Dancus, R. Freeman, H. Geissel, P. Ghenuche, L. Gizzi, F. Gobet, G. Gosselin, M. Gugiu, D. Higginson, E. D'Humieres, C. Ivan, D. Jaroszynski, S. Kar, L. Lamia, V. Leca, L. Neagu, G. Lanzalone, V. Meot, S. R. Mirfayzi, I. O. Mitu, P. Morel, C. Murphy, C. Petcu, H. Petrascu, C. Petrone, P. Raczka, M.Risca, F. Rotaru, J. J. Santos, D. Schumacher, D. Stutman, M. Tarisien, M. Tataru, B. Tatulea, I. C. E. Turcu, M. Versteegen, D. Ursescu, S. Gales, N. V. Zamfir, Romanian Reports in Physics **68**, Supplement, S37-S144 (2016)
3. I. V. Panov, H.-Th. Janka, Astr. Astroph., **494**, 829 (2009)
4. R. Surman, G.C. McLaudhlin, M. Ruffert, H.-Th. Janka, W. R. Hix, The Astrophysical Journal **679**, L117-L120 (2008)
5. C. Danson, D. Hillier, N. Hopps, D. Neely, High Power Sci. Engineering **3**, e3 (2015)
6. D. Habs, P. G. Thirolf, M. Gross, K. Allinger, J. Bin, A. Henig, D. Kiefer, W. Ma, J. Schreiber, Appl. Phys. B **103**, 471 (2011)
7. J. Badziak, S. Jabłoński, P. Parys, M. Rosiński, J. Wołowski, A. Szydłowski, P. Antici, J. Fuchs, A. Mancic, J. Appl. Phys. **104**, 063310 (2008)
8. J. Badziak, P. Antici, J. Fuchs, S. Jabłoński, A. Mancic, P. Parys, M. Rosiński, R. Suchańska, A. Szydłowski, J. Wołowski, AIP Conf. Proc. **1024**, 63-77 (2008)
9. A. Sgattoni, S. Sinigardi, A. Macchi, Appl. Phys. Lett. **105**, 084105 (2014)
10. T. Esirkepov, M. Borghesi, S.V. Bulanov, G. Mourou, T. Tajima, Phys. Rev. Lett. **92**, 175003 (2004)
11. A. P. L. Robinson, M. Zepf, S. Kar, R. G. Evans, C. Bellei, New J. Phys. **10**, 012021 (2008)
12. O. Klimo, J. Piskal, J. Limpouch, V.T. Tikhonchuk, Phys. Rev. ST-Accelerators and Beams **11**, 031301 (2008)
13. T. Schlegel, N. Naumova, V.T. Tikhonchuk, C. Labaune, I. V. Sokolov, G. Mourou, Phys. Plasmas **16**, 083103 (2009)
14. A. Macchi, M. Borghesi, M. Passoni, Rev. Mod. Phys. **85**, 751 (2013)
15. J. Domanski, J. Badziak, S. Jabłonski, Laser Part. Beams **35**, 286-293 (2017)
16. J. Domanski, J. Badziak, S. Jabłonski, Proc. of SPIE **10445**, 1044543-1 (2017)
17. J.-L. Dubois, F. Lubrano-Lavaderci, D. Raffestin, J. Ribolzi, J. Gazave, A. Compant La Fontaine, E. d'Humieres, S. Hulin, Ph. Nicolai, A. Poye, V.T. Tikhonchuk, Phys. Rev. E **89**, 013102 (2014)
18. A. Poye, J.-L. Dubois, F. Lubrano-Lavaderci, E. d'Humieres, M. Bardon, S. Hulin, M. Bailly-Grandvaux, J. Ribolzi, D. Raffestin, J.J. Santos, Ph. Nicolai, V.T. Tikhonchuk, Phys. Rev. E **92**, 043107 (2015)

* Corresponding author: jaroslaw.domanski@ifpilm.pl

Why do Electrons with "Anomalous Energies" appear in High-Pressure Gas Discharges?

Andrey Kozyrev[1,2*], *Vasily* Kozhevnikov[1,2], and *Natalia* Semeniuk[1]

[1] Institute of High Current Electronics, Laboratory of Theoretical Physics, 634055 2/3 Akademichesky ave, Tomsk, Russia
[2] National Research Tomsk State University, Faculty of Physics, 634050 36 Lenin ave. Tomsk, Russia

Abstract. Experimental studies connected with runaway electron beams generation convincingly shows the existence of electrons with energies above the maximum voltage applied to the discharge gap. Such electrons are also known as electrons with "anomalous energies". We explain the presence of runaway electrons having so-called "anomalous energies" according to physical kinetics principles, namely, we describe the total ensemble of electrons with the distribution function. Its evolution obeys Boltzmann kinetic equation. The dynamics of self-consistent electromagnetic field is taken into the account by adding complete Maxwell's equation set to the resulting system of equations. The electrodynamic mechanism of the interaction of electrons with a travelling-wave electric field is analyzed in details. It is responsible for the appearance of electrons with high energies in real discharges.

1 Introduction

This paper deals with the phenomenon of runaway electrons generation in fast gas discharges under high pressures. The generation of runaway electrons in high-pressure pulsed discharges is a fundamental physical phenomenon. The possibility of runaway electrons appearing in the Earth atmosphere was predicted at first in 1925 [1], and much later the prediction was successfully proven in laboratory experiments [2].

At present the fact of fast (runaway) electrons detection can be firmly established at the initial stage of high-pressure gas breakdown in discharge gaps with strongly non-uniform electric field. At the same time, various researchers obtain fast electron current pulses with largely spread parameters: amplitudes from 0.1 up to tens of amperes with durations from tens picoseconds to nanoseconds [3, 4]. As the number of runaway electrons is strongly depends on several critical parameters (gas type, geometric enhancement of the electric field near the sharp edges of electrodes, scales of field-enhancement regions, time-resolution of the experimental equipment), so experimental results of fast electrons detection will also differ considerably.

Owing to modern experimental equipment and novel methods of experimental data analysis, it was shown that the runaway electron beams have a broadband spectrum containing several groups of electrons with different mean energies [5, 6]. The main amount of runaway electrons constitutes the group of particles with mean energy corresponding to the voltage amplitude at the gap. However, it was discovered that power spectrum also includes a small group of electrons with mean energy much greater than the applied values. Such electrons represent a group of electrons with so-called "anomalous energies".

For high-pressure discharges, it was shown that number of electrons with "anomalous" energies usually does not exceed ten percent of the total number of runaway electrons. On the other hand, electrons with "anomalous energies" do not yield to the equilibrium Maxwellian distribution, i.e. pointed out to the nontrivial mechanism of the formation of a runaway electron beam. These experimental results showed the need to formulate new theoretical models allow explaining the appearance of the "anomalous" runaway electron beam component self-consistently.

Essential non-stationarity and spatial three-dimensionality in real experiments represent a great challenge for the theoretical modeling. So simple zero-dimensional and one-dimensional plate theoretical models allowing to understand the mechanism of runaway electron beam formation, do not provide ideal correlations with the existing experimental data [7, 8].

A more realistic one-dimensional coaxial model provides better agreement between the theoretical and experimental data [9, 10]. Simulation show that the breakdown of the coaxial gas-filled gap occurs due to the rapid propagation of the ionization wave front between inner electrode (cathode) and outer electrode (anode). Namely, the necessary conditions for the generation of runaway electrons with "anomalous energies" are formed at the front of the ionization wave.

The aim of this paper is to describe in details the physical mechanism of electron acceleration to "anomalous energies".

* Corresponding author: kozyrev@to.hcei.tsc.ru

2 The pilot problem of electron acceleration by travelling electric field pulse

2.1 Statement of the model problem

We consider the one-dimensional interaction of a travelling-wave longitudinal electric field pulse with a single electron (initially at rest), as shown in Fig. 1. For the simplicity reason, it is convenient to consider quite simple rectangular shape of field strength profile. Let the amplitude of the pulse is equal to E, and the pulse width is equal to L.

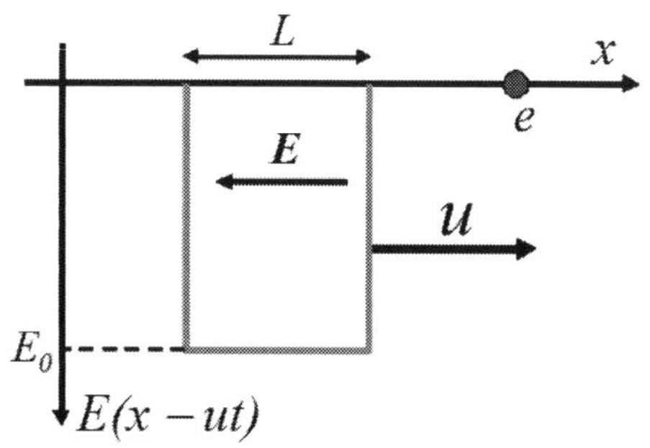

Fig. 1. Schematics of the model problem.

The field pulse moves from left to right with a constant velocity, u, and at some time point overlaps the rest electron. Electron starts to move accelerated if the electric field acts on it. In Fig. 2 the area where the electron interacts with travelling-wave field is shown with shaded region at the coordinate-time plane.

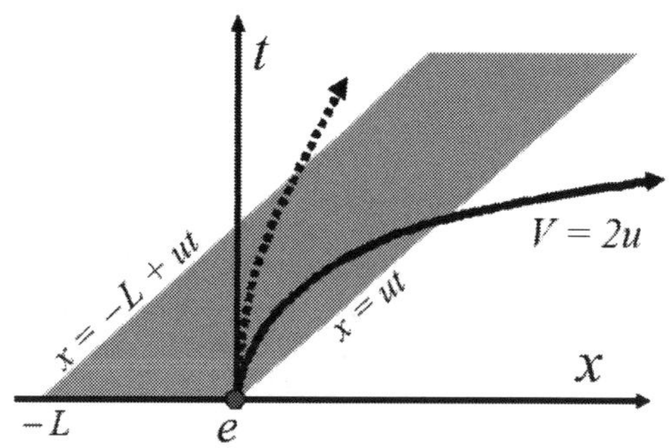

Fig. 2. Two modes of interaction of a field pulse and an electron. The solid line describes the "reflecting mode", the dashed line corresponds to the "escaping mode".

2.2 Mathematical description of the rectangular field pulse profile

For simplicity, let us consider the non-relativistic case $u \ll c$ (c is a speed of light). The generalization to the case of relativistic velocities is straightforward, it is only slightly complicates the mathematical formulation. While the electron is in the scope of travelling-wave, it moves with the constant acceleration (e is elementary charge, and m is mass of electron):

$$x(t) = \left(eE_0 / 2m\right)t^2. \tag{1}$$

As can be seen in Fig. 2, there are two fundamentally different scenarios of such interaction.

2.2.1 Escaping mode of interaction

If the rear edge of the electric field pulse advances the electron, the latter gains lower velocity. This mode is shown in Fig. 2 with the dashed curve. The phase trajectories of the electron and the trailing edge of the field pulse will intersect if the quadratic equation has a real root (when its discriminant D is positive):

$$\begin{gathered} x(t) = \left(eE_0 / 2m\right)t^2 = -L + ut, \\ D = u^2 - \left(2eE_0 L / m\right) > 0, \\ \frac{mu^2}{2} > eE_0 L. \end{gathered} \tag{2}$$

If the last condition of (2) is satisfied, then the final velocity of the electron remains less than the velocity of the field pulse, u. In this interaction mode, the electron escapes through the traveling electric field area. Therefore, it can be called the "escaping mode". This mode takes place at a low voltage drop across the pulse, $eU_0 = eE_0L$, as compared to the "relative kinetic energy" of the electron, $mu^2/2$.

2.2.2 Reflecting mode of interaction

To understand the cause of the appearance electrons with anomalous energies more important is another mode of interaction, when the condition (2) is not satisfied. In this case, the accelerating electron eventually outpaces the field pulse and leaves ahead (solid curve in Fig. 2). We can say that the electron is reflected from the incoming high-voltage pulse, and this mode can be called the "reflecting mode".

It is easy to show that the velocity of the electron that leaves forward is exactly twice the velocity of the field pulse. Indeed, the interaction time t_0 is determined by the equation: $x(t) = \left(eE_0 / 2m\right)t_0^2 = ut_0$. Therefore, the final velocity of the electron will be equal to $V = \left(eE_0 / m\right)t_0 = 2u$. The kinetic energy of the electron, ε, is appropriate to be written together with the condition of the reflecting mode:

$$\varepsilon = \frac{mV^2}{2} = 2mu^2, \quad \frac{mu^2}{2} < eU_0. \tag{3}$$

In accordance with (3), the kinetic energy of the electron will be limited from above:

$$\varepsilon = 2mu^2 = 4\frac{mu^2}{2} \le 4eU_0. \tag{4}$$

The last inequality shows that the electron passed forward may have kinetic energy exceeding the value of eU_0.

2.3 Generalization for general cases

After analyzing the simplest problem, we can extend our conclusions to more general cases. In particular, it is possible to extend the expressions (3) to more general profiles of the field pulse and to the relativistic energies.

2.3.1 Extension to more general pulse profiles

The fact that in the reflecting mode the electron acquires an exactly double value of the velocity of the traveling field pulse occurs for any bell-shaped pulse profiles *E(s)*. The restriction applies only to the function *E(s)* was single-humped and did not change the sign of the field.

For such a pulse profile, it is possible to determine the voltage drop

$$U_0 = \int_{-\infty}^{+\infty} E(s)ds \,. \qquad (5)$$

If we go over to the reference frame in which the pulse profile is stationary, then the electron bumps into the potential barrier with the same velocity u. If the height of the potential barrier will be greater than kinetic energy of the incident electron, it is reflected from it in the opposite direction at the same speed. With the reverse transformation to the laboratory frame, we obtain exactly twice the value of the velocity $V = 2u$.

Thus, expression (3) is completely valid for a more general field-pulse profile. The Fig. 3 shows the final kinetic electron energy from the velocity field pulse.

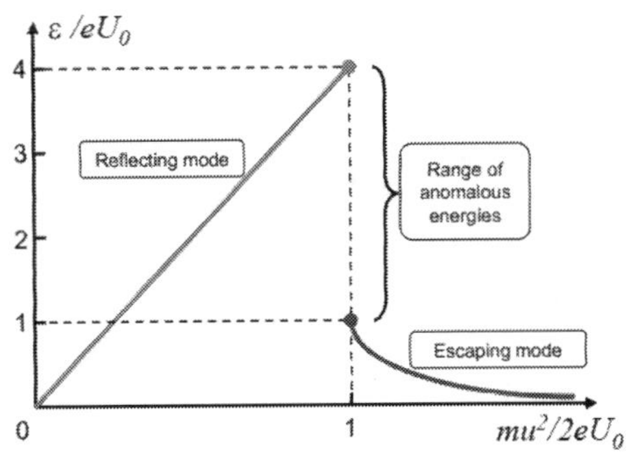

Fig. 3. Dependence of the final kinetic energy of the electron from the speed of motion of the field pulse at $u \ll c$.

2.3.2 Extension to relativistic energies

Finally, it is possible to extend the calculations to the relativistic energy range. In relativistic mechanics, it is customary to express the kinetic energies of particles ε using the relativistic factor γ.

Turning to the reference system, tied to a field pulse, and then back to the laboratory frame of reference, one can obtain the following expression for the velocity and kinetic energy of the electron interaction in the reflectance mode. The condition of the "reflecting mode" (3) is written in the following form:

$$mc^2(\gamma - 1) < eU_0, \quad \gamma = \frac{1}{\sqrt{1-u^2/c^2}}. \qquad (6)$$

The electron velocity, *V*, in the laboratory system after its reflection from the traveling field pulse will be determined by the law of velocity composition:

$$V = \frac{2u}{1+u^2/c^2}. \qquad (7)$$

The kinetic energy of the reflected electron, ε, is uniquely determined by the field pulse factor γ:

$$\varepsilon = mc^2\left(\frac{1}{\sqrt{1-V^2/c^2}} - 1\right) = mc^2 \cdot 2\left(\gamma^2 - 1\right). \qquad (8)$$

Formulas (5), (6) and (8) generalize two formulas (3) to arbitrary pulse profile and relativistic energies. For low potential barriers $eU_0 \ll mc^2$ from (8) we obtain the classical formulas (4) for the maximum electron energy: $\varepsilon_{max} \approx 4 \cdot eU_0$. But, in the ultra-relativistic limit, when $eU_0 \gg mc^2$, $\gamma \gg 1$ take place, we obtain expression for the maximum electron energy: $\varepsilon_{max} \approx 2\gamma \cdot eU_0$.

3 Kinetic model of high-pressure gas discharge in coaxial gap

Let us show that in the nanosecond breakdown of a gas-filled coaxial gap there are conditions for the generation of electrons with "anomalous energies". The theoretical analysis is based on the model of the kinetic description of the electronic component of gas-discharge plasma [11, 12].

We are going to describe the discharge process by using of the kinetic Boltzmann relativistic equation for electron distribution function (EDF) with the model right-hand part:

$$\gamma\left(\frac{\partial f}{\partial t} + \frac{p}{m\gamma}\frac{\partial f}{\partial r} - eE(r,t)\frac{\partial f}{\partial p}\right) = -Q_- + Q_+ + S_{sc}. \qquad (9)$$

Here $f(r, p, t)$ is the EDF depending on time t and radial coordinate r, $p = mv\gamma$ is the relativistic momentum, $\gamma = \sqrt{1+(p/mc)^2}$ is the relativistic factor, E is the electric field strength .

The first term Q_- on the right-hand side of (9) describes the loss of electrons in a given element of the phase space due to inelastic electron-atom collisions, the second term Q_+ describes the production of electrons in an element of the phase space in the same inelastic processes, and the third term S_{sc} describes the effect of scattering, in particular, elastic collisions on the EDF.

In the framework of the kinetic approach, we must write the integral expression for the convective electron current in any cross section of the discharge gap:

$$J_e(r,t) = -\frac{e}{m}\int_{-\infty}^{\infty}(p/\gamma)f(r,p,t)dp. \quad (10)$$

A coaxial diode (cathode radius r_c, anode cylinder radius r_a, discharge tube length L) is chosen as a simplified model of a discharge with nonuniform discharge-gap geometry. An electric pulse $U_0(t)$ with a steep leading edge and an amplitude by far exceeding the static gas breakdown voltage is applied to the gap switched in series with a voltage supply and a ballast resistance R. This equivalent power-supply circuit matches well with the discharge of the pulse forming line with impedance R loaded onto a diode, which is used in most experiments.

Using the Kirchhoff rule for the complete circuit, we can relate the voltage drop across the gap to the value of the total current in the discharge circuit J_{tot}:

$$J_{tot}(t) = -\frac{1}{2\pi rLR}\left(U_0(t) + \int_{r_c}^{r_a} E(r,t)dr\right). \quad (11)$$

Finally, using the conservation law for the total current (convective current of electrons plus Maxwell's displacement current), we can close our system by the evolution equation for the intensity of the electric field in the gap (ε_0 is vacuum permittivity):

$$\varepsilon_0\frac{\partial E(r,t)}{\partial t} = J_{tot}(t) - J_e(r,t). \quad (12)$$

The above system of equations (9)-(12) represents mathematical equations of the kinetic model of a fast gas discharge where the electron component of the discharge plasma is described from the first principles in terms of the EDF evolution.

4 Results of the discharge simulation in nitrogen

A discharge in a coaxial diode filled with nitrogen (cathode radius r_c = 1 mm, anode radius r_a = 10 mm, length L = 1 cm) is developed at a pressure of 760 Torr. A voltage pulse with leading edge duration 300 ps, in its level 0.1–0.9 of the amplitude 200 kV, was fed to the diode switched in series with ballast resistance R = 75 Ω.

To simplify the problem, we took into account only two types of electron-atom collisions, namely, impact ionization and elastic scattering collisions. The experimental data for these cross-sections are given in [13] for electron energies up to 1 keV. We extrapolate the dependence of the cross-sections to the high-energy region, using the asymptotic dependences [14]. The dependences of the cross section on the energy of the electron are shown in Fig. 4 (solid lines show experimental values from [13] and dashed lines correspond to high-energy extrapolations).

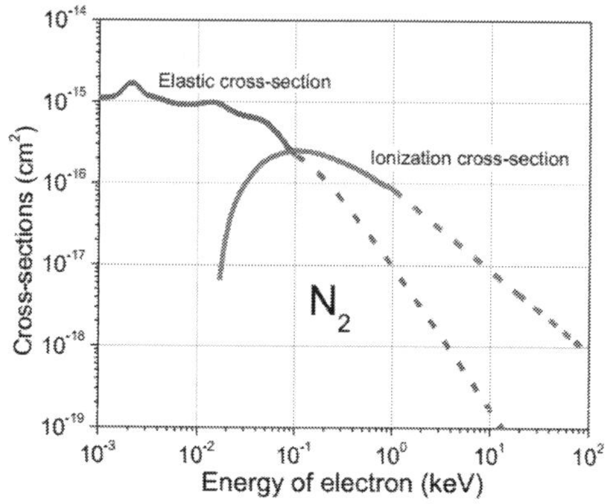

Fig. 4. Dependences of collision cross section on the energy of an electron in nitrogen.

The system of equations (9)-(12) is solved numerically with uniform initial conditions for the EDF and the electric field. We choose the initial Maxwellian EDF, which determines the initial electron density of $n_0 \sim 10^3$ cm^{-3} that homogeneously fills the discharge gap with the thermal spread $kT \sim 5$ eV.

Calculations show a detailed picture of the breakdown development. The current switching in the gap occurs due to its rapid filling with dense plasma, the front of which moves from the cathode (internal electrode) to the anode. The dynamics of the electric field distribution in the gap during the development of breakdown is shown in Fig. 5. It can be seen the field strength at the ionization front has a local maximum, which moves with increasing speed to the anode.

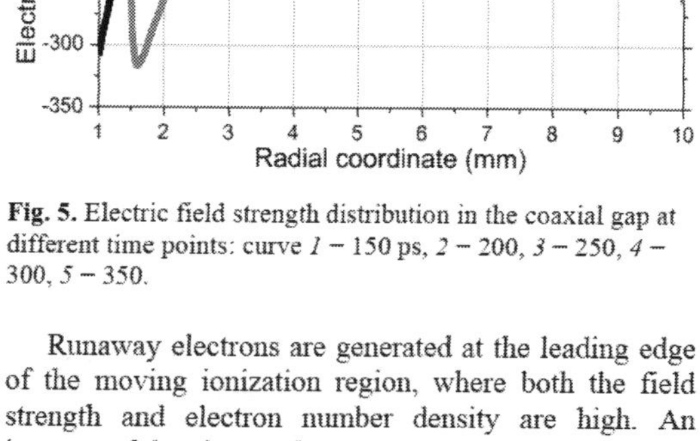

Fig. 5. Electric field strength distribution in the coaxial gap at different time points: curve *1* – 150 ps, *2* – 200, *3* – 250, *4* – 300, *5* – 350.

Runaway electrons are generated at the leading edge of the moving ionization region, where both the field strength and electron number density are high. An increase of the plasma electrons energy in the EDF tail is accompanied by a decrease of the collision cross section, and a group of fast electrons is formed.

This process is clearly shown in the phase portraits shown in Fig. 6. These functions are obtained as a solution to the Boltzmann equation (9). We pay attention to a group of runaway electrons that is separated from the total plasma group under the impact of a strong electric field. Here, some instantaneous EDF in the vicinity of the maximum discharge voltage are shown. The dashed lines show the level of the electron momentum corresponding to the kinetic energy equal to the instantaneous gap voltage.

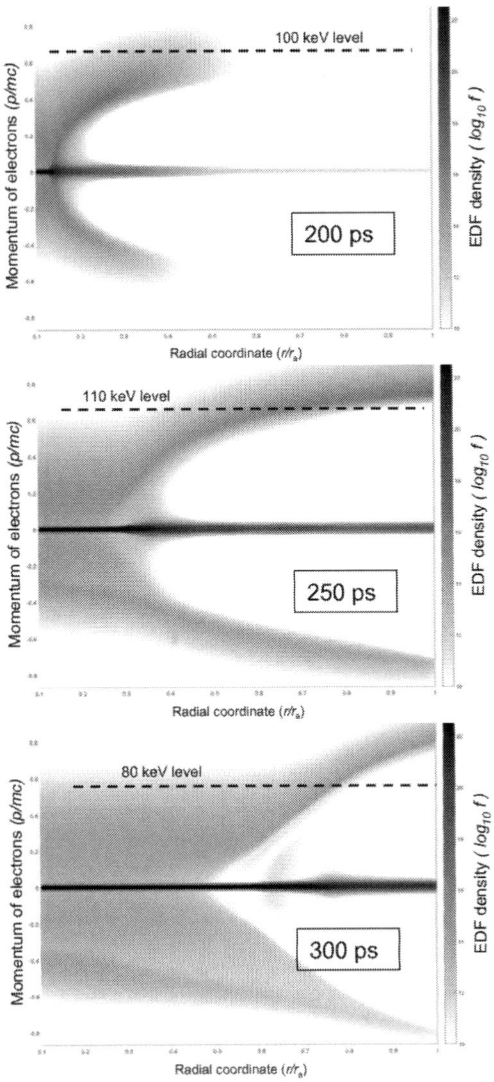

Fig. 6. Phase portraits EDF for various time points of the breakdown in nitrogen (cathode left, right anode). Shaded region represents a logarithmic scale.

Here the gap voltage is maximal (~110 kV) at time point 240 ps. The phase diagrams show how an electron beam with "anomalous energy" is formed at the anode. The travelling-wave electric field profile forms in the discharge (see Fig. 5) providing extra acceleration to the electrons beyond the maximum level of the potential drop at the gap. This represents the same situation that was discussed in Section 2.

In order to obtain the fast electrons current, which values could be compared with the experimental data, in the calculations we use an "energy filter" – an $d = 10$ µm Al foil, which partially cuts off slow plasma electrons coming to the anode from the discharge column. The spectral attenuation factor of the foil, $\varphi(d, \varepsilon)$, is calculated using semi-empirical formulas [15]. Thus, the current, $J_{fast}(t)$, and the total spectrum of fast electrons per pulse behind the anode, $N(\varepsilon)$, are given with

$$J_{fast}(t) = 2\pi r_a L \cdot e \int_0^\infty \varphi(d,\varepsilon) f(r_a,\varepsilon,t) v(\varepsilon) d\varepsilon,$$

$$N(\varepsilon) \propto \varphi(d,\varepsilon) \int_0^\infty f(r_a,\varepsilon,t) dt.$$

It should be noted, the bulk of the electron beam has an "anomalous energy" as shown in phase pictures Fig. 6. The integral spectrum of electrons per pulse, shown in Fig. 7, of course, contains electrons with lower energies also.

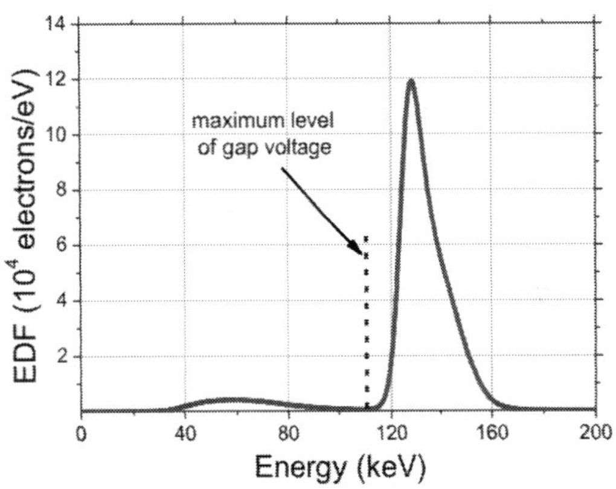

Fig. 7. Integral spectrum of electron beam behind the anode foil.

The amplitude of the runaway electron beam current $J_{fast}(t)$ at 25 A is reached at the 250 ps time point. Current pulse duration at half maximum is equal to 12 ps. Thus, the total yield is estimated to be $2 \cdot 10^9$ fast electrons per pulse. This value agrees well with typical experiment data [5, 16, 17].

5 Summary

This work presents an original theoretical model of a non-stationary high-pressure discharge. The physical kinetics of electrons is described using the Boltzmann kinetic equation for the electron momentum distribution function, and the collisions are described on the basis of known sections of elementary processes.

Self-consistent descriptions of both the process of forming total EDF in the discharge, and generate a group of runaway electrons. It does not use any semi-empirical coefficients and the functional dependence of the type: the critical field, the field enhancement factor, the drift velocity, etc.

The physical mechanism responsible for the appearance in the discharge of electrons with "anomalous energies" is convincingly demonstrated.

However, we must mention that in actual discharges of the three-dimensional geometry, the relative fraction of electrons with "anomalous energies" should be significantly lower than our calculations showed. This is clear because in our model of axisymmetric discharge all the electrons generated in the amplified field on the ionization wave front, while in real discharges only a small group of particles participate in this interaction with field wave [5].

We believe that such a mechanism for electron acceleration can also take place in other phenomena, for example, during the expansion of dense plasma initiated by a focused laser beam.

This work has been supported by Russian Foundation for Basic Research (projects 15-08-023983).

References

1. C. T. R. Wilson, *Mathematical Proceedings of the Cambridge Philosophical Society*, **22**, 4, 534 (1925)
2. S. Frankel, V. Highland, T. Sloan, O. van Dyck, and W. Wales, Nuclear Instruments and Methods, **44**, 2, 345 (1966)
3. T. Shao, V.F. Tarasenko, C. Zhang, E.Kh. Baksht, P. Yan, Y.V. Shut'ko, Laser and Particle Beams, **30,** 369 (2012)
4. V. F. Tarasenko et al., *Generation of Runaway Electron Beams and X-Rays in High Pressure Gases* (Nova Science Publishers Inc., NY, USA, 2016)
5. E.H. Baksht, A.G. Burachenko, V.Y. Kozhevnikov, A.V. Kozyrev, I.D. Kostyrya, and V.F. Tarasenko, Journal of Physics D: Applied Physics, **43**, 30, 30520 (2010)
6. L. Babich, T.V. Loiko, V.A. Tsukerman, Sov. Phys. Usp., **33**, 521 (1990)
7. V. Yu. Kozhevnikov, A. V. Kozyrev, N. S. Semeniuk, IEEE Trans. on Plasma Science, **43,** 4077 (2015)
8. V.Y. Kozhevnikov, A.V. Kozyrev, and N.S. Semeniuk, EPL, **112**, 15001 (2015)
9. A.V. Kozyrev, V.Yu. Kozhevnikov, M.I. Lomaev, D.A. Sorokin, , and V.F. Tarasenko, EPL, **114**, 45001 (2016)
10. A.V. Kozyrev, V.Yu. Kozhevnikov, N.S. Semeniuk, Matter and Radiation at Extremes, **1**, 264 (2016)
11. V.Y. Kozhevnikov, A.V. Kozyrev, and N.S. Semeniuk, Russian Physics Journal, **60** (2017) (in press).
12. V.Y. Kozhevnikov, A.V. Kozyrev, N.S. Semeniuk, E.M. Baranova, E.Kh. Baksht, V.F. Tarasenko, M.I. Lomaev, D.A. Sorokin, A.G. Burachenko, International Conference on Electromagnetics in Advanced Applications (ICEAA), Verona, Italy, (2017), pp. 489-492.
13. H. Bethe, Annalen der Physik, **397**, 325 (1930)
14. T. Tabata and R. Ito, Nucl. Instrum. Methods, **127**, 429 (1975)
15. T. Shao, V. F. Tarasenko, C. Zhang, et al., Laser and Particle Beams, **30**, 369 (2012)
16. G. A. Mesyats, M. I. Yalandin, A. G. Reutova, *et al.*, Plasma Phys. Rep., **38**, 29 (2012)

Spectral gain investigation of large size OPCPA based on partially deuterated KDP

Marco Galimberti*, *Alexis* Boyle, *Ian O.* Musgrave, *Pedro* Oliveira, *Dave* Pepler, *Waseem* Shaikh, *Trevor B.* Winstone, *Adam* Wyatt, *Cristina* Hernandez–Gomez

Central Laser Facility, STFC Rutherford Appleton Laboratory, Chilton, Didcot, OX11 OQX, UK

Abstract. The Optical Parametric Chirped Pulse Amplification is one of the most promising techniques to deliver 20PW laser system. The already available KD*P in large size is a good candidate as nonlinear crystal. In this article we report the experimental analysis of the spectral small signal gain for KD*P at 70% deuteration level for different phase matching and non-collinear angle. The data is also compared with a theoretical model.

1 Introduction

The Central Laser Facility has made leading contributions to the development of Optical Parametric Chirped Pulse Amplification. Through a series of projects the technique has been developed from pre-amplifiers to feasibility tests at large aperture. The culmination of this work is the proposal for and development of a 20PW upgrade to the current Vulcan laser facility [1]. The new beam line will be fully integrated with the existing 1PW beam line enabling new types of experiments 1PW + 20PW. In addition to the peak power the requirement on the laser pulse will be high contrast, good focusability and able to reach intensities greater than 10^{23} W cm^{-2}.

Whilst funding is being sought to fully deliver this goal a technology development programme for multi-PW OPCPA systems has been established to ensure that the relevant components required are available when funding is identified.

Key components in the OPCPA are the nonlinear crystals, responsible to the amplifications. To de-risk the project, but also to improve the knowledge and investigate ways to increase the project specifications, a dedicated laboratory was setup to investigate large size OPCPA system.

The deuteration level of KD*P has an impact in the absorption in the NIR: while the signal is not impacted from it, the idler could be affected reducing the gain and the efficiency of the amplification. Different deuteration change also the phase matching of the process, having an impact on the amplification bandwidth achievable.

While there is a great variety of nonlinear crystals, only a small number are available in large size (>100mm). One of the most used crystal already available in large size is the KD*P, used in large fusion laser system, like NIF. KD*P is similar to the KDP, where the hydrogen has been partially substituted by deuterium. The degree of substitution is called deuteration level and it has influence on the optical properties of the crystal, mainly on the refractive index and the absorption in the near infrared.

A detailed theoretical and numerical study on the influence of the deuteration level on the spectral gain of an OPCPA is available in literature [2], where an optimal spectral gain was obtained for different deuteration level, by changing the phase matching angle and the pump non-collinear angle. The result are shown in fig. 1.

From this study, partially deuterated KD*P could be used in the 20PW project. While the results are promising, an experimental study of the spectral gain of an OPCPA is required to validate the theoretical model.

2 Experimental Investigation

Based on the 20PW project [1], the high energy OPCPA stages required around 600J of pump laser and large beam size (20cm), not easy available. Instead of the final

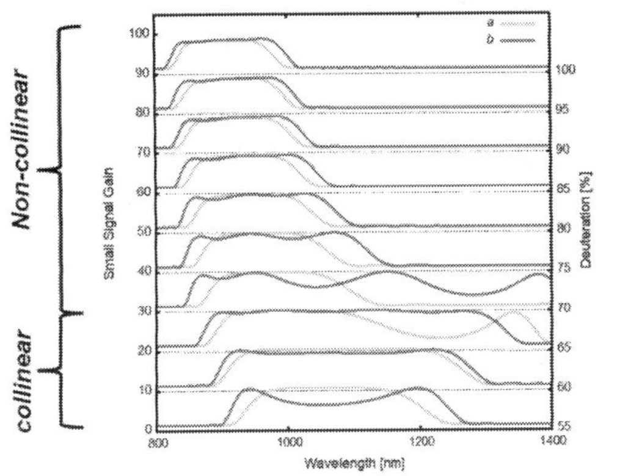

Fig. 1. Optimal spectral small signal gain for KD*P with different deuteration level [2].

* Corresponding author: marco.galimberti@stfc.ac.uk

https://doi.org/10.1051/epjconf/201816701006

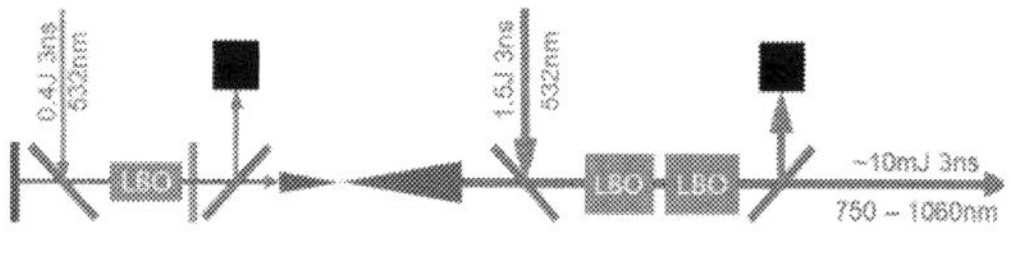

Fig. 2. Optical layout of the OPG source

amplifier, the study was based on the booster stage, where the pump beam size is 35mm with energies of 15J at 526.5nm. The main required to perform the study are:

- Crystal samples;
- Broad band signal, to be used as a seed;
- Pump laser and OPCPA setup;
- Procedure to find the phase matching.

Different deuterated KD*P were obtained. For this article we chose the 70% deuterated KD*P, more readily available in large size. The test crystal was 50mm x 50mm, 64mm long.

2.1 Signal generation

The required seed need to cover a large spectral range, from 750nm to 1050nm, comparable pulse duration and synchronizable with the pump laser and of energies of the order of few mJ.

Different techniques could be used to generate this seed. Typically fs laser system are used, followed by OPG or white light generation. Unfortunately, those systems are complex and require a stretcher to match the pulse length to the pump laser.

In our case we opted for a simpler approach to directly generate a seed pulse with a suitable pulse duration and energy, based on optical parametric oscillator and OPA. A schematic of our scheme is shown in fig. 2. The seed generation was performed by a resonant cavity around a 38mm long Type I LBO crystal, pumped by a 0.4J 3ns at 532nm providing with high single pass gain in excess of 10·. The cavity length allowed 3 round trips during the pump pulse duration, generating laser pulses exceeding 30mJ. The central wavelength could be tuned by varying the phase matching angle of the crystal [3].

The generated beam was spatially filter and expanded to 10mm before be amplified by an additional dual stage OPA. After the spatial filter the spectrum was maintained but the energy was reduced to 10 mJ.

The second OPA amplification stage was constituted by two 13mm thick LBO crystal pumped by 1.5 J at a fluence of 1.9 J/cm^2. At the output a pulse energy of 90 mJ was obtained, with a stability of 6% rms. By changing the phase matching angles, a tuneable range of 300 nm from 750 to 1050 nm was achieved as shown in fig. 3.

2.2 Pump laser and OPCPA setup

To deliver a pulse energy of the order of 15J at 526.5nm in 3ns needed for the test, a Nd:Glass laser system was setup, basing the design on the rod chain of the Vulcan laser system. Due to the properties of the OPCPA,

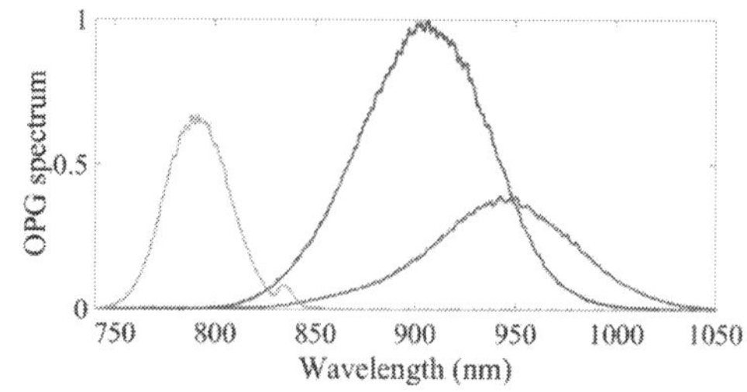

Fig. 3. OPG spectra, centred at 790, 910 and 950nm obtained by changing the phase matching angle.

temporal and spatial properties of the laser pulse are crucial to achieve the required amplification. More in detail, a flat top uniform beam profile, single frequency (single longitudinal mode) and temporal square pulses is needed. A layout of the pump laser is shown in fig. 4 [3, 4].

The laser pulse was generated by a long pulse front end with a programmable temporal pulse shape, capable to deliver around 1mJ on a single frequency at 1053nm.

The pulse was spatially filtered by an diffraction limited pinhole and amplified by a series of flash pumped phosphate Nd:Glass amplifiers, with the final amplifier was based on a 45mm rod in a double pass configuration. The system was capable to deliver reliable pulses of 30J at 1053nm with pulse length of 3ns.

The infrared pulse was frequency doubled by a type I KDP 22mm long. The resulting green pulses were sent to the OPCPA crystal by a vacuum telescope, with energy of 15J.

The signal, generated by the OPG, was expanded to 35mm by an achromatic vacuum telescope and sent to the OPCPA crystal. A full set of diagnostics, including NF, FF and spectrum, were setup to investigate the amplification process.

Because the repetition rate of the pump laser was low, with a shot every 5 minutes, a procedure to tune the phase matching of the OPCPA crystal was necessary. To investigate different phase matching angle in one shot, using the spatial extension of the beam, a cylindrical negative lens was inserted just before of the OPCPA crystal. In this way, the divergence of the pump and the signal beams was resulting in different incidence angles

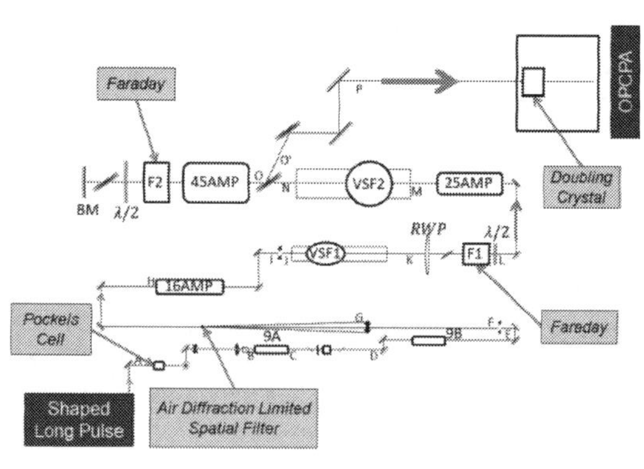

Fig. 4. Layout of the pump laser.

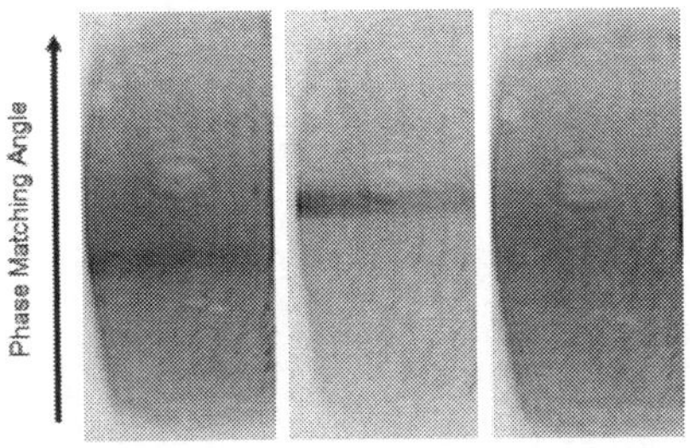

Fig. 5. Near field images during phase matching tuning. The three images were acquired at different phase matching angles.

on the crystal along one spatial direction. In case of phase matching, a line in the near field image, related to the region with gain, should appear.

A set experimental images is shown in fig. 5. The line where there was gain is clearly visible in the first and second image. Each image are related to different angular position of the crystal, and it clearly visible the line moving with the crystal.

3 Results and discussion

A complete phase matching scan was performed for four non-collinear angles of the pump. Due to the characteristics of the seed, for a single position of phase marching and non-collinear angle various shots were taken, to cover all the required spectral region. The full set of data was acquired over three month with hundreds of shots.

The data was analysed numerically to join the different acquisition and to extract the spectral gain. The results are shown in fig. 6. The phase matching angle θ is experimental and it is subjected of a systematic error of

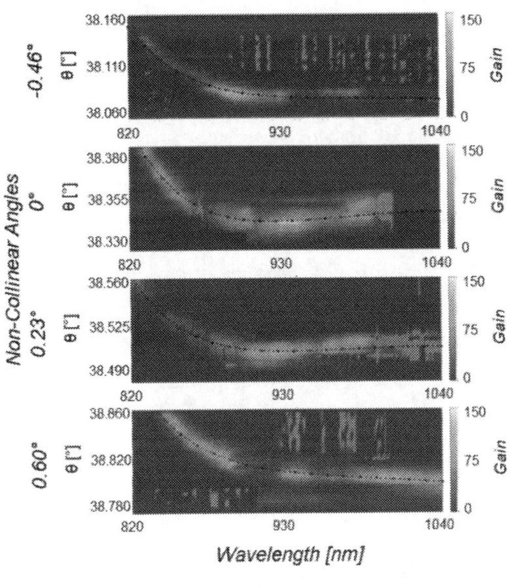

Fig. 6. Measured spectral small signal gain for different non-collinear angles and phase matching angles θ.

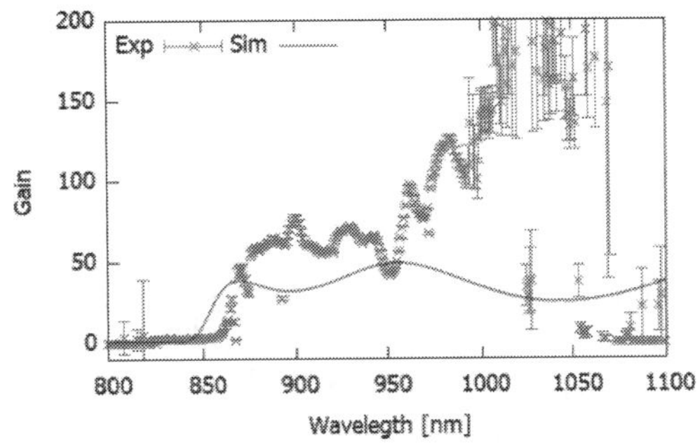

Fig. 7. Comparison between experimental and simulated small signal gain.

about 0.5°, due to the precision of the cut of the crystal. The measured gain was approximately in agreement with the theoretical model.

Selecting the data with maximum spectral bandwidth for the non-collinear angle 0.23°, the data could be more easily compared with the theoretical model. Best match was found for a phase matching angle of 38.32°. From the comparison (fig. 7), theoretical model predicted gain at lower wavelength than the experimental data.

Reconsidering the entire dataset, a possible explanation of this mismatch was found analysing the phase matching at different deuteration level. As shown in fig. 6, the data was in agreement with the phase matching curve, shown in dashed line, for deuteration level of 66%.

4 Conclusion

The small signal gain of an OPCPA based on partially deuterated KD*P at 70% was investigate. To perform the study a broadband tuneable source was setup within a 15J pump laser and a laboratory.

A complete dataset of phase matching and non-collinear angles was acquired. A preliminary analysis showed a good agreement between the experimental data and the theoretical model if a slightly different deuteration level is considered (66%).

Future work will be focused on investigating different deuteration level and to improve the data analysis.

Bibliography

1. C. Hernandez-Gomez et al, 'The Vulcan 10 PW Project: Facility Design Report', Technical Report, Central Laser Facility, STFC, 2010.
2. M. Galimberti et al, Optics Coms. **309**, 80-84 (2013)
3. I. Musgrave et al, Proceedings of ICUIL conference 2014, GOA, India
4. M. Galimberti et al. The Review of Laser Engineering: vol. 42(2), 137-140 (2014)

 https://doi.org/10.1051/epjconf/201816702001

Escaping Electrons from Intense Laser-Solid Interactions as a Function of Laser Spot Size

Dean Rusby[1,2], Ross Gray[2], Nick Butler[2], Rachel Dance[2], Graeme Scott[1], Vincent Bagnoud[3], Bernhard Zielbauer[3], Paul McKenna[2], and David Neely[1,2*]

[1]STFC Rutherford Appleton Laboratory, Oxfordshire OX11 0QX, UK
[2]SUPA Department of Physics, University of Strathclyde, Glasgow G4 0NG, UK
[3]PHELIX Group, Gesellschaft fur Schwerionenforschung, D-64291 Darmstadt, Germany

Abstract. The interaction of a high-intensity laser with a solid target produces an energetic distribution of electrons that pass into the target. These electrons reach the rear surface of the target creating strong electric potentials that act to restrict the further escape of additional electrons. The measurement of the angle, flux and spectra of the electrons that do escape gives insights to the initial interaction. Here, the escaping electrons have been measured using a differentially filtered image plate stack, from interactions with intensities from mid 10^{20}-10^{17} W/cm^2, where the intensity has been reduced by defocussing to increase the size of the focal spot. An increase in electron flux is initially observed as the intensity is reduced from 4x10^{20} to 6x10^{18} W/cm^2. The temperature of the electron distribution is also measured and found to be relatively constant. 2D particle-in-cell modelling is used to demonstrate the importance of pre-plasma conditions in understanding these observations.

1 Introduction

When a high intensity laser ($I\lambda^2 > 10^{18}$ W/cm^2 μm^2) interacts with a solid target, multi-MeV electrons are accelerated and travel into the target. These so called `hot electrons' will generate x-rays via bremsstrahlung inside the target and, upon reaching the rear surface, create a sheath field that can accelerate protons to mega electron Volt energies. The initial electron heating caused by the laser absorption is a crucial aspect of the interaction as it dictates all subsequent particle acceleration/generation. At intensities > 10^{16} W/cm^2, resonance absorption usually plays a dominant role. This mechanism accelerates the electrons perpendicular to the target or critical surface. For higher intensities (>> 10^{16} W/cm^2), the ponderomotive force becomes relevant. This force accelerates the electrons along the laser axis. The dominance of each of these processes depends on the interaction conditions, such as the incident laser intensity and the pre-plasma density scale length on the front surface.

The incident laser intensity depends on three laser parameters; the laser energy, pulse duration and focal spot size. Many investigations have been conducted to show how the laser energy and pulse duration affects the emitted protons[1] and x-rays[2].

It has been demonstrated in the past that the size of the focal spot changes the dynamics of the laser interaction with the target. This has been investigated with regard to the proton acceleration mechanism numerous times. Coury et al. [3] showed that as the laser was defocused from a peak intensity of 4x10^{20} W/cm^2, the maximum energy of the protons decreased. However, this reduction was smaller for a given intensity than that demonstrated when the target was kept at best focus and the laser energy reduced. This measurement of the maximum proton energy is consistent with the data found by Brenner et al. [4,5] for shorter pulse laser conditions. Measurements of the divergence of the electrons were also conducted by Coury et al. using the K-alpha x-rays from the rear surface. This found no significant trend for the angular divergence of the electrons as a function of intensity for variation of energy or focus. The primary observable for both of these studies, and others, is the effect that the laser spot size has on the accelerated protons. No studies have examined in detail the escaping electron population as a function of focal spot size.

Electrons can be measured using FUJI-film image plate which is highly sensitive to ionising radiation [6]. Hidding et al. [7] showed that interleaving multiple layers of image plate with filtering material in a stack behind a laser-solid interaction enables, the temperature and flux to be extracted from simple Monte Carlo modelling. Curving the image plate and also using multiple layers provides angular information on the flux and temperature [8,9]. In this investigation, the so called `wraparound' diagnostic is deployed on an experimental campaign at the PHELIX laser. During this campaign the laser was defocused by up to 400 μm, decreasing the incident intensity onto the target from 4x10^{20} - 4x10^{17} W/cm^2. Previous studies have employed wraparound diagnostics to examine the rear surface escaping

* Corresponding author: david.neely@stfc.ac.uk

https://doi.org/10.1051/epjconf/201816702001

electrons [9,10] whilst simultaneously investigating the electron flux emitted the front surface [8]. The experimental results from this campaign are accompanied by two dimensional particle-in-cell (PIC) simulations, investigating the influence of a front surface pre-plasma on a defocused pulse.

2 Experimental Method

The experimental campaign was conducted on the 1.05 μm PHELIX glass laser facility at GSI. The maximum laser energy post-compressor was 140 J and the pulse length measured using an auto-correlator was 700±60 fs. The laser is focused using an F/1.5 copper parabolic mirror, which is capable of delivering a laser focal spot with a Full Width Half Maximum (FWHM) of approximately 4 μm. This leads to a peak intensity of $4x10^{20}$ W/cm^2. The laser was s-polarised and incident at 20 degrees relative to the target normal. The target was a 100 μm thick copper foil.

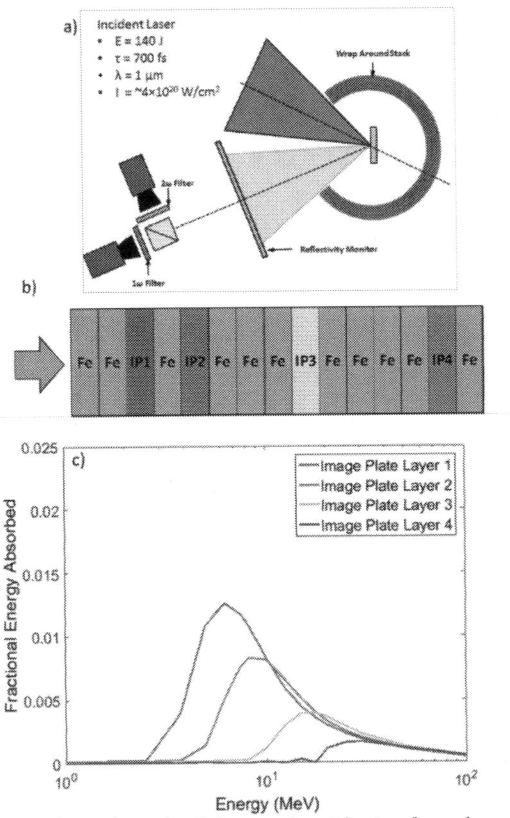

Fig 1. a) A schematic of the experimental setup from above with the aproximate laser parameters. A cross section of the wraparound diagnostic is shown in b) where the image plate (IP) layers are separated by iron (Fe) filters. Each of the IP layers is coloured to correspond with the response shown in c).

A wraparound diagnostic was deployed to measure the escaping electrons. A cross-section of the diagnostic is shown in Figure 1 b). As electron pass through the 0.85mm thick iron filtering layers, they lose energy. Therefore, the deeper layers will only detect the higher energy electrons. The response function of each layer of image plate, as calculated by Rusby et al. [9], was calculated using the Monte Carlo code GEANT4; this is shown in Figure 1 c). The first layer of image plate will only detect electron energies greater than 3 MeV. The copper target was placed directly in the middle of the wraparound diagnostic so it provides angular information about the escaping electrons. It is biased vertically such that it only observes the lower half of the escaping electron beam. The specular reflectivity was also monitored using two 12-bit CCDs; one that observes the 1 μm laser reflection and the other the emitted 2nd harmonic light from the interaction. A basic schematic of the setup with the previously mention laser parameters is shown in Figure 1 a).

3 Results

Polar plots of the escaping electrons from a series of shots of varying laser defocus are shown in Figure 2. These are taken from a lineout across the most intense part of the signal. On each plot the incident angle of the laser and the target plane is shown. The angular distribution of the escaping electrons appears to either be pointed towards the laser axis and target normal, with some smaller features also observed occasionally (see Figure 2 d) and f)). Although there is a shot to shot variation, collecting the angular profiles demonstrates the benefits of using such a diagnostic to measure the electrons and not relying on a standard electron spectrometer, which only samples a very small solid angle. The integrated and peak flux extracted from the wraparound diagnostic is shown in Figure 3 a), as a function of incident laser intensity. It is clear that there is an increase in both integrated and peak flux when the laser is defocused; with the optimum occurring for a defocus of 100 μm which corresponds to an intensity of approximately $6x10^{18}$ W/cm^2. This is an increase of a factor of ~x2 from the tight-focus shot. When the intensity is reduced further by increasing the focal spot size, the signal reduces rapidly.

An estimate of the electron temperature inferred by the wraparound diagnostic is made by comparison of the measured signal with a simulated output calculated using the response curve shown in Figure 1)c. As image plate is sensitive to ionising radiation, the filtering is selected to optimise the electron response. Protons up to energies of <33 MeV are filtered out using the first 1.8 mm of Fe. However, x-rays produced inside the target will be sufficiently energetic to penetrate to the IP. An estimation of this flux was calculated using GEANT4. Various electron temperature distributions were directed onto a 100 μm thick Cu target in GEANT4 and a detector positioned behind the target to record all the electrons and x-rays that pass through it. Thus, the x-ray flux and spectra as a function of electron temperature were estimated.

https://doi.org/10.1051/epjconf/201816702001

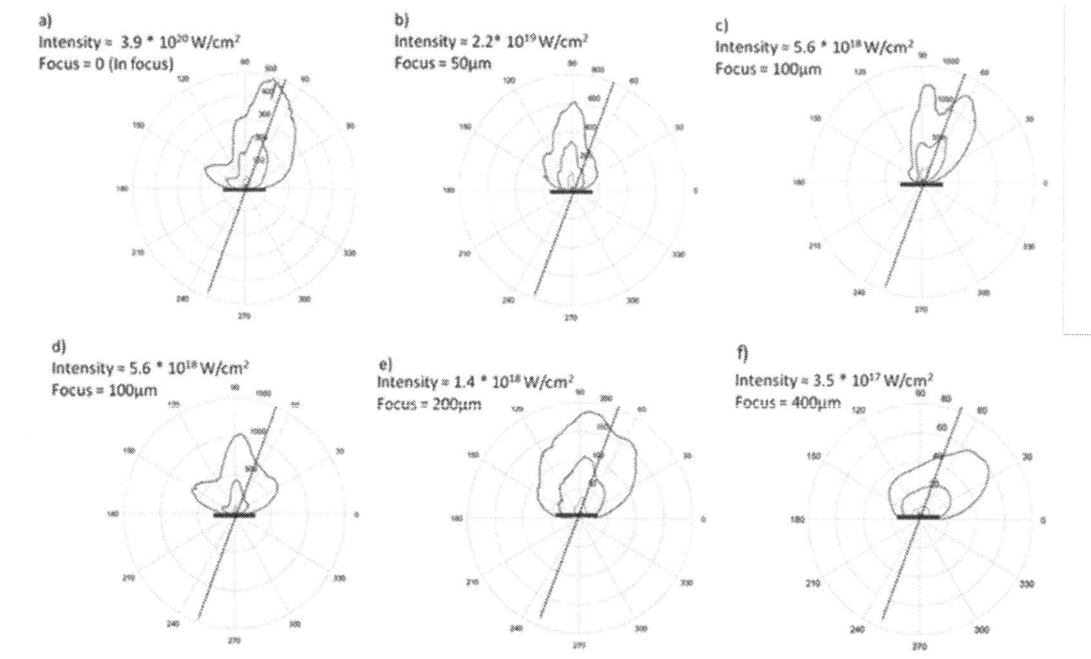

Fig 2. Polar plots of the angularly-resolved electron varied by laser defocus. The laser intensity is reduced from a) to f). The data is taken from a line-out across the most intense region of signal

Experimentally, not all the electrons will escape the target as they experience a rear surface sheath that restricts them. The escaping percentage of electrons is estimated to be between a few percent and 30% [11] depending on laser and target parameters.

The experimental data is compared to various electron temperatures and escaping fractions. The comparison is conducted using the least-squares method; with the most suitable temperature being determined when the residuals between the two sets of data is reduced to a minimum. Varying the escaping electron fraction changes the measured temperature slightly; these changes are taken as the largest uncertainty in the temperature extraction.

The temperatures extracted from the peak of the angular data on the wraparound diagnostic are shown in Figure 3 b) as a function of calculated laser intensity for a given defocus. Also plotted are 3 estimations of the hot electron temperature from Beg, Haines and Wilks [12-14]. Figure 3 b) clearly shows that the escaping hot electron temperature is very similar as the laser is defocussed over 3 orders of magnitude of incident intensity.

To better understand why the flux peaks for a defocused laser and the temperature remains constant, 2D PIC simulations were conducted.

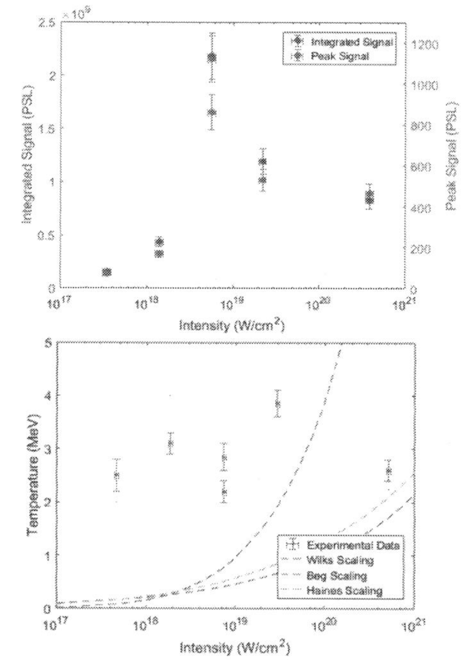

Fig 3 a) Integrated and peaks counts taken from the wraparound diagnostic as a function of incident laser intensity. The full energy shots have a peak in flux when the laser is defocused to 100 μm. b) The temperatures extracted from each focus as a function of incident laser intensity. The temperature remains constant as a function of intensity, which is contrary to the three temperature models also plotted.

https://doi.org/10.1051/epjconf/201816702001

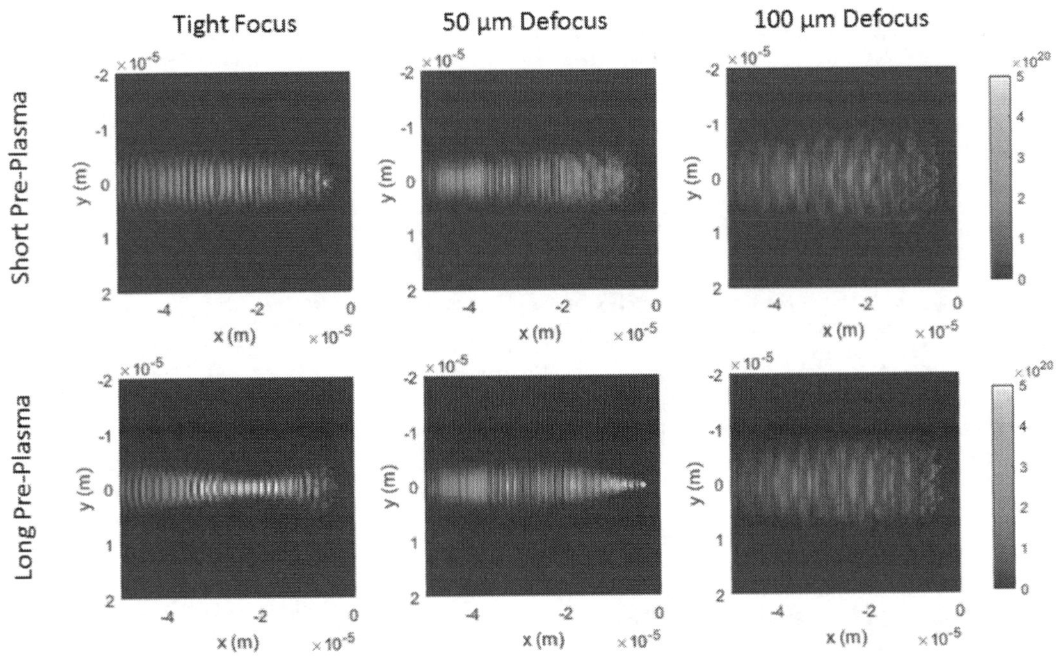

Fig 2. Two dimensional PIC simulation results of the laser incident onto the target for different focus and plasma conditions. The short pre-plasma case (top row) shows that the spot size increases as the focus is moved further away from the target surface. However, for the longer pre-plasma, the laser self focuses, causing the beam to achieve a higher intensity on target for a 50 µm defocus than for the tight focus case.

4 2D EPOCH Simulations

The PIC code EPOCH [15] was used to simulate a defocused laser onto a planar solid target. The spatial size of the simulation box is set to 200 × 120 µm, which is made up of 8000×4800 cells. This leads to a resolution of 25 nm. To reduce the computational resources required to conduct the simulations, the target thickness and pulse duration was reduced by a factor of 4 to 25 µm and 175 fs respectively.
Simulations corresponding to the experimental campaign were performed in which the laser was defocused relative to the target surface. Rather than making the beam laterally larger, curvature was added to the phase of the beam, φ; this ensures that the same amount of energy enters the simulation by keeping the peak intensity constant. The phase of the beam is given as,

$$\varphi = -\frac{-k_L x^2 - k_L y^2}{2R_C} + \operatorname{atan}\left(\frac{x}{R_r}\right) \quad (1)$$

where $k_L = 2\pi/\lambda$, R_r is the Rayleigh range $R_r = \pi w_0^2/\lambda$, $R_C = -x\left(1+\left(\frac{R_r}{x}\right)^2\right)$ is the radius of curvature and w_0 is the diffraction limited width of a Gaussian beam. Experimentally an F/1.5 off-axis parabola was employed, which yields an ideal waist size of ~3.7 µm. The beam focus is offset in the positive x direction to simulate the defocus of the laser.

The experimentally measured contrast of 10^{-7} would form a preplasma on the front surface. Although, this was not measured on this experiment, the extent of the expansion can be estimated using previously measured pre-plasmas. McKenna and Carroll et al. [16,17] show a long, shallow density gradient measured using interferometry of a probe beam across the front surface of the target. Wagner et al. [18] characterised the contrast of the PHELIX laser and its effects on the pre-plasma [19] by measuring the expansion of the plasma using shadowgraphy of a probe beam. These two sources provide sufficient information to make a first estimate of the pre-plasma. Coupled with this, additional simulations are conducted with a short scale length.

The intensity of the laser beam as it is incident onto the target surface for the different defocus and pre-plasma parameters is shown in figure 4. As expected, for the case of the short pre-plasma, the size of the laser in the y-direction increases as a function of the distance of best focus from the target surface. However, in the case of the longer pre-plasma, the laser undergoes relativistic self-focusing, which for the best focus case causes the laser to focus prior to the target. For the 50 µm defocus case, the laser is kept from spreading and finally focuses to a very tight, high-intensity focal spot. This leads to a higher energy population of electrons accelerated into the target similar to that observed experimentally. These simulations clearly demonstrate the need to further investigate and characterise the plasma density scale length present near the peak of the pulse in order to understand the real laser intensity distribution present on full shots, which could be significantly different from the intensity which would be present without a pre-plasma.

5 Conclusion

Understanding the internal electron distribution present during laser solid target interactions is important to enable a greater degree of control and optimisation of secondary sources such as Bremsstrahlung x-rays, ions and neutrons. The entire angular distribution of the forward escaping electrons in the 3-20 MeV energy region has been observed as a function of defocus distance from copper foil targets. A 2 times enhancement of the escaping electron flux at a defocus of 100 μm from the optimum pre shot focal position was measured. As the laser is moved further away from focus, the escaping electrons numbers drop rapidly. Surprisingly, the temperature of the escaping electrons remains almost constant even though the flux changes considerably. The angular distribution of the electron beam does vary from shot to shot, though it generally points in the laser forward direction. These measurements highlight the importance of observing the entire escaping electron distribution, as a single conventional electron spectrometer would not observe the shot to shot beam distribution changes. PIC simulations clearly demonstrate that the presence of a pre-plasma plays a significant role in determining the real intensity conditions achieved on experimental shots. Future work will investigate this relationship in more detail and also the role that the-pre plasma has on modifying the angular distribution and spectra of the escaping electrons.

The authors gratefully acknowledge the expert assistance of the PHELIX operations team, targets from Scitech and funding from Laserlab grant 284464 and EPSRC (Grant Nos. EP/K022415/1, EP/M018091/1 and EP/R006202/1). Data associated with this paper can be accessed at http://dx.doi.org/10.5286/edata/709

References

1. Diado, H. et al. Review of laser-driven ion sources and their applications, Reports on Progress in Physics, 75,5, 2012
2. Fiorini, F. et al. Characterization of laser-driven electron and photon beams using the Monte Carlo code FLUKA, Laser and Particle Beams, 32, 233-241, 2013
3. Coury, M. et al. Influence of laser irradiated spot size on energetic electron injection and proton acceleration in foil targets. Appl. Phys. Lett. 100, 074105 (2012).
4. Brenner, C. et al. Dependence of laser accelerated protons on laser energy following the interaction of defocused, intense laser pulses with ultra-thin targets. Laser Part. Beams 29, 345–351 (2011).
5. Brenner, C. M., McKenna, P. & Neely, D. Modelling the effect of laser focal spot size on sheath- accelerated protons in intense laser–foil interactions. Plasma Phys. Control. Fusion 56, 084003 (2014).
6. Bonnet, T. et al. Response functions of imaging plates to photons, electrons and 4He particles. Rev. Sci. Instruments 84, 103510 (2013).
7. Hidding, B. et al. Novel method for characterizing relativistic electron beams in a harsh laser-plasma environment. Rev. Sci. Instruments 78 (2007).
8. Gray, R. J. et al. Surface transport of energetic electrons in intense picosecond laser-foil interactions. Appl. Phys. Lett. 99, 171502 (2011).
9. Rusby, D. R. et al. Measurement of the angle, temperature and flux of fast electrons emitted from intense laser–solid interactions. J. Plasma Phys. 81, 475810505 (2015).
10. Ge, X. L. et al. Directed fast electron beams in ultraintense picosecond laser irradiated solid targets. Appl. Phys. Lett. 107, 091111 (2015).
11. Link, A., Freeman, R. R., Schumacher, D. W. & Van Woerkom, L. D. Effects of target charging and ion emission and the energy spectrum of emitted electrons. Phys. Plasmas 18 (2011). DOI 10.1063/1.3587123.
12. Beg, F. N. et al. A study of picosecond laser–solid interactions up to 10^{19} W/cm^2. Phys. Plasmas 4, 447–457 (1997).
13. Haines, M. G., Wei, M. S., Beg, F. N. & Stephens, R. B. Hot-electron temperature and laser-light absorption in fast ignition.Phys. Rev. Lett. 102, 1–4 (2009). DOI 10.1103/PhysRevLett.102.045008.
14. Wilks, S. C. & Kruer, W. L. Absorption of Ultrashort, Ultra-Intense Laser Light by Solids and Overdense Plasmas. 33, 1954–1968 (1997).
15. Arber, T. D. et al. Contemporary particle-in-cell approach to laser-plasma modelling. Plasma Phys. Control. Fusion 57, 113001 (2015).
16. McKenna, P. et al. Effects of front surface plasma expansion on proton acceleration in ultraintense laser irradiation of foil targets. Laser Part. Beams 591–596 (2008). DOI 10.1109/PLASMA.2009.5227320.
17. Carroll, D., Quinn, M., Yuan, X. & McKenna, P. Effects of front surface plasma expansion on proton acceleration driven by the Vulcan Petawatt laser. clf.rl.ac.uk 19–22 (2008).
18. Wagner, F. et al. Temporal contrast control at the PHELIX petawatt laser facility by means of tunable sub-picosecond optical parametric amplification. Appl. Phys. B: Lasers Opt. 116, 429–435 (2014).
19. Wagner F., et al, Pre-plasma formation in experiments using petawatt lasers, Opt. Express 22, 29505-29514 (2014)

Laser contrast and other key parameters enhancing the laser conversion efficiency in ion acceleration regime

Lorenzo Torrisi

Dipartimento di Scienze Fisiche, MIFT, Università di Messina,
V.le F.S. D'Alcontres 31, 98166 S. Agata, Messina, Italy

Abstract. Measurements of ion acceleration in plasma produced by *fs* lasers at intensity of the order of 10^{18} W/cm^2 have been performed in different European laboratories. The forward emission in target-normal-sheath-acceleration (TNSA) regime indicated that the maximum energy is a function of the laser parameters, of the irradiation conditions and of the target properties.In particular the laser intensity and contrast play an important role to maximize the ion acceleration enhancing the conversion efficiency. Also the use of suitable prepulses, focal distances and polarized laser light has important roles. Finally the target composition, surface, geometry and multilayered structure, permit to enhance the electric field driving the forward ion acceleration.Experimental measurements will be reported and discussed.

1 Introduction

More and more progresses are achieved in the field of the ion acceleration using high intensity pulsed lasers, thank to many experiments and results obtained using different laser parameters. The high intensity lasers using *fs* pulse duration are accelerating ions above 1 MeV per charge state and, by optimizing some important parameters of the used laser, irradiation conditions and target properties, recently are reaching values higher than 10 MeV per charge state [1, 2].

The main mechanism used to accelerate ions at high kinetic energy is the target normal sheath acceleration (TNSA) occurring at the rear surface in forward direction [3]. The dominant accelerated ion species are protons, due to their high charge-to-mass ratio, which are always present as taking part of the target composition or as contaminant layers at the surfaces and interfaces.

The highest electron and ion density at the target rear surface, and consequently highest acceleration field obtained, is expected for ultrathin targets of a conducting material. The electric field of ion acceleration depends on the following parameters:

$$E = \sqrt{\frac{kT_e n_e}{\varepsilon_0}} \qquad (1),$$

where kT_e and n_e represent the plasma electron temperature and density, respectively, and ε_0 is the vacuum permittivity [4].

Generally, as the target thickness is decreased, the transverse spreading of the relativistic electrons inside the target is reduced, resulting in a higher rear surface charge density and consequently in an higher ion acceleration and directionality along the normal to the target surface. For very high intensities and ultrathin targets, relativistic transparency allows part of the laser pulse to be transmitted through the target and contribute to increase the electron heating.

On the other hand, the thinning of the thickness of the target produces an increase in energy of the ions but also to a reduction of their yield, which therefore could decrease to such values as to be practically unusable.

One of the most severe limiting factors in laser-plasma ion acceleration is the presence of a laser pedestal of amplified spontaneous emission (ASE) or laser prepulses to the main pulse, which affects the laser-solid and laser-plasma interaction [5]. A high pedestal, in facts, may destroy the solid target prior to the arrival of the main laser pulse, thus deteriorating the conditions to have better TNSA regime. From this point of view thin foils can be destroyed also by relatively low energetic pedestals with respect to thick foils, thus TNSA conditions and ion energy maximization need of a compromise between pedestal and main pulse intensities to obtain the best acceleration conditions.

The laser contrast, in terms of ratio between the intensity of the main pulse to that of the pedestal or prepulse, in fact, plays a fundamental role in the forward TNSA ion acceleration mechanism and the pedestal must be controlled accurately in duration and in intensity before to be applied to a given target irradiation.

I some cases the laser prepulse or the use of a laser focal positioning in front of the target may produce a pre-plasma in front of the solid target very useful to induce self-focusing effect [6], consisting in the increment of the refractive index in which the laser light may be further focalized up to spot dimensions comparable with the laser wavelength, so that the intensity increases, the

* Corresponding author: lorenzo.torrisi@unime.it

accelerated electron energy also and consequently the final ion acceleration enhanced.
However the key parameters to enhance the particle acceleration are many and some of them concern the target composition and structure, that must be employed to improve the TNSA mechanism, to enhance the yield of the ions that we want accelerate and to improve their ion energy and angular emission distributions.
In this paper the influence of the laser parameters, such as the contrast, of the irradiation conditions, as the focal distance from the target surface, and of the target properties, such as the composition and geometry, will be presented, studied and discussed in order to enhance the laser-to-protons and ions energy conversion efficiency and energy distribution.

2 Materials and methods

Measurements using *fs* Ti:sapphire laser have been performed at CELIA-Bordeaux, in France and at IPPLM-Warsaw, in Poland. Further measurements have been performed in PALS-Prague in Czech Republic and useful comparisons of experimental results were performed with literature data by experiments performed in other facilities.
Generally the used lasers are similar, operating at a wavelength of 800 nm, with intensities of the order of 10^{18-19} W/cm^2, with main pulse duration of 30-40 *fs*, a pulse energy of about 300 mJ and a focal spot diameter of 10 microns. However, two different experiments were obtained using a contrast of 10^5 at CELIA Lab and of 10^9 at IPPLM Lab. In the two cases the pedestal duration is about 10 ps.
Experiments were performed using different focal position (FP) distance, which was moved in vacuum with micrometrical step motors, from about -200 microns (in front of the target) up to +200 microns (inside the target). Laser beams used P-polarized radiation with a Gaussian beam distribution.
Many thin targets were used with different composition, multilayers and thickness. Polymers, semiconductors and metals were employed with thicknesses from about 0.1 micron up to about 20 micron. Advanced targets were prepared using thin film of graphene, conductive, transparent to the IR radiation and with high mechanical resistance. Graphene thickness was 10 nm, in order to reduce significantly the prepulse laser reflection effect, increasing the laser-matter energy transfer to the target. Graphene, in facts, has peculiar optical properties, such as high transmittance in the visible and IR regions, above 90%, and very low reflectance, less than 1% in the visible and near IR regions. It is worthy highlighting graphene's unique optical properties such as its low reflectivity and absorption and high transmission as well as that it produces an unexpectedly high opacity for an atomic monolayer in vacuum, absorbing about 2.3% of red light [7].
A typical example of reflectance measurements is reported in Fig. 1a for copper and graphene. The plot shows that the copper reflectivity is high, and of the order of 90% for wavelengths of about 800 nm, while the graphene reflectivity is very low, below 1%, at the used laser wavelengths of 800 nm. Thus, while the direct irradiation of the metallic copper in the first instant of the laser pulse, before all target is vaporized, produces high laser reflections with significant energy loss, the use of a copper surface covered with some graphene monolayers reduces such reflections enhancing the laser energy penetrating the solid target. The trasmittivity of 3 monolayers of grahene, in facts, is high, above 90%, as reported in the plot of Fig. 1b.
Thus we expect that the plasma generated by targets with the graphene covering layers should be more energetic than in the case of copper surface directly irradiated in the same experimental conditions.

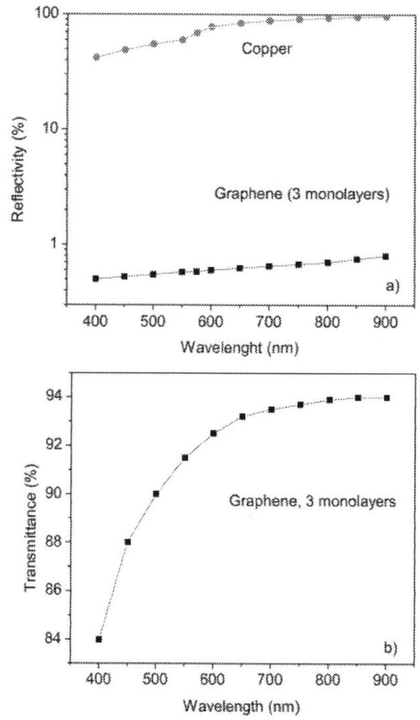

Fig. 1. Reflectivity vs. wavelength for copper and graphene (a) and transmittance vs. wavelength for 3 monolayers of graphene (b).

The plasma diagnostics uses mainly ion collectors (IC), consisting in Faraday cups with the suppressor grid, connected in time-of flight (TOF) configuration with a fast storage oscilloscope and semiconductor detectors, such as SiC diodes with different active region adapt to detect high energetic ions emitted from plasma using TOF approach [8]. IC and semiconductor detectors are placed at different angles with respect to the normal target surface and at different known distances from the target, in order to detect the forward and the backward ion acceleration.
To use TNSA regime driving targets thinner than 1 μm maintaining high contrast it is necessary to
maintain undisturbed the target rear surface, other while only ponderomotive ion acceleration occurs in plasma

and no so coordinate manner and a lower forward ion acceleration can be recorded.

A typical experimental set-up scheme is reported in Fig. 2a, in which different diagnostics are employed in TOF approach using diamonds, silicon, SiC and IC detectors. Fig. 2 b shows the typical *fs* laser pulse and its pedestal relative to the CELIA Lab laser.

3 Results

In order to understand the experimental differences obtained irradiating the same target with two lasers having different contrast, a target constituted by graphene/Cu layers, 10 nm in thickness deposited on a Cu substrate (14 μm in thickness), was irradiated from the graphene face and the Cu as rear surface. In both cases the laser focal spot was 10 microns in diameter focused on the target surface (FP=0).

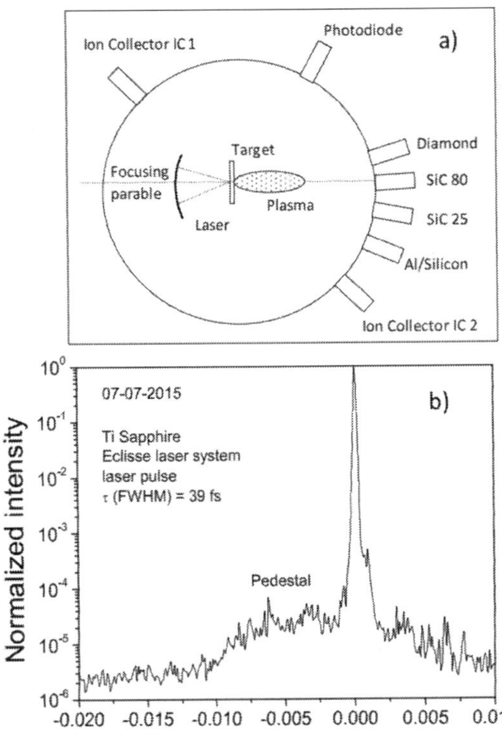

Fig. 2. Scheme of the experimental set-up (a) and pulse shape intensity vs. time of the Eclipse laser in Celia Laboratory (b).

Fig. 3a reports the SiC spectrum (SiC80) of the plasma forward (0°) radiations detected at a flight distance of 72.5 cm using a contrast of 10^5, while Fig. 3b that of SiC, in forward emission (0°), at a flight distance of 78.6 cm using a contrast value of 10^9.

The elements expected from the ion acceleration, due to the target composition, are protons (as contamination), carbon and copper, of which these last must be predominant. In fact, the spectra shows such ions with a main peak due to Cu ions but which energy, calculate from TOF approach, is very different in the two cases.

At low laser contrast (Fig. 3a) the protons are detected at about 92 ns, thus by considering the flight distance of 74.5 cm their maximum kinetic energy is about 330 keV, the carbon are detected at about 2.0 MeV and the copper at about 9.2 MeV. Such measurements are agree with an ion acceleration of about 330 keV per charge state, in fact, assuming the carbon and the copper atoms to be all ionized, the more energetic C^{6+} and Cu^{29+} ions should have an energy of 6x330 keV~2 MeV and 29x330 keV~ 9.5 MeV, respectively, in good agreement with that detected. Moreover, in this case the maximum spectrum ion yield is low, of about 1.5 mV.

At high laser contrast (Fig. 3b) the protons are detected at a maximum kinetic energy of about 600 keV, the carbon at about 3.6 MeV and the copper at about 16 MeV. Such measurements are agree with an ion acceleration of about 600 keV per charge state, in fact, assuming the carbon and the copper atoms to be all ionized, the more energetic C^{6+} and Cu^{29+} ions should have an energy in agreement with that detected. Moreover, in this case the maximum ion yield of the spectrum is high, of about 2.3 V.

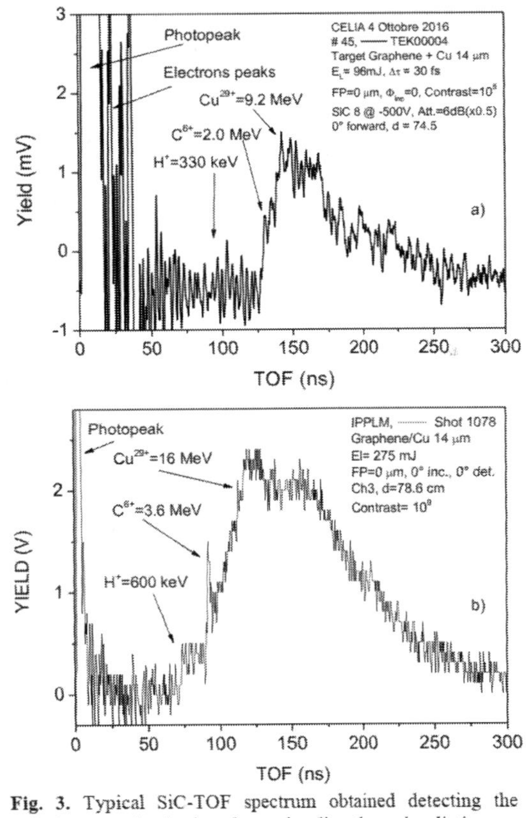

Fig. 3. Typical SiC-TOF spectrum obtained detecting the radiations emitted in forward direction irradiating a graphene/Cu target at low contrast (a) and at high contrast (b).

Thus the higher contrast permits to maintain low the pedestal that limits the ablation of the target ensuring that the rear side of the target remains solid and that TNSA regime may occur. The higher contrast produces higher electron density and temperature and

 https://doi.org/10.1051/epjconf/201816702002

consequently, according to eq. (1), higher electric field driving the forward ion acceleration. Not only the ion energy is increased by about a factor two but also the ion yield enhances strongly and of about a factor higher than 1500, by considering that the same detector in this second case is more distant form the target. The use of different focal positions generates different plasmas due to the different spot diameter that changes the laser intensity and to the possibility to induce preplasma in which not linear self-focusing effects can be produced. The properties of the produced plasma change by irradiating the graphene/copper target, 14 micron in copper thickness, by changing the focal position with respect to the target graphene surface.

Fig. 4a and 4b report two typical SiC spectra (SiC80) obtained using two different focal positions of FP = -100 microns (in front of the target) and FP = +100 micron (inside the target), respectively. In the first case the maximum ion energy is measured at 275 keV, 1.75 MeV and 7.6 MeV for protons, carbon ions and copper ions, respectively, indicating a mean charge acceleration of about 275 keV per charge state. In the second case the maximum ion energy is measured at 1.6 MeV, 9.7 MeV and 40 MeV for protons, carbon ions and copper ions, respectively, indicating a mean charge acceleration of about 1.6 MeV per charge state.

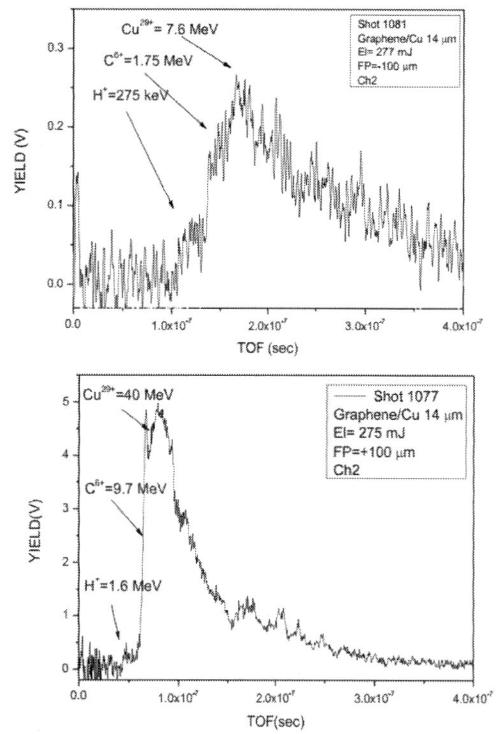

Fig. 4. Typical SiC-TOF spectrum obtained irradiating the graphene/Cu foil with high contrast laser with a focal position of -100 microns (in front of the target surface) (a) and +100 microns (inside the target surface) (b).

This result is particularly interesting because the only focal shift of 200 microns determines an ion energy increase of about a factor 5.8, demonstrating that the focalization on the copper foil inside the target, instead that on the superficial graphene layers, increases the relativistic electron emission from the target and consequently the forward electric field driving the ion acceleration.

The plots of Fig. 5a and 5b report the results about the proton accelerations obtained using a laser contrast of 10^5 and 10^9, respectively, irradiating the same target of graphene/copper at different focal positions. The fist plot demonstrates that the maximum proton acceleration is obtained at FP=0 microns, decreasing at the other focal positions, while the second plot demonstrated that the maximum proton acceleration is obtained focusing at FP=+100 microns, confirming a different process occurring in the target. In this second case, thank to the high laser contrast, the laser pedestal don't destroy the graphene superficial layer that transmits the laser pulse and hits the copper surface producing TNSA ion acceleration from the rear face of the copper. In the first case, instead, due to the low laser contrast the pedestal destroys the graphene layer and part of the copper surface, lower laser energy accelerates electrons and TNSA ion acceleration occurs also but with a less ion acceleration process.

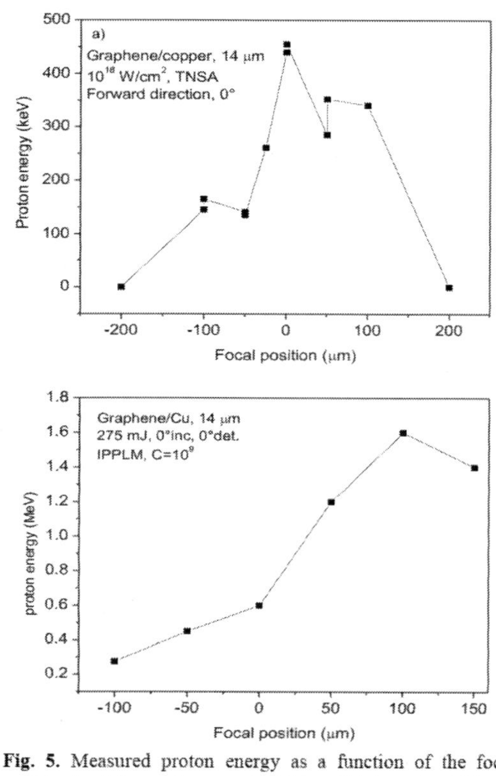

Fig. 5. Measured proton energy as a function of the focal position for *fs* laser at low contrast (a) and high contrast (b).

The maximum ion acceleration exhibits a clear trend with the target thickness influencing the accelerated electron transmission from the thin foil in term of quantity, energy and angular spreading. The number of accelerated protons and the proton energy changes with the target composition and thickness. According to eq.

(1), by increasing the rear hot electron temperatures and densities enabling the higher protons and ions acceleration. Generally using thinner targets more energetic electrons are transmitted from the target surface to its rear side and the electric field driving the Generally the proton acceleration is too low if the target is thick or is very thin, and assumes a maximum value for an optimal thickness that maximize the electron transmission to the rear side of the foil. The accelerated surface electrons, their energy loss and transversal spreading inside the target modify the ion acceleration and permit to generate a maximum electrical field if their emission has spatial and temporal coherence form the rear face of the target. The proton yield instead depends mainly on the target composition, being high for hydrogenate targets, such as polymers and hydrates, and low for hydrogen contamination of surfaces.

At high laser contrast, for low target thickness, the maximum proton energy increases because more electrons are injected in the rear surface and a better charge separation occurs increasing the forward electric field driving the ion acceleration. Thus, in such conditions it is possible to obtain high proton energies for submicrometric target thicknesses.

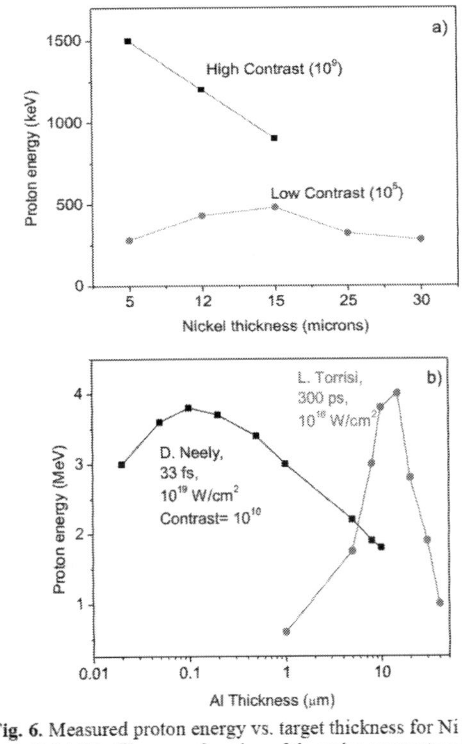

Fig. 6. Measured proton energy vs. target thickness for Ni (a) and Al (b) thin films as a function of the pulse contrast and of the pulse duration, respectively.

Of course too low foil thickness reduce the number of particles to negligible values and determine a very low flux of relativistic electrons up to a reduced charge separation and a limit in the ion acceleration is obtained. As reported in literature [9] sometimes at high laser contrast, of the order of 10^{10} the proton energy continues to increase also up to sub-micrometric thicknesses, possibly due to an increasing role of electron recirculation with decreasing the target thickness. Below about 100 nm, we find a decrease in the maximum proton acceleration, resulting in a clear optimum in thickness.

This decrease could be due to the finite temporal contrast on the picosecond time scale, given by the shoulders of the main pulse rather than to ASE or separate prepulses.

Fig. 6b reports the result of the dependence on the protons energy accelerated from a thin Al target irradiated at 10^{16} W/cm^2 using the PALS iodine laser operating at 1315 nm wavelength and 300 ps, described by L. Torrisi [10] reaching about 4 MeV energy at the optimum target thickness of 10 microns Al in TNSA regime. For comparison the same plot shows the results obtained by D. Neely [9] using the LLC laser which delivers 33 fs pulses with a 1 ns duration ASE, at an intensity contrast of 2×10^8 and a contrast of 10^{10}, when it is focused to an intensity of 10^{19} W/cm^2 on Al foils. In this case the optimal thickness to reach the maximum proton energy of about 4 MeV is obtained at 0.1 microns. The comparison shows that the different laser intensity and temporal pulses optimize the proton acceleration in two different thickness regions, of the order of tens microns in the first case and of 100 nm in the second case, respectively.

The laser conversion efficiency in proton acceleration can be calculated accurately knowing the ion current, the ion energy distributions and the angular emission distribution from the target. The SiC detector spectra, showing the ion current versus time, permits to give an evaluation of the conversion efficiency taking in consideration the average current of the diode, $I=V/R$. For example from the spectra of Fig. 3b referred to high contrast, it is possible to evaluate the real proton current by the equation:

$$I_p = \frac{V}{R}\cdot\frac{E_c}{E_p} = \frac{0.4V}{50\Omega}\cdot\frac{7.78eV}{3.10^5 eV} = 0.21\mu A \quad (2),$$

where V is the ion yield in Volt, R is the input resistance of the storage oscilloscope of 50Ω, E_c represents the energy to produce a pair in the semiconductor (7.78 eV) and E_p is the mean energy of the protons, supposed to be 300 keV.

The number N of protons produced by a single laser shot can be calculated as follows:

$$N = I_p \frac{\Delta t}{e}\cdot\frac{\Omega}{\Delta\Omega} = \quad (3),$$
$$= 0.21\mu A\cdot\frac{25ns}{1.6\cdot10^{-19}C}\cdot\frac{1.75sr}{6.5\mu sr} = 8.8\cdot10^9$$

where Δt is the proton current duration, e is the electron charge, Ω is the total solid angle of proton emission, assumed of 1.75 *sr*, and $\Delta\Omega$ is the solid angle of detection (6.5 μsr).

Assuming these ions have a mean kinetic energy (E_{max}) of 300 keV and the laser pulse energy to be 275 mJ (E_L), thus the laser conversion efficiency in protons acceleration is:

$$\eta(\%) = \frac{N \cdot E_{\max}}{E_L} \cdot 100 = \qquad (4).$$

$$\frac{8.8 \cdot 10^9 \cdot 3 \cdot 10^5 \cdot 1.6 \cdot 10^{-19} J}{275 \cdot 10^{-3} J} \cdot 100 = 0.15\%$$

Of course the conversion efficiency of proton acceleration using low laser contrast will be lower, as immediately observable from their yield of about a factor 1500 lower and minor current and energy.

4 Discussion and Conclusions

The process of TNSA ion acceleration depends on many parameters connected to the used laser system, to the irradiation conditions and to the target properties.

The laser contrast is fundamental because if it is low it means that the pedestal pulse contains high laser energy which will hit the target destroying it totally or partially, depending on its thickness and composition, before that the main pulse arrives after tens *ps*. To use TNSA acceleration mechanism the rear side of the target must be maintained solid during the main pulse arrive and the contrast must be more high possible.

The irradiation conditions are also essential to be controlled accurately, as demonstrated, for instance, from the different laser focal position with respect to the target surface depend the ion acceleration. The optimal acceleration condition must be searched changing the focal position and maximizing the proton energy.

The target properties are also very important because from its composition depends the main ion stream to be accelerated, from its thickness depends the maximum ion acceleration and from its properties depends the maximum laser energy transferred to the target. For example the use of thin graphene layers reduces the reflectivity maintaining high its transmission, permitting high energy transfer to the substrate on which the film is deposited. Its use is recommended when high laser contrast is employed.

Although many factors are under control, high intensity laser are used and advanced target permit to transfer high energy from photons to the matter, the ion beam conversion energy remains low, as reported in many literature [11, 12].

A laser-to-proton beam energy conversion efficiency of 0.1%, for protons with maximum energy of 600 keV is obtained from our calculations. This efficiency is based on the use of advanced targets with absorbent graphene superficial layers and very short pulses from a Ti:sapphire laser operating at about 0.3 J. With a contrast of 10^9, the optimum target thickness is found to be of about 5 microns or lower, in agreement with literature reporting the optimal thickness at 10^{10} contrast of about 0.1 micron. The results show a significant enhancement in proton beam scaling with a variation in target thickness and clearly demonstrate the benefits of using an ultrahigh contrast laser to drive thin targets.

This work was supported by the "Research and Mobility" project of Messina University No. 74893496, scientifically coordinated by Professor L. Torrisi. It has been implemented at the CANAM (Center of Accelerators and Nuclear Analytical Methods) infrastructure LM2015056 and has been supported by project No. P108/12/G108.

References

1. L. Torrisi, *Physics of Plasma* **24**, 023111 (2017)
2. R. Prasad, S.Ter-Avetisyan, D. Doria, K.E. Quinn, L. Romagnani, P.S. Foster, C.M. Brenner, J.S. Green, P. Gallegos, M.J.V. Streeter, D.C. Carroll, O. Tresca, N.P. Dover, C.A.J. Palmer, J. Schreiber, D. Neely, Z. Najmudin, P. McKenna, M. Zepf, M. Borghesi, *Nucl. Instr. and Methods in Physics Res. A* **653**, 113 (2011).
3. L. Torrisi, M. Cutroneo, and J. Ullschmied, *Physics of Plasmas* **24**, 043112 (2017).
4. S. Eliezer (Ed). The interaction of high power lasers with plasmas. Bristol: Institute of Physics Publishing, (2012).
5. I.O. Musgrave, C. Hernandez-Gomez, D. Canny, R. Heathcote, R.J. Clarke, J.L. Collier, S. Bandyopadhyay, *Central Laser Facility Annual Report*, 197 (2004/2005).
6. L. Laska, K. Jungwirth, J. Krasa, E. Krousky, M. Pfeifer, K. Rohlena, J. Ullschmied, J. Badziak, P. Parys, J. Wolowski, S. Gammino, L. Torrisi, F.P. Boody *Laser and Particle Beams* **24**(1), 175 (2006).
7. M. Aliofkhazraei, N. Ali, W.I. Milne, C.S. Ozkan, S. Mitura, J.L. Gervasoni, Graphene Science Handbook, Electrical and Optical Properties *CRC Press , Taylor & Francis*, London (2016).
8. M. Cutroneo, P. Musumeci, M. Zimbone, L. Torrisi, F. La Via, D. Margarone, A. Velyhan, J. Ullschmied, L.Calcagno, *J. Mat. Research* **28**(01), 87 (2012).
9. D. Neely, P. Foster, A. Robinson, F. Lindau, O. Lundh, A. Persson, C.G. Wahlström, P. McKenna, *Applied Physics Letters* **89**, 021502 (2006).
10. L. Torrisi and M. Cutroneo, *Journal of Applied Physics* **115**, 083105 (2014).
11. S. Kar, K.F. Kakolee, B. Qiao, A. Macchi, M. Cerchez, D. Doria, M. Geissler, P. McKenna, D. Neely, J. Osterholz, R. Prasad, K. Quinn, B. Ramakrishna, G. Sarri, O. Willi, X.Y Yuan, M. Zepf, M. Borghesi, Phys. Rev. Lett. 109, 185006 (2012).
12. M. Roth, A. Blazevic, M. Geissel, T. Schlegel, T.E. Cowan, M. Allen, J.C. Gauthier, P. Audebert, J. Fuchs, J. Meyer-ter-Vehn, M. Hegelich, S. Karsch, A. Pukhov, *Phys. Rev. Spec. Topics* **5**, 061301 (2002).

Ion energy distributions from laser-generated plasmas at two different intensities

Giovanni Ceccio[1,*], *Lorenzo* Torrisi[1], *Masahiro* Okamura[2], *Takeshi* Kanesue[2] and *Shunsuke* Ikeda[2]

[1]Dottorato di Ricerca in Fisica-Dipartimento MIFT, V. le F.S. d'Alcontres 31, Messina 98166, Italy
[2]Brookhaven National Laboratory, Upton, NY 11973, USA

Abstract. Laser-generated non-equilibrium plasmas were analyzed at Brookhaven National Laboratory (NY, USA) and MIFT Messina University (Italy). Two laser intensities of 10^{12} W/cm^2 and 10^9 W/cm^2, have been employed to irradiate Al and Al with Au coating targets in high vacuum conditions. Ion energy distributions were obtained using electrostatic analyzers coupled with ion collectors. Time of flight measurements were performed by changing the laser irradiation conditions. The study was carried out to provide optimum keV ions injection into post acceleration systems. Possible applications will be presented.

1 Introduction

In order to produce high energy particles, traditional accelerators or high intensity lasers can be used. From interactions of laser with solid thin target, in high vacuum condition, it's possible to obtain energies in the order of keV at about 10^{10} W/cm^2, while using 10^{20} W/cm^2 it is possible to reach energies of MeVs [1].
The most diffused regime to induce high ion acceleration, using sub-nanosecond laser pulses interacting with thin foils, it's known as Target Normal Sheath Acceleration (TNSA), in which the plasma is produced mainly in forward direction with high energy ions above 1 MeV per charge state [2].
Another acceleration regime is the Backward Plasma Acceleration (BPA), that generally occurs during the interaction with thick targets with low intensity lasers (10^{9-13} W/cm^2) generating high ion yields and low energetic ions, of the order of 1-10 keV per charge state [3]. The latter can be used for the generation of plasma that find uses in several applications, for example Laser Ion Source (LIS) and post ion acceleration to generate implantable ions [4, 5]. After the laser interaction, it is possible to observe the erosion of surface with the presence of a crater in the case of thick target and a micrometric hole in thin foils. The eroded material, converted in plasma and from which ions are produced, depends on the target composition and laser characteristics [6].
In order to study the plasma properties generated by low laser intensity and composed targets, we investigated some Al substrates targets with Au coating using different plasma diagnostics.
The plasma characterizations were performed using time-of-flight (TOF) techniques both at Brookhaven National Laboratory (BNL) and plasma laser laboratory of University on Messina. The interest for these plasmas comes for LIS applications in accelerator systems, such as the superconducting cyclotrons, and radio frequency quadrupole (RFQ) accelerator in which the laser-plasma produced ions can be injected.

2 Experimental Setup

2.1 Laser System

At Brookhaven National Laboratory a nanosecond Nd:YAG laser beam, with 10^{12} W/cm^2 intensity, was used at fundamental frequency with energy of 1094 mJ measured on the target surface, time per pulse of 6 ns and focalized spot of 100 μm in diameter. The laser-target incident angle was 20°. In University of Messina was employed also a Nd:YAG laser system at fundamental frequency, but with a laser energy of 100 mJ and 3 ns pulse duration, focalized with a spot of 1 mm in diameter in order to obtain a laser intensity of about 10^9 W/cm^2. In both cases the interaction chamber was placed at 10^{-6} mbar of vacuum.

2.2 Al coated with Au film targets

Thin gold films on the aluminium substrates were prepared using physical vapour deposition (PVD) technique. For plasma production, we have employed a different film thicknesses, from 250 nm up to 750 nm, of Au deposited on Al substrate. In addition, pure Al substrate was also prepared, in order to understand the differences with the addition of Au layers. The thicknesses of gold films were measured during the deposition process with a resonant quartz crystal monitor. The surface roughness, measured with a surface profiler, was very low and in the order of 1 micron. The prepared targets were mounted on a 2-axis linear stage,

* Corresponding author: gceccio@unime.it

moved after every laser shot; in order to irradiate only fresh surfaces in single laser pulse and high vacuum conditions.

2.3 Faraday Cup and Ion Energy Analyser

The plasma diagnostics in BNL are based on Faraday Cup (FC) and Ion Energy Analyser (IEA) both in Time of Flight (TOF) configuration [7]. Thanks to FC, located at 2.4 m away from the target, we can get the first informations on laser-generated plasma, like the maximum ion energy and the yield of plasma. The FC is located in normal direction of expansion of plasma and after the measures, can be removed to allow at the particles to travel to the IEA. The IEA is based on the electrostatic deflection of the ions, emitted from the plasma, due to an electrostatic field. IEA allows to determine the ion energy and charge state distributions by varying the voltage of the deflection plates. A symmetric potential to deflect of 90° the ions towards the electron multiplier is used, according to the following relation:

$$E/z=ekU \qquad (1)$$

where E is the ion energy, z is the charge state, e is the electron charge, k is the geometrical factor of IEA and U is the potential gap between the plates. Only the ions with a given energy-to-charge state ratio, E/z, can pass through the IEA and arrive to the electron multiplier, thus IEA operates such as an energy filter [8].

IEA detector is placed at 1.4 m after the FC position for a total flight distance of 3.8 m from target. In Fig. 1a is plotted the experimental setup present in BNL where are indicates the laser incidence direction, the vacuum value of operation and the diagnostic position. The same diagnostics were employed in Messina with the same configuration but different distances, 0.88 m and 1.45 m for IC Ring and IEA, respectively.

The particles that pass through the IEA reaching the detector, was in an Hamamatsu R2362 Secondary Electron Multipliers (SEM). Electron multipliers are mainly used as high efficient ion detectors. Hamamatsu electron multipliers have a high gain (multiplication factor) and a low dark current, allowing operation in photon counting mode, to detect and measure extremely small low energy incoming particles. Fig. 1b reports the electron emission ratio of SEM as function of ion acceleration voltage.

In Fig.1c is reported a typical IEA spectrum obtained irradiating 750 nm of Au on Al bulk; it is possible to observe several charge states of gold from 1+ up to 5+, the resolution of device allows to observe the separation between two charge states.

3 Results

3.1 BNL measurements

Fig. 2 shows the time of flight spectra of radiation emitted from laser generated plasma irradiating Al pure target and Al with Au films of different thicknesses: 250 nm, 400 nm, 500 nm and 750 nm. The insert shows a zoom of the peaks of last two thicknesses (500 nm and 750 nm of gold), the bigger quantitive of gold reduced the yield value up to the value of few mV. All signals show a photopeak due to photoelectric effect, that gives the start time, and the ion peak located at different TOF times from which is possible to evaluate the ion energy knowing the flight distance.

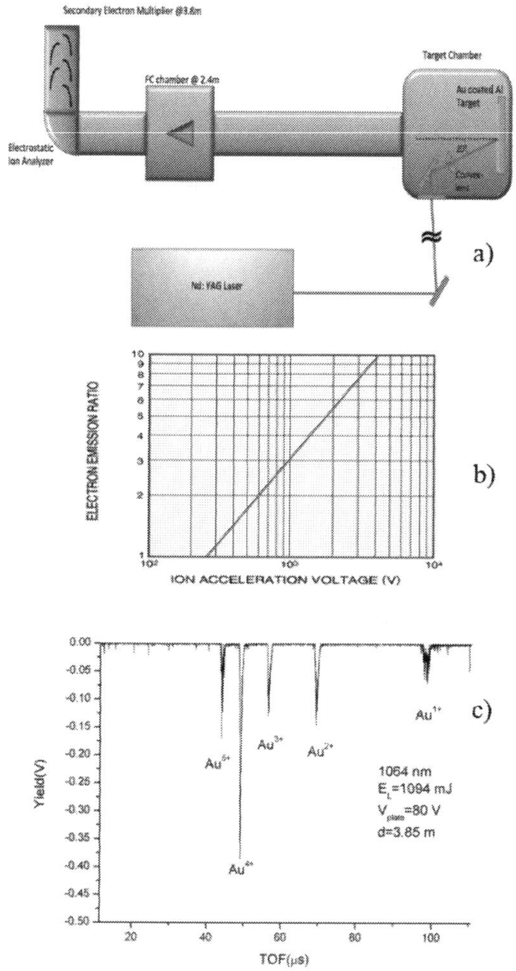

Fig. 1. Experimental setup of BNL laboratory for Plasma diagnostics. b) The Electron emission ratio for Hamamatsu SEM R2362. c) Example of IEA spectra obtained at BNL.

Observing the ions TOF values and the yields for each target it is possible to note that these values have a trend depending on the Au thickness. In particular the TOF value for Al pure was 5.76 μs, for 250 nm of Au on Al was 6.55 μs, for 400 nm of Au on Al 10.83 μs, for 500 nm of Au on Al 14.13 μs and for 750 nm of Au on Al 14.8 μs. From these values, it is possible to deduce that the target with 250 nm of gold film has behaviour similar to the Al pure target but with lower energy, instead the 750 nm target shows a result similar to Au bulk, in terms of times of flight and yield, the 400 and 500 nm targets have an intermediate behaviour.

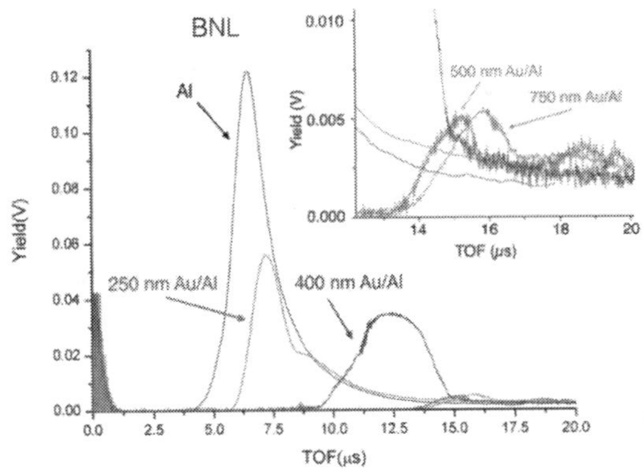

Fig. 2. ToF spectra for coated target irradiated at BNL, the reduction of energy and yield follow the thicknesses of gold films.

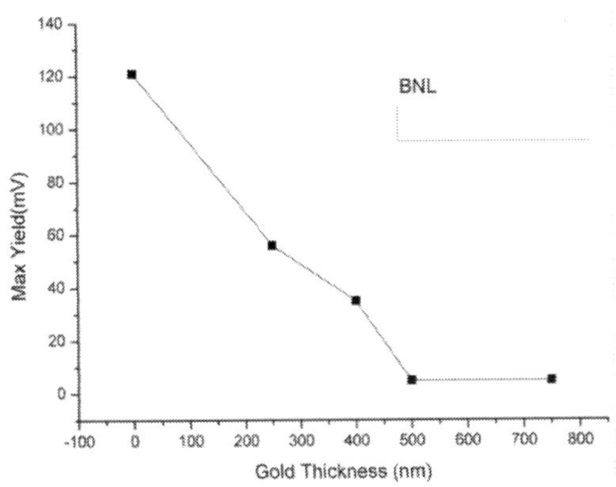

Fig. 3. Measured plasma yield with FC.

In Fig. 3 is reported the maximum yield for each target as a function of the gold thickness.

The total ion yield shows a maximum value of 120 mV in pure Al, 56 mV for 250 nm of Au on Al and decreases with the thickness until to reach the value of 5 mV for the higher gold thickness (750 nm). This result shows that the Al ion yield are reduced when are covered by Au thin films, as expected.

Additional information about the maximum energy, can be obtained from TOF spectra making the deconvolution operation using the Coulomb-Boltzmann Shifted (CBS) function [9]:

$$f(t) = A\sqrt{\left(\frac{m}{2\pi k_B T}\right)^3}\frac{L^4}{t^5}exp\left[-\frac{m}{2k_B T}\left(\frac{L}{T}-\sqrt{\frac{\gamma k_B T}{m}}-\sqrt{\frac{2zeV_0}{m}}\right)^2\right]$$

where A represents the normalization constant, m the atomic mass, L the distance between the target and the ion collector, γ the adiabatic coefficient (1.67 for monoatomic species), ze the ion charge, V_0 the equivalent voltage developed in the non-equilibrium plasma, and $k_B T$ the equivalent plasma temperature in (eV).

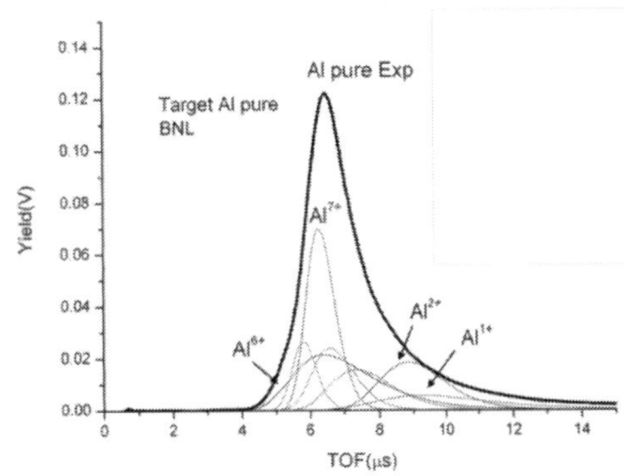

Fig. 4. Result of deconvolution on Al pure target using CBS function.

Fig. 4 reports a typical deconvolution process using the CBS function, for the pure Al spectrum. The value of 4.5 μs corresponds to a kinetic energy of 38.5 keV obtained from the higher charge state that in this case is 6+.

The slower part of curve consists in lower charge states, in particular to 1+ and 2+. Observing the faster Al and Au ions TOF values in all targets, we obtained the values of energies reported in Fig. 5, where it is clear that the trend of ion energy is similar at the trend of the ion yield and decreases with the gold thickness from a value of 38.5 keV to 2 keV for Al pure target and from 51 keV up to 28 keV for 500 nm and 750 nm of Au film thickness, respectively.

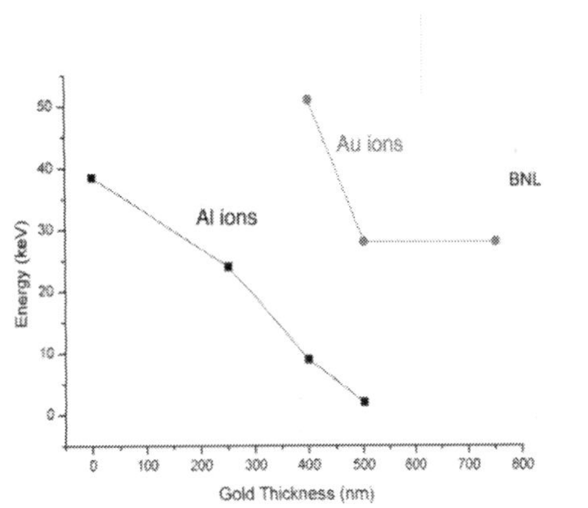

Fig. 5. Maximum energy value for Al and Au ions obtained from deconvolution of TOF spectra.

Another parameter that change with the thickness of the gold coverage is the plasma temperature. For a calculation of the plasma temperature we have used the energy distribution obtained from IEA measurements, taking in consideration the charge states and making a fit operation using the CBS function in energy. From these deconvolutions, performed for Al and for Au in the produced plasmas, we obtained the plasma temperature for our targets reported in Fig. 6. This plot shows that the temperature is constant for Al pure and less than 1 keV but for the other targets covered with Au films, it increases with the charge state and gold thickness reaching values up to 4.5 keV for the charge state 5+ of Au with 750 nm of film thickness. The results demonstrate that temperature gradients characterize the produced plasma depending on the kind of irradiated target. With IEA measurements one parameter that can be possible to calculate is the ion energy acquired per charge state.

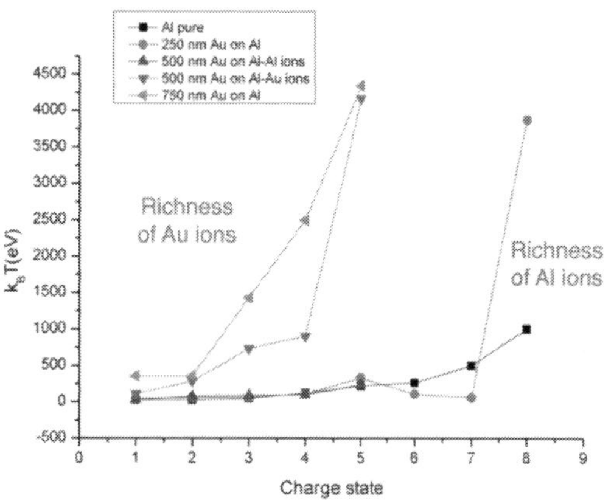

Fig. 6. Plasma temperature obtained from fit operation using CBS

3.2 Messina measurements

The results obtained in Messina laboratory showed similar results. The maximum peak yield and the peak energy versus the gold thickness are shown in Fig.7: full symbols indicate the results for particles energy, while the empty symbols the yield of FC signal. The yield of FC signal decreases from a value of 112 mV for Al pure target up to 20 mV for 750 nm of gold on Al. The ion energy increases with the film thickness reaching the values of 5.5 keV and 1.6 keV for Au and Al ions, respectively. Of course, the lower intensity laser reduces the charge states and the ion energy and the plasma temperature, as expected.

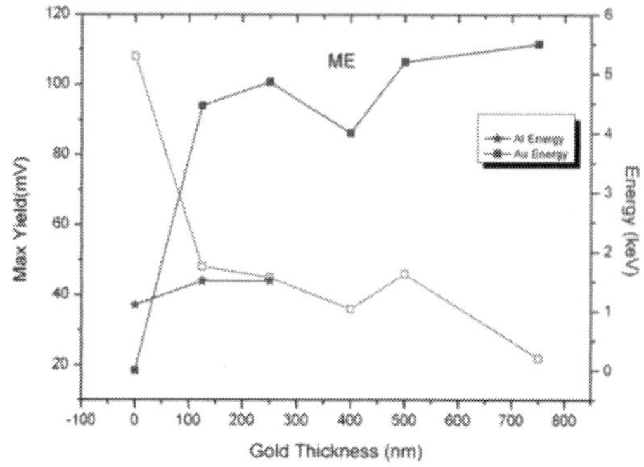

Fig. 7. Experimental results obtained in Messina

4 Conclusions

The reported data indicate that using a laser intensity of about 2.3 10^{12} W/cm^2 irradiating pure Al the plasma temperature is under 1 keV. Using the same laser intensity irradiating 750 nm Au/Al increase up to 4.5keV, demonstrate that the increment of the electron density of the plasma, due to the presence of gold, enhances the plasma temperature. The gold in the plasma produces also an increment of the charge separation and an increment of the ion acceleration that from the 900 eV per charge state grows up to the value of 7.5 keV per charge state. The use of the 750 nm Au on Al produces a plasma very rich in gold and in electrons, and very poor in Al, demonstrating that this thickness is the minimum value to produce ablation from gold similar to a bulk gold target. Using lower laser intensity of 10^9 W/cm^2 the plasma yield decreases from a value of 112 mV for pure Al up to 20 mV with the cover of 750 nm of gold, with all intermediate thicknesses that follow the trend. The plasma energy increases its value with the presence of electrons from the gold films. Considering the results obtained about the charging states in the two different laboratories using two different laser intensities, we can state that at a laser intensity of 2.3×10^{12} W/cm^2 it is possible to obtain higher charging states, for aluminium from 1+ to 8+ whereas for Au from 1+ to 12+, while at lower intensity the maximum charge states for both species are 3, but with more energy constants than for the high intensity. Furthermore, the studies have shown that the plasma temperature is constant for the various charge states in plasmas obtained in Messina laboratory. These obtained results suggest different applications. Direct Plasma Injection Scheme (DPIS) is a possible application where coated target can be used to produce high charge states. The DPIS in the new method combining a Laser Ion Source and a Radio Frequency Quadrupole (RFQ) Linac without a beam transport. In the DPIS, multiple charge state ions are extracted from plasma at the entrance of RFQ without selection of charge state [10-11]. In the DPIS scheme a target is located in a vacuum chamber directly connected to an RFQ Linac. In this way, even if there is an angular distribution of emitted ions a large fraction of plasma, it goes directly into the RFQ channel. Typically, parameters of RFQ are 100 MHz of operating frequency and input energy of 17 keV/amu. This means that our measures can be used to increasing the charge state inside the plasma and coupling with extraction voltage. Future investigation will be done to characterize the best operation frequency and the best output current beam with coated target. Using low intensity laser, it would be possible to use the coupling of an ion source that uses the coated targets with an Electron Cyclotron Resonance (ECR) that allows to enhance charge state producing ions with charge state up to 30+ and with extraction voltage of about 20-30 kV as performed at INFN-LNS experiments in Catania (Italy). [12]

This work was supported by the "Research and Mobility" project of Messina University No. 74893496, scientifically coordinated by Professor L. Torrisi.

Bibliography

1. L. Torrisi, NUKLEONIKA; **60**(2): 207-212 (2015)
2. J. Badziak et al., Plasma Phys. Contr. Fusion **46**, B541 (2004)
3. J. Badziak, P. Parys, A. B Vankov, J Wolowski, and E. Woryna, Appl. Phys. Lett. **79**, 21 (2001)
4. K. Sakakibara, M. Okamura, S. Kondrashev, T. Hattori, H. Kashiwagi, T. Kanesue, Rev. Sci. Instrum **77**, 03B304 (2006)
5. K. Kondo, T. Kanesue, J. Tamura, M. Okamura, Rev. Sci. Instrum. **81**, 02A511 (2010)
6. D. Margarone, L. Laska, L. Torrisi, et al., Applied Surface Science, **254**, 2797–2803, (2008)
7. L. Laska et al. Czechoslovak Journal of Physics, **46**, 1099–1115, (1996)
8. E. Woryna et al., Laser and Particle beams, **13**, 293-321, (1996)
9. L. Torrisi, Radiation Effects and Defects in Solids, **171**:1-2, 34-44, (2016)
10. J. Tamura. M. Okamura, et al., Joint Accelerator Conference Website, HIAT09, 2009
11. Y. Fuwa, M. Okamura, et al., Joint Accelerator Conference Website, LINAC14, 2015
12. S. Gammino, L. Torrisi, et al,. Proceeding of EPAC 2002

Graphite oxide based targets applied in laser matter interaction

Mariapompea Cutroneo[1,2*], *Lorenzo* Torrisi[2], *Jan* Badziak[3], *Marcin* Rosinski[3], *Vladimir* Havranek[1], *Anna* Mackova[1,4], *Petr* Malinsky[1], *Zdenek* Sofer[5], *Jan*Luxa[5], *Antonino* Cannavò[2], *Jan* Lorincik[6]

[1]Nuclear Physics Institute, AS CR, 25068 Rez, Czech Republic
[2]Department of Physics and Earth Sciences, Messina University, V.le F.S. d'Alcontres 31, 98166 S. Agata, Messina, Italy
[3]Institute of Plasma Physics and Laser Microfusion, P.O. Box 49, 00-908 Warsaw, Poland
[4]Department of Physics, Faculty of Science, J.E. Purkinje University, Ceske mladeze 8, 400 96 Usti nad Labem, Czech Republic
[5] Department of Inorganic Chemistry, University of Chemistry and Technology Prague, Technická 5, 166 28 Prague 6, Czech Republic
[6]Analytical Laboratory NFC, Husinec-Rezc.p. 130 25068 Czech Republic

Abstract. In the present work, we propose the production of a hybrid graphene based material suitable to be laser irradiated with the aim to produce quasi-monoenergetic proton beams using a femtosecond laser system. The unique lattice structure of the irradiated solid thin target can affect the inside electron propagation, their outgoing from the rear side of a thin foil, and subsequently the plasma ion acceleration. The produced targets, have been characterized in composition, roughness and structure and for completeness irradiated. The yield and energy of the ions emitted from the laser-generated plasma have been monitored and the emission of proton stream profile exhibited an acceleration of the order of several MeVs/charge state.

1. Introduction

Graphene exhibits unique electronic band structure, thermal conductivity and excellent electronic transport properties indeed electrons in graphene often are indicated as " massless"[1]. Since the first publication in 2004 [1], when Geim's group plucked monolayer graphene sheets from graphite by adhesive tape, one of the foremost challenge is to functionalize its band gap as it is crucial for a controllable manipulation of its atomic and electronic properties. Extensive research had been conducted to investigate the applicability of graphene based material in medicine[2], energy storage[3], fundamental research [4-5], and others, but its suitability in laser matter interaction, in order to improve the ion acceleration from plasma, has been about ignored in literature despite the manufacturing of graphene sheets in a composite material is an indicated way to enhance its properties [6]. Recently, graphene has been produced on a sufficient scale and homogeneously incorporated in a polymeric matrix [7] to be irradiated by laser, exploiting its properties, optical, conductive and mechanical, with the aim to generate hot and dense plasmas in vacuum from which protons and carbon ion beams can be extracted and accelerated at high energy for many applications: hadrontherapy [8], astrophysical studies, laser-driven ion acceleration [9], inertial confinement fusion [10], and others. In the present work, we followed up the increasing demand of advanced target engineering in laser matter interaction, proposing a hybrid graphene based material fabricated to be irradiated by sub-picosecond high intensity laser, taking into account the significance of target composition, geometry, resistivity, roughness and thickness for the electron propagation and the production of sharp proton bunches; we paid attention to the close dependence between the electron spatial distribution and the features of emitted ions plasma laser-generated. It is well established that the conductivity of a material is strictly connected to its lattice. In the case of high current of electrons propagating in a solid target, as a consequence of high energy sub-picosecond laser pulse, due to charge neutrality, an "inverse" current of thermal electrons will induce the heating of a wide volume of the target generating a non equilibrium state of warm dense matter consisting of hot electrons and cold ions which are temporarily in the same position where they were in the condensed matter state. In this situation the target conductivity depends on its ability to maintain its lattice structure. Over a plethora of members in the graphene-like materials family, the one, emerging for the ability of tailoring its mechanical, electrical and optical properties, is graphite oxide. It is a layered material produced by oxidation of graphite and consisting of hydroxyl and epoxide functional groups as basal planes and carbonyl and carboxyl groups at the edges [11] and, it is an insulating material [12] as most of the carbon atoms are sp3-hybridized. A systematic study of compositional, structural, optical properties of graphene based material, correlated to its response, is essential for understanding as various topological defects and

* Corresponding author: cutroneo@ujf.cas.cz

chemical functionalization can modify the properties of the initial graphene system and for reproducibility of the sample production. To understand the potentiality of the prepared target, the emission of energetic ions induced under laser bombardment was detected online by semiconductor SiC detectors. The structural, compositional and optical properties of the samples have been performed at Tandetron laboratory of the Nuclear Physics Institute, CAS, in Rez (Czech Republic).

2. Experimental setup

Despite of the Hummers method is one of the oldest preparation techniques, it is one of the most suitable methods for the formation of bulk graphene.

In this study, fresh graphene oxide (GO) foil was synthesized like-Hummers method as reported in Ref. [13] to prepare GO thin foils. The evaluation of GO density was carried out by a Mettler Toledo Micro-Balance with ± 1μg absolute accuracy in standard microbalance weighing technique. A small square of GO foil was cut and its area was accurately determined by through a digital microscope image analysis. In Fig.1 the typical image of the cross sectioned GO observed by optical microscopy reports the measures of the thickness foil in several positions: 5.95μm, 5.98μm, 5.90μm, 6.20 μm. The evaluated thickness was 6 μm and the density was 0.72 gr/cm^3, with an error of 2%.

Two samples have been fabricated, both 1x1 cm^2 of size: graphene oxide foil 6 μm thick (labeled as GO) and graphene oxide covered by 200 nm gold film (labeled as Au+GO) by PVD technique.

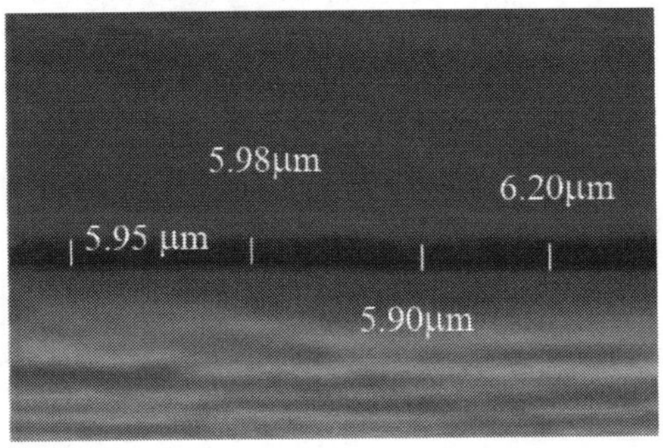

Fig. 1. Optical microscope image of the graphene oxide foil section.

The AvaSpec-2048 spectrophotometer with UB-600 lines/mm grating and a bandwidth of 195–757 nm has been employed to carry out optical measurements in reflectance mode. The adopted incidence angle and the investigated wavelength range were 15° and 175–2500, nm respectively. The source designed by Avantes was a deuterium halogen lamp AvaLight-DH-S with an integrated TTL-shutter and filter holder for filters of up to 50x50x5.0 mm. An area of 1.8 mm^2 is analysed using a probes of 1.5 mm in diameter and equipped with a setscrew to place the probes into suitable positions.

The morphology of the substrates was studied by atomic force microscopy (AFM) analysis [14] using a NTEGRA AURA instrument (NT-MDT) in the tapping mode in the ambient atmosphere, and a chip (NSG01) with the radius less than 10 nm. The arithmetic average height (R_a) and the mean roughness (RMS) for pristine and irradiated polymers with three different ion fluences were estimated and their error does not exceed 10%. Scanning Electron Microscope (SEM) and Energy Dispersive Spectroscopy (EDS) measurements have been conducted using both a FIB-SEM system LYRA3 (TESCAN ORSAY HOLDING a.s.) and an EDS Aztec X-MAX 80 (Oxford Instruments plc). In order to probe the relative amounts of carbon and oxygen present in the sample, 5 kV and 20 kV voltages electronic microscope [15] and the detectors of secondary electrons (SE) and of the backscattered electrons (BSE) for a better z material contrast have been adopted.

An InVia Raman microscope (Renishaw, England) supplied with a spectrometer operating in backscattering geometry has been employed to estimate the structural properties of the substrates. The system consists of a CCD detector and a Nd-YAG laser operating at 532 nm of wavelength, 50× of magnification objective and at no more than 5% of the max power output (50 mW) to avoid the sample damage. In this context, the characteristic Raman band of Silicon at 520 cm^{-1} has been adopted as calibration reference.

Preliminary ion beam irradiation of GO foil was performed at Helmholtz-Zentrum Dresden-Rossendorf to characterize the changes in a set of GO foils under ion irradiation in different conditions. The composition of the foils in the most superficial layer (~1 μm) has been evaluated by Rutherford Backscattering Spectrometry(RBS) and Elastic Recoils Detection(ERD) analyses at Tandetron laboratory in the Czech Republic. In both cases, a beam of 2.9 MeV protons has been used as projectile at incident angles of 0°. Solid state detectors placed at 160° and 35° for each analysis, monitored the backscattered and forward recoiled atoms.

After the characterization, thy have been laser irradiated by a multi-TW femtosecond laser at the Institute of Plasma Physics and Laser Microfusion (IPPLM) in Warsaw. The central laser wavelength is 810 nm, the pulse duration is ~ 40 fs, the maximum energy of the output pulse is 500 mJ, the contrast is 10^{-9} , the power density is 10^{18} W/cm^2, 10 μm focal spot diameter, and operating in single-pulse mode or at 10 Hz repetition rate. A high resolution CCD camera evaluates the laser focal position (FP) on the surface of the target and presently was +150 μm (i.e. inside the target) and the main plasma stream was generated close to the target surface. The laser-target interaction give rise to complex phenomena and ends with the emission of protons and heavier ions driven by electrons. Because of the ion acceleration is strictly connected to electrons, the fast electron transport has been analyzed by the monitoring of the proton bunch profile as a function of both space and intensity in the rear side of the target[16]. The online bunches monitoring was performed in Time of Flight (TOF) approach where ions propagate over a known distance and then revealed by Ion Collectors and robust semiconductor SiC detectors[17] placed at 0° with respect to the normal to the target surface, in the forward direction.

3. Results

SEM analysis conducted at 20 keV electron beam on GO foil in the Au side (b, c) as well as in the GO side (a) reveals in the latter case (Fig. 2a) as main elements, for C, O and S, the mass concentrations of 0.62, 0.35, 0.02 respectively and also minor elements in traces of Cl, Al, Mn and K.

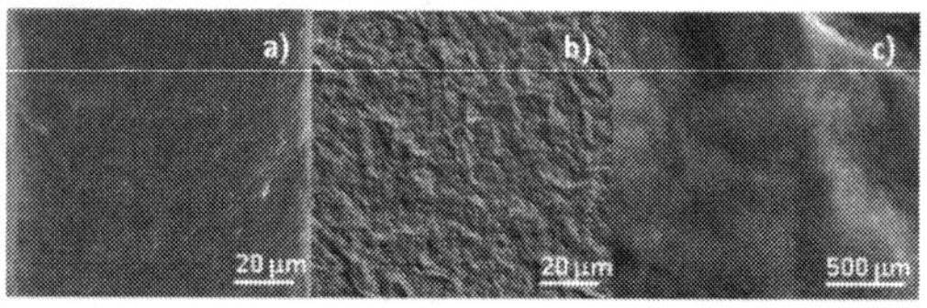

Fig. 2. SEM images of virgin GO foil 100 µm field view a) Au on GO foil 100 µm b) and 2 µm field views c).

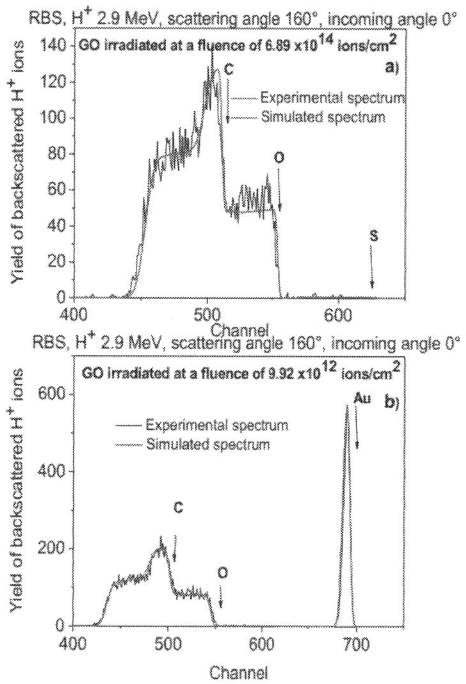

Fig. 3. RBS spectra for GO a) and for GO-Au b).

In the former (b, c), Au as main element and C as contaminant have been revealed with the morphology reported in Fig. 2b. Au microstructures on top of the underlying surface are observed from Fig. 2c. RBS and ERDA analyses have been performed to evaluate the composition of GO and GO covered by gold and typical spectra are reported in Fig. 3 and Fig. 4.

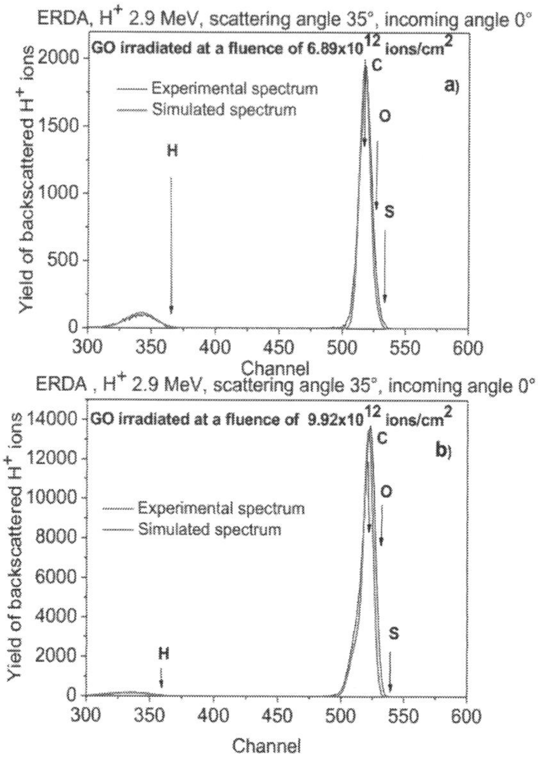

Fig. 4. ERDA spectra for GO a) and for GO-Au b).

In both samples, the main components are carbon, oxygen, hydrogen and trace elements, such as sulphur. This last component is present as a consequence of the chemical synthesis of GO. The ERDA spectrum reports a value of hydrogen content of 17.4% for GO foil and 22.6% for GO Au foil, gathering the final increment of 23 %, as reported in Fig. 3b. carbonization effects occurring during the GO foil irradiation. Because of RBS provides information about the presence of heavy ions, whereas ERDA about the presence of light ions, by the comparison between the obtained atomic concentrations we obtained the overall consistency of analyses. From such ion beam analyses was possible to find out the hydrogen/oxygen (H/O) and the C/O ratios, of about 1.68 and 0.572 for GO and 1.59 and 0.766 for GO Au, respectively.

Table I reports the values of the C, O, H, C/O and H/O atomic value determined by the ion beam analyses. The enhancement of C of about 9.1 % (0.47-0.513)/0.47) was a consequence of the thin gold coverage of the GO substrate.

Table I. The composition of GO and GO Au foils.

COMPOSITION OF THE GO FOILS [atm%]			
Sample/ Ions	C	O	H
GO	51.3± 0.5	30.4± 0.3	17.4± 0.2
GO_Au	47± 0.5	29.5± 0.3	22.6± 0.2

The optical characterization of the prepared thin foils was studied by the reflectivity-curves for all the samples as a function of the wavelength, as reported in Fig. 5. It shows a reflectivity decreasing going from Au+GO to Go virgin and to GO+Au sample with seven main peaks at 370.7, 465.7, 486.4, 581, 610.3, 656, 674.8 and 691 nm. The direct evaluation of the reflectivity of the samples, performed at room temperature and in the range of wavelength of 300-760 nm, reports a good agreement between the experiment and literature. The highest reflectivity is presented for Au+GO foil illuminating the Au side, due to the contribution of the gold film which gives high an high reflectivy of about 98-99% in the near infrared region, but a limited and very low reflectivity at wavelengths shorter than ~ 400 nm.

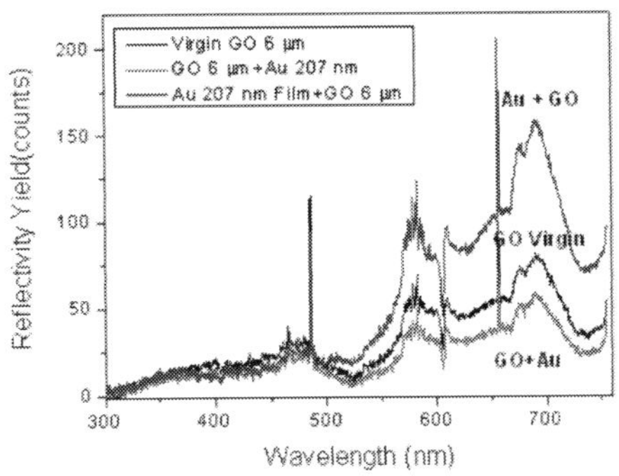

Fig. 5. Optical measurements in reflection mode for graphene oxide, graphene oxide coated with thin gold film illuminated on the graphene oxide surface side and then, on the gold surface side.

The reflectivity of GO is slightly higher in the near UV region and markedly lower in the near IR region due to its unique electronic structure [18]. For completeness the reflectivity of the Au+GO foil illuminating the GO face (i.e. GO+Au) has been evaluating, exhibiting the lowest value caused by the opacity of the foil, which increases with the membranes' thickness. Thus, also if the graphene is constituted by a monoatomic layer, it is found that it absorbs a significant ($\pi\alpha$= 2.3%) fraction of the incident visible light, according to literature [18].

Returning to the application of the thin GO foil to prepare advanced target for laser irradiation in target-normal-sheath acceleration (TNSA) regime, there is a reason of the correlation between proton stream properties and electron beam transport. Protons have been exploited to diagnose the behaviour of electron beam laser accelerated into the target. Fig. 6 shows a typical SiC spectrum recorded in time-of-flight (TOF) approach [17] for ions emitted irradiating with the IPPLM of Warsaw terawatt laser the GO+Au target. The focal position was optimized to obtain dense and hot plasma from the target and to maximize the ion acceleration in the forward direction. The data were collected using a fast storage oscilloscope (TekTronix DPO2024, 200 MHz, 1GS/s).

UV and X-ray laser plasma generated are indicated by the first peak, 3 ns wide in Fig. 6, followed by faster ions at 34.4 ns, 47 ns and 50 ns corresponding to protons, carbon and oxygen ions, with kinetic energies of about 2.7 MeV, 17 MeV and 20.5 MeV, respectively. The wide peak starting at 50 ns to 76 ns is due to the emission of low energetic Carbon and Oxygen. The peak at 81 ns is due to the contribution of gold ions. This acceleration increment is due to the thin Au coverage enhancing the electron density of the plasma and the electric field driving the ion acceleration regime[19].

This interesting result can be ascribed also to the reduction of the target reflectivity, the lattice of the GO foil, the enhancement of the electron density due to the presence of gold all together providing better transferring of the laser energy to the target and decreasing of the electron energy loss through the target. Since a high intensity laser at 10^{18} W/cm^2 and 39 fs pulse duration interacts with a thin solid target, a high fraction of laser energy is absorbed by the target surface electrons, producing high current of relativistic accelerated electrons.

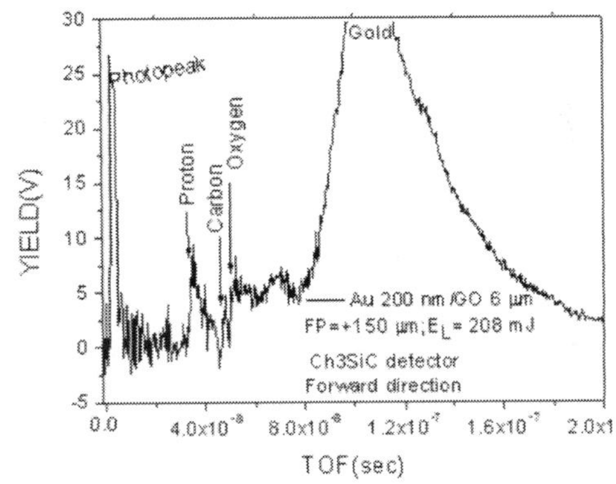

Fig. 6. SiC spectrum in the forward emission irradiating GO foil with a gold film 200 nm thick

This effect induces a significant enhancement of the target temperature (~ 10 keV) and an high electric field developed in the rear side of the thin irradiated foil. In this step, in facts, it is generated a "non-equilibrium state of dense matter", consisting of hot electrons and cold ions, generating an electron cloud out of the target and a high positive target surface in which ion acceleration take place. The developed electric field, direct orthogonally to the target surface, is of the order of tens GV/m, as reported in literature [20].

Conventionally, the ion acceleration per charge state can be evaluated by the maximum proton energy detectable from TOF measurements, reaching a value of about 3.2 MeV relatively to the TOF spectrum of Fig. 6, in well agreement with literature [20].

4. Summary and conclusions

Universally is recognised that conductive targets, laser irradiated at high intensities, induce the emission of smooth proton stream profiles, while insulator generate non-homogeneities in the proton density. The electron current is undergone to electromagnetic instabilities which split the flow of hot electrons into filamentations [22]. In this work, hybrid GO based materials have been employed to be laser irradiated and their response were surprising if compared with experiments conducted in the same laser facility, at optimized laser conditions, but using different targets. GO is an insulator material and it has been simply coated with gold to enhance its electron density. The main purpose of this work is to perform a comparative study of structural, compositional and optical of graphite oxide based materials properties to potentially implement design and fabrication of advanced targets usable in laser matter interaction for too long overlooked in literature. Certainly the achieved result can be maximized changing the laser parameters (focal position, laser energy, contrast, pulse duration..) and this is our proposed approach. The emerging preparation of sophisticated targets leads to a high level development of ions sources laser plasma generated useful for many applications and the use of graphene based materials arises several issues such as: what does occur increasing the number of the graphene layers? would be better to use lower thickness of metallic substrate to enhance the electron density and to reduce the electron energy loss? How the optical properties of graphene control the laser-plasma interaction?. Work is in progress to reply to these questions.

The research has been realised at the CANAM (Center of Accelerators and Nuclear Analytical Methods) infrastructure LM 2015056 and has been supported by project No. P108/12/G108. Z.S. and J.L. were supported by Czech Science Foundation (GACR No. 16-05167S) and by specific university research (MSMT No 20-SVV/2017). This work was supported by the "Research and Mobility" project of Messina University No. 74893496, scientifically coordinated by Professor L. Torrisi..

References

1. K.S. Novoselov et al., Science **306**, 666-669 (2004)
2. Y. Yang, A. Mohamed Asiri, Z. Tang, D. Du, and Y. Lin, Materials Today **16**, 10 (2013)
3. W. Lv, Z. Li, Y. Deng, et al., Energy Storage Materials **2**, 107-138 (2016)
4. K. S. Novoselov, A. K. Geim, S. V. Morozov, D. Jiang, M. I. Katsnelson, I.V. Grigorieva, S. V. Dubonos, A. A. Firsov, Nature **438,** 197-200 (2005)
5. Y. B. Zhang, Y. W. Tan, H. L. Stormer, P. Kim, Nature**438**, 201-204 (2005)
6. S. Stankovich, D. A. Dikin et al., Nature **442**, 282-286 (2006)
7. L. Torrisi, L. Torrisi, G. Ceccio, N. Restuccia, E. Messina, P. G. Gucciardi, M. Cutroneo, Laser and ParticleBeams 1-10,(2017)
8. F. Schillaci, A. Anzalone, G.A.P. Cirrone , M. Carpinelli, G. Cuttone, M. Cutroneo, et al., ELIMED, MEDical and multidisciplinary applications at ELI-Beamlines, Journal of Physics: Conference Series **508**, 012010 (2014)
9. E. Clark et al., Phys. Rev. Lett. **84**,670 (2000)
10. M. Tabak et al., Phys. Plasmas **1**, 626 (1994)
11. H. He, J. Klinowski, J. Foster & A. Lerf, Chem. Phys. Lett. **287**, 53-56 (1998)
12. C. Punckt, F. Muckel, S. Wolff et al., Applied Physics Letters **102**, 023114 (2013)
13. P. Malinsky, A. Mackova et al., Physical Chemistry Chemical Physics **19**, 10282-10291 (2017)
14. V. Lavrientev, A. Stupakov, J. Pokorny, I. Lavrentieva, J. Vacik, J., et al., J. Phys. D: Appl. Phys. **48**, 335002 (2005)
15. J. Fuchs, T. E. Cowan, P. Audebert, H. Ruhl, L. Gremillet, et al., Phys. Rev. Lett. **91**, 255002 (2003)
16. Fuch et al., Phys. Rev. Lett. 91 (2003)255002
17. M. Cutroneo, P. Musumeci, M. Zimbone et al., J. Material Research **28**, (01) 87-93 (2012)
18. R. R. Nair, P. Blake, A. N. Grigorenko, K. S. Novoselov, T. J. Booth, T. Stauber, N. M. R. Peres, A. K. Geim, Science **320**, 1308-1308 (2008)
19. L. Torrisi, L. Calcagno, D. Giulietti, M. Cutroneo, M. Zimbone, J. Skala, Nucl Instr. and Methods B **355**, 221-226 (2015)
20. L. Torrisi, Nukleonika **60**(2), 207-212 (2015)
21. L. Torrisi, Rad. Eff. and defects in Solids **171**, (1-2), 34-44 (2016)
22. A. Macchi, M. Borghesi, M. Passoni, Reviews of Modern Physics **85**, 751 (2013)

Investigation of Plasma Formation with ns Laser by Using Focused Sub-ns Laser Probe

Shunsuke Ikeda*, *Takeshi* Kanesue, and *Masahiro* Okamura

Collider-Accelerator Department, Brookhaven National Laboratory, NY, USA

Abstract. A laser ablation plasma formation process was investigated by shooting another laser beam. The initial plasma was created by a mildly focused nanosecond laser beam and the probe laser, which has about ten times shorter pulse length with tightly focussed condition, was irradiated on it. By analysing high temperature plasma created by the probe laser, we could detect an interaction between the initial low temperature plasma and the probe laser. The interaction caused that the velocity of high temperature plasma generated by the sub-ns laser became smaller and the amount of the highly charged ions decreased. We found that the interaction does not occur during the irradiation of the first half of the ns laser. This fact indicates that the plasma is not produced during the first half of the ns laser.

1 Introduction

Laser ablation plasma is suitable for providing intense ion flux with various charge states ranging from singly charged to fully stripped. The advantage of the ion production scheme is that a large number of ions can be produced from most of the solid materials under high vacuum condition. The induced ions can be applied to many fields such as an ion source for particle accelerators[1].

For production of low charge state ions, laser intensity on a target surface is close to the laser ablation threshold. Typically, the threshold is 10^8 W/cm^2 ~ 10^9 W/cm^2 for nanosecond and sub-nanosecond lasers [2-4]. Most of the ions are singly charged for the laser intensity [5, 6]. In the intensity region, the plasma is not uniform but evolves temporally and spatially. To investigate the plasma formation, we are proposing to use a focused another laser pulse with a shorter pulse width as a probe. If one irradiates a laser target with a focused sub-nanosecond laser during the irradiation of a mildly focused ns laser pulse, the interaction of the sub-ns laser and the ns-laser plasma depends on the property of the ns-laser plasma. The interaction makes the difference of the plasma property after the expansion. Therefore, one can investigate the property of the ns-laser plasma with temporal and spatial resolutions by analysing the plasma after the expansion with different spot positions and timings of the sub-ns laser.

As a first step of the investigation, we measured ion current and charge state distribution of plasma with different timings of the sub-ns laser focused on the center of the ns laser spot. Based on the results, we discussed the production process of the ns-laser plasma.

2 Experimental setup

Figure 1 shows the experimental setup. An aluminium plate was irradiated with a nanosecond and a sub-nanosecond lasers. The target plate was placed on linear stages and moved every laser shot. The target system was installed in the grounded chamber evacuated to less than 4×10^{-4} Pa.

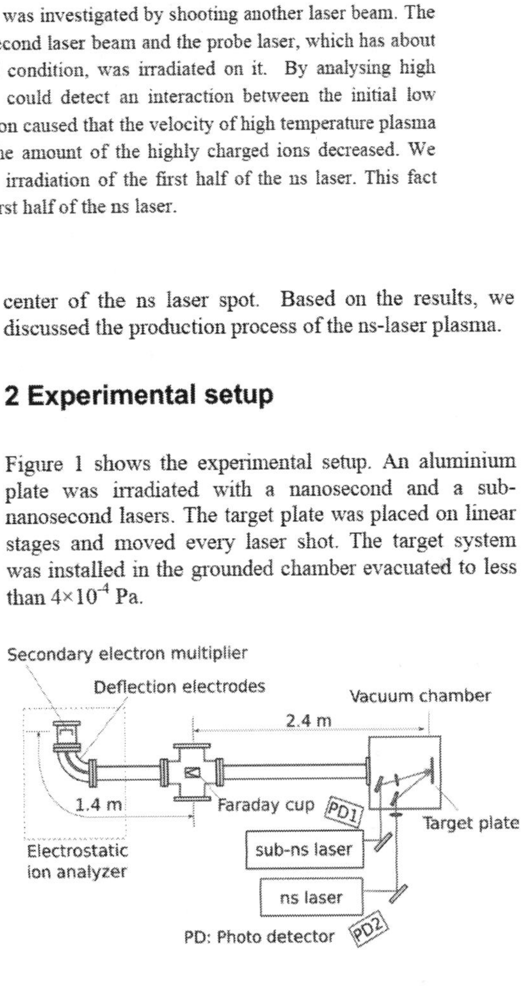

Fig. 1. Experimental setup.

The nanosecond laser pulse was generated with a Nd:Yag laser (THALES SAGA 230). The pulse duration was 6 ns and the wave length was 1064 nm. The energy, the area, and the intensity of the laser pulse on the target were 750 mJ, 0.16 cm^2, and 8×10^8 W/cm^2, respectively. The incident angle was 40 degree with respect to the target normal. The sub-nanosecond laser pulse was

* Corresponding author: SIkeda@bnl.gov

generated with a Nd:Yag laser (Ekspla SL230). The pulse duration, the wave length, and the energy on the target were 170 ps,1064 nm, and 300 mJ, respectively. The laser pulse focused by a plano-convex lens with the focal length of 120 mm. The distance between the target and the lens was adjusted to obtain a minimum time of flight of the plasma traveling from the target to a Faraday cup. The laser intensity was less than 1×10^{14} W/cm^2 that was estimated by assuming a Gaussian beam. The incident angle was 20 degree with respect to the target normal.

During the experiment, we watched the laser pulses near the lasers with photo detectors, PD1 and PD2. The ns and sub-ns laser pulses arrived at the target with an interval time as illustrated in Fig. 2. We define interval time, dt = (arrival time of sub-ns laser peak) – (arrival time of ns laser peak). For dt =0, the peaks of the ns-laser pulse and the sub-ns laser pulse arrived simultaneously. For dt >0 (or <0), the sub-ns laser arrived after (or before) the ns laser.

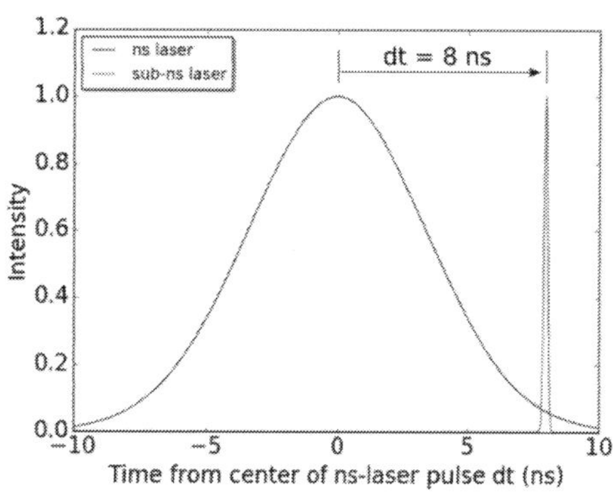

Fig. 2. Concept of time sequence of two laser pulses to explain the definition of interval time between two lasers, dt. The figure shows the case for dt = 8 ns.

A current detector at 2.4 m far from the target was used to measure the plasma ion current. It was composed of a grounded aperture with the diameter of 10 mm, negatively biased metal mesh, and Faraday cup. The mesh bias was determined to be -5 kV to obtain saturated ion current.

The charge state distribution was measured with an electrostatic ion analyser [5, 7]. During the charge state analysis, the Faraday cup was retracted.

3 Results and discussion

Figure 3 shows the plasma ion current waveforms. The horizontal axis is the time from the laser irradiation. The red, blue, and green curves correspond with the waveforms with the sub-ns laser, ns laser, and both lasers with a certain interval time, respectively. Because laser ablation plasma expands rapidly after the production, one can assume that the ions travel without collisions to the detector far from the target compared with the laser spot [8]. Under the assumption, the velocities of the ions are proportional to the inverses of the time of flights. The figure shows that the ns laser generated larger and slower plasma while the sub-ns laser generated smaller and faster plasma. When both lasers were used, the current waveform had two peaks. The time at the faster and slower peaks were around the waveforms with the sub-ns and ns lasers. This shows that the faster and slower peaks were generated mainly by the sub-ns and ns lasers, respectively.

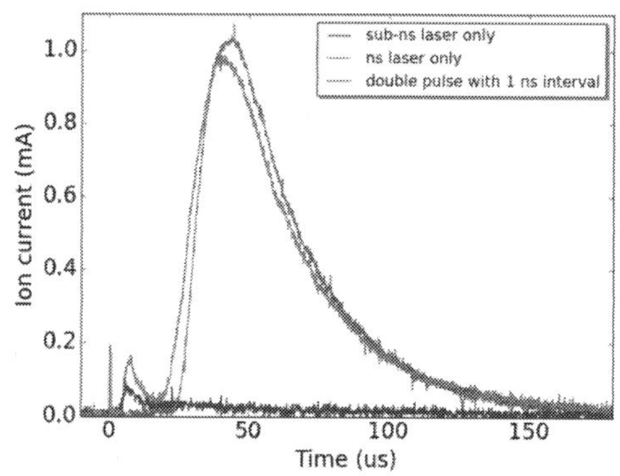

Fig. 3. Plasma ion current waveforms measured by a Faraday cup with sub-ns laser (blue), ns laser (green), and both lasers (red).

To investigate the influence of the time interval on the faster plasma, we measured plasma ion current with different the time intervals, dt. Figure 4 displays the wave forms around 10 μs for the time intervals dt = -1 ns (blue), 0 ns (green), and 8 ns (red). For comparison, the waveform only with the sub-ns laser is also shown (black). For dt < -1 ns, the ion current did not change as shown in the similarity between the waveforms only with the sub-ns laser and with dt = -1 ns interval. For larger dt, the height of the current peak became larger and the peak time also became larger.

 https://doi.org/10.1051/epjconf/201816702005

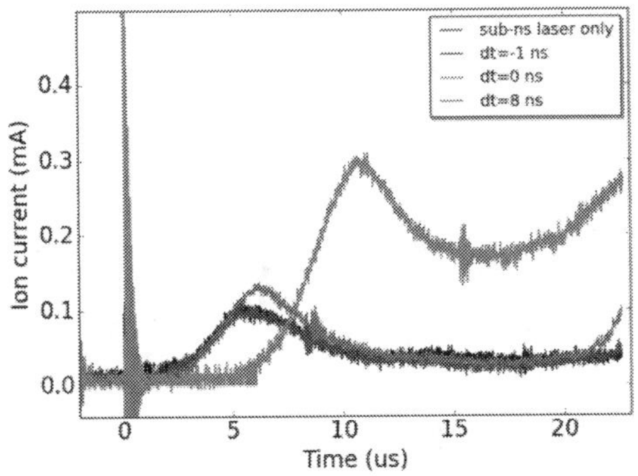

Fig. 4. Plasma ion current waveforms measured around 10 μs for the time intervals dt = -1 ns (blue), 0 ns (green), and 8 ns (red). For comparison, the waveform only with the sub-ns laser is also shown (black), which is overlapped with the blue curve.

Figure 5 shows the veocity estimated from the peak time and the height of the peak current as a function of *dt*. The graph shows clearly that the peak velocity decreased and the peak current increased monotonically with the increase of dt for $dt > -1$ ns.

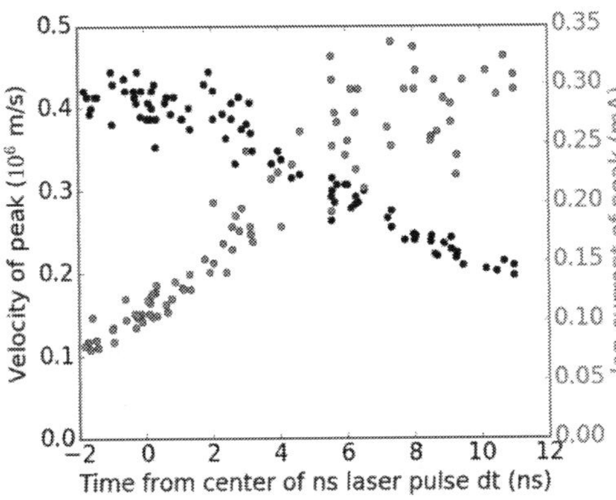

Fig. 5. Velocity estimated from the peak time (black) and height of the peak ion current (red) as a function of *dt*.

Figures 6(a) and 6(b) show the time structures of (a) Al^{11+} and (b) Al^{8+} generated only with the sub-ns laser (black) and the both laser pulses for $dt = 8$ ns (red). The horizontal axis is the time at the Faraday cup estimated by assuming the time of flight is proportional to the traveling distance of the ions. The vertical axis is the voltage of the secondary electron multiplier used in the electrostatic ion analyser. Although calibrations are needed to compare the voltage among different species, one can compare the voltages for the same species directly. As shown in the figures and the similar figures for the other charge states that are not shown here, it was found that the amounts of the highly charged ions (12+ ~ 9+) and the ions with lower charge states (< 8+) decreased and increased, respectively, when the both pulsed laser was used. The decrease of the highly charged ions around 5 μs and the increase of the lower charge states for the measurement time later than 10 μs are consistent with the difference between the ion current waveforms shown in Fig.4.

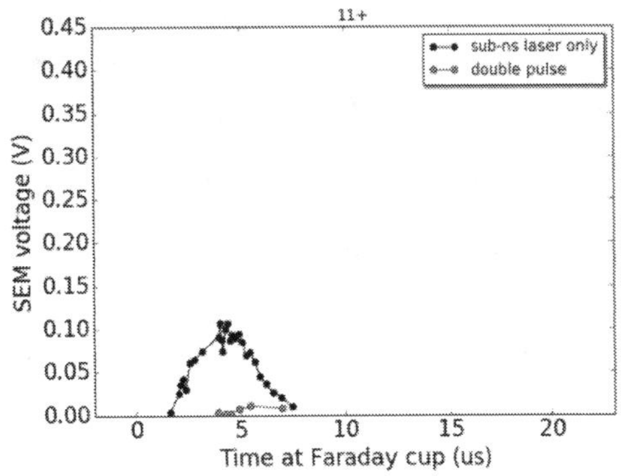

(a)

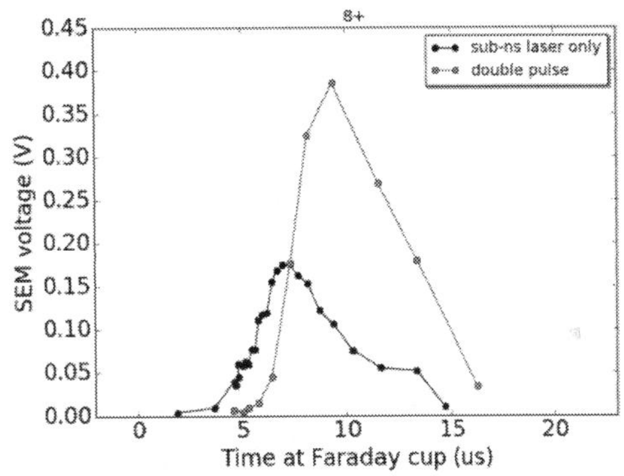

(b)

Fig. 6. Time structures of (a) Al^{11+} and (b) Al^{8+} generated only with the sub-ns laser (black) and the both laser pulses for $dt = 8$ ns (red) as a function of time at the Faraday cup.

We observed that the velocity of the faster plasma decreased and the highly charged ions decreased while the ions with lower charge states increased when the both laser pulses were irradiated. The changes indicate that the plasma temperature was lower. This shows the influence of the ns-laser plasma. By the first ns laser, large volume of the plasma was generated. The volume was larger than the plasma generated only with the sub-ns laser as guessed from the larger plasma ion current shown in Fig. 3. Then, the energy of the sub-ns laser was distributed to the larger ns-laser plasma, resulting in the lower energy per ion and temperature.

The influence of the ns-laser plasma was observed only for $dt > -1$ ns. This shows that the ns-laser plasma dense enough to interact with the second sub-ns laser pulse was produced after the first half of the ns laser pulse. In other word, the plasma was not produced before the first half of the laser pulse.

During this experiment, the focused sub-ns laser was irradiated on the center of the ns laser spot. If one scans the sub-ns within the ns laser spot, one can obtain the spatial information about the plasma production.

4 Conclusions

We investigated the evolution of the plasma generated with mildly focused ns laser by using sub-ns laser focused on a target within the ns laser spot. We measured the plasma ion current with a Faraday cup with the different interval time between the ns and sub-ns lasers. Then we measured the charge state distribution with an electrostatic ion analyser. We observed that the velocity of fast plasma generated mainly by the sub-ns laser became smaller and the amount of the highly charged ions decreased. This resulted from the interaction between the ns-laser plasma and the sub-ns laser. We found that the interaction did not occur during the irradiation of the first half of the ns laser. This fact indicates that the plasma is not produced during the first half of the ns laser.

This work was supported by the US Department of Energy under contract number DE-SC0012704.

References

1. T. Kanesue, et al., "The Commissioning of the Laser Ion Source For RHIC EBIS", in proceedings of IPAC2014, Dresden, Germany, 2014, 1890-1892.
2. S. Amoruso, et al., J. Phys. B-At. Mol. Opt., **32**(14), R131 (1999).
3. R. B. Stuart, et al., J. Opt. Soc. Am. B, **13**(2), 459 (1996).
4. R. B. Stuart, et al., Phys. Rev. B, **53**(4), 1749 (1996)
5. S. Ikeda, et al., Rev. Sci. Instrum. **85**(2), 02B913 (2014).
6. T. Kanesue, et al., "Measurement of Ion Beam from Laser Ion Source for RHIC EBIS", in proceedings of EPAC2008, Genoa, Italy, 2008, 421-423.
7. E, Woryna, et al., Laser Part Beams, **14**(3), 293 (1996).
8. R. Kelly and R. Dreyfus, Surf. Sci., **198**(1), 263 (1998)

Very Fast Current Diagnostic for Linear Pulsed Beams

Vincenzo Nassisi*, *Domenico* Delle Side and *Vito* Turco

Dipartimento di Matematica e Fisica, Lab. LEAS – Università del Salento and I.N.F.N. sect. of Lecce, Italy

Abstract. Fast current pulses manage lasers and particle accelerators and require sophisticate systems to be detected. At today Rogowski coils are well known. They are designed and built with a toroidal structure. In recently application, flat transmission lines are imploded and for this reason we develop a linear Rogowski coil to detect current pulses inside flat conductors. To get deep information from the system, it was approached by means of the theory of the transmission lines. The coil we build presents a resistance but it doesn't influence the rise time of the response, instead the integrating time. We also studied the influence of the magnetic properties of coil support. The new device was able to record pulses of more hundred nanoseconds depending on the inductance, load impedance and resistance of the coil. Furthermore, its response was characterized by a sub-nanosecond rise time (~100 ps), The attenuation coefficient depends mainly on the turn number of the coil, while the quality of the response depends both on the manufacture quality of the coil and on the magnetic core characteristics. In biophysical applications often, a double line is employed in order to have a sample as control and a sample stressed by a light source. So, in this case we build two equal plane lines by 100 Ω characteristic resistance connected in parallel. We diagnosed the current present in a line. The attenuation factor resulted to be 11,5 A/V.

1 Introduction

Rogowski coil is a well-known electric current detector since about a century[1]. Nevertheless, today its capabilities are still of great interest. TEA lasers and accelerator system need to detect fast current[2]. Also flat transmission lines utilised for radiofrequency biophysical experiments[3] need to control the current.

It was developed to detect variable currents fluxing in a conductor, which is responsible of the magnetic field around the conductor proper axis. In principle, the voltage provided by the coil is proportional to the input pulse current derivative and in order to obtain a sound measure of the originating signal, an integrating circuit is applied[4]. For high frequency signals, its response becomes self-integrating, providing an attenuation factor independent of the time duration or signal frequency. Moreover, the behaviour of the system is more complex for fast pulses due to the creating a capacitance with the metallic shield.

In the past, this detail has been analysed and developed by a few authors [5-7]. Under a few restrictions, the proposed device allows to justify its behaviour for fast pulses, large dimensions of the detector and manufacture quality.

To study our detector, we build two equal plane lines by 100 Ω characteristic resistance connected in parallel. We diagnosed the current present in a line.

2 Theory

The theory of the Rogowski coil is based on the main laws of electromagnetism.

2.1. Rogowski Coil

A current fluxing in a conductor generates a magnetic field dependent on distance r from it and a toroidal inductor is influenced by the magnetic flux derivative. *Fig. 1* shows a sketch of the system.

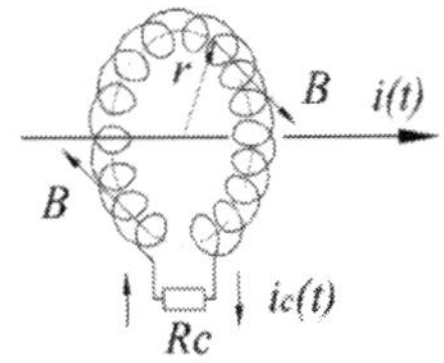

Fig. 1: Electrical circuit of a toroidal Rogowski coil with the input current in the centre.

Applying the Kirchhoff's equation [1, 8, 9], in the Laplace domain we have,

$$pI_i(p)\frac{L_c}{n} = pL_cI(p) + R_cI(p) \quad (1)$$

with $I_i(p)$ and $I(p)$ are the Laplace transforms of the input current $i(t)$ and of the coil current $i_c(t)$, respectivevely,e

* Corresponding author: vincenzo.nassisi@unisalento.it

$L_c = \mu_o n^2 A/2\pi r$ is the inductance of the coil, n is the number of loops and p is the Laplace variable. Now, hypothesizing a Heaviside function as current signal, i.e. $i(t) = I_o u(t)$, its transform is $I_i(p) = I_o/p$. Then the real output current results

$$i_c(t) = \frac{I_o}{n} e^{-\frac{t}{\tau}} \cdot u(t) \qquad (2)$$

where $\tau = L_c/R_c$. This shows that the overall attenuation obtained in the detector with respect to the input current strongly depends on time due to the exponential factor. If we consider an observation time $T \ll \tau$ (more in general, T is the pulse length of the input pulse), then the response on resistive load becomes

$$i_c(t) = \frac{I_o}{n} \cdot u(t) \qquad (3)$$

The above result implies that the detector is also able to detect high frequency currents. Indeed, the steady-state response can be obtained from Eq. 1 substituting p with $j\omega$, where j is the imaginary unit and ω the frequency of the input current. The attenuation factor is

$$\frac{i_c(t)}{i(t)} = \frac{1}{n} \frac{j\omega}{j\omega + \frac{R_c}{L_c}} \qquad (4)$$

then, to get a frequency independent response, in the steady-state case we observe that the attenuation is constant if $R_c/L_c \ll \omega$.

We stress that the results obtained do not depend on the shape of the coil loops neither on the area of each turn. Indeed, such assumptions are not satisfactory for short pulses.

The Rogowski coil we developed is devote for planar currents. This model is very useful, apart for laser discharges, also plane transmission lines. This last allows us to perform experiments of radiofrequency biophysical stresses [7, 9-11]. A sketch of a planar current is shown in Fig. 2. A current $i(t)$ in the conductor (width a) determines a current density that we call $j(t) = i(t)/a$.

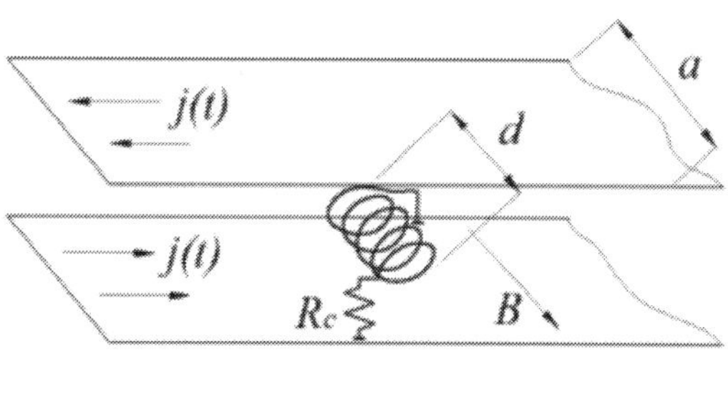

Fig. 2: Sketch of a Rogowski coil employed in a planar conductor.

Again, applying the Kirchhoff's equation, in the Laplace domain we have

$$pI_i(p)\frac{L_c}{2n}\frac{d}{a} = pL_c I(p) + R_c I(p) \qquad (5)$$

where d is coil length and again $I_i(p)$ is the Laplace transform of the input current $i(t)$.

For an Heaviside-like input current, the corresponding Laplace transform is $I_i(p) = I_o/p$, and the final form of Eq. 5 becomes

$$I_o \frac{L_c}{2n}\frac{d}{a} = pL_c I(p) + R_c I(p) \qquad (6)$$

whose real solution is

$$i_c(t) = \frac{1}{n}\frac{d}{a} I_o e^{-\frac{R_c}{L_c}t} \qquad (7)$$

and for $t \ll L_c/R_c$, the response will be time independent:

$$i_c(t) = \frac{1}{n}\frac{d}{a} I_o \qquad (8)$$

This result is comparable to that obtained for the toroidal Rogowski coil described by Eq. 3, apart from the correction factor $d/2a$. Again, utilizing $j\omega$ instead of p it is possible to get the steady-state response.

2.2 Rogowski coil as distributed elements

For fast pulses the above circuits are insufficient because they are unsuitable for transmit fast signals. So we consider the Rogowski coil as composed by distributed elements. It allows us to justify the presence of anomalies on the signal response generally due to imperfect manufacture.

To avoid interferences of stray signals, the coil is enclosed in a shielding box. It is a cracked metallic box operated on the ground conductor. The crack is realized by operating a slim slit on the conductor. The slit changes the current direction which wraps the coil, inducting again the same electromotive force, see Fig. 3.

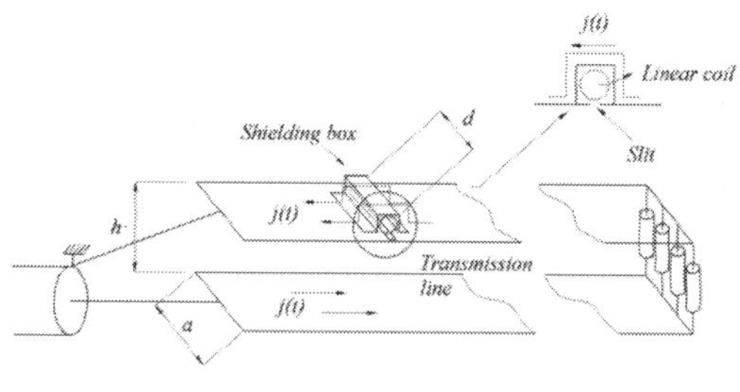

Fig. 3: Sketch of a flat transmission line with a linear Rogowski coil closed in the shielding box.

Now, the electromotive force can be considered distributed along the line and it can depend, apart from t, even from position, x (with x directed along the coil axis).

So, to study it, let us to call $f(t, x)$ the electromotive force per unit length. Now, the presence of the metallic box exhibits a distributed capacitance along the coil. Let us to call $C(x)$ the capacitance per unit length. Even the inductance of the coil is distributed along the device and per unit length, it is $L(x)$, as well as that of the ground conductor, namely $L_g(x)$. Therefore, the device we analyse can be considered exactly as a transmission line, Fig 4.

Let us consider the electric circuit of this line by means of an infinite number of cells. A typical cell is illustrated in Fig. 4 where $f(t, x)$, $L(x)$ and $C(x)$ are the voltage, the inductance and the capacitance per unit length, respectively. Besides, $L_g(x)$, $R(x)$ and $G(x)$ are the inductance of the box conductors, the resistance of the coil wire and the conductance of the insulator per unit length, respectively.

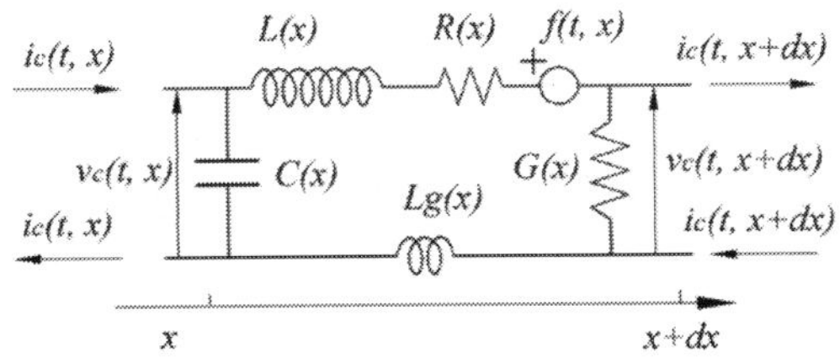

Fig. 4: Electric sketch of a Rogowski coil closed into the metallic box.

Then, applying Kirchhoff's equations, the differential of voltage and current for every cell to differential of position depend through the following equations:

$$\frac{\partial v_c(t,x)}{\partial x} = -\{Z(t,x)\} i_c(t,x) - f(t,x)$$
$$\frac{\partial i_c(t,x)}{\partial x} = -\{Y(t,x)\} v_c(t,x) \qquad (9)$$

with $\{Z(t,x)\}$ and $\{Y(t,x)\}$ the impedance and admittance operators of the circuit per unit length. They correspond to $\{Z\} = \left\{R + L\frac{d}{dt} + L_g\frac{d}{dt}\right\}$ and $\{Y\} = \{G + C\int dt\}$. Next, to simplify the solution, the resistance R, the inductance L_g and the conductance G are neglected and the line is represented by the circuit of Fig. 5. Operating the Laplace transform of Eq. 9, the impedance operator becomes $Z(p) = pL(x)$, the admittance one becomes $Y(p) = pC(x)$ and the electromotive force one becomes $F(p, x) = pI_i(p)Ld/na$.

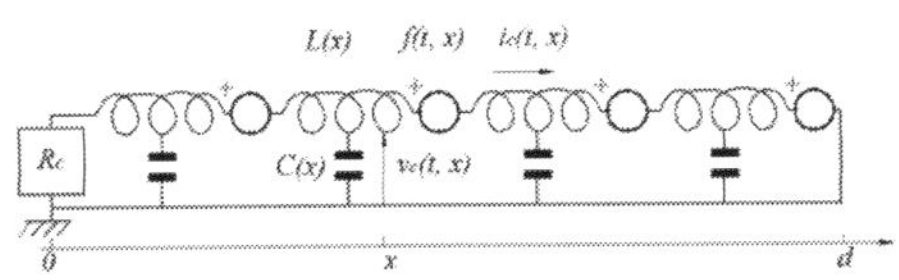

Fig. 5: Electric circuit of the Rogowski coil closed inside the metallic box. The voltage and the current along the coil are $v_c(t, x)$ and $i_c(t, x)$, respectively.

Theoretically, the last three parameters should be x-independent and this depends on the coil manufacture quality. Therefore, under the above constraints, the Laplace transform of Eq. 9 is:

$$V_c(x,p) = \frac{F(p)R_cR_0}{pL\,(R_c+R_0)} \cdot \frac{e^{-\tau_0 px} - e^{-\tau_0 p(2d-x)}}{1+\vartheta e^{-2\tau_0 pd}} \qquad (10a)$$

$$I_c(x,p) = \frac{F(p)\,R_c}{pL\,(R_c+R_0)} \frac{e^{-\tau_0 px} + e^{-\tau_0 p(2d-x)}}{1+\vartheta e^{-2\tau_0 pd}} - \frac{F(p)}{pL} \qquad (10b)$$

where we have considered the line short-circuited at $x = d$ and closed on resistive load R_c at $x = 0$. The parameter ϑ represents the reflection coefficient, namely $(R_c - R_o)/(R_c + R_o)$, where R_o is the characteristic resistance of the line whose expression is $R_o = \sqrt{Z(p)/Y(p)}$, while the propagation function per unit length is $-\tau_0$. In our conditions, the latter is a constant, $\tau_0 = \sqrt{Z(p)Y(p)} = \sqrt{LC}$ and it expresses the delay per unit length. Because it is known to be of the order of 10^{-9} s/m let us suppose that $|\vartheta e^{-2\tau_0 pd}| < 1$ and we can operate the expansion

$$|\vartheta e^{-2\tau_0 pd}| \cong 1 - \vartheta e^{-2\tau_0 pd} + \vartheta^2 e^{-4\tau_0 pd} - \vartheta^3 e^{-6\tau_0 pd} + \cdots \qquad (11)$$

Substituting the Eq. 11 in the first Eq. 10, the voltage across R_c is obtained placing $x = 0$

$$V_c(x,p) = \frac{F(p)\,R_cR_0}{pL(R_c+R_0)} \left[1 - \vartheta_V e^{-2\tau_0 pd} + \vartheta\vartheta_V e^{-4\tau_0 pd} - \vartheta^2\vartheta_V e^{-6\tau_0 pd} \ldots\right] \qquad (12)$$

with $\vartheta_V = (1 + \vartheta) = 2R_c/(R_c + R_0)$. Instead, substituting the Eq. 11 in the second Eq. 10, the current in R_c is obtained again placing $x = 0$

$$I_c(0,p) = \frac{F(p)\,R_c}{pL\,(R_c+R_0)}\left[1 + \vartheta_I e^{-2\tau_0 pd} - \vartheta\vartheta_I e^{-4\tau_0 pd} + \vartheta^2\vartheta_I e^{-6\tau_0 pd} \ldots\right] - \frac{F(p)}{pL} \qquad (13)$$

with $\vartheta_I = (1 - \vartheta) = 2R_0/(R_c + R_0)$.

For an Heaviside-like input current, the corresponding electromotive force per unit length is constant

$$F(p) = \frac{I_0 L}{n}\frac{d}{a} \qquad (14)$$

Using Eq. 14, for $R_c = R_0$, we have $\vartheta = 0$, and consequently $\vartheta_V = 1$ as well as $\vartheta_I = 1$. The response at $x = 0$ becomes:

$$V_c(0,p) = \frac{I_0 R_0}{2pn}\frac{d}{a} \cdot \left[1 - e^{-2\tau_0 pd}\right] \qquad (15a)$$

$$I_c(0,p) = \frac{I_0}{2pn}\frac{d}{a} \cdot \left[1 + e^{-2\tau_0 pd}\right] - \frac{I_0}{pn}\frac{d}{a} \qquad (15b)$$

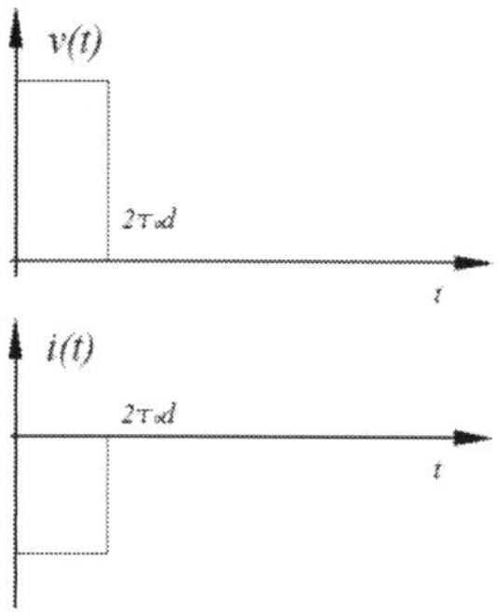

Fig. 6: Pictorial representation of the response waveforms for voltage and current with $x = 0$ and $R_c = R_0$.

In the time domain, the result is a voltage and current pulse $2\tau_0 d$ long that corresponds to the time of a signal to propagate and came back for the whole line, Fig. 6. Nevertheless, this response does not match the input signal we want to detect. Instead, closing the Rogowski coil on a load resistor of low value such that $R_c < R_0$, it results that $-1 < \vartheta < 0$, $0 < \vartheta_V < 1$ and $1 < \vartheta_I < 2$. At $x = 0$ the response is a voltage signal that decreases every $2\tau_0 d$ seconds of a little quantity, while the current signal is initially negative with the overlapping of a small positive signal every $2\tau_0 d$ seconds. Fig.7 shows a pictorial representation of the situation.

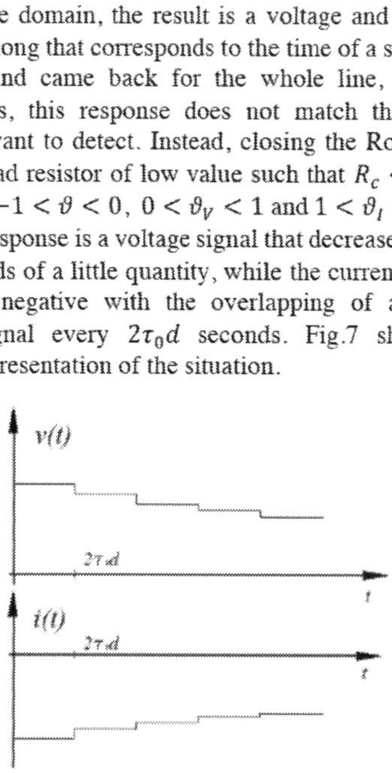

Fig. 10: Pictorial representation of the waveforms for voltage and current with $x = 0$ and $R_c < R_0$.

This result is very interesting particularly if $R_c \ll R_0$ because we can have: $\vartheta \cong -1$, $\vartheta_V \cong 0$ and $\vartheta_I \cong 2$ and by the Eqs. 12 and 13 the response is approximately:

$$V_c(0,p) \cong \frac{I_0 R_c}{pn}\frac{d}{a} \quad (16a)$$

$$I_c(0,p) \cong -\frac{I_0}{pn}\frac{d}{a} \quad (16b)$$

Performing the inverse transformation in order to get the time dependent relations, the results are:

$$v_{Rc}(t) \cong \frac{I_0 R_c}{n}\frac{d}{a}u(t) \quad (17a)$$

$$i_{Rc}(t) \cong \frac{I_0}{n}\frac{d}{a}u(t) \quad (17b)$$

The relations obtained show that the theory developed give results very similar to the one expressed by Eq. 8 but now they deliver more information. Indeed, if the coil is not uniformly shaped, a different electromotive force will be induced at different positions and different values of inductance and capacitance are exhibited. In this case the propagation function per unit length isn't constant and an overlapped signal propagate along the line generating ripples on R_c. To attenuate the ripple, authors have suggested to apply some resistors along the coil in order to damp these undesirable signals [13]. In this case, the behaviour of the system improves but the value of the integrating time, that theoretically is Ld/R_c, decreases. However, the insertion of the resistors must be realized without creating new loops and/or modify the capacitance values because they could introduce new elements that should modify the behaviour.

2.3 Influence of the coil resistance

During the development of the previous theory, the resistance of the coil wire has been neglected. The response of the Rogowski coil for a Heaviside-like signal is principally characterised by an attenuation coefficient and by a damping time. Then, considering a no-null value for R, the voltage on the load R_c becomes

$$V_c(x,p) = \frac{F(p)R_c}{(pL+R)\left(1+\frac{R_c}{pL+R}p\tau_0\sqrt{1+\frac{R}{Lp}}\right)} \left[\frac{e^{-\tau p x}}{1-\frac{Z-R_c\tau p}{Z+R_c\tau p}e^{-2\tau p d}} - \frac{e^{+\tau p x}e^{-2\tau p d}}{1-\frac{Z-R_c\tau p}{Z+R_c\tau p}e^{-2\tau p d}}\right] \quad (18)$$

where the parameter $\tau = \tau_0\sqrt{1+R/pL}$. Then, before operating the inverse transformation, it is interesting to note that the part of Eq. 18 in square brackets governs the propagation of the signal. If we consider again the load resistance $R_c \sim 0$, condition necessary to get easily the result we expect, at $x = 0$ the value of this part is 1. Instead, the first part governs the attenuation factor. The inverse transformation of only this first part is very difficult, but the solution can be reached imposing that the coil resistance is generally small. Then, the expansion of the square root, considering only the first two terms is $\sqrt{1+\frac{R}{Lp}} \cong 1 + R/2Lp$ and Eq. 18 then reads

$$V_c(0,p) \cong \frac{F(p)R_c}{(pL+R)\left[1+\frac{R_c}{pL+R}p\tau_0\left(1+\frac{R}{2Lp}\right)\right]} \quad (19)$$

Always for a Heaviside current signal as input, we have

$$V_c(0,p) \cong \frac{LI_0 R_c}{n\,(pL+R)\left[1+\frac{R_c}{pL+R}p\tau_0\left(1+\frac{R}{2Lp}\right)\right]}\frac{d}{a} \quad (20)$$

The expression for $V_c(0,p)$ has got a pole that for $R_c \sim 0$ it amounts to $-R/L$. Therefore, performing the inverse transformation, the real voltage on R_c exhibits an exponential behaviour, namely:

$$v_{Rc}(t) = \frac{I_0 L R_c}{n(L+R_c\tau_0)} e^{-\frac{Rt(2L+R_c\tau_0)}{2L(L+R_c\tau_0)}} \frac{d}{a} \cdot u(t) \quad (21)$$

which for $R_c \sim 0$ becomes

$$v_{Rc}(t) \cong \frac{I_0 R_c}{n} \frac{d}{a} e^{-\frac{R}{L}t} \cdot u(t) \quad (22)$$

Comparing the Eq. 21 to Eqs. 17, we observe that a non-negligible R does not affect the attenuation coefficient but changes only the time behaviour. Therefore, during the design of a Rogowski coil, for high-precision measurements, it is necessary to consider both the values of R and R_c in order to find the current, while the voltage is still simply the current multiplied by only R_c.

2.4 Influence of core characteristics

Utilizing magnetic materials for the core of the coil its inductance changes as well as the electromotive force. An inductor with a core having only the real component of permeability, μ_r, larger than 1, has a μ_r times increased inductance value, but also the electromotive force increases of the same factor μ_r. This behaviour could be considered as an advantage, since it reduces the effect of construction anomalies. In fact, loops of different cross section but with the same magnetic properties by constant diameter exhibit very near inductance values. Nevertheless, the use of a magnetic core involves a constant value of the permeability coefficient and of magnetic flux. In this way, a linear response is ensured.

In principle, the time response is not influenced by the permeability μ_r, provided that the latter is constant. Instead, if the permeability has also an imaginary component, the risetime response is increased limiting the applications on frequency range, particularly the maximum frequency value.

3 Experimental Material and method

3.1 Material

The device presented in the above sections has been designed and built for recording fast currents fluxing in a plane line. It is made by a Cu wire wound along a Lucide rod of 4 mm diameter 27 mm long. The wire has a 0.11 mm in diameter which provides a maximum value of turn density of 80 $rings/cm$ with a total of 216 turns. The coil load R_C was the characteristic impedance of a 50 Ω cable, while the value of R has been measured to be of 4 Ω.

Fig. 11: Double plane transmission line with the linear Rogowski probe on the ground conductor. The transverse dimension is 6 cm.

The transmission line we used in this work was a double transmission line of 100 Ω characteristic impedance connected in parallel as can be seen in Fig. 11. Both the lines are composed by two plane Cu conductors 6 cm wide, spaced 2 cm. The end of each line is closed on a resistive load by means of 4 parallel resistors of 400 Ω, equal distributed. Similar configurations are present in the laser system MOPA and in biophysical applications.

Under this configuration, the current in each line is the half of the total current and Eq. 22 is modified in

$$v_{Rc}(t) \cong \frac{I_0 R_c}{2n} \frac{d}{a} e^{-\frac{R}{L}t} \cdot u(t) \quad (23)$$

In practice the expression (23) might contain the value of R_c and, being the value of L of the order of 500 nH, the time duration of the signals we can record may be of at max 200 ns.

The Rogowski coil is applied on the ground conductor of the plane transmission line. The slit operated in correspondence of the coil was 35 mm long in order to guide the input current around the coil.

3.2 Results

The behaviour of the probe was studied by means of a current pulser made by a high voltage power supply connected to a coaxial cable (triax cable Suhner 50 Ω) and a fast switch (Reed relay h12). The output of the pulser was applied to the double transmission line and the provided current was monitored by means a high impedance resistive probe having an attenuation of 46, see Fig. 12. Instead, the Rogowski coil output was closed to a 50 Ω coaxial cable utilised also to connect the output to oscilloscope, 2 GHz (20GS/s).

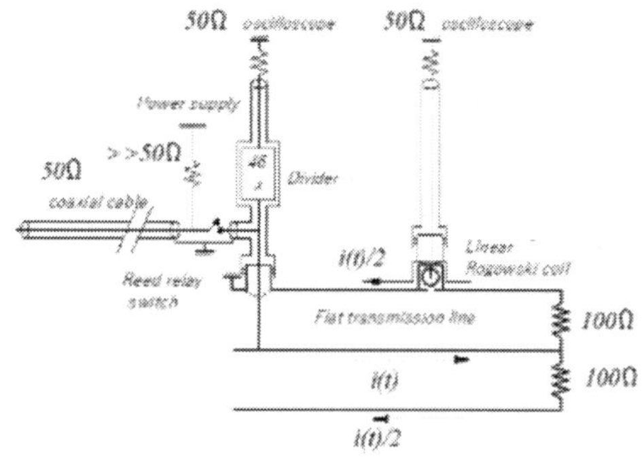

Fig. 12: Sketch of the electric circuit used to calibrate the linear probe.

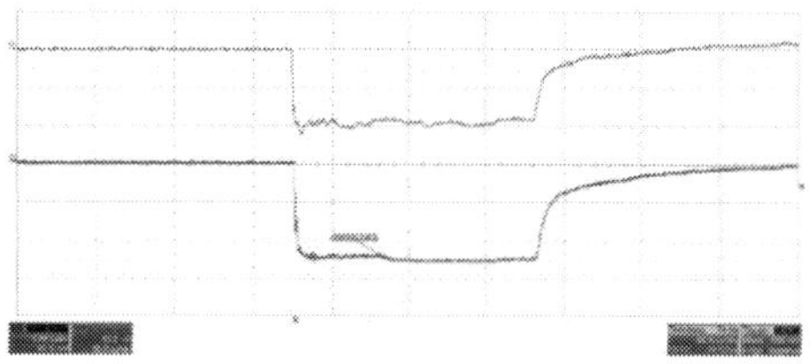

Fig. 13: Waveform of the input and output signals. Upper lower is the input current; upper trace is the output signal. Time scale 50 ns/div.

The waveform response is very near to the input one, as can be seen in Fig 13, where we have used one pulse forming line of 16 m.

The lower trace scale is 1 V/div which correspond to $1\,\text{V}\times\frac{46}{50\Omega}\frac{1}{2} = 0.46$ A/div, while the upper trace is 50 mV/div while correspond to $50\ \text{mV/div}$. The attenuation factor is about $12\,A/V$, while theoretically one expect $216\times\frac{60}{27}\times\frac{1}{50} = 9.6\,\text{A/V}$.

The discrepancy we found is mainly due to the distributed current inside the line owing to the slit that we have operate only for 50 mm. The maximum time duration of the detection signal is about 157 ns, as can be seen in Fig. 13.

In order to test the behaviour of our device, we perform an experiment with very short pulse utilising as pulse forming line of about 5 cm. The result of the latter test are illustrated in Fig. 14, where the time scale is 1 ns/div.

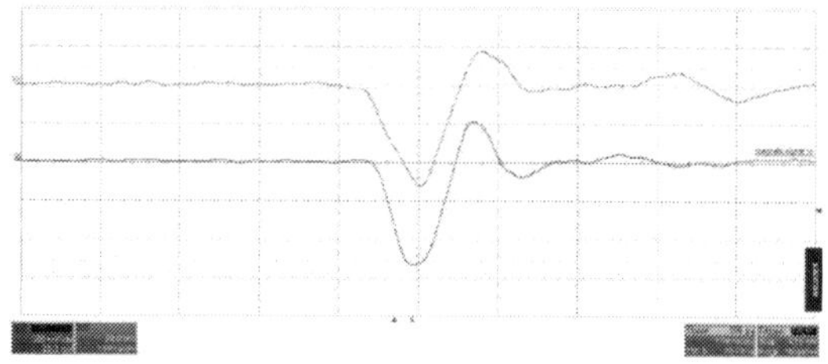

Fig. 14: Experimental result: Upper trace is the input signal of 650 ps FWHM (0.5 V/div); Lower trace is the output signal of 700 ps (20 mV/div). Time scale 1 ns/div.

If we turn our attention to the fast rise time of the response, we find very interesting results, represented in Fig. 15, with a 500 ps/div time scale. From Fig. 15, the input signal presents a rise time of about 500 ps, while the output signal of about 600 ps. Therefore, the discrepancy is of about 100 ps, that we can ascribe to response quality of the Rogowski coil.

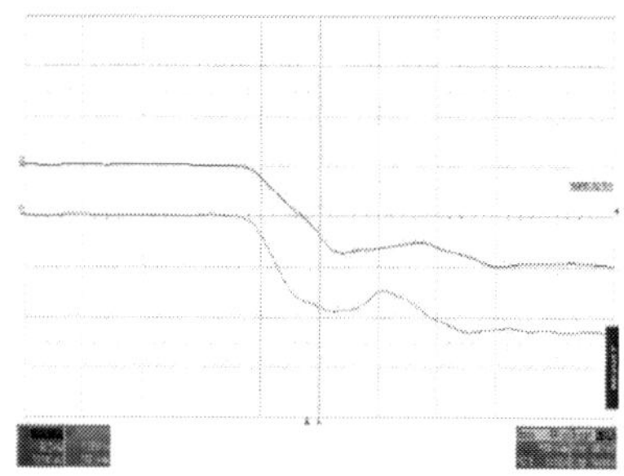

Fig. 15: Experimental result: Upper trace is the input signal (1 V/div); Lower trace is the output signal (50 mV/div). Time scale 500 ps/div.

If we turn our attention to the fast rise time of the response, we find very interesting results, represented in Fig. 15, with a 500 ps/div time scale. From Fig. 15, the input signal present a rise time of about 500 ps, while the output signal of about 600 ps. Therefore, the discrepancy is of about 100 ps, that we can ascribe to response quality of the Rogowski coil.

4 Conclusion

By this device it is possible to diagnose fast current pulses in plane conductors. The response quality might be improved using ferromagnetic cores. The attenuation coefficient is independent by the magnetic permittivity and the disagreement we found between the expected and measured attenuation factor is mainly ascribed to the length of the slit we operated on the line conductor. The calculations developed above show how the wire resistance and the load resistance do not affect the rise time.

References

1. W. Rogowski and W. Steinbhaus, Arch. Electrotech. **1**, 141-50 (1912)

2. A. Luches, V. Nassisi and M.R. Perrone, J. Phys. E: Sci. Intrum. **20**, 1015-1018 (1987)
3. F. Belloni, V. Nassisi, P. Alifano, C. Monaco, A. Talà, M. Tredici and A. Rainò, Rev. Sci. Instrum. **76**, 0543021-6 (2005)
4. D. A. Ward and J. L. T. Exon, Eng. Sci. Edu **2**, 105-13 (1993)
5. J. Cooper, Plasma Physics. **5**, 285-89 (1963)
6. V. Nassisi and A.Luches, Rev. Sci. Instrum. **50**, 900-02 (1979)
7. R.Y. Han, W.D. Ding, J.W. Wu, H.B. Zhou, Y. Jing, Q.J. Liu, Y.C. Chao, A.C. Qui, Rev. Sci. Instrum. **86**, 035114 (2015)
8. R.H. Uddlestone, S.L. Leonard, *Plasma diagnostic technique*, Academic Press, New York (1965)
9. D. G. Pellinen, Rev. Sci. Instrum. **42**, 667-9 (1971)
10. V. Nassisi, P. Alifano, A. Talà and L. Velardi, J. Appl. Phys. **112**, 014702 (2012)
11. D. Delle Side, L. Velardi, V. Nassisi, Appl. Surf. Sci. **272**, 124-127 (2013)
12. S. Agosteo, M.P. Anania, M. Caresana et al., Nucl. Instr. Meth. Phys. Res. **B 331**, 15-19 (2014).
13. J.M. Anderson, Rev. Sci. Instrum. **42**, 915-30 (1971)

Surface modifications of AISI 420 stainless steel by low energy Yttrium ions

Vincenzo Nassisi[1], *Domenico* Delle Side, *Vito* Turco, *Luigi* Martina

Dipartimento di Matematica e Fisica "E. De Giorgi", Università del Salento - Lecce, Via Provinciale Lecce-Monteroni, 73100 Lecce – Italy, Tel. +39 0832 297495, Fax. +39 0832 297482, E-Mail: vincenzo.nassisi@le.infn.it

Abstract. In this work, we study surface modifications of AISI 420 stainless steel specimens in order to improve their surface properties. Oxidation resistance and surface micro-hardness were analyzed. Using an ion beam delivered by a Laser Ion Source (LIS) coupled to an electrostatic accelerator, we performed implantation of low energy yttrium ions on the samples. The ions experienced an acceleration passing through a gap whose ends had a potential difference of 60 kV. The gap was placed immediately before the samples surface. The LIS produced high ions fluxes per laser pulse, up to $3x10^{11}$ ions/cm^2, resulting in a total implanted flux of $7x10^{15}$ ions/cm^2. The samples were characterized before and after ion implantation using two analytical techniques. They were also thermally treated to investigate the oxide scale. The crystal phases were identified by an X-ray diffractometer, while the micro-hardness was assayed using the scratch test and a profilometer. The first analysis was applied to blank, implanted and thermally treated sample surface, while the latter was applied only to blank and implanted sample surfaces. We found a slight increase in the hardness values and an increase to oxygen resistance. The implantation technique we used has the advantages, with respect to conventional methods, to modify the samples at low temperature avoiding stray diffusion of ions inside the substrate bulk.

1. Introduction

Ion-implantation is an established method of surface modification of materials leading to an improvement of their mechanical proprieties e.g. micro-hardness, wear resistance and corrosion resistance [1, 2]. Alloys undergoing high temperatures are frequently used in power plants, gasification systems, petrochemical industry, combustion processes and in aerospace applications [3]. Depending on the application, materials are exposed to corrosive atmospheres and/or thermal cycling.

High temperature resistant alloys have to combine with two main requirements: low scale growing rate and adequate scale adherence on the alloy, especially if thermal cycles are considered. According to the literature, the control of the oxidation is particularly due to formation of Cr_2O_3 and Al_2O_3[4–5].

The beneficial effects of active element additions on the oxidation resistance of heat resistant alloys are well known. Small amounts (usually below 1%) of reactive elements (Sc, Ti, Y, Zr, Ce, La, ect.) clearly improve the oxidation behaviour of chromia- and alumina-forming alloys [6–7]. Several explanations are given about this effect, usually called reactive element effect (REE), namely a modification of the diffusion mechanisms, a reduction of vacancies condensation at the internal interface, a formation of reactive element oxide inclusions trapping alloy impurities [8-10].

In this work, thermal treatment was carried out in order to study the influence of implanted yttrium on the oxide scale adherence on 410 steel specimens oxidized at 1123K. We have chosen the yttrium, in particular, for its availability and its low cost.

X-ray diffraction analyses indicate that the oxides formed near 1000 K are different on blank compared to implanted samples. Measurements of XRD allow analyzing the oxide scale composition after cooling to room temperature. Consequently, we also focus on the possible correlation between the presence of implanted yttrium and the oxide structure formed at high temperature.

2. Experimental apparatus

The experimental apparatus utilized for performing implantation of samples is an accelerator composed by a KrF excimer laser (Lambda Physics, Compex), operating at 248 nm wavelength (5 eV photon energy, total energy fixed at 11 mJ/pulse), 23 ns pulse duration to induce plasmas by PLA and a vacuum chamber made of stainless steel [11]. Inside the vacuum chamber, an expansion chamber is placed tightly closed around the target support. It was 18 cm in length and 8 cm in diameter. The target and the expansion chamber are connected to a power supply of positive polarity. The

* Corresponding author: vincenzo.nassisi@unisalento.it

expansion chamber end has a 1.5 cm hole to allow ion extraction. A ground electrode was placed at 3 cm distance from the expansion chamber. After this electrode, a third electrode at 2 cm distance from ground electrode, connected to a power supply of negative polarity, acted as samples holder. The same electrode can also work as Faraday cup. In this configuration, the electrode connects to an oscilloscope using a high voltage capacitor (2 nF), in order to separate the oscilloscope from high voltage, and a voltage attenuator, in order to decrease the electric signal . In Fig. 1 the stretch of the device is shown.

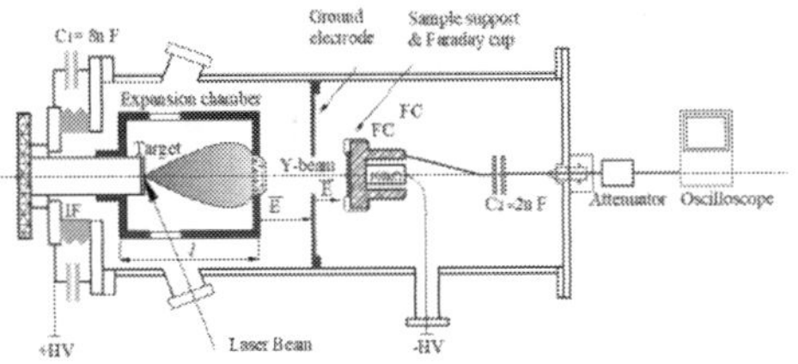

Fig. 1: Schematic drawing of the experimental apparatus.

During the experiments, we fixed the positive and the negative voltage respectively at +40 kV and -20 kV. This corresponds to a maximum acceleration voltage of 60 kV. In this condition, the ion flux resulted of $3x10^{11}$ ions/cm^2. For a deeper characterization of the ion beams, we remand the reader to the analysis in [12] and references therein.

The samples we used to perform oxidation experiments were foils of AISI 420 stainless steels. This is a widely used alloy, for example for surgical tools. Therefore, increasing the oxidation resistance of such material is very useful. The composition of the AISI420 is given in Table 1. All the samples utilized had a square shape (15 mm x 15 mm) and a thickness of 2 mm.

Table 1: Chemical composition (wt%) of the stainless steel: AISI 420

Chemical composition (wt%)	
Fe	86.7
Cr	13.0
C	0.3

In the described experimental conditions, we obtain mainly yttrium ions with charge states +1 and +2, with a small presence of +3 states. This translates to an ion range inside the samples ranging from 15 to 40 nm.

The scratch tests were performed with a diamond tip on which a 0.60 N perpendicular force was applied. This produced a scratch on the samples surface and we measured the scratch width and depth. For the roughness measurements, we used the stylus surface profiler (Tencor Instruments ALPHA-STEP 200), which determined the arithmetic surface roughness using the graphical-centerline method. The stylus surface profiler was set with an applied force of 0.12 mN, a scan length of 2000 μm and a scan speed of 100 μms-1.

Oxidation behavior was investigated by utilized a classic muffle furnace with controller (GEFRAN 400, GEFRAM S.p.A.) at atmospheric pressure in air. Even treatment was performed putting the sample in a porcelain crucible at a temperature of 1123 K for 24 hours. After the thermal treatment and cooling to room temperature the samples were analyzed by an XRD (Rigaku D/MAX Ultima+).

3. Experimental results

Before implantation, we operated an accurate cleaning of samples in ethanol. The samples were then implanted on one face, which corresponds to approximately 70% of the total surface area due to the mask utilized to hold the samples on the support. The final implanted flux was reached applying 22000 laser pulses, resulting in $7x10^{15}$ ions/cm^2.

After implantation, we measured the micro-hardness characteristics, using the "scratch test" method. Fig. 2 shows the scratch test results of our samples. The scratch test results shows that implanted samples exhibit scratches with lower depth (~ 0.2 μm) with respect to blanks (~ 0.5 μm). Therefore, the smaller scratch points out a lower tip depth in the sample surface, i.e. a sensible, although little, increment of the surface micro-hardness occurred in the implanted sample.

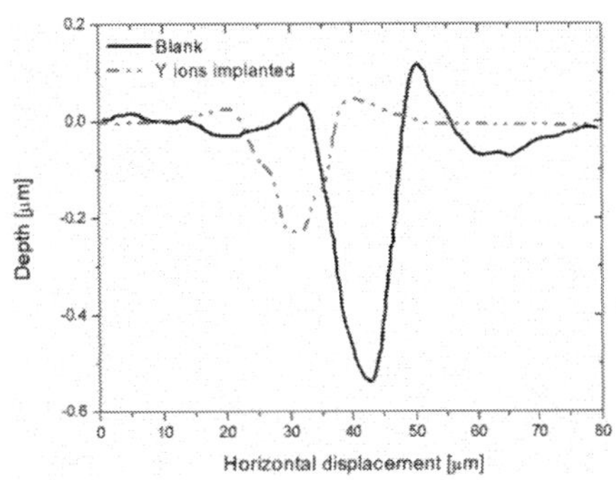

Fig 2: Profile of the scratch on the blank (black solid line) and on the Y-implanted sample (red un-solid line).

The XRD spectra of the blank and Y-implanted sample did not show significant differences and both exhibited mainly iron peaks, Fig. 3. Instead, the thermally processed samples (blank and Y-implanted) exhibited, apart the iron peaks, also many peaks related to the oxide families. In fact, by close examination, it can be concluded that Cr_2O_3 (Joint Commitee Powder Diffraction Standard file: JCPDS no. 38-1479) is the most probable oxide in the scale. Before furnace treatment, we observed also presence of Fe_2O_3 nuclei. Other corrosion products, mainly Cr_2O_3 and $Mn_{1.5}Cr_{1.5}O_4$, then covered them. The signal intensities of Cr_2O_3 were different for the two cases analyzed. Fig 4a shows the spectrum of the XRD analysis for the blank sample, while Fig. 4b shows the spectrum of the XRD analysis for the Y-implanted sample. The intensity (cps) of the Cr_2O_3 signal for the blank sample was higher than

https://doi.org/10.1051/epjconf/201816702007

the implanted one. In fact the signal of Cr_2O_3 at $2\theta=24.38°$ is 209 for the blank sample, while it is 43 for the Y-implanted sample.

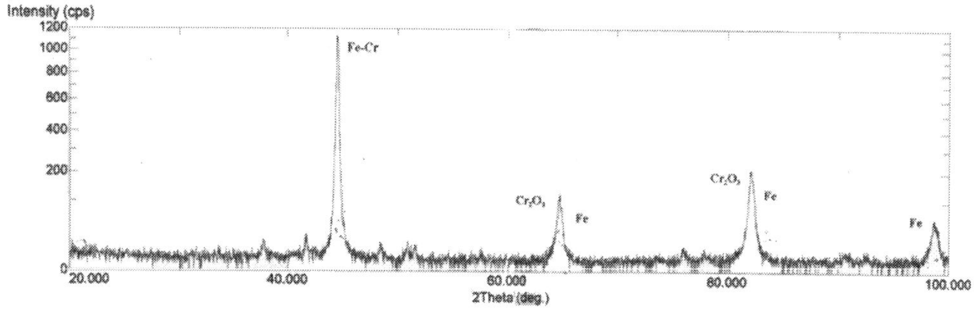

Fig. 3: XRD spectrum of the blank and Y-implanted samples before thermal treatment

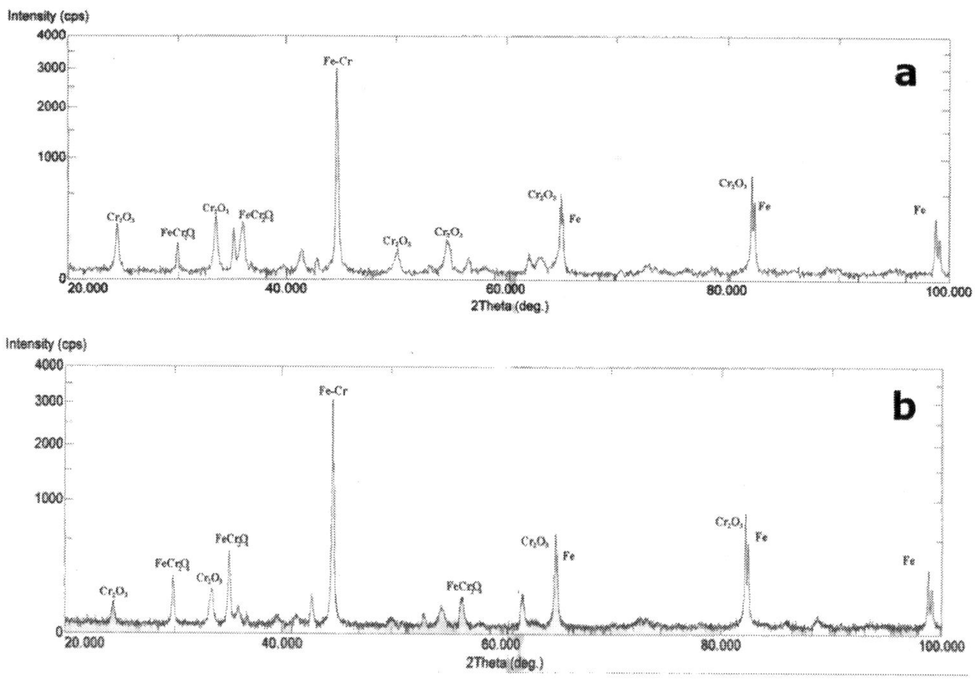

Fig. 4: XRD spectrum after thermal treatment. a) blank sample; b) Y-implanted sample.

We can observe that the implantation of Y ions modify the characteristic of the stainless steel samples. As proposed by other authors [13-16], we have shown that yttrium implantation also induce some surface alloy structural modification promoting partial phase transformation of austenite into bcc structure (α-ferritic) near the surface. Even if any yttrium containing oxides have not been observed by XRD (Fig. 4b), results show that yttrium reduces the oxidation rate.

4. Conclusions

We found that implantation of low energy yttrium ions is able to increase hardness of AISI 420 stainless steel samples. XRD measurements further pointed out that an enhancement of the oxidation resistance is also present. In the experiment we used just one dose. We can suppose that at higher doses hardness and oxidation resistance could improve. This treatment is simple and ensures effective results on the AISI 420 surfaces.

https://doi.org/10.1051/epjconf/201816702007

Consequently, it could represent a valuable tool for improvements of AISI 420 surface properties.

References

1. G. Dearneley, Nucl. Instr. and Meth. **182-183** (1981) 899.
2. M.J. Bennett, A. T. Tuson, Mat. Sci. Eng. A **116** (1989) 79.
3. F.J. Perez, M.J. Cristobal ,M.P. Hierro ,F. Pedraza Surf. & Coat. Tecnol. **120** (1999) 442.
4. C. Wagner, Atomic Movements, American Society of Metallurgy, Cleveland, 1951, p. 153.
5. M.J. Bennet, D.P. Moon, in: E. Lang (Ed.), The Role of Active Elements in the Oxidation Behaviour of High Temperature Metals and Alloys, Elsevier, Amsterdam, 1989, p. 111.
6. M. Landkof, A.V. Levy, D.H. Boone, R. Gray, E. Yaniv, Corros.-NACE **41** (1985) 344.
7. D.P. Whittle, J. Stringer, Philos. Trans. R. Soc., Lond. Ser. A **295** (1980) 309.
8. F.I. Wei, F.H. Stott, Corros. Sci. **29** (1989) 839.
9. P.Y. Hou, J. Stringer, Oxid. Met. **34** (1990) 299.
10. E.A. Polman, T. Fransen, P.J. Gellings, Oxid. Met. **33** (1990) 135-155.
11. Velardi, L., Siciliano, M. V., Delle Side, D., & Nassisi, V., Rev Sci Instrum **83**(2) (2012) 02B717.
12. Nassisi V., Velardi L. and Delle Side D., Appl Surf Sci, **273** (2013) pp. 114-118.
13. R. Guillamet, J. Lopitaux, B. Hannoyer, M. Lenglet, J. Phys. IV **C9** (3) (1993) 349.
14. M. Guemmaz, A. Mosser, J.J. Grob, R. Stuck, Surf. Coat. Technol. **100–101** (1998) 353.
15. A. Gutierrez, M.F. Lopez, F.J. Perez-Trujillo, Surf. Interface Anal. **30** (2000) 130.
16. G. Xie, M. Song, K. Mitsuishi, K. Furuya, J. Nucl. Mater. **281** (2000) 80

Fabrication and characterization of porous opaque PMMA foils to be laser irradiated producing ion acceleration

Mariapompea Cutroneo[1,2*)]

[1]Nuclear Physics Institute, AS CR, 25068 Rez, Czech Republic
[2]Department of Physics and Earth Sciences, Messina University, V.le F.S. d'Alcontres 31, 98166 S. Agata, Messina, Italy

Abstract. In this study, the effect of pore size in the opaque poly(methyl methacrylate) and its composition is investigated by optical measurements as well as Rutherford Backscattering Spectroscopy and Elastic Recoil Detection Analyses. The enhancement of the absorption coefficient induced by the presence of micrometric beads makes these porous thin foils high absorbent to IR radiation and suitable to be laser irradiated in order to generate a hot plasma rich in proton emission. The presented results indicate that the high optical transparency of PMMA foils can be strongly reduced by the presence of the micrometric acrylic beads and that the presence of high Z-metallic nanoparticles, such as gold, embedded in the polymer enhances the acceleration of emitted ions. The fabricated advanced targets have been irradiated by lasers at low intensity (Messina University) and at high intensity (PALS Research Infrastructure in Prague) generating plasma accelerating high proton yield and energy.

1. Introduction

Recently advances in polymeric processing have attracted growing interest from fundamental and experimental point of view, especially concerning the fabrication of micro and submicro-scale materials. By varying the micro-lattice geometry, in fact, it is possible to modify the base material physical and chemical properties: damage tolerance, transparency, strength, stiffness, optical and electrical characteristics etc...

Special attention to the polymer structure comes from their high content of hydrogen and carbon atoms, and by the optical properties, which play an important role in case thin polymeric foils are used to generate non-equilibrium plasma by laser pulse and consequent ion acceleration. To this, coherent photon energy must be converted into ion kinetic energy. Hot plasma rich in protons and carbon ions can be obtained irradiating in vacuum thin polymeric foils by high intensity laser pulses. Thin polymeric films can be used to accelerate protons by laser impact from target-normal-sheath-acceleration (TNSA) regime using intensities higher than 10^{15} W/cm^2 [1]. The drawback of this procedure consists in the too low laser absorption due to the limited thickness and high transparency of the polymer foil. In this way, a significant advantage of its use consists in the transformation of the polymer structure in a high absorbent opaque film to generate high laser absorption, transforming the energy of the deposited coherent photons into kinetic energy of electrons, protons and other particles.

Beside the strong metallic or ceramic microlattices, the polymerization in polymers allows the improvement of their mechanical and optical behaviour [2]. The main parameter playing a key role in both the optical and mechanical performances of the microlattice material is its effective density. In this work the fabrication of clear and opaque thin layer of PMMA (Polymethylmethacrylate) by spin coating method is reported. The present study is focused on the investigation of optical polymer properties changing by their degree of polymerization. Typically, semi-crystalline polymers, consisting of ordered (crystalline phase) and disordered (amorphous phase) molecular chains, are usually opaque due to the scattering of the light at the boundaries between the crystalline and amorphous regions as well as the light absorption at the nucleation centres. On the other hand, decreasing the density of the boundary and the assembly of the nucleation centres, the semi-crystalline polymer can become translucent. It is worthy to mention as the growth of the crystalline regions, induced by temperature gradient can be responsible for the generation of quasi-spherical aggregates with size ranging between 1 μm and 10 μm [3].

Among the wide range of polymer, PMMA has been proposed for its high number of applications, such as sensors, optical fibers [4], resist for ion lithography [5] and high absorption coefficient to specific wavelength bands. PMMA has stoichiometry $(C_5O_2H_8)_n$, with an high content in hydrogen, and a density of 1.18 g/cm^3, with a fusion pint at 160°C.

The use of metallic nanoparticles embedded in PMMA has been also proposed to induce Surface Plasmon Resonance (SPR) effects [6], with electric dipole absorption at specific wavelength bands

* Corresponding author: cutroneo@ujf.cas.cz

depending on the nanoparticles size, shape and matrix nature in which they are embedded.

The thin foils, prepared as advanced target for high intensity laser irradiation, with micrometric thickness, are suitable to induce high electron and ion acceleration in the forward direction from TNSA regime, promising to accelerate protons and carbon ions, which are useful for many applications, from protontherapy to nuclear physics, from nuclear fusion to laser ion source, from astrophysical investigations, to photochemical reactions [7, 8].

Moreover, PMMA foils will be presented as useful target to produce protons and carbon ions acceleration using laser pulse durations from *ns* up to *fs* scales and high intensities ranging from 10^{10} up to 10^{19} W/cm^2.

2. Experimental setup

The fabrication route consists of different stages. In the first stage 300 mg of PMMA in form of powder (99.9% of purity) with size of 600 μm and density of 1.18 g/cm^3, supplied by Goodfellow, was dissolved in 1.5 ml of chloroform, supplied by Aldrich, and mixed for 6 hours at room temperature. Afterward 1.5 ml of solution was poured on a clean polyethylene substrate to fully cover it while a spin coater was used at 25°C selecting a spin coating duration of 3 minutes with an angular velocity of 1000 rpm. Successively, the PMMA foils were peeled from the substrate. Going through the same procedure, and adding in the former solution 0.1 wt% Au NanoParticles (AuNPs), a foil of PMMA containing AuNPs has been produced. The AuNPs were produced by laser in water at a concentration of the order of 1 mg/ml, with a spherical size and a mean diameter of 50 nm, as reported in literature [9]. The nanoparticle coalescence was avoided adding a little amount of sodium citrate to the solution containing Au NPs.

Chloroform absorbs heat from the solution producing polymer cooling [10]. The optical characteristics of the PMMA foils, in terms of translucent or opaque appearance in the visible region, depends on the fast or slow cooling of the chloroform solution as a consequence of the solvent evaporation. A fast cooling as consequence of the fast solvent evaporation induces the production of high number of assembled nuclei/volume and small beads dimension (translucent PMMA foil i.e. clear foil).

A slow cooling during the solidification of the foil, induces the production of low number of assembled nuclei/volume and big dimension of the beads, as reported in Fig. 1a (opaque PMMA foils) of the order of microns and observed with higher optical magnifications in Fig. 1b,1c and 1d as will be explained in the following.

The control of the cooling is managed using substrates with different thermal conductivity. The key parameters in the production of clear or opaque PMMA foils are based on the cooling rate, temperature gradient, solution properties and adopted chemical solvent.

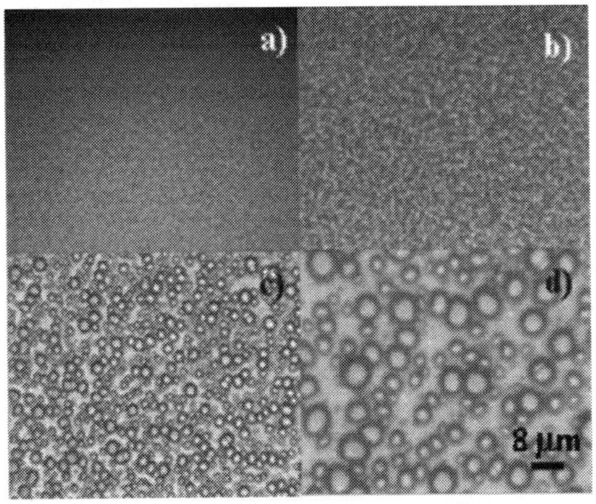

Fig. 1. Optical images of opaque PMMA at different magnifications

The PMMA effective density was changed using suitable chemical approaches. The density of the solid samples was measured by the weight method consisting of two stages: a Mettler-Toledo balance with ±1 μg absolute accuracy to measure the weight of a well-defined surface area and thickness of the foils. Then, the foil was gripped between two polymeric slices with a contrasted colour obtaining a single block. Afterward, the block was polished and the cross-section of the foil was measured by optical microscope, obtaining a final thickness of 9 μm±2 μm. By the mass-to-area ratio of the foil, an areal density of 1.18 g/cm^3 for the clear PMMA and of 0.75 g/cm^3 for the opaque PMMA were evaluated. In Fig.1 are reported the optical image of opaque PMMA magnified x5 (Fig. 1a), x10 (Fig. 1b), x40 (Fig. 1c) and x100 (Fig. 1d) times. The average dimension of the beads is 3.5 μm while the average thickness of the cell-wall is 0.8 μm evaluated in an area of 3.8x3.8 μm^2.

A FLUKE Ti100 CCD thermocamera was employed during the preparation of the clear and opaque PMMA foils for monitoring the temperature gradient. By the thermocamera images, it is possible to observe in both cases an initial temperature of about 24°C. During the preparation of the clear PMMA, using the chemical solution and a metallic holder, the solution temperature reaches a minimum value of 16.5°C. During the preparation of the opaque PMMA, rich in beads, using the chemical solution and a plastic holder, the solution temperature reaches a minimum temperature of 5.1°C. Fig. 2a shows the images obtained by a thermo-camera during the preparation of clear PMMA foil (inset) and a plot of the temperature of the solution as a function of the time. Fig. 2b shows the images obtained by a thermo-camera during the preparation of the opaque PMMA foil (inset) and a plot of the temperature of the solution as a function of the time.

Moreover, the stabilizing temperature after the evaporation of chloroform has been reached after 38 seconds and 1 minute and 30 seconds for clear and opaque foils respectively.

Optical measurements of the prepared polymers were performed using different lamp sources with characteristic light peaks from near UV region, to Vis and to near IR region, and using a high resolving monocromator spectrophotometer system (Lynear-Horiba Jobin Yvon) using optical fibers.

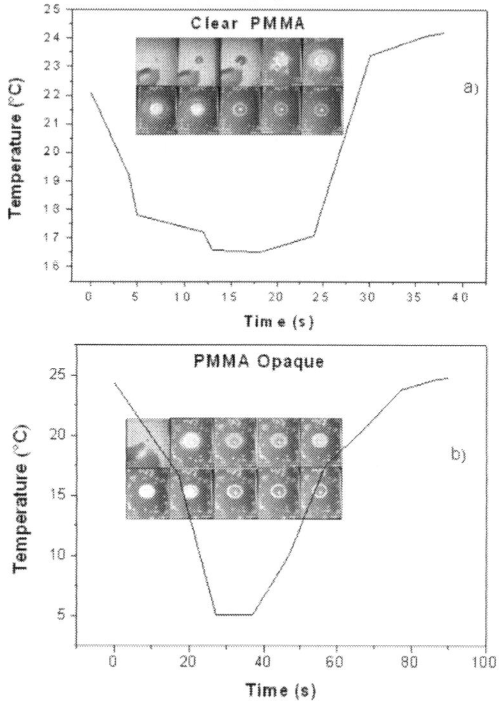

Fig. 2. Images obtained by a thermo-camera during the preparation of clear PMMA foil (a) and during the preparation of opaque PMMA foil (b).

To characterize the composition of the foils the different elements were evaluated by Rutherford Backscattering Spectrometry (RBS) and Elastic Recoil Detection Analysis (ERDA), for hydrogen measure component, using 2.0 MeV He^+ ions at Tandetron Lab [11], Nuclear Physics Institute of the ASCR, in the Czech Republic.

Incoming helium ions arrive on the substrate orthogonally then they are backscattered and detected at an angle of 170° by Ultra-Ortec PIPS detector placed below the incoming beam in the Cornell geometry for RBS analysis [12], whereas during ERDA the incoming beam angle is 75° in glancing and the scattering angle is 30° where the forward recoiled atoms (light elements, such as protons) are detected by a Canberra PD-25-12-100 AM detector fixed in plane with the incoming ion beam and covered by a 12 µm Mylar foil to avoid the detection of back-scattered He^+ ions [13].

Typically, RBS and ERDA spectra were evaluated and transformed into concentration-depth profiles taking into account all the parameters involved in particles-solid interaction and experimental configuration (stopping power, elastic recoil-cross section, kinematic and geometric factors, energy spreading). Spectra were studied comparing with simulation data by utilizing SIMNRA 6.06 code [14] and cross-sections data from IBANDL [15]. The typical He^+ ion current for each analysis, measured with a suitable Faraday cup, was about 5.3 nA and the irradiation time was about 5 minutes for the clear PMMA and 10 minutes for the opaque PMMA, no more prolonger times were adopted to avoid the thermal degradation of the samples due to ion beam irradiation dose.

Successively, clear PMMA and opaque PMMA, with and without AuNPs, have been irradiated using a Nd:YAG laser, 3 ns of duration, 1064 nm of wavelength, 10^{10}W/cm^2 of intensity, delivering an energy of 100 mJ at Messina University. Moreover they were irradiated using an Iodine laser at 1315 nm wavelength, 300 ps pulse duration, 600 J pulse energy and 10^{16} W/cm^2 intensity in PALS facility (Czech Republic).

3. Results

Optical measurements of 8 µm PMMA absorption at 1064 nm and 1315 nm were performed in order to investigate on the effect of near IR irradiation of the prepared polymer to be laser irradiated. The optical measurements were performed using a Nd:YAG laser operating at very low energy, under the ablation threshold. The absorption measurements of the prepared samples at 1064 nm and 1315 nm, the wavelength of the two lasers used in order to produce plasma and accelerated protons, and the absorbance increase using embedded AuNPs is reported in Fig.3.

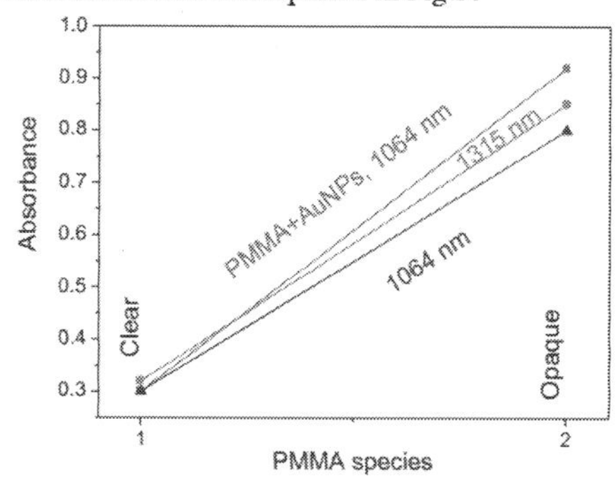

Fig. 3. Absorbance at 1064 and 1315 nm for PMMA clear, opaque and opaque containing Au NPs 1% wt (b) .

The transmission curve in ref [16] shows two absorption peaks located at 1100 nm and 1300 nm at which transmission is about 70% and 68% having high transparency for the clear PMMA. Fig. 3b reports the absorbance of PMMA clear and opaque at two different wavelengths, and the absorbance value at 1064 nm for PMMA containing Au NPs 1% wt.

RBS and ERDA analyses have been evaluated qualitatively and quantitatively both the elements and the hydrogen concentrations contained into the prepared polymeric foils.

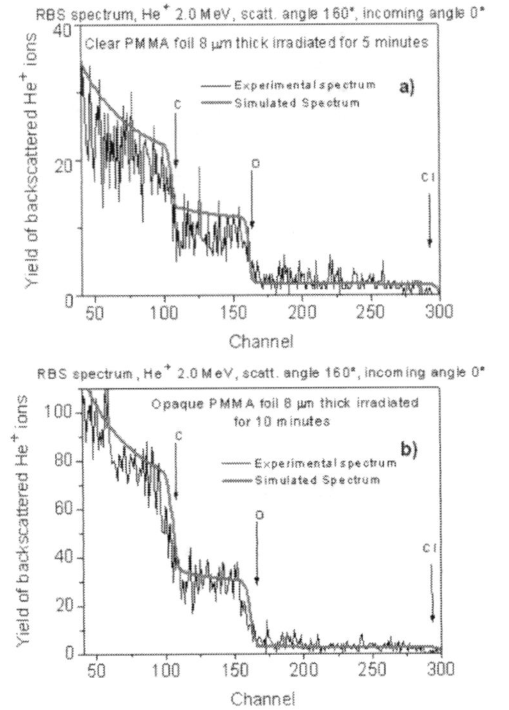

Fig. 4. RBS spectra achieved irradiating clear (a) and opaque (b) PMMA foils.

In Fig. 4a is reported the RBS analysis measurement performed in clear PMMA foil. The maximum atomic concentrations of C, O, and Cl into clear PMMA found at a depth of 400 nm are 60%, 36% and 0.8%, respectively. The spectra show the experimental data and the simulation data (solid red line) comparison from which the quantitative concetration have been evaluated.

In Fig. 4b is reported the RBS analysis measurement performed in opaque PMMA foil. The maximum atomic concentrations of C, O, and Cl into opaque PMMA found at a depth of 400 nm appears to be 70%, 26% and 0.5%, respectively.

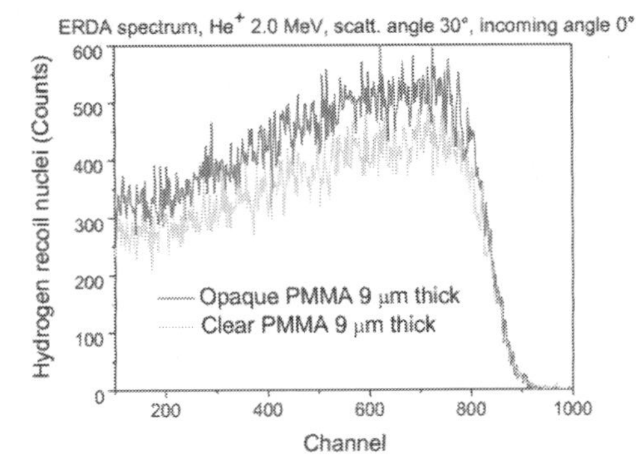

Fig. 5. ERDA analysis for the clear and opaque PMMA foils

Thus, in this case an increment of carbon and chlorine and a decrement of the oxygen are measured with respect to the clear polymer.

Fig. 5 reports the comparison between the ERDA analysis in clear and opaque PMMA foils. These results are in agreement with that showed by the thermocamera: the evaporation in the clear PMMA is lower than in the opaque PMMA as well as the oxygen concentration. The hydrogen content in both polymer species is comparable. In Fig. 6 are reported three spectra revealed by SiC detectors, at a flight distance of 74.5 cm from the target, in TOF approach during the laser irradiation of clear PMMA, opaque PMMA and opaque PMMA containing 1% of embedded AuNPs (50 nm size).

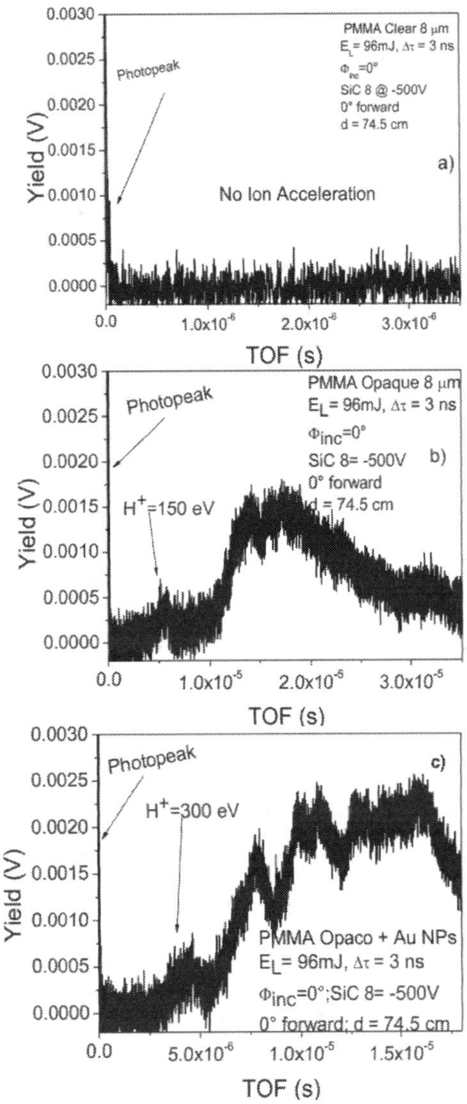

Fig. 6. TOF spectra obtained by SiC detector irradiating at low laser intensity clear (a), opaque (b), and opaque PMMA containing gold NPs (c).

The used laser was a Nd:YAG operating at 1064 nm wavelength, with 3 ns pulse duration and 10^{10} W/cm^2 intensity.

In the first case the clear PMMA is transparent to the laser wavelength, thus the released laser energy is minimal and no ion acceleration occurs (Fig. 6a). In the second case the opacity of the polymer enhances the laser absorption and more energy is transformed in ion acceleration, detecting maximum proton energy of 150 eV (Fig. 6b). In the third case the presence of the Au NPs permits to increase the laser absorption and to enhance the electron density of the plasma, thus the maximum proton energy reaches 300 eV (Fig. 6c).

The enhancement of 50% of proton energy is due to the increment of the electric field [17].

Similar results were also obtained irradiating the polymer with high intensity laser at PALS at 10^{16} W/cm^2 using 1315 nm wavelength and pulse duration of 300 ps. The acceleration was performed using thin PMMA foils, 8 micron in thickness, in the forward direction in TNSA regime. In these experiments SiC was placed in forward direction at 60 cm distance from the target and laser was focalized on the target with 70 micron spot diameter. X-ray, electrons and ions were produced by the hot and dense plasma and protons were detected at high kinetic energy, as expected.

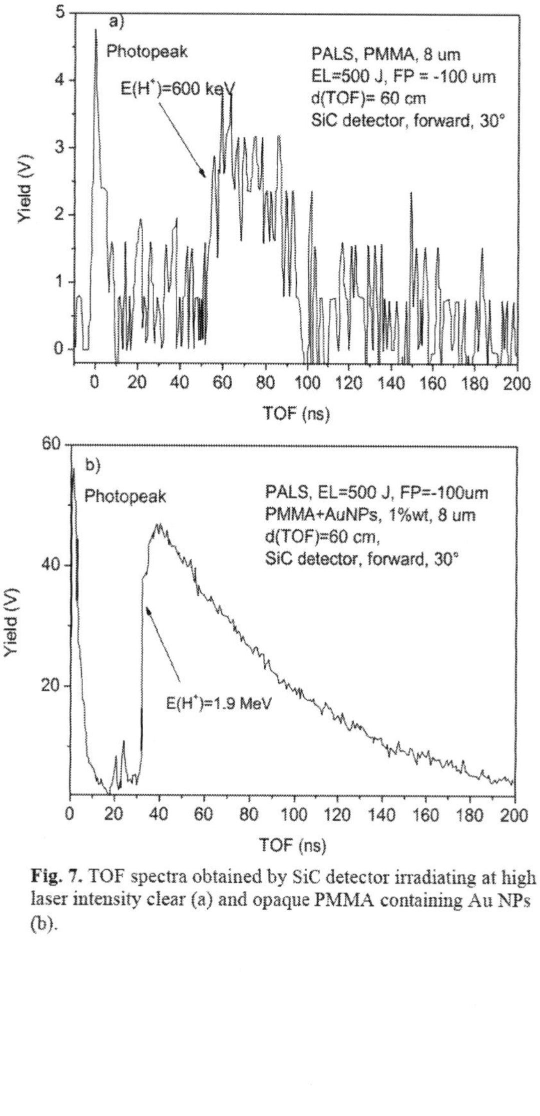

Fig. 7. TOF spectra obtained by SiC detector irradiating at high laser intensity clear (a) and opaque PMMA containing Au NPs (b).

The laser irradiation of the clear PMMA is lower absorbed and a cold plasma is obtained accelerating a low proton yield at a kinetic energy of 600 keV (Fig. 7a). The laser irradiation of the opaque PMMA in which Au NPs are embedded produces hot plasma and a very good forward acceleration in TNSA regime, accelerating protons up to 1.9 MeV (Fig. 7b). It means that the high intensity laser produces high relativistic electron acceleration at the polymer surface, thanks to the presence of high Au nanoparticle concentration. Such electrons are transmitted to the back face of the thin film producing high positive charge concentration in the polymeric film. Coulomb explosion is generated in the solid and ions are emitted. The high electric field generated in the back face of the target (forward direction) drives the ion (protons, carbon and oxygen) acceleration. The high proton energy is due to the increment of the electric field driving the acceleration which depends on the temperature and the electron density of the produced plasma, according to literature [17].

4. Summary and conclusions

Hydrogenated polymeric targets, such as PMMA, can be prepared as advanced targets to be have low absorption or high absorption in the near IR wavelengths. The study as the aim to use thin PMMA foils to be laser irradiated at high intensity and to produce plasma in which high proton acceleration can occurs.

Particularly the use of porous opaque polymer, containing microstructured beads, or metallic nanoparticles embedded inside it permit to enhance both the laser absorption coefficient, inducing surface plasmon resonance (SPR) effects [6], and the plasma electron density with a significant increment of the resulting ion acceleration energy.

The maximum acceleration in the polymer isobtained using opaque PMMA as a matrix in which Au NPs have been embedded. At relatively low laser intensities, of the order of 10^{10} W/cm^2, the maximum proton acceleration is 300 eV, demonstrating that the PMMA with Au NPs could be a good candidate target to accelerate protons at high yield and energy, useful for many applications.

The use of opaque PMMA containing micrometric beads confers high laser absorption and a good plasma production, but the addiction of Au NPs permits to enhance the electron plasma density and the plasma temperature, i.e. the electric field driving the proton acceleration from TNSA regime.

Possible applications of such opaque PMMA with Au nanoparticles concern the proton acceleration for protontherapy [18], the proton and deuterium acceleration to induce nuclear reaction and fusion processes [19], the simulation of astrophysical environment, the use of ion beams for material treatments and the use of ion implantation to modify the chemical and physical properties of many elements [20].

The research has been realised at the CANAM (Center of Accelerators and Nuclear Analytical Methods) infrastructure LM 2015056 and has been supported by project No. P108/12/G108.

References

1. L. Torrisi, M. Cutroneo, L. Calcagno, M. Rosinski and J. Ullschmied, Journal of Physics: Conference Series **508** 012002 (2014).
2. I.G. Jiang, Y.S. Zhou, W. Xiong et al., Opt. Lett. **39** 3034-3037(2014).
3. L. Torrisi, M. Cutroneo, V. Semian, G. Ceccio, Applied Surface Science **351** 580-587 (2015).
4. I. Capan, R. Capan, T. Tanrisever, S. Beyaz, Mater. Lett. **59** 2468-2471 (2005).
5. M. Cutroneo, V. Havranek, A. Mackova, V. Semian, L. Torrisi, L. Calcagno, Nuclear Inst. and Methods in Physics Research, B, **371** 344-349 (2016).
6. M.A. Garcia, J. Phys. D: Appl. Phys: **44** 283001 (2011).
7. L. Torrisi, Molecules **19** 17052-17065 (2014).
8. C.B. Simone and R. Rengan, Cancer J. **20**(6):427-32 (2014).
9. L. Torrisi, Recent Patents on Nanotechnology **9**(1),51-60 (2015).
10. N. Patra, A. C. Barone, M. Salerno, Adv. Polym Technol. **30** 1 12-20 (2011).
11. Canam Website 2017: http://canam.ujf.cas.cz/en/laboratory/lt-laboratory-of-tandetron
12. E. Kotai, AIP Conference Proceedings **392** 631 (1997).
13. L. Torrisi and M. Cutroneo, Surface and Interface Analysis **48** (1) 10-16 (2016).
14. M. Mayer, SIMNRA version 6.06, Max-Planck-Institut fur Plasmaphysik, Garching, Germany, 2011, available at: http://home.rzg.mpg.de/~mam/Download.html
15. IBANDL actual website 2017: http://www-nds.iaea.org/ibandl/
16. I. Jones and J. Rudlin, Joining Plastics 2006, London, National Physical Laboratory (NPL), 25-26 April 2006
17. L. Torrisi, M. Cutroneo, and J. Ullschmied Physics of Plasmas **24** 043112-043123 (2017).
18. L. Torrisi, Recent Patents on Nanotechnology **9**(1), 51-60 2015.
19. L. Torrisi, S. Cavallaro, M. Cutroneo, J. Krasa and D. Klir, Phys. Scr. T **161** 014026 (2014).
20. M. Cutroneo, L. Torrisi, J. Ullschmied and R. Dudzak, Nukleonika **61** 2, 109-113(2016).

Nanoimaging using soft X-ray and EUV laser-plasma sources

Przemyslaw Wachulak [1,*], *Alfio* Torrisi [1], *Mesfin* Ayele [1], *Andrzej* Bartnik [1], *Joanna* Czwartos [1], *Łukasz* Węgrzyński [1], *Tomasz* Fok [1] and *Henryk* Fiedorowicz [1]

[1] Institute of Optoelectronics, Military University of Technology, 2, Kaliskiego Street, 00-908 Warsaw, Poland

Abstract. In this work we present three experimental, compact desk-top imaging systems: SXR and EUV full field microscopes and the SXR contact microscope. The systems are based on laser-plasma EUV and SXR sources based on a double stream gas puff target. The EUV and SXR full field microscopes, operating at 13.8 nm and 2.88 nm wavelengths are capable of imaging nanostructures with a sub-50 nm spatial resolution and short (seconds) exposure times. The SXR contact microscope operates in the "water-window" spectral range and produces an imprint of the internal structure of the imaged sample in a thin layer of SXR sensitive photoresist. Applications of such desk-top EUV and SXR microscopes, mostly for biological samples (CT26 fibroblast cells and Keratinocytes) are also presented. Details about the sources, the microscopes as well as the imaging results for various objects will be presented and discussed. The development of such compact imaging systems may be important to the new research related to biological, material science and nanotechnology applications.

1 Introduction

Use Recent advancements in nanoscience and nanotechnology will surely not be possible without nanometer scale resolution imaging tools and techniques, such as extreme ultraviolet (EUV) and soft X-ray (SXR) [1] microscopy. The EUV/SXR microscopy proved to be a useful tool for imaging of objects with nanometer spatial resolution and allows to obtain additional information about the objects investigated, providing at the same time high optical contrast in the specific wavelength ranges. The EUV radiation is strongly absorbed in thin layers of materials, so it is highly suitable for investigations of thin films and layers of materials, while the SXR radiation, specifically in the "water-window" (λ = 2.3 - 4.4 nm), is particularly suitable for high resolution biological imaging, due to high achievable contrast between carbon and water – the main constituents of biological material.

Most of studies in this field are conducted using large scale facilities, such as synchrotron or free-electron laser installations [2,3]. Synchrotron and FEL facilities are being used for cutting-edge scientific experiments and provide highest available photon flux, tunability and spatial and temporal coherence. However, they also have some disadvantages (high maintenance cost, require highly trained staff due to the complexity of those sources and have limited user access). Recent progress in development of a new generation of compact EUV and SXR compact sources, especially laser-plasma sources permits to overcome some of these limitations and allow for imaging experiments in the laboratories worldwide.

So far, many efforts have been made to perform nanometer spatial resolution imaging in the EUV and SXR spectral ranges over the last few years employing both large scale and compact sources. Some examples of such works are listed as follows. The synchrotron radiation at λ = 2.4 nm was used for imaging frozen-hydrated samples at atmospheric pressure [4] to observe internal details of algae cells with a spatial resolution of ~35 nm. A SXR source, emitting at λ = 2.88 nm, based on a liquid jet nitrogen target was employed recently for microscopy in the "water-window" range with a sub-50 nm spatial resolution, but this system present a very complicated hardware [5]. A compact high order harmonic generation (HHG) source was employed for sub-100 nm spatial resolution imaging [6]. However, the HHG systems require a femtosecond laser as a driver for HHG process with conversion efficiency (~10-6-10-4) which often results in long exposures or does not allow a proper reconstruction of the image. Ptychographic techniques, usually based on employment of hard X-ray beams, although provide very high spatial resolution, are extensively time consuming during the reconstruction process [7]. Imaging in the EUV range permits to analyse very thin samples, nanofilms and nanostructures, because the EUV radiation is absorbed by solid materials with thicknesses of the order of 100 nm [8] and by gaseous materials with thicknesses of the order of a few millimetres [9]. Employing such radiation allows to visualize in a direct way through absorption contrast mechanism the flow of the gas, usually investigated with other techniques, such as the interferometry [10]. It was demonstrated that the radiation from a capillary discharge laser operating at a wavelength of = 46.9 nm EUV images permits to obtain a spatial resolution better than 55 nm [11, 12], and that the spatial resolution of holographic images [13], employing the same

* Corresponding author: : wachulak@gmail.com

wavelength, can be improved up to sub-50 nm [14]. A table top EUV laser emitting 13.2 nm wavelength radiation from Ni-like Cd ions was employed for imaging of with a sub-38 nm spatial resolution and a picosecond time resolution [15]. Using xenon gas discharge source a sub-100 nm zone-plate based zooming EUV microscope was demonstrated [16]. Very nice results were also recently demonstrated [17] with liquid nitrogen target based system and 1.3 kHz repetition rate Nd:YAG laser to record images of biological samples and nanostructures with half-pitch spatial resolution of 40-50nm, however, the laser driver was quite complicated and large comparing to other, more compact ones. Another interesting and already well established technique for obtaining high resolution images of samples is contact microscopy. It uses the SXR radiation transmitted through the sample to expose a high resolution photoresist in contact with the sample. Using this method human platelets [18], fibroblasts [19] and hydrated biological cells [20] were imaged.

It is necessary, however, to find a good compromise in the desk-top SXR/EUV imaging systems, between the performance (high spatial resolution, low exposure time provided by large scale facilities) and the complexity, size, cost of these setups. Some of these problems can be addressed employing laser plasma sources based on double stream gas puff target for microscopy purposes. These sources represent a useful alternative for lab-scale analysis in order to perform high spatial resolution imaging by using short wavelength radiations.

In this article we would like to show our recent developments of three simple and compact SXR/EUV microscopes, capable of resolving 50-80 nm features that require short exposure times and have a desk-top footprint. The laser-plasma source employed allows the possibility to generate the plasma efficiently and without debris production, they are also robust and easy to operate, providing relatively high EUV and SXR emission flux [21]. Additionally, they have compact and easy construction. Moreover, the EUV/SXR microscopes, based on those sources [22] do not require sample preparation such as gold coating for SEM microscopes for example or marking/staining the samples for STED microscopy. In such systems the gas puff target is produced by injection of a small amount of high-Z gas – working gas, into a stream of low-Z gas – outer gas, by a fast electromagnetic double valve system, as depicted in Fig. 1a). The gas puff target is then irradiated by focused laser pulses from Nd:YAG laser, which allows generation of the EUV and SXR radiation. Both full-field microscopes use reflective optics to focus the short wavelength radiation onto a sample and FZPs to obtain magnified images of the sample with high spatial resolution, while the contact microscope uses the radiation directly, without pre-focusing, to form an imprint of the sample in the photoresist layer.

2 Microscopy systems

Soft X-ray contact microscope and EUV/SXR microscopes based on Fresnel optics are schematically depicted in Fig. 1b) and Fig. 1c), respectively. The sources for those microscopes were developed using a compact Nd:YAG laser, λ = 1064 nm, 0.5-0.7 J pulse energy, 3 ns pulse duration and 1-10 Hz repetition rate.

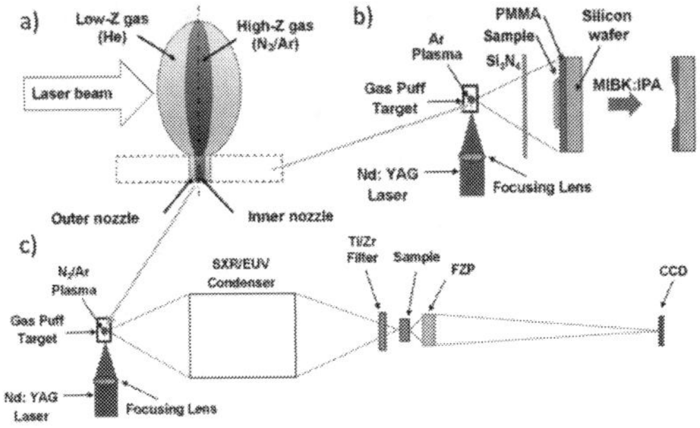

Fig. 1. a) Scheme of the double stream gas puff target employed in laser-plasma EUV/SXR sources, b) scheme of the contact SXR microscope and c) general scheme of the SXR/EUV full-field microscopes.

The laser beam is focused onto a double-stream gas puff target, schematically depicted in Fig. 1a), produced by an electromagnetic valve with double radially-symmetric nozzles. Laser pulses irradiate the gaseous target producing plasma, which efficiently emits radiation from various spectral ranges, including the EUV and SXR regions of electromagnetic spectrum. The gas puff target source was optimized for emissions in the "water window" (both quasi-monochromatic for the SXR full-field microscope and broadband for contact SXR microscope) and the EUV spectral ranges using spectral narrowing with thin filters, depending on the type of the microscope. A nitrogen plasma emission, filtered with Ti foil, provides λ =2.88 nm wavelength radiation and an Ar plasma emission, filtered with Mo/Si multilayer mirror and Zr filter, provides radiation at λ =13.84 nm. However, for the argon/helium plasma if silicon nitride filter is used, it provides emission from λ =2.8-4 nm, well within the "water window" spectral range with much larger number of photons, suitable for contact microscopy. Such sources were employed for SXR and EUV experiments, respectively. The gas pressures, nozzle position in respect to the laser focus, valve timing in respect to the laser pulse were optimized in order to maximize the photon flux at the sample plane.

The inner nozzle injects a small amount of working gas (high Z gas - N_2 at 8 bar, in case of SXR full-field microscope, and Ar, at 10 bar, in case of the EUV full-field microscope and contact SXR microscope), while the outer nozzle injects an outer gas (low Z gas, He, with a pressure of 6 bar), to narrow down the flow of the working gas, reducing its density gradient along the nozzle axis. This allows obtaining higher target density at 1-2 mm from the nozzle exit and, in turn, allows for higher photon yield in the EUV/SXR region. Interaction of Nd:YAG laser pulses with gaseous target produces plasma. The radiation from the plasma is focused using condenser optics and illuminates the sample, located in a second focal point of a condenser in the case of full-field microscopes. Depending of the type of microscope, plasma radiation is collected and focused by a different condenser and filter, and images of the objects are obtained with different zone plates using a back-illuminated CCD camera. In the case of contact SXR microscope the scheme is much simpler, because the radiation from the plasma is transmitted through a 200 nm thick Si_3N_4 filter and illuminated the sample directly, to form its imprint in the surface of the photoresist. In the following paragraphs the systems will be described in more detail.

In the full field SXR microscope to focus the SXR radiation from the nitrogen plasma an ellipsoidal, axi-symmetrical SXR condenser (Rigaku, Czech Republic) coated with nickel film was used. A titanium filter (200 nm thick, Lebow), positioned downstream the condenser, selects the He-like nitrogen line at λ=2.88 nm from nitrogen plasma emission, suitable for the objective optic. Filtered SXR radiation illuminates the sample/object, positioned 140 mm downstream the condenser, in the second focal point. A FZP objective (Zoneplates Ltd., UK) was used to form a magnified image onto a back-illuminated SXR-sensitive CCD camera (Andor, iKon-M DO-934-BN, 1Mpixel, 13 x 13 μm^2 pixel size) in transmission mode. A silicon nitride FZP (400 nm thick, 250 µm in diameter, outer zone width of 30 nm) with a focal length f =2.6 mm at 2.88 nm wavelength, was used as an objective. The numerical aperture of the zone plate was matched to the numerical aperture of the condenser, providing incoherent illumination [23]. The geometrical magnification of the system was ~220x. It was already demonstrated that such system is capable to achieve a half-pitch spatial resolution of ~60nm [24]. More details about this work can be found in [25].

In the full field EUV microscope an Ar plasma radiation was collected and spectrally narrowed by an ellipsoidal off-axis mirror coated with a Mo/Si multilayer (MLM), employed as a condenser (Reflex, Czech Republic, and IOF, Germany). The condenser was designed and manufactured in order to reflect the radiation at the wavelength of 13.5±0.5 nm at 45° incidence angle. To eliminate longer wavelengths, emitted from the Ar plasma (at wavelengths λ 16 nm), a 250 nm thick zirconium filter (Lebow) was used. The sample was imaged using a FZP objective (diameter 200 µm, number of zones 1000 and outer zone width Δr = 50 nm, made from PMMA, from Zoneplates Ltd., UK) onto the CCD camera (Andor, iKon-M DO-934-BN camera). The FZP was fabricated using e-beam lithography in a 200 nm thick PMMA layer deposited on top of a 50 nm thick Si3N4 membrane. As in case of the SXR microscope, an incoherent illumination was provided by matching of the numerical aperture of the condenser and of the FZP objective. The geometrical magnification of the objective was 410x, however, it can be easily changed by adjusting the FZP-CCD distance. Such system, in this configuration, is capable to achieve a half-pitch spatial resolution of 48 nm [26]. More details about this system can be found in [27].

In the contact SXR microscope the Ar plasma emission was narrowed down to the "water-window" spectral range by employing 200 nm thick Si3N4 filter. Using that filter the most energy will reside in the wavelength range from 2.8 nm to 4 nm, well within the "water-window" range. Such broad band SXR radiation from Ar/He gas puff target has approximately one order of magnitude more photons than from N2/He gas puff target, which allows in this case for irradiation and exposure of the high resolution photoresist (500 nm thick PMMA on top of a silicon wafer). The object is then placed in contact with the PMMA. The light that is locally transmitted by the object's structure illuminates the photoresist and changes its physical and chemical structure. After the irradiation the photoresist is chemically developed in solution of methyl isobutyl ketone and isopropyl alcohol (MIBK:IPA 1:2 v/v) for 90 seconds and modulation of the light intensity absorbed by the object is converted in this process to a modulation of the thickness of the resist. This creates a relief-like structure in its surface. The height of the relief structure is directly proportional (however, not necessarily linearly), to the dose of radiation absorbed in the resist volume. In such process a high resolution imprint of the internal structure of the object can be stored in the near-surface of the photoresist and later converted to an image using for example AFM or SEM microscope. More details about this system can be found in [28].

3 Results

An example of SXR "water window" images of organic sample, acquired with the full-field SXR microscope is depicted in Fig. 2.

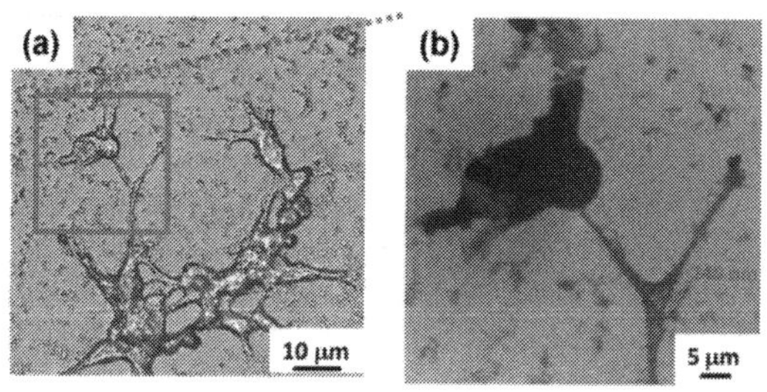

Fig. 2. CT 26 fibroblast cells. Comparison of optical image, 40x objective, (a) and detail imaged (b) with the SXR "water window" microscope (red square in (a)) that shows small features of the order of 240 nm (indicated with red arrow). Exposure of 200 EUV pulses.

Fig. 2a) and b) shows a sample of CT 26 fibroblast from Mus musculus colon carcinoma (strain BALB/c), prepared on top of a 30 nm thick Si_3N_4 membrane. A direct comparison between the image acquired with a traditional optical microscope (Fig. 2a)) and the SXR microscope image (Fig. 2 b)), acquired with 200 SXR pulses, at a source repetition rate of 10 Hz, exposure time of 22 seconds and detector (CCD) temperature of -

https://doi.org/10.1051/epjconf/201816703001

20 °C, are shown. The sample was prepared with a gradual dehydration in ethanol series (final concentration 70% EtOH), without any fixation procedure. The SXR image shows improved spatial resolution due to the employment of shorter wavelength, beyond the diffraction limit of the optical-visible light microscopes. Some internal and external structures can be distinguished due to phase contrast in the visible light microscopy images and due to the modulation in the absorption of the SXR light through the sample in the SXR images.

The full field EUV microscope was employed for imaging of CT26 fibroblast cells. The EUV images were acquired with 200 EUV pulses - 22 sec. exposure time and detector temperature of −20°C. A sample of CT 26 cells fixed with 30% hexamethildisilazane (HDMS) in absolute EtOH on top of 30 nm Si_3N_4 membrane was imaged with the optical microscope, equipped with 40x objective, as depicted in Fig. 3a) and at 13.84 nm wavelength with the EUV microscope - Fig. 3b). It can be seen that the EUV image shows a very high contrast and resolution enhancement and permits to investigate features of the order of 100 nm or smaller in size. The EUV images exhibit superior spatial resolution and much higher optical magnification (410x) comparing to visible light microscopy - Fig. 3a). Moreover, although the EUV radiation does not allow to see the internal structure of the cells, it shows in detail the fine structures outside the cell membrane (interconnections between the cells, flagella, etc.), which are not visible in the SXR range due to the lack of optical contrast in those very thin features.

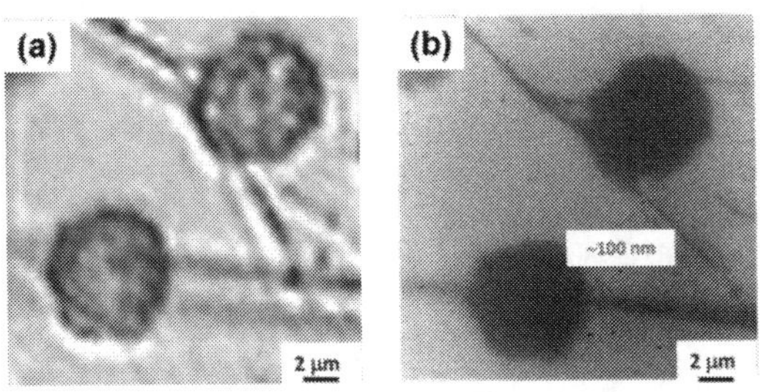

Fig. 3. Comparison of CT 26 fibroblast cells sample acquired with optical microscope (small section from the optical image acquired with 40x objective) (a) and EUV microscope (b). In this last image it is possible to appreciate small features of the order of 100 nm (as indicated in red arrows). Exposure of 200 EUV pulses.

The contact microscope has been used for imaging fixed epidermal cells (Keratinocytes), as depicted in Fig. 4. The sample was prepared in collaboration with the Institute of Biotechnology, Warsaw University of Technology, Poland. The sample was cultured on PMMA photoresist in a controlled culture conditions. A Dulbecco's Modified Eagle's Culture Medium (DMEM, Sigma-Aldrich) supplemented with 1% L-glutamine, 1% penicillin/streptomycin and 10% heat inactivated foetal bovine serum (Celbio) was used as a culture medium. The culture medium was washed with PBS buffer solution followed by fixation and dehydration of the cells using paraformaldehyde 3.7% in phosphate buffered saline at 35 °C. The exposure of 200 SXR pulses, 20 second exposure time, was required to expose the PMMA photoresist and reach spatial resolution of approximately 80 nm half-pitch.

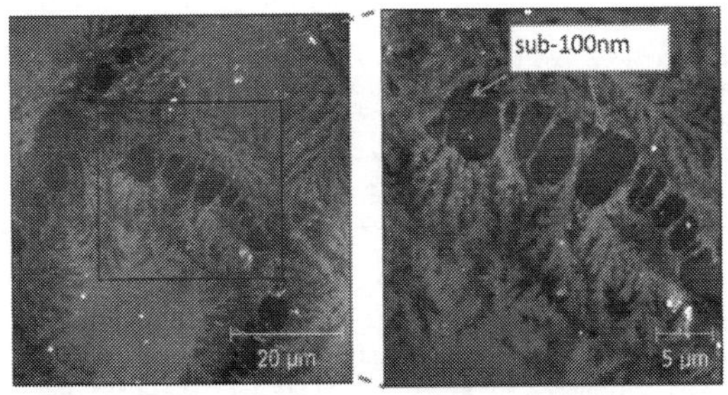

Fig. 4. SXR contact microscopy images of fixed epidermal cells (Keratinocytes). Images were obtained with SXR exposure of 200 pulses, 20 seconds and the photoresist was scanned using AFM microscope in semi-contact mode. Structures with sizes below 100 nm are visible.

The photoresist was treated with previously mentioned procedure and later scanned with AFM (AFM, NT-MTD, Russia) operating in semi-contact mode. Each image was 512x512 points over 60x60 μm^2 and 25x25 μm^2 in left and right images, respectively. The structures with sizes below 100 nm are clearly visible. More details about this work can be found in [27]

4 Summary and conclusions

The SXR/EUV desk-top microscopes based on laser plasma sources with double stream gas puff target allow capturing magnified images of the samples, with 50-80 nm half-pitch spatial resolution and exposure time as low as few seconds. The microscopes are compact and permit to obtain images with a nanometer spatial resolution and short exposure time. Such characteristics represent key parameters which open possibility of a future commercialization of such systems. Those systems employ photons and not electrons, like in SEM, allow obtaining additional/complementary information about the sample. Their spatial resolution is comparable to what is possible to obtain with Stimulated Emission Depletion microscopy (STED) [29], however, such table-top systems allow for a direct acquisition of a full field images, in contrary to STED acquisition, operating in scanning mode and do not require fluorescent bio markers or staining that can modify the morphology of the sample. Additionally, short wavelengths provide high optical contrast, either in biological (SXR microscope) or in all matter (EUV microscope).

These microscopes may represent an important alternative to perform experiments in small academic laboratories and could have an impact on the nanotechnology in the near future. The goal of developing of these microscopes is to show feasibility to achieve high resolution imaging, low exposure times, together with a compact footprint, which may, in the future, open a possibility for commercialization.

This work is supported by the National Science Centre, Opus programmes, grant agreement number UMO-2015/17/B/ST7/03718 and UMO-2015/19/B/ST3 /00435, the Education, Audiovisual and Culture Executive Agency (EACEA) Erasmus Mundus Joint Doctorate Programme Project No. 2012-0033 and from the European Union's Horizon 2020 research and innovation program, under Laserlab-Europe IV, grant agreement No. 654148. We would like to thank to Tomas Parkman, Šarka Salacová and Jana Turňová from Faculty of Biomedical Engineering, Czech Technical University in Prague for biological samples and Andrzej Kowalik and Mrs. Anna Rojek, from Institute of Electronic Materials Technology (ITME) for the preparation of the photoresist in their laboratory. We also acknowledge Dr. Tomasz Kobiela and Mrs. Anna Sobiepanek, from the Faculty of Chemistry Warsaw University of Technology for providing biological samples used for contact microscopy experiments.

References

1. The International Organization for Standardization, "ISO 21348 Definitions of Solar Irradiance Spectral Categories," 6–7 (2007).
2. R.N. Wilke, M. Priebe, M. Bartels, K. Giewekemeyer, A. Diaz, P. Karvinen, T. Salditt, Optics Express 20, 19232 (2012).
3. M. A. Le Gros, G. Mcdermott, B. P. Cinquin, Journal of Synchrotron Radiation 21 (6), 1–8 (2014).
4. G. Schneider, Ultramicroscopy 75, 85–104 (1998)
5. K.W. Kim, Y. Kwon, K. Nam, J. Lim, K.W. Kim, K. Chon, B. Kim, D.E. Kim, J. Kim, B.N. Ahn, H.J. Shin, S. Rah, K.W. Kim, J.S. Chae, D.G. Gweon, D.W. Kang, S.H. Kang, J.Y. Min, K. Choi, S.E. Yoon, E. Kim, Y. Namba, K. Yoon, Physics in Medicine and Biology 51, N99–N107 (2006)
6. H. T. Kim, Springer Proceedings in Physics 136, 323-328 (2010)
7. M. Dierolf, New J. Phys. 12, 035017 (2010)
8. P. W. Wachulak, R. A. Bartels, M. C. Marconi, C. S. Menoni, J.J. Rocca, Y. Lu, B. Parkinson, Optics Express 14, 9636 (2006)
9. M.C. Marconi, P.W. Wachulak, Progress in Quantum Electronics 34, 173–190 (2010)
10. P.W. Wachulak, L. Wegrzynski, A. Bartnik, T. Fok, R. Jarocki, J. Kostecki, M. Szczurek and H. Fiedorowicz, Laser and Particle Beams 31, 2, 219-227 (2013)
11. C. Brewer, F. Brizuela, P. Wachulak, D.H. Martz, W. Chao, E.H. Anderson, D.T. Attwood, A. V Vinogradov, I. a Artyukov, A.G. Ponomareko, V. V Kondratenko, M.C. Marconi, J.J. Rocca, C.S. Menoni, Optics Letters 33, 518–520 (2008)
12. P.W. Wachulak, C.A. Brewer, F. Brizuela, W. Chao, E. Anderson, R. A. Bartels, C.S. Menoni, J.J. Rocca, M.C. Marconi , Journal of the Optical Society of America B 25, B20, (2008)
13. E. B. Malm, N. C. Monserud, C. G. Brown, P. W. Wachulak, H. Xu, G. Balakrishnan, W. Chao, E. Anderson, M. C. Marconi, Optics Express21, 8, 9959-9966 (2013)
14. P.W. Wachulak, M.C. Marconi, R. a Bartels, C.S. Menoni, J.J. Rocca, Journal of the Optical Society of America B 25, 1811–1814 (2008)
15. G. Vaschenko, F. Brizuela, C. Brewer, M. Grisham, H. Mancini, C.S. Menoni, M.C. Marconi, J.J. Rocca, W. Chao, J. Liddle, E.H. Anderson, D.T. Attwood, V Vinogradov, I. Artioukov, Y.P. Pershyn, V. V Kondratenko, Optics Letters 30, 2095–7 (2005)
16. L. Juschkin, Journal of Physics: Conference Series 186, 012030 (2009)
17. H. Legall, G. Blobel, H. Stiel, W. Sandner, C. Seim, P. Takman, D. H. Martz, M. Selin, U. Vogt, H. M. Hertz, D. Esser, H. Sipma, J. Luttmann, M. Höfer, H. D. Hoffmann, S. Yulin, T. Feigl, S. Rehbein, P. Guttmann, G. Schneider, U. Wiesemann, M. Wirtz, and W. Diete, Optics Express 20, 16, 18362 (2012)
18. J. Kirz, Rev. Sci. Instr. 56, 1-13 (1985)
19. P. C. Cheng, NIMA 246, 668-674 (1986)
20. M. Kado, M. Kishimoto, S. Tamotsu, K. Yasuda, M. Aoyama and K. Shinohara, Proc. of SPIE 8849, 88490C-1 (2013)
21. A. Bartnik, W. Lisowski, J. Sobczak, P. Wachulak, B. Budner, B. Korczyc, H. Fiedorowicz, Appl Phys A 109, 39-43 (2012)
22. P. W. Wachulak, A. Bartnik, M. Skorupka, J. Kostecki, R. Jarocki, M. Szczurek, L. Wegrzynski, T. Fok and H. Fiedorowicz, Applied Physics B 111, 2, 239-247 (2013),
23. J.M. Heck, D.T. Attwood, W. Meyer-Ilse, E.H. Anderson, J. X-ray Sci. Technol. 8, 95 (1998)
24. P. Wachulak, A. Torrisi, M.F. Nawaz, A. Bartnik, D. Adjei, Š. Vondrová, J. Turňová, A. Jančarek, J. Limpouch, M. Vrbová, H. Fiedorowicz, Microscopy and Microanalysis 21, 1214–1223 (2015)
25. P. W. Wachulak, A. Torrisi, A. Bartnik, D. Adjei, J. Kostecki, L. Wegrzynski, R. Jarocki, M. Szczurek, H. Fiedorowicz, Applied Physics B 118, 573-578 (2015)
26. P. W. Wachulak, A. Torrisi, A. Bartnik, Ł. Węgrzyński, T. Fok and H. Fiedorowicz, Applied Physics B 123:25, 1-5 (2017),
27. A. Torrisi, P. Wachulak, Ł. Węgrzyński, T. Fok, A. Bartnik, T. Parkman, Š. Vondrová, J. Turňová, B. J. Jankiewicz, B. Bartosewicz and H. Fiedorowicz, Journal of Microscopy 265, 2, 251-260 (2017),
28. M.G. Ayele, J. Czwartos, D. Adjei, P. Wachulak, I.U. Ahad, A. Bartnik, Ł. Wegrzynski, M. Szczurek, R. Jarocki, H. Fiedorowicz, M. Lekka, K. Pogoda and J. Gostek, Acta Physica Polonica 129, 2, 237-240 (2016),
29. K. Otomo, T. Hibi, Y. Kozawa T. Nemoto, Microscopy (Oxf) 64, 4, 227-36 (2015)

Development and optimization of a "water window" microscope based on a gas-puff target laser-produced plasma source

Alfio Torrisi[1,2,*], *Przemyslaw* Wachulak[2], *Andrzej Bartnik*[2], *Łukasz* Węgrzyński[2], *Tomasz* Fok[2] and *Henryk* Fiedorowicz[2]

[1]Dept. of Medical Physics & Biomedical Engineering, University College London, Malet Place, Gower Street, London WC1E 6BT
[2]Institute of Optoelectronics, Military University of Technology, Kaliskiego 2 Str., 00-908 Warsaw

Abstract. A laser-plasma double stream gas-puff target source coupled with Fresnel zone plate (FZP) optics, operating at He-like nitrogen spectral line λ=2.88nm, is capable of acquire complementary information in respect to optical and electron microscopy, allowing to obtain high resolution imaging, compared to the traditional visible light microscopes, with an exposition time of a few seconds. The compact size and versatility of the microscope offers the possibility to perform imaging experiments in the university laboratories, previously restricted to large-scale photon facilities. Source and microscope optimization, and examples of applications will be presented and discussed.

1 Introduction

Since the '90s, significant efforts were made to bring technical capabilities of synchrotron-based microscopes to the development of compact sources [1]–[8], because that the small size and versatility of compact microscopy system offers the possibility to perform imaging experiments, previously restricted to large-scale facilities, in university laboratories. It was already demonstrated that spatial resolution can be increased in nanoscale imaging systems employing short wavelength radiation such as soft X-ray (SXR) [9]. Such radiation has been successfully employed in different microscope techniques, mainly in transmission mode, including scanning microscopy [10], 3D tomography [11] and contact microscopy [12]. Nanosecond lasers operating in the near IR range can generate hot plasma in vacuum irradiating solid and gas targets, from which extreme ultraviolet (EUV) and SXR are produced. In particular, the "water window" spectral range (λ = 2.3 - 4.4 nm) is suitable for biological imaging, due to a significant difference in absorption of water (oxygen) and carbon - biological sample constituents, allowing the possibility to obtain high contrast images. Although synchrotrons allow imaging acquisition in the "water window" spectral range, with a spatial resolution of ~10 nm [13], a compact high resolution imaging systems is of major importance from the point of view of a worldwide spread and possible commercial applications. The imaging in short wavelength ranges requires a high photon flux. For this reason, until now, most of studies in this field are conducted using large scale facilities, such as synchrotron or free-electron laser [14]–[16]. As it is possible to understand, the trade-off between the exposure time and the size and complexity of the system represents a deep challenge in the development of compact systems. In this work, a table-top SXR microscope based on a double stream nitrogen gas puff target, working at the monochromatic wavelength of λ=2.88 nm, including its characterization, parameters optimizations and few applications will be presented and discussed.

2 Experimental setup

A picture showing the overall features of the SXR microscope and the main components inside the SXR source chamber are depicted in Fig.1 a) and b), respectively. The microscope was built at the Institute of Optoelectronics, Military University of Technology, in Warsaw (Poland). The system is located inside a vacuum chamber, 60 cm in diameter and 35 cm in height and the entire system, including the pumping laser, fits on top of a single 1.8 x 1.2 m^2 optical table. The chamber source has a diameter of 22 cm. A nitrogen plasma was produced by focusing Nd:YAG pumping laser pulses (NL303HT, from EKSPLA), with duration of 4 ns, 10 Hz rep. rate and energy of 0.74 J by a focusing lens with a f number of 25 mm onto a double-stream gas puff target. The gases are injected through the nozzles driven by separate electromagnetic valve systems located in a single body. The electromagnetic valve, made of a stainless steel, is driven by electric pulses synchronously with the laser driving pulse. The gaseous target is formed by two circularly symmetric nozzles (0.4 mm and 0.7–1.5 mm in diameter for inner and outer nozzle, respectively). The inner nozzle injects a small amount of nitrogen (working gas) into the vacuum, while the outer nozzle injects a low Z-number gas of helium (outer gas), to confine the nitrogen gas flowing through the inner nozzle, which makes it possible to obtain higher working gas density further away from the nozzle. This improves

* Corresponding author: alfio.torrisi@ucl.ac.uk

the SXR emission yield and avoids nozzle ablation by repetitive plasma formation. Further details about the laser pulse emission synchronization with the gas emission can be found in [17].

The gas pressures were set to 8 and 6 bar for nitrogen and helium, respectively, to ensure achievement of the highest possible photon flux at the CCD camera. Moreover, the pumping system allows to maintain a constant pressure of $2 \cdot 10^{-2}$ mbar in the vacuum chamber during the 10 Hz laser source operation. The gas-puff target source is very compact, robust and easy to operate and, moreover, being gaseous, is "debris-free", eliminates debris production.

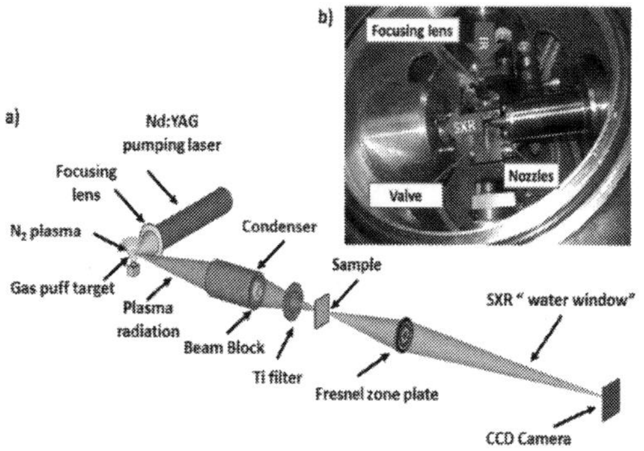

Fig. 1. a) Overall 3D scheme of the "water window" microscope and b) Top view of the gas-puff target source chamber.

2.1 Condenser optics for SXR microscopy

The SXR radiation was collected by a nickel-coated condenser mirror, manufactured by Rigaku Innovative Technologies (Czech Republic, Prague). The condenser efficiently reflects radiation from the SXR region at a small angle, focusing the radiation on the sample plane, with the energy cut-off of ~800 eV (~1.5 nm critical wavelength), above $K_\checkmark$-O edge (E = 543 eV). The condenser was mounted on a three-axis motorized translation stages, enabling a smooth adjustment of the position in three directions (xyz) and the on manual stage to change tilt angle in two directions. The distance source-condenser was equal to 60 mm, while the distance condenser-sample was equal to 140 mm. Entrance and exit diameters of the condenser were equal to 11.7 and 14 mm, respectively, which results in entrance numerical aperture of $NA_{C\_in} = 0.09$ and exit numerical aperture of $NA_{C\_out} = 0.05$. For SXR, high reflectivity is possible employing small grazing angles. Fig. 2 a) shows the reflectivity p-polarization for different metallic surfaces at 2.88 nm (data from [18]). For higher incidence angles the major part of the radiation is absorbed in the mirror's surface. However, it is possible to observe that the reflectivity at small grazing angles is sufficiently high for such mirror to be employed as a grazing incidence condenser. Moreover, out of presented, commonly used materials the nickel exhibit one of the highest reflectivity. It is also relatively easy to deposit it using i.e. electroplating.

2.2 Fresnel zone plate objective for SXR microscopy

A 200 nm thick Ti filter (Lebow Company, California, USA) allows to cut off H-like line almost completely, so the main part of the energy is in a single line (λ= 2.88 nm), in the "water window" spectral range. A 400 nm thick Si_3N_4 Fresnel zone plate (FZP) objective (from Zoneplates Ltd., UK) was used to magnify image of the sample onto a back-illuminated SXR-sensitive CCD camera in transmission mode. The FZP, 250 μm in diameter, has the outer zone width of 30 nm and the focal length at 2.88 nm wavelength f = 2.6 mm. The magnified image was acquired with a back-illuminated CCD camera, sensitive to SXR (Andor, i-Kon). The numerical aperture of the zone plate, NA_{ZP} = 0.048, is matched to the exit numerical aperture of the condenser, NA_{C_out}, limiting the loss of photons and providing incoherent illumination, since σ = NA_{C_out}/NA_{ZP} = ~1. The image plane is located 574 mm from the zone plate. Thus, geometrical magnification of the system is ~220× allowing to achieve a half-pitch spatial resolution of ~60 nm [19].

The diffraction efficiency of a diffractive lens can be calculated from the refractive index, *n*. Fig. 2b) reports different diffraction efficiency curves in the first diffraction order for different materials at λ=2.88 nm. The calculation is based on [20]. As it is possible to observe, that lenses made of silicon and nickel give the highest efficiency of the diffraction grating in the first order. However, silicon is a very soft material, extremely difficult to manufacture in thin layers for a lens, whereas the procedure for making nickel lens require additional steps in the lithographic process, such as sputtering and etching. Taking into account these motivations, a Si_3N_4 lens having a thickness of 400 nm is a good candidate for zone plate dedicated for the SXR microscope, giving a maximum diffraction efficiency, which at λ= 2.88 nm is of about 15%.

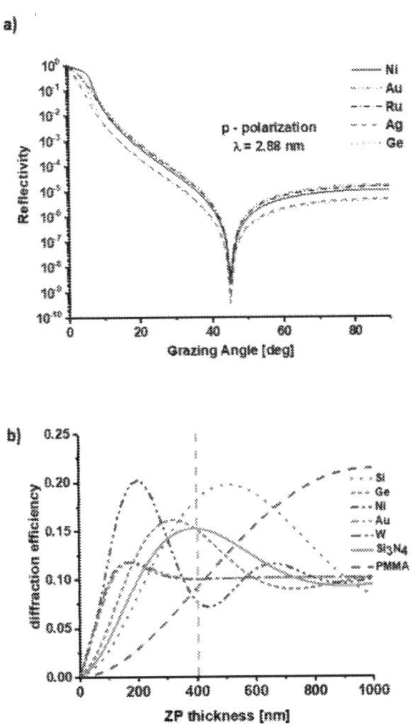

Fig. 2. a) Typical reflectivity values for p (parallel) polarization of $\lambda = 2.88$ nm radiation incident on different surface and b) Fresnel zone plate diffraction efficiency in the first diffraction order for different materials commonly used for its fabrication, as a function of the thickness of the lens, at $\lambda = 2.88$ nm.

3 Results

3.1 Source optimization for SXR radiation

3.1.1 Measurement of the plasma spectral distribution

The SXR spectrum obtained with and without a Ti filter was measured using a transmission grating spectrometer equipped with a free-standing grating (5000-lines/mm), 33 μm entrance slit and the CCD camera. The grating was positioned ~770 mm from the plasma, the entrance slit was placed 4 mm from the grating, and distance between the grating and the CCD was 425 mm. To improve the spectrometer resolution, a 20 μm wide auxiliary slit was placed 300 mm from the plasma. The inverse relative bandwidth of the spectrometer was estimated from spectrum to be $\lambda/\Delta\lambda \sim 70$ at $\lambda = 2.88$ nm wavelength. The N_2 measured spectrum (solid line in Fig.3a) was quasi-monochromatic, consisting of N^{5+}:1s2-1s2p He-like nitrogen line at $\lambda = 2.88$ nm and N^{6+}1s-2p H-like nitrogen line at $\lambda = 2.48$ nm, according to L-α Ti edge at $\lambda \sim 2.7$ nm allows to filter out H-like line almost completely; thus, most energy of the beam illuminating the sample resides in a single line at $\lambda = 2.88$ nm in the "water window" spectral range, that represents the N_2 plasma radiation employed for imaging. The plasma distribution was acquired, for comparison, employing three different thicknesses of a free-standing Ti filter, i.e. 100 nm (dashed line), 200 nm (dotted line) and 500 nm (solid line). Results are plotted in Fig. 3a). As it is possible to observe, the 500 nm of Ti filters out completely the N_2 emission at 2.48 nm. However, subsequent acquisitions of images using this filter, have demonstrated that the photon flux is too low to obtain a good signal to noise ratio. A reasonable compromise in order to achieve optimal imaging conditions is given by the employment of a 200 nm thick Ti filter.

3.1.2 Photon flux measurements

The source photon flux was measured at the sample plane using an AXUV100 silicon p-n junction photodiode (from, IRD Inc., California, USA), active area 10 x 10 mm^2. The detector does not have any filters deposited, thus, due to broad spectral emission from the plasma, a 200 nm Ti filter was used, placed 31 cm from the detector. The detector signal was stored using a 4 GHz bandwidth Tektronix Inc. (Oregon, USA), DPO 70404 oscilloscope and later integrated numerically. The circuit was used to reverse bias the detector up to a voltage of ~ - 28 V. The photon flux acquired was equal to $(7.9 \pm 0.2) \cdot 10^9$ photons/pulse, which corresponds to energy of 561 ± 17 nJ/pulse at 2.88 nm wavelength.

3.1.3 Visualization of the plasma spatial distribution

The spatial distribution of nitrogen plasma in the "water window" spectral range was obtained using a pinhole camera equipped with 32 μm in diameter pinhole. A 200 nm Ti filter was mounted in order to assess the plasma size in the "water window" spectral range. The nitrogen plasma FWHM size was measured to be (HxV) 0.47×0.31 mm^2. The plasma profile and its distribution are shown in Fig. 3b).

3.1.4 Condenser Alignment

In order to illuminate the test sample uniformly and symmetrically, the condenser was initially aligned employing the radiation of a laser diode (from the visible spectrum, 650 nm wavelength), focused by a spherical lens, in order to simulate the plasma generated by the Nd:YAG laser radiation. Because of the dispersion of the source lens, which has been adjusted to the wavelength of the Nd:YAG laser emission (1064 nm), it is necessary take into account a small shift of approx. 0.3 mm, between the position in which the plasma is generated by the laser and the focus position of the radiation from the laser diode. Such shift might affect the intensity and spatial distribution of the SXR radiation emerging out of the condenser. To avoid that, the condenser illumination pattern was imaged with a two-zone-plate arrangement. Such system allows to project images of the objects onto an image plane by adjusting the distance between the two FZP, although it requires large photon flux due to low throughput of the imaging system ~0.41%. As a result, the distance between two conjugated planes may be adjusted, as schematically depicted in Fig. 4a), overcoming geometrical restrictions of the vacuum system and allowing for visualization of the illumination pattern without the need to convert short wavelength radiation to the visible range using scintillators. Much more details about this techniques can be found in [21].

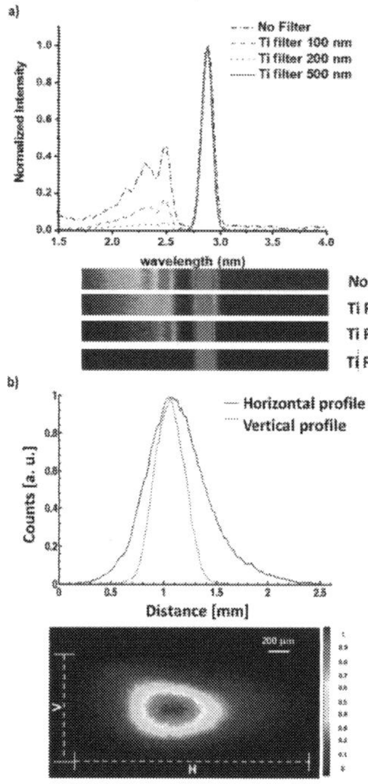

Fig. 3. a) Nitrogen plasma emission in the "water window" spectral range, unfiltered (dash-dotted line) and filtered with 100 (dashed line), 200 (dotted line) and 500-nm (straight line) thick Ti filter and b) SXR emission from nitrogen plasma generated by focused Nd:YAG laser pulses irradiating a double stream gas-puff target in the "water window" spectral range.

Condenser images, acquired at different positions along x and y axes, (with 20 SXR pulses, exposure time of 20 s and a bin of 4×4 pixels, to increase the signal-to-noise ratio) are shown in Fig. 4 b). The central picture (indicated with (0,0)) shows the condenser in an axial symmetric position and uniform illumination. It is possible to observe how the illumination becomes less uniform moving away from the central position.

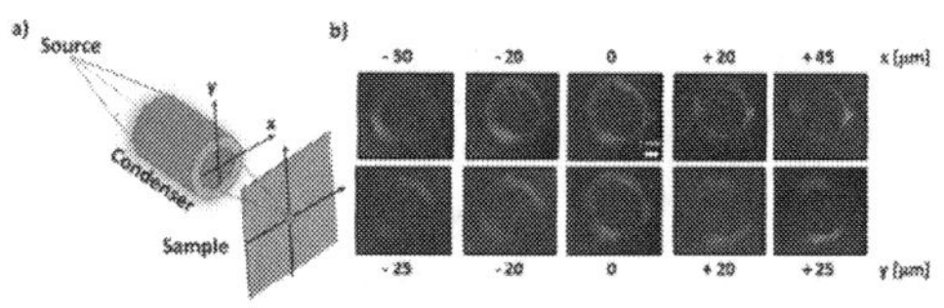

Fig. 4. a) Scheme of the condenser alignment and b) Condenser illumination patterns acquired at different distances from the central position (uniform illumination). The condenser was placed in a three-step motor and moved along x and y axes, with different shifts.

3.2 Imaging Results

Herein few applications employing the "water window" microscope are presented and discussed. The first sample in Figs.5 a-b) shows hippocampal neurons from E17 mouse embryos, cultured for on poly-D-lysine coated, 50 nm thick Si_3N_4 substrate. The neurons were imaged using a visible light microscope (40× objective, 400× magnification), Fig. 5 a) and the SXR microscope (sample to CCD magnification of 410×, exposure of 200 SXR pulses, 20 s), Fig. 5 b). The comparison shows a significant improvement of the spatial resolution employing the "water window" radiation where it is possible to observe that the high absorption coefficient at SXR wavelengths enhances the optical contrast, allowing to observe some features that are barely visible in Fig. 5 a). Indeed, in the SXR image it is possible to distinguish the neuron and dendrites that are blurred due to inferior resolution and phase-type imaging using the optical microscope. Same considerations can be made for the sample presented in Fig. 5 c-d), showing diatom cells deposited on top of 30 nm thick Si_3N_4 membrane. Also in this case small internal features are visible in the SXR image, in which is possible to obtain a better contrast respect to the optical image, distinguishing small features that are difficult to observe by visible light.

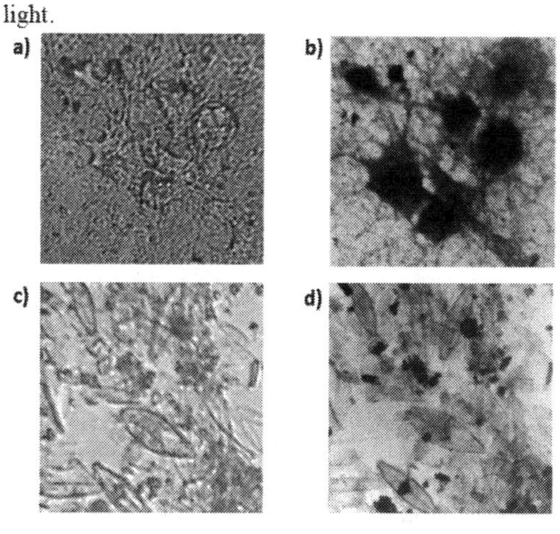

Fig. 5. Comparison of a) optical and b) SXR image of a hippocampal neurons from E17 mouse embryos; Diatoms sample imaged with c) visible light and d) SXR radiation.

4 Discussion and conclusions

We presented a SXR microscope, working in the "water window" spectral range. The system is compact in size and makes possible wider collaboration in a multidisciplinary context with biologists, chemists, and scientists from other disciplines locally, in the university laboratories or companies. Moreover, such system offers higher spatial resolution to the visible light microscopes [17], [22] and do not require sample preparation like SEM microscopes. Although the spatial resolution achieved cannot compete with the SEM, it has to be noted that the SXR microscope still operates using photons instead of electrons. This is an important characteristic, because electrons interact with matter differently than the photons, thus providing the possibility to obtain different information from the samples. Moreover, although their resolution is similar to STED in the visible light region [23], [24], they enable the acquisition of full field images, contrary to STED acquisition, which operates in scanning mode, and does not require any fluorescence markers or staining, which modify the sample to be imaged.

In conclusion, the SXR compact microscopes can be regarded as complementary imaging tools to already existing, well-established ones, providing additional information about investigated samples. Compact overall dimension, easy accessibility and simple operation are key parameters for compact microscopes, such as fast exposure and tens of nanometers spatial resolution. This open the possibility of widespread commercialization of such systems, in the near future. Moreover, different samples were imaged in the SXR spectral range, microscope, including material science, nanotechnology and biology [19], [25], showing applicability and potential usability of such a "water window" microscope to different areas of research.

This work is supported by the National Science Centre, Opus programmes, grant agreement number UMO-2015/17/B/ST7/03718 and UMO-2015/19/B/ST3/00435, the Education, Audiovisual and Culture Executive Agency (EACEA) Erasmus Mundus Joint Doctorate Programme Project No. 2012-0033 and from the European Union's Horizon 2020 research and innovation program, under Laserlab-Europe IV, grant agreement No. 654148. We would like to thank Dr. Michal Odstrcil from Paul Scherrer Institut, Villigen, Switzerland, for the hippocampal neurons from E17 mouse embryos sample preparation and Dr. E. Viaggiu, University of Tor Vergata, Rome (Italy), AlgaRes srl, for the preparation of diatoms sample.

References

1. G. Vaschenko, F. Brizuela, C. Brewer, M. Grisham, H. Mancini, C. S. Menoni, M. C. Marconi, J. J. Rocca, W. Chao, J. Liddle, E. H. Anderson, D. T. Attwood, A. V. Vinogradov, I. A. Artioukov, Y. P. Pershyn, V. V. Kondratenko, Opt. Lett. **30**, 2095 (2005).
2. E. S. Wyndham, M. Favre, M. P. Valdivia, J. C. Valenzuela, PPC2009 - 17th IEEE Int. Pulsed Power Conf. **688** (2009).
3. R. Lebert, W. Neff, D. Rothweiler, J. Xray. Sci. Technol. **6**, 107 (1996).
4. M. Benk, K. Bergmann, D. Schäfer, T. Wilhein, Opt. Lett. **33**, 2359 (2008).
5. C. Brewer, F. Brizuela, P. Wachulak, D. H. Martz, W. Chao, E. H. Anderson, D. T. Attwood, A. V Vinogradov, I. a Artyukov, A. G. Ponomareko, V. V. Kondratenko, M. C. Marconi, J. J. Rocca, C. S. Menoni, Opt. Lett. **33**, 518 (2008).
6. M. Berglund, L. Rymell, M. Peuker, T. Wilhein, H. M. Hertz, J. Microsc. – Oxford **197**, 268 (2000).
7. H. Legall, G. Blobel, H. Stiel, W. Sandner, C. Seim, P. Takman, D. H. Martz, M. Selin, U. Vogt, H. M. Hertz, D. Esser, H. Sipma, J. Luttmann, M. Hofer, H. D. Hoffmann, S. Yulin, T. Feigl, S. Rehbein, P. Guttmann, G. Schneider, U. Wiesemann, M. Wirtz, W. Diete, Opt. Express **20**, 18362 (2012).
8. M. Müller, T. Mey, J. Niemeyer, K. Mann, Opt. Express **22**, 23489 (2014).
9. The International Organization for Standardization, "ISO 21348 Definitions of Solar Irradiance Spectral Categories," **6–7** (2007).
10. C. Jacobsen, J. Kirz, Nat. Struct. Biol. **5**, 650 - 653 (1998)
11. G. McDermott, M. A. Le Gros, C. G. Knoechel, M. Uchida, C. A. Larabell, Trends Cell Biol. **19**, 587 (2009).
12. G. Poletti, F. Orsini, D. Batani, Solid State Phenom. **107**, 7 (2005).
13. W. Chao, P. Fischer, T. Tyliszczak, S. Rekawa, E. Anderson, P. Naulleau, Opt. Express **20**, 9777 (2012).
14. H. N. Chapman, A. Barty, M. J. Bogan, S. Ebastien, M. Frank, S. P. Hau-riege, S. Marchesini, B. W. Woods, S. Bajt, W. H. Benner, R. A. London, E. Plonjes, M. Kuhlmann, R. Treusch, S. Düsterer, T. Tschentscher, Jochen R. Schneider, E. Spiller, T. Möller, C. Bostedt, M. Hoener, D. A. Shapiro, K. O. Hodgson, D. van der Spoel, F. Burmeister, Magnus Bergh, Carl Caleman, G. Huldt, M. Marvin Seibert., F. R. N. C. Maia, R. W. Lee, A. Szöke., N. Timneanu, J. Hajdu, Nat. Phys. **2**, 839-843 (2006).
15. M. A. Le Gros, G. Mcdermott, B. P. Cinquin, J. Synchrotron Radiat. **21**, 1 (2014).
16. R. Ortega, S. Bohic, R. Tucoulou, A. Somogyi, G. Devèst, Anal. Chem. **76**, 309 (2004).
17. P. W. Wachulak, A. Torrisi, A. Bartnik, D. Adjei, J. Kostecki, L. Wegrzynski, R. Jarocki, M. Szczurek, H. Fiedorowicz, Appl. Phys. B **118**, 573 (2015).
18. CXRO X-Ray Interactions With Matter - database online: http://henke.lbl.gov/optical_constants/.
19. P. W. Wachulak, A. Bartnik, M. Skorupka, J. Kostecki, R. Jarocki, M. M. Szczurek, L. Wegrzynski, T. Fok, H. Fiedorowicz, Appl. Phys. B Lasers Opt. **111**, 239 (2013).
20. J. Kirz, J. Opt. Soc. Am. **64**, 301 (1974).
21. P. W. Wachulak, A. Torrisi, A. Bartnik, Ł. Węgrzyński, T. Fok, R. Jarocki, J. Kostecki, M. Szczurek, H. Fiedorowicz, J. Opt. **17**, 55606 (2015).
22. P. Wachulak, A. Torrisi, M. F. Nawaz, A. Bartnik, D. Adjei, Š. Vondrová, J. Turňová, A. Jančarek, J. Limpouch, M. Vrbová, H. Fiedorowicz, Microsc. Microanal. **21**, 1214 (2015).
23. K. Otomo, T. Hibi, Y. Kozawa, Microscopy **64**, 1 (2015).

24. J. N. Farahani, M. J. Schibler, L. A. Bentolila, Microsc. Sci. Technol. Appl. Educ. **2**, 1539 (2010).
25. A. Torrisi, P. Wachulak, M. Fahad Nawaz, A. Bartnik, D. Adjei, Š. Vondrová, J. Turňová, A. Jančarek, H. Fiedorowicz, Acta Phys. Pol. A **129**, 169 (2016).

Photoionized plasmas induced in molecular gases by extreme ultraviolet and X-ray pulses

Andrzej Bartnik*, *Wojciech* Skrzeczanowski, *Henryk* Fiedorowicz, *Przemysław* Wachulak, *Tomasz* Fok, *Łukasz* Węgrzyński, *Roman* Jarocki

Institute of Optoelectronics, Military University of Technology, Kaliskiego 2, 00-908 Warsaw, Poland

Abstract. In this work a laser-produced plasma (LPP) source was used to create low temperature plasmas. An extreme ultraviolet and soft X-ray (EUV/SXR) radiation pulse was used for ionization of molecular gases, injected into a vacuum chamber synchronously with the EUV/SXR pulse. Energies of photons exceeding 100 eV were sufficient for dissociative ionization, ionization of atoms or even ions. The resulting photoelectrons had also enough energy for further ionizations or excitations. Time resolved UV/VIS spectra, corresponding to single charged ions, molecules and molecular ions, were recorded. For spectral lines, corresponding to radiative transitions in F II and S II ions, electron temperature was calculated based on a Boltzmann plot method. Numerical simulations of the molecular spectra were fitted to the experimental spectra allowing for determination of vibrational and rotational temperatures.

1 Introduction

Low temperature plasmas are widely used in different technologies [1]. Probably the most important is application of plasma processes in the microelectronics industry. For the microfabrication of integrated circuits, several fabrication steps are typically plasma based [2]. The other very important application is deposition of thin films [3,4]. This type of deposition allows for production of layers having unique properties like hydrogenated silicon layers or diamond-like carbon. Another important process is plasma etching [5,6]. Reactive etching is connected with formation of volatile products like FW_6, GeF_4 or SiF_4. This effect can be thus utilized for reactive plasma etching of tungsten, germanium, silicon or other materials containing these elements like SiC [7] or glass [8]. Plasma etching is widely used for manufacturing of micro-electromechanical systems (MEMS), microfluidic devices, optical diffractive components, X-ray diffractive or refractive optics or in different kinds of nanotechnologies. Plasma treatment can be also employed for surface activation and functionalization of polymers [9]. Careful and appropriate selection of the types of plasma source could introduce various functional groups on polymer surface in order to change wettability, adhesion characteristics, and biocompatibility of polymers by directly changing the surface chemical compositions [10-12]. Crosslinking of the polymer chains results in increasing of the molecular weight of polymers and could cause polymer surface to establish three-dimensional molecular networks, by which the polymer layer became an insoluble material [13]. Fluorination of polymers by plasma treatment results in hydrophobic surfaces without degradation of their bulk properties [14,15]. The SF_6 plasma treatment of silicon surface can convert it to so called black silicon with minimized reflection losses. Low reflectance of the black silicon is very important for optoelectronic devices, in particular, for high efficiency solar cells [16]. Fluorination was also employed to improve properties of amorphous carbon films resulting in increase of the film roughness and hydrophobicity [17]. Apart from that, SF6 plasma treatment can be employed for functionalization of graphene and carbon nanotubes [18,19]. Low temperature plasmas are also utilized for spectrochemical analysis. In particular it was shown by various authors that the excitation of atomic or molecular species in cold plasmas is a powerful method for such analysis. The method was used for a large variety of analytical tasks such as the determination of noble metals [20], the determination of the rare earth elements [21] or the determination of Hg with the aid of continuous cold vapour generation [22]. In this case the sample aerosol was introduced into the low temperature plasma and the emission spectra of the excited species were measured and analysed.

For creation of plasmas used in these technologies plasma generators based on electrical discharge, working in a stationary regime, are used. They allow to create plasmas of low temperature ($T_e \sim 10^4$K) and low electron density ($n_e \sim 10^{11} - 10^{13}$cm^{-3}) [23]. Even in plasmas produced under atmospheric pressure the electron density, hence, also ion density (n_i) does not exceed these values. In our previous works, an alternative method of low temperature plasmas creation, by irradiation of gases with extreme ultraviolet (EUV) or soft X-ray (SXR) pulses, was proposed. A possibility to create plasmas of significantly higher electron density,

* Corresponding author: andrzej.bartnik@wat.edu.pl

comparing to plasmas produced using standard generators, even exceeding $10^{18}cm^{-3}$ [24], was demonstrated. All these works concerned spectral investigations of atomic or ionic species. In this work EUV induced plasmas, created by irradiation of molecular gases, were investigated. Time resolved spectral measurements in the optical range were performed. Based on ionic and molecular spectra, the electron, vibrational and rotational temperatures were estimated.

2 Experimental arrangement

In our experiments, a 10-Hz laser produced plasma (LPP) EUV/SXR source, based on a double-stream gas-puff target, irradiated with the 10ns/5J Nd:YAG laser pulses, was used. The target was formed by pulsed injection of xenon gas into a hollow stream of helium by employing an electromagnetic valve system, equipped with a double nozzle set-up. The laser focusing conditions and plasma parameters were adjusted such as to obtain maximum emission in the EUV spectral region. The most intense emission was in the relatively narrow spectral region centred at λ ~ 11 ± 1 nm, however, significant part of the radiation was emitted in a broad spectral range λ ~ 1- 20 nm. The LPP source was used to ionize a small portion of gas injected into the vicinity of the laser plasma. The gas was injected using an auxiliary gas puff valve synchronized with laser pulses. Its density in the interaction region could be regulated within the range of the order of 1 ÷ 10% of the atmospheric density. Ionization of the gas resulted in creation of low temperature plasmas. Spectral investigation of these plasmas were performed in a UV/VIS range (λ = 200-780 nm), using an Echelle Spectra Analyzer ESA 4000. To avoid a strong radiation from the laser plasma, it was screened by a specially prepared metallic plate mounted between the plasma and the spectrograph. Apart from that the measurements were performed with a temporal resolution which allowed to register exclusively spectral lines originating from the EUV induced plasmas, and not from the expanding laser plasmas.

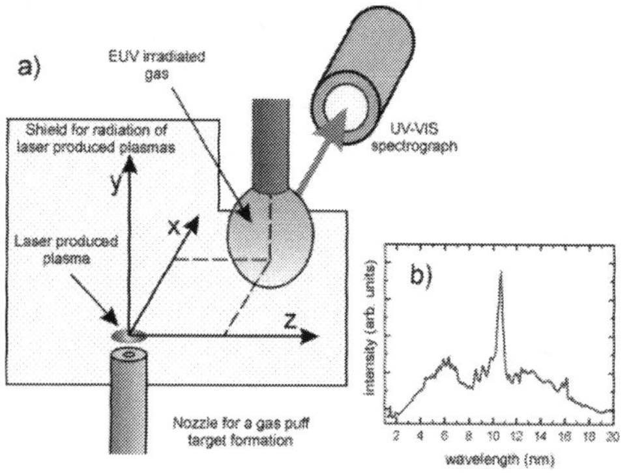

Fig. 1. Creation of EUV induced low temperature plasmas: a) experimental arrangement, b) spectrum of the LPP radiation pulses

A schematic view of the experimental set-up together with the spectrum of the LPP source is presented in Fig.1.

3 Experimental results

Irradiation of molecular gases, injected into the interaction region, by EUV/SXR photons resulted in numerous processes including dissociative ionization or creation of molecular ions. Photoelectrons released during these processes have sufficient energies for further ionization or excitation of atoms, ions and molecules. This work is devoted to spectral investigation of low temperature plasmas induced in a nitrogen and SF6 gas. Spectra were acquired for different delays according to the ionizing pulse. The most intense spectrum, recorded at a time delay $t = 150\,ns$ after the gas irradiation by the EUV/SXR pulse, contains series of lines corresponding to vibrational emission bands of N_2 and N_2^+ species. In Fig.2 two parts of the N_2, $C^3\Pi_u - B^3\Pi_g$ electronic system, recorded in the UV range, are presented.

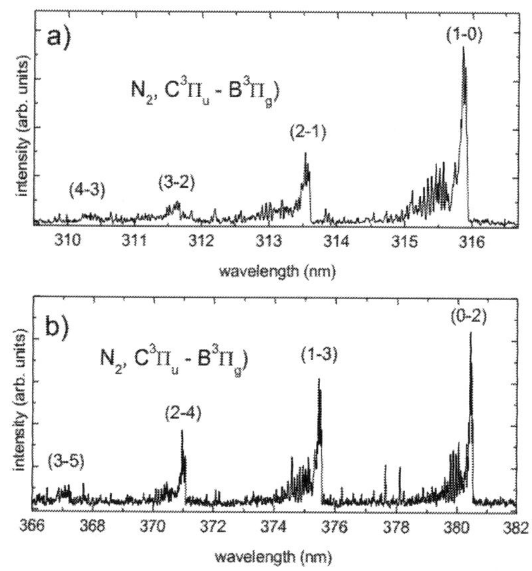

Fig. 2. Vibrational-rotational N_2 spectra corresponding to the $C^3\Pi_u - B^3\Pi_g$ electronic system: a) $\Delta\upsilon = -1$, b) $\Delta\upsilon = 2$

Due to sufficiently high resolution of the spectrometer, $\lambda/\Delta\lambda \approx 20000$, not only the vibrational but also rotational structure of the spectra corresponding to the transitions within the $C^3\Pi_u - B^3\Pi_g$ system, are visible. The band heads were identified based on the Ref. 25. Several other heads with the most intense one corresponding to the (0-0) transition at the wavelength λ = 337.1 nm were also detected.

As mentioned above, the spectrum contains also lines corresponding to transitions in the molecular ions N_2^+. Their spectral intensity is similar or even higher comparing to the N_2 lines. Several vibrational emission bands of the $B^2\Sigma_u^+ - X^2\Sigma_g^+$ electronic system with a change of the vibrational quantum number $\Delta\upsilon =$

$-1, 0, 1$, were recorded. Vibrational bands with $\Delta\upsilon = -1$ of the $B^2\Sigma_u^+ - X^2\Sigma_g^+$ electronic system and bands with $\Delta\upsilon = 1$ of the $C^3\Pi_u - B^3\Pi_g$ system are partially overlapping, hence, not very convenient for analysis. On the other hand, two other bands of the $B^2\Sigma_u^+ - X^2\Sigma_g^+$ system, with $\Delta\upsilon = 0, 1$ are well separated from the N_2 spectrum. Hence, this part of the N_2^+ spectrum could be used for estimation of the vibrational and rotational temperatures. The estimation was based on simulation of the spectrum employing the LIFBASE code [26]. Simulations were performed for several vibrational and rotational temperatures (T_v and T_r respectively). The best fit was obtained for $T_v = 3000\ K$ and $T_r = 200\ K$. The corresponding experimental and calculated spectra are presented in Fig.3.

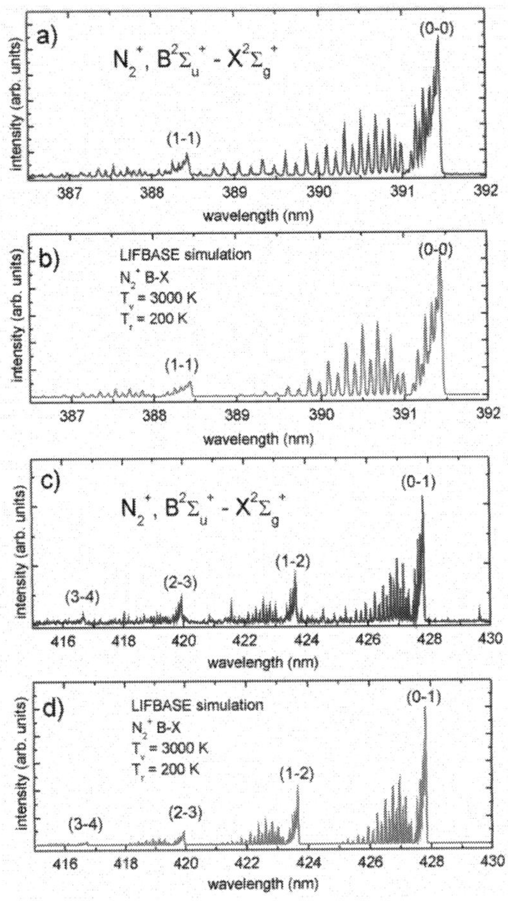

Fig. 3. Vibrational-rotational N_2^+ spectra corresponding to the $B^2\Sigma_u^+ - X^2\Sigma_g^+$ electronic system: a) experimental, $\Delta\upsilon = 0$; b) $\Delta\upsilon = 0$, numerical simulation ; c) experimental, $\Delta\upsilon = 1$; d) $\Delta\upsilon = 1$, numerical simulation

Spectra recorded for different experimental conditions were dominated by the lines corresponding to the N_2 and N_2^+ molecular bands. Lines corresponding to radiative transitions in nitrogen atoms were not detected. On the other hand two spectral lines corresponding to singly charged ions $2s^22p3s(^1P_1^o) - 2s^22p3p(^1D_2)$ and $2s^22p3s(^3P_2^o) - 2s^22p3p(^3P_2)$ at the wavelengths $\lambda = 399.5\ nm$, $\lambda = 463.05\ nm$ respectively, were revealed. Their intensities, however, are an order of magnitude lower comparing to the most intense molecular lines. According to the NIST database [27] these lines are the most intense among the N I and N II lines. It can be thus concluded that the other lines, including the N I lines, were too weak to be detected. It means that except the molecular ionization and excitation, the dissociative ionization or dissociation with further ionization of the nitrogen molecules also took place. It should be expected due to high energy of the driving photons, capable even for the inner-shell ionization.

In principle, based on the intensity ratio of two N II lines it would be possible to calculate the plasma electron temperature. On the other hand intensities of the lines can be affected by self-absorption, especially in case of the strongest lines. It means that an error of the temperature estimation, performed this way, could be significant. Much better is to utilize a Boltzmann plot method [28,29] based on relative intensities of multiple lines corresponding to radiative transitions between the levels i and k. The plot is constructed from individual points with the values of $\ln(I_{ki}\lambda_{ki}/A_{ki}g_k)$ versus the upper level energy E_k. Quantities I_{ki}, λ_{ki}, A_{ki}, g_k are as follows: the line intensity, its wavelength, the radiative transition probability and the statistical weight of the upper level, respectively. In case of LTE (local thermodynamic equilibrium) or at least partial LTE (pLTE) and spectral lines originating from the same ion species a straight line can be fitted to the points and the slope $1/kT$ yields the electron temperature. The method could not be employed for the nitrogen plasma, due to lack of multiple N I or N II lines, hence, the measurements were performed for the plasma induced in other molecular gas, namely SF_6. In this case numerous lines originating from F II and S II ions were detected. In Fig.4 Boltzmann plots constructed for these lines are presented. The electron temperature estimated from these plots is of approximately $T_e \sim 2 \cdot 10^4 K$.

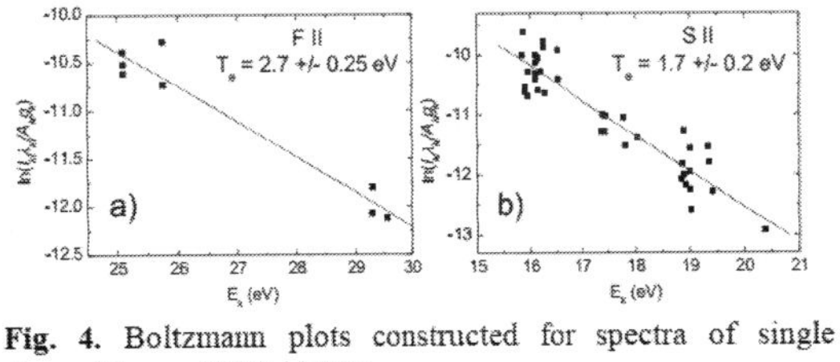

Fig. 4. Boltzmann plots constructed for spectra of single charged ions: a) F II, b) S II

Similar values were obtained in our earlier experiments with plasmas induced in rare gases [30,31]. It can be thus assumed that also for the nitrogen plasma the electron temperature is of the same order. In this case the electron temperature is almost an order of magnitude higher comparing to the vibrational temperature.

In Fig.5 a temporal dependence of relative intensities of the (0-0) emission bands corresponding to the $C^3\Pi_u - B^3\Pi_g$ and $B^2\Sigma_u^+ - X^2\Sigma_g^+$ systems are presented. At a sub-image a time profile of a typical EUV/SXR pulse, used for creation of the low temperature plasmas, was added. An FWHM (full width at half maximum) of the EUV pulse, measured using a fast AXUV HS1 detector, is $\tau = 11\,ns$. This value is comparable to the time duration of the driving laser pulse.

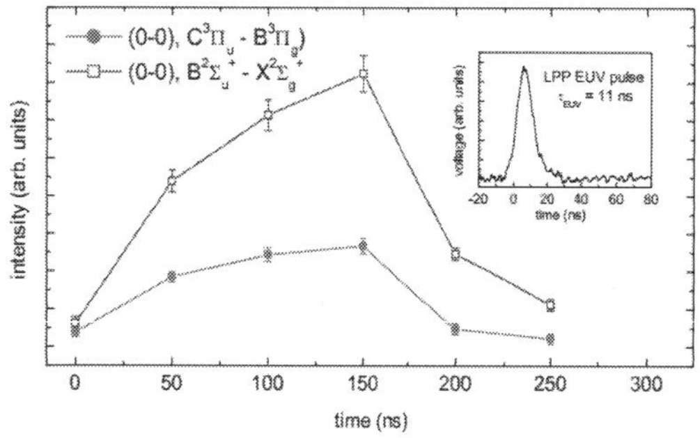

Fig. 5. Temporal dependence of relative intensities of emission bands corresponding to N_2 and N_2^+molecules. Time profile of the driving LPP EUV pulse at the sub-image.

As can be noticed from Fig.5 a time of the optical emission from the nitrogen plasma is much longer, over an order of magnitude longer comparing to the driving pulse. Apart from that, intensity of the UV-VIS radiation emitted from the plasma increases within a time of approximately $t_{N_2}{\sim}150\,ns$. It means that the molecular electronic excitations are driven mostly by collisional processes following the photoionization phase. Similar temporal dependence, presented in Fig.6 was obtained for the emission line corresponding to the $2s^22p3s(^1P_1^o) - 2s^22p3p(^1D_2)$ transitions.

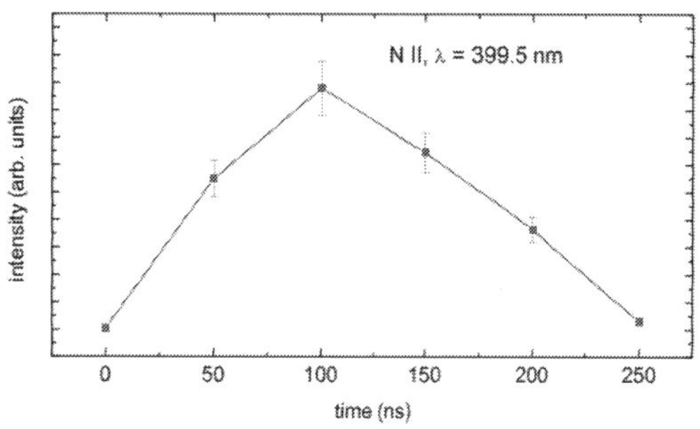

Fig. 6. Temporal dependence of relative intensities of the $2s^22p3s(^1P_1^o) - 2s^22p3p(^1D_2)$ emission line

In this case, however, the intensity maximum is for the time $t_{N\,II}{\sim}100\,ns$. It is reasonable taking into account that the dissociative ionization potential of the nitrogen is relatively high, $I_{N\,I} \sim 24.3\,eV$ and an additional energy exceeding 21 eV has to be delivered for creation of the $2s^22p3p(^1D_2)$ excited state. These processes require a presence of electrons with relatively high energy, an order of magnitude exceeding the average electron energy corresponding to the electron temperature ($T_e{\sim}2\,eV$). Number of such electrons initially increases as a result of photoionization, Auger emission and further collisional ionization but after a finite time decreases, due to thermalization and non elastic collisions with atoms or molecules. From the temporal dependence in Fig.6 this time can be estimated to approximately 100 ns.

4 Summary

In this work first results of investigations of low temperature plasmas produced by irradiation of molecular gases with intense EUV/SXR ionizing pulses, are presented. The gases were delivered into the interaction region synchronously with the ionizing pulses. The resulting low temperature plasmas were investigated using the UV/VIS spectroscopy. Emission lines corresponding to excited states in singly charged ions, molecular ions and neutral molecules were detected. Numerical simulations of the molecular spectra allowed to determine vibrational and rotational temperature of the nitrogen plasma. For spectral lines corresponding to radiative transitions in F II and S II ions, the electron temperature was calculated, based on the corresponding Boltzmann plots. Additionally temporal evolution of the emission from ionic and molecular species was presented. It was shown that the lifetime of the EUV induced plasma is an order of magnitude longer comparing to the time duration of the driving pulse.

This work was supported by the National Science Centre, Poland, grant agreement no. UMO-2016/23/B/ST7/00949, and partially by European Union's Horizon 2020 Programme (LASERLAB-EUROPE) grant agreement no. 654148

References

1. A. Bogaerts, E. Neyts, R. Gijbels, J. van der Mullen, Spectrochimica Acta Part B **57,** 609-658, (2002)
2. M.A. Lieberman, *Advanced Technologies Based on Wave and Beam Generated Plasmas*, NATO Science **67,** 1–22, (1999),
3. M. Kumar, Y. Ando, JNN **10**, 3739-3758 (2010)
4. M. Konuma, *Film Deposition by Plasma Techniques*, Springer, New York, 1992
5. D.M. Manos, D.L. Flamm, *Plasma Etching: An Introduction*, Academic Press, New York, 1989
6. S. Samukawa, T. Mieno, Plasma Sources Sci. Technol. **5,** 132–138, (1996)
7. N.O.V Plank, M.A. Blauw, E.W.J.M. van der Drift, R. Cheung, J. Phys. D **36**, 482–487, (2003)
8. L. Lallement, C. Gosse, C. Cardinaud, M.-C. Peignon-Fernandez, A. Rhallabi, Journal of Vacuum Science & Technology A **28**, 277 (2010)
9. S. Yoshida, K. Hagiwara, T. Hasebe, A. Hotta, Surface & Coatings Technology **233**, 99–107 (2013)
10. K.R. Kull, M.L. Steen, E.R. Fisher, J. Membr. Sci. **246**, 203, (2005)

11. J.N. Lai, B. Sunderland, J.M. Xue, S. Yan, W.J. Zhao, M. Folkard, B.D. Michael, Y.G. Wang, Appl. Surf. Sci. **252,** 3375, (2006)
12. D. Hegemann, H. Brunner, C. Oehr, Nucl. Instrum. Methods Phys. Res. B Beam Interact. Mater. Atoms **208**, 281, (2003)
13. S. Tajima, K. Komvopoulos, Appl. Phys. Lett. **89**, 124102 (2006)
14. E.C. Rangel, W.C.A. Bento, M. E. Kayama, W.H. Schreiner, N.C. Cruz, Surf. Interface Anal. **35**, 179–183 (2003)
15. R.Y. Korotkov, T. Goff, P. Ricou, Surface & Coatings Technology **201**, 7207–7215 (2007)
16. Y. Xia, B. Liu, S. Zhong, C. Li, Journal of Electron Spectroscopy and Related Phenomena **184**, 589–592 (2012)
17. N.M.S. Marins, R.P. Mota, R.Y. Honda, P.A.P. Nascente, M.E. Kayama, K.G. Kostov, M.A. Algatti, N.C. Cruz, E.C. Rangel, Surface & Coatings Technology **206**, 640–645 (2011)
18. M. Baraket, S. G. Walton, E. H. Lock, J. T. Robinson, and F. K. Perkins, Applied Physics Letters **96**, 231501 (2010)
19. A.J. Barlow and J.S. Quinton, *ICONN 2010 - Proceedings of the 2010 International Conference on Nanoscience and Nanotechnology*, IEEE Xplore, 120-123 (2010)
20. F. Liang, D.X. Zhang, Y.H. Lei, H.Q. Zhang, Q.H. Jin, Microchem. J. **52**, 181-187 (1995)
21. Y. Duan, Y. Li, X. Tian, H. Zhang, Q. Jin, Anal. Chim. Acta **295**, 315-324 (1994)
22. J.F. Camun-Aguilar, R. Tereiro-Garcia, J.E. Sanchez-Uria, Spectrochim. Acta Part B **49**, 475-484 (1994)
23. K. Ogura, H. Yamada, Y. Sato, Y. Okamoto, Appl. Spectrosc., **51**, 1496–1499 (1997)
24. A. Bartnik, P. Wachulak, T. Fok, H. Fiedorowicz, T. Pisarczyk, T. Chodukowski, Z. Kalinowska, R. Dudzak, J. Dostal, E. Krousky, J. Skala, J. Ullschmied, J. Hrebicek, T. Medrik, Physics of Plasmas **22**, 4, 043302 (2015)
25. A. Lofthus, P.H. Krupenie, J. Phys. Chem. Ref. Data, **6**, 113-307 (1977)
26. J. Luque and D.R. Crosley, *SRI International Report MP 99-009* (1999)
27. http://physics.nist.gov/PhysRefData/ASD/lines_form.html
28. C. Aragón, J.A. Aguilera, Spectrochimica Acta Part B **63**, 893–916 (2008)
29. S. S. Harilal, Applied Optics, 43, 3931-3937 (2004)
30. A.Bartnik, P. Wachulak, H. Fiedorowicz, W. Skrzeczanowski, Physics of Plasmas **23**, 043512 (2016)
31. A. Bartnik, W. Skrzeczanowski, P. Wachulak, I. Saber, H. Fiedorowicz, T. Fok, Ł.Węgrzyński, Laser and Particle Beams **35**, 42-47 (2017)

Characterization of X-ray emission from laser generated plasma

Antonino Cannavò[1,*], *Lorenzo* Torrisi[1], *Giovanni* Ceccio[1], *Mariapompea* Cutroneo[2], *Lucia* Calcagno[3], *Antonella* Sciuto[4] and *Massimo* Mazzillo[5]

[1]Dottorato di Ricerca & Dip.to di Scienze Fisiche MIFT, Università di Messina, V.F.S. D'Alcontres 31, 98166 S. Agata(ME), Italy
[2]Physics Institute, ASCR, 25068 Rez, Czech Republic
[3]Dipartimento di Fisica & Astronomia, Università di Catania, V. S. Sofia 64, 95123 Catania, Italy
[4]CNR-IMM, VIII strada n , 95121 Catania, Italy
[5]ST-Microelectronics, Str.le Primosole 50, 95121 Catania, Italy

Abstract. X-ray emission from laser generated plasma was studied at low (10^{10} W/cm^2) and high (10^{18} W/cm^2) intensity using ns and fs laser, respectively. Plasma characteristics were controlled trough the laser parameters, the irradiation conditions and the target properties. The X-ray spectra were acquired using fast detection technique based on SiC diodes with different active regions. The X-ray yield increases with the atomic number of the target, both at low and high intensity, and a similar empirical law has been obtained. The X-ray emission mechanisms from plasma are correlated to the plasma temperature and density and to the Coulomb charge particle acceleration, due to the charge separation effects produced in the non-equilibrium plasma. Functional dependences, theoretical approaches and interpretation of possible mechanism will be presented and discussed.

1 Introduction

When a high intensity laser pulse (I = 10^{10}-10^{18} W/cm^2) is focused on the surface of a target placed in vacuum the generated plasma acts as source of photons, from IR to X-rays regions, high energetic and thermal electrons and ions. So the possibility to obtain very compact X-rays source with high time and space resolution has given a great impulse in this research field.
From the theoretical point of view the laser-matter interaction and the consequent X-rays production has been widely reported in literature [1]. There are three fundamental emission processes due to free-free, free-bound and bound-bound electrons transitions. In the first case, the emission in a continuum spectrum is due to the interaction between electrons and the Coulomb potential of ions, the so called bremsstrahlung effect. The second process leads again to the emission of a continuum spectrum and is due to the re-combination phenomena through the passage between an initial free electron state to a bound electron state. The last mechanism produces line spectrum due to transition between discrete energy levels [2]. The ratio in which Bremmstrahlung or recombination phenomena occurs, is strongly influenced by the atomic number of the target. In particular, the ratio of the spectrally integrated recombination (W_r) and Bremsstrahlung (W_B) intensities:

$$\frac{Wr}{W_B} = 2.4\,\frac{Z^2}{E_H}k_BT \qquad (1)$$

in which E_H=13.6 eV is the Hydrogen ionization energy, Z is the atomic number and k_BT is the plasma temperature. For low Z and/or high temperature plasmas, the Bremsstrahlung emission overcomes the recombination emission. Concerning the X-rays pulse duration, the overall emission is comparable with the laser pulse but changes according to the considered spectral range: the higher is the energy of the considered photons the shorter is the duration of the plasma emission. Finally the peak of X-ray emissivity is located in the plasma region which combines both high temperature and high electron density, typically just beyond the critical surface density.

2 Experimental set-up

The experiments were carried out using the so called time of flight technique (TOF) in which the radiations energy can be determined by the time to cross a known distance. The electronic chain is reported in Fig. 1a and consists in a fast storage oscilloscope (Tektronix TDS5104B, 5 Gs/s sampling rate) connected to the detector through a coupling capacitance C while the detector is biased independently via the resistance R.
Choosing the appropriate values for R and C, which determine the decay constant of the electronics, it is possible to put in evidence the very fast component of the spectrum, essentially due to the prompt photon and electron emissions, or to the slow component due to ions detection. There are many advantages in the use of the TOF technique. First of all, it gives the possibility to

* Corresponding author: acannavo@unime.it

 https://doi.org/10.1051/epjconf/201816703004

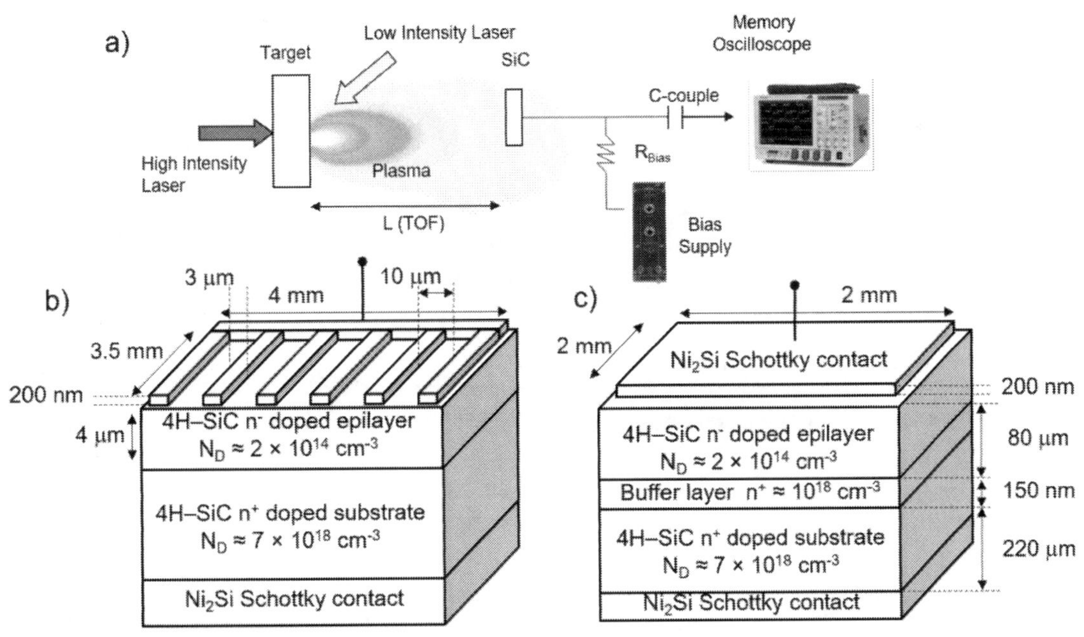

Fig. 1. Description of the TOF technique set-up a) and of the geometry of the SiC AZ 80 b) and SiC IAZ4 c) detector.

monitor very well high particle flux, without any problems concerning pile-up effects. This technique gives also time information because it is based on the time of arrival of radiation to the detector.

Finally it can be applied in a wide range of laser intensities and conditions. In fact the results presented in this work have been acquired using three lasers operating in Italy, in Czech Republic and in France:

- Nd:YAG laser, 1064 nm, 3 ns, 10^{10} W/cm^2 @ UniME;
- Iodine laser, 1315 nm, 300 ps, 10^{16} W/cm^2 @ PALS;
- Ti Sapphire laser, 800 nm, 10^{18} W/cm2 @ CELIA.

The disadvantage to use TOF approach for photon measurements consists in the photon energy evaluation which is not immediate. In fact, while the energy of the particles can be determined with high precision from their time of flight from the target to the detector, photons arrive almost instantaneously to the detector, travelling with a velocity c along limited flight distances of the order of 1 m. However, some estimation about the mean energy of photon can be done by using different absorbers placed in front of the detector, as described in [3]. The used detectors are based on Silicon Carbide (SiC) material. Thanks to the physical properties of this semiconductor, the detectors have a high temperature operability (up to 1240 °C), fast response (900 cm^2/Vs electron mobility and 110 cm^2/Vs holes mobility), and they can be used with high electrical field having a breakdown electric field of 3 MV/cm. Moreover they have got peculiar characteristics to be employed in harsh environment such that which characterize the diagnostic of plasma generated by laser. SiC detectors have an high resistance to damage due to radiations thanks to the higher displacement energy (25 eV) with respect to Silicon (13.5 eV). But the key point is the fact that because of the high value of the bandgap these detectors cut the very high intensity visible radiation emitted from the laser-generated plasmas, which represents the main source of noise.

The used devices, schematized in Fig. 1b and 1c, were designed and fabricated in the STMicroelectronics and CNR-IMM clean room facilities and are widely described in previous paper [4].

They have different geometry of the top metallization layer and different thickness of the n-doped 4H-SiC sensible layer.

They give complementary information being the first device, labeled SiC IAZ4, able to detect low energy ions, electrons, soft X-rays and UV [5] due to the active region directly exposed to radiation and to the low thickness of the active region (4 µm). On the other hand the second detector, labeled SiC AZ 80, presents a thick top metallization (Ni_2Si, 200 nm thick) and a deep active region about 80 µm applying 600 V and for these reasons is very efficient in detecting hard X-rays and high ions energy [6].

3 Results

A typical TOF spectrum, acquired at PALS facility, is shown in Fig. 2.

It was obtained irradiating a polyethylene target (10 µm thick) and detecting through the SiC AZ 80 placed in forward direction, at 0° with a flight distance of 72 cm.

It presents a fast intense peak, called photopeak, due to hard X–rays and UV.

It is followed by an edge due to hot relativistic electrons and a little tail due to cold electrons.

Instead the last wide structure is due to protons, to carbon ions and at time of flight higher than 150 ns to the massive slower CxHy groups.

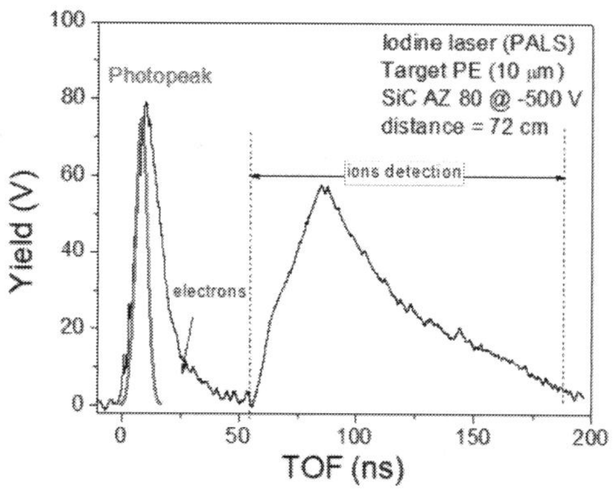

Fig. 2. TOF spectrum obtained at PALS facility irradiating PE target.

The two electron components can be distinguished using a Maxwell-Boltzmann distribution with two different temperatures:

$$f(v) = \sqrt{\left(\frac{m_e}{2\pi k_B T}\right)}\, 4\pi v_e^2 \exp\left[-\frac{m_e v_e^2}{2k_B T}\right] \quad (2)$$

where m and v_e are the electron mass and velocity, respectively.

Considering the ion acceleration due to charge separation effects in the plasma, the ions follow a Coulomb Boltzmann Shifted functions [7]:

$$f(v) = A\sqrt{\left(\frac{m}{2\pi k_B T}\right)^3}\, v^3 \exp\left[-\frac{m}{2k_B T}(v - v_k - v_c)^2\right] \quad (3)$$

where v represents the total velocity along the normal to the target surface and is due to the sum of thermal velocity v_{th}, adiabatic expansion velocity v_k and Coulomb velocity v_c.

3.1. Target parameters

The first analysis concerns the dependence of X-rays emission on the target composition. In the graphs of Fig. 3 are plotted the maximum yield as a function of the atomic number.

The spectra regard measurements performed both at low intensity (Nd:YAG laser in Messina) and at high intensity (Ti Sapphire laser at CELIA), were acquired using SiC IAZ 4 and SiC AZ 80, both in backward direction. In both situations a similar behavior was obtained and the experimental results follow the empirical law:

$$Y = 15.47 \times [1 - \exp(-0.0245 \times Z)] \quad (4)$$

$$Y = 3.40 \times [1 - \exp(-0.02 \times Z)] \quad (5)$$

These results can be explained taking into account that increasing the atomic number of the target more electrons are injected into the plasma as a consequence of the ponderomotive force. The higher electron density promotes the particles acceleration with a consequent increase of the photon emission and of the plasma temperature.

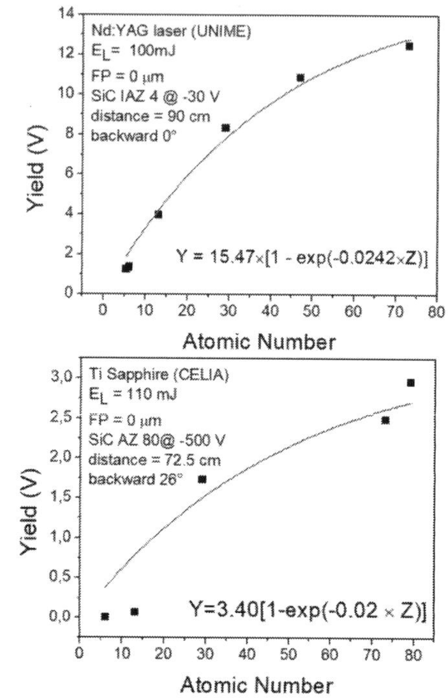

Fig. 3. Dependence of X-rays emission on the atomic number of the target.

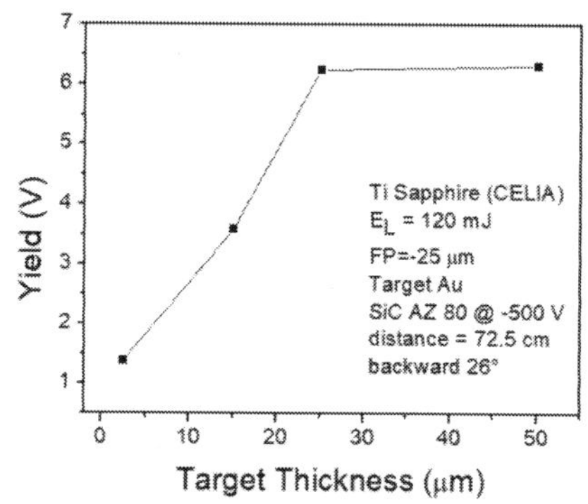

Fig. 4. Dependence of X-rays emission on the thickness of the target.

Also the influence of the target thickness was evaluated and the experimental results were reported in Fig. 4.

In particular these data were obtained at high intensity laser, detecting the photons emitted in backward direction from Au targets (2.48, 15, 25 and 50 μm thick). The graph shows an initial increase and then saturation.

In fact, increasing the thickness of the target the amount of material involved in the laser ablation processes also increases until, for thickness of about 30 μm, the saturation is reached. The saturation occurs when the target thickness is higher than the penetration depth of the laser into the target.

3.2. Laser parameters

Many plasma parameters depend on the $I\lambda^2$ factor, being I the laser intensity and λ the laser wavelength [8]. In order to test the influence also on the X-rays emissivity, the laser intensity has been modified changing separately the laser energy, the focal position and the pulse duration, as reported in Fig. 5, Fig. 6 and Fig. 7, respectively. The spectra were obtained irradiating, at low laser intensity, a multilayer target consisting in Carbon and Nickel, while at high laser intensity a thin target of Gold was used. In both cases the detectors were placed in backward direction. As expected by increasing the laser power, a linear increase in photons yield is also observed.

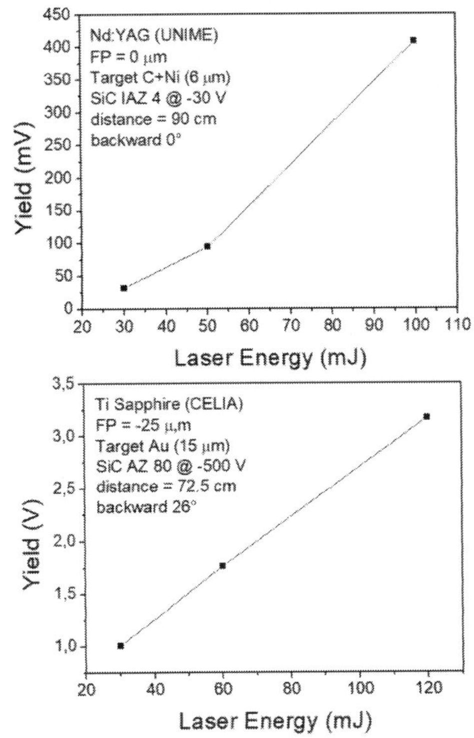

Fig. 5. Dependence of X-rays emission on the laser energy.

In fact, increasing the laser intensity, electrons can gain more energy leading to a greater emission of photons. Also the focal position plays an important role as can be seen in Fig. 6: the position in which laser is focused, with respect the target surface, not only determines the size of the laser spot but a proper choice of this distance can induce self-focusing effects and non-linear acceleration enhancement.

In this way it is possible to explain the presence of a second peak for FP = 50 μm, in addition to the maximum obtained for FP = 0 μm.

A similar dependence on the focal position was observed also for the ion yield as reported in literature [9].

Finally also the effects of pulse duration have been analyzed.

For low intensity laser, graph placed in the top of Fig. 7, were taken in consideration data from literature [10], in which the conversion efficiency was investigated as a function of the laser pulse duration, using a Nd:YAG laser

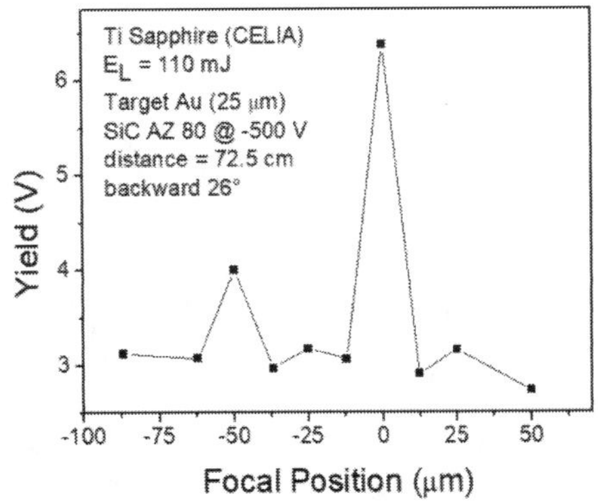

Fig. 6. Dependence of X-rays emission on focal position of the laser.

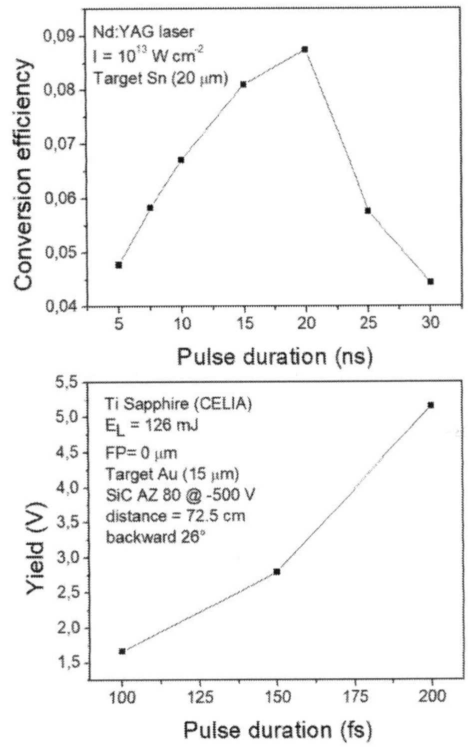

Fig. 7. Dependence of X-rays emission on the laser pulse duration.

with an intensity of 10^{13} Wcm^{-2}, focalized on a bulk Sn target. The photon emission increases with the pulse duration, because electron can interact for longer times with the laser pulse, gaining more energy. However, when the laser pulse becomes too long, the overall laser intensity decreases leading to the reduction of photon emission. For high intensity laser only the first region was investigated due to the constrains imposed by the laser configuration parameters. The lower and the higher pulse duration cannot be investigated with our method of analysis.

4 Conclusion

In this work the X-rays emission of laser generated plasmas was characterized, analyzing how it is influenced by the target composition and laser parameters. Concerning the target influence, our data performed on X-rays can be combined with other works [11]-[13] about ions. Also in this case the same behavior was obtained: higher is the atomic number and higher is the total yield. Moreover, the analysis of experimental data shows also an increase of plasma temperature and of equivalent accelerating voltage for ions [14], confirming the considerations done about the increase of the electron current injected into plasma.

Data concerning the laser influence can be interpreted considering the $I\lambda^2$ factor. For example the ponderomotive energy that for high laser intensities represents the energy transferred to electrons is given by [8]:

$$\varepsilon_p = \frac{e^2 E_0^2 \lambda^2}{16\pi^2 m_e c^2} = 9.33\times10^{-14}(I\lambda^2) \text{ in (eV)} \quad (6)$$

where e is the electron charge, E_0 the electric field of the electromagnetic wave, m_e the electron mass and c the speed of light in vacuum.

Many other plasma parameters depend on the $I\lambda^2$ factor, such as the shockwave pressure, the ion temperature, the emitted ion current and the electric field generated as a result of charge separation effects or the hot electron temperature, kT_{hot}, given by the relation [8]:

$$kT_{hot} = \left(\sqrt{1 + \frac{I\lambda^2}{2.8\times10^{18} W\mu^2/cm^2}} - 1\right) \cdot 511 \text{ (keV)} \quad (7)$$

All the presented results show that the target parameters as well as the laser parameters play a fundamental role in the generation of X-ray from laser-target interaction.

This work was supported by the "Research and Mobility" project of Messina University No. 74893496, scientifically coordinated by Professor L. Torrisi.

References

1. D. Giulietti and L. A. Gizzi, La rivista del Nuovo Cimento, **21**, (1998).
2. H.R. Griem, *Plasma Spectroscopy* (McGraw-Hill, New York, 1964).
3. D. Margarone, L. Torrisi, S. Cavallaro, E. Milani, G. Verona-Rinati, M. Marinelli, C. Tuvè, L. Làska, J Kràsa, M. Pfeifer et al., ECA F **31**, (2007) P-5.011 .
4. M. Mazzillo, G. Condorelli, M. E. Castagna, G. Catania, A. Sciuto, F. Roccaforte, V. Raineri, IEEE Photonics Technology Letters, **21** 23, (2009) 1782-1784.
5. A. Sciuto, L. Torrisi, A. Cannavò, G. Ceccio P. Musumeci, M. Mazzillo, L. Calcagno, Vacuum **131**, (2016) 170-179.
6. A. Cannavò, L. Torrisi, JINST **11**, (2016) C09005.
7. L. Torrisi, Radiations & Defects in Solids, **171** 1-2; (2016) 34-44.
8. L. Torrisi, Nukleonika, **60** (2), (2015) 207-212.
9. L. Làska, S. Cavallaro, K. Jungwirth, J. Kràsa, E. Krousky, D. Margarone, A. Mezzasalma, M. Pfeifer, K. Rohlena, L. Ryc, et al., Eur. Phys. J. D **54**, (2009) 487-492.
10. B. Atalay, N. Kenar, A. Demir, Turk J Phys **33**, (2009) 363-368.
11. J. Badziak, P. Parys, J. Wolowski, H. Hora, J. Kràsa, L. Làska, K. Rohlena, Optica Applicata, XXXV,(2005) 1.
12. G. Ceccio, JINST **11**, (2016) C04017.
13. L. Torrisi, F. Caridi, L. Giuffrida, Laser and Particles Beams, **29**, (2011) 29-37.
14. P. Musumeci, M. Cutroneo, L. Torrisi, A. Velyhan, M. Zimbone, L. Calcagno, Phys. Scr. T **161** (2014) 014021.

Effects induced by high and low intensity laser plasma on SiC Schottky detectors

Antonella Sciuto[1*], *Lorenzo* Torrisi[2] , *Antonino* Cannavò[2], *Massimo* Mazzillo[3] and *Lucia* Calcagno [4]

[1]CNR-IMM, VIII strada n°5, 95121 Catania, Italy
[2]Dip.to Scienze Fisiche-MIFT, Università di Messina, Agata (ME),Italy
[3]STMicroelectronics, Str.le Primosole 50, Catania, Italy
[4]Dip.to di Fisica ed Astronomia, Università di Catania, Italy

Abstract. Silicon-Carbide detectors are extensively employed as diagnostic devices in laser-generated plasma, allowing the simultaneous detection of photons, electrons and ions, when used in time-of-flight configuration. The plasma generated by high intensity laser (10^{16} W/cm^2) producing high energy ions was characterized by SiC detector with a continuous front-electrode, and a very thick active depth, while SiC detector with an Interdigit front-electrode was used to measure the low energy ions of plasma generated by low intensity laser (10^{10} W/cm^2). Information about ion energy, number of charge states, plasma temperature can be accurately obtained. However, laser exposure induces the formation of surface and bulk defects whose concentration increases with increasing the time to plasma exposure. The surface defects consist of clusters with a main size of the order of some microns and they modify the diode barrier height and the efficiency of the detector as checked by alpha spectrometry. The bulk defects, due to the energy loss of detected ions, strongly affect the electrical properties of the device, inducing a relevant increase of the leakage (reverse) current and decrease the forward current related to a deactivation of the dopant in the active detector region.

1 Introduction

Silicon Carbide (SiC) detectors are now-days widely employed for the monitoring of ionizing radiations. Due to the physical properties of the material, these detectors can be used at high temperatures and in high level radioactive ambient [1-3]. Irradiation of SiC detectors with neutrons, protons and heavy ions evidenced the higher radiation hardness of these devices with respect to Si detectors, principally thanks to higher atomic displacement energy of the SiC. It has been shown [4,5] that SiC detectors retain their detection properties (such as efficiency and resolution) up to irradiation doses as high as 10 MGy [6,7]. These remarkable results allowed to widely use SiC detectors in nuclear and plasma physics [8,9].
In the last decade, they are extensively used as diagnostic devices in laser generated plasma for both high and low intensity laser. In this field they found large application thanks to the blindness to visible light emitted from pulsed laser and to the visible component of the plasma itself. In particular, they are used in Time of Flight (TOF) configuration as this technique allows the simultaneous detection of photons, electrons and ions emitted from laser plasma; the energy of radiation can be measured with high accuracy and the detector show a high efficiency and sensitivity together with a fast response [10-12].
In our recent works, we have shown [13] that SiC detectors with two different front-electrode geometries allow to detect ions in a wide range of energy from few keV to 10 MeV. The detectors with a continuous front-electrode and a thick active region are able to measure the high energy ions generated by high intensity (10^{16} W/cm^2) *ns* or *fs* pulsed laser. Instead SiC devices with an interdigit front-electrode can be employed to detect the low energy of the ions emitted by low intensity (10^{10} W/cm^2) laser.
Some of us demonstrated, in a recent paper [14], that by using both these detectors we can realize a telescope ion detector to measure the energy and the energy loss of ions generated by high intensity lasers. The proposed telescope system was based on the use of a M-SiC device with thick active detection layer to measure the total ion energy and of an I-SiC device with thinner active detection layer to measure the ion's energy loss. By plotting the ion energy versus the ion energy loss, it is possible to distinguish the different detected ion species.
Furthermore, due to their micrometric size, these detectors are very promising for the realization of

* Corresponding author: antonella.sciuto@imm.cnr.it

https://doi.org/10.1051/epjconf/201816703005

pixilated sensors for the measurements of radiation spatial distribution [15,16].

Although SiC detectors are very useful for the laser plasma characterization, their use could be limited by the damage produced during long-time exposure to plasma. In particular, we must distinguish between a bulk damage consisting of defects created in the detector active depth and due to the ion energy loss and a surface damage due to the debris and defect clusters generated by plasma at the device surface [17].

In the present paper we report about damage produced in both type of detectors (M-SiC and I-SiC) after plasma exposure. The surface damage is monitored by optical microscopy and the increases of defect concentration with laser exposure time was monitored. Moreover, the effect of damage generated by low and high intensity laser on the electrical characteristics of the devices and on the detector performances was examined.

2 Experimental

SiC detectors based on Schottky diode technology with different active volume and with two different front electrode structures were examined in this work. They were fabricated at CNR-IMM-Catania (M-SiC) and STMicroelectronics-Catania (I-SiC) facilities and are widely described, from the electrical and the detection point of view, in previous work of authors of the present manuscript [13]. The first proposed detector (M-SiC), is a continuous metal vertical Schottky device, fabricated on *n-type* 4H-SiC epitaxial layer, 80 μm thick with a 2×10^{14} cm^{-3} *n*-dopant concentration, grown on *n-type* heavily doped (~ 6×10^{18} cm^{-3}) substrate.

The second type (I-SiC), is an interdigit metal vertical Schottky device, fabricated on *n-type* 4H-SiC epitaxial layer, 4 μm thick with a 1×10^{14} cm^{-3} *n*-dopant concentration grown on *n-type* heavily doped substrate.

In both devices, the ohmic contacts on the sample back side were formed by sputtering of a 100 nm thick nickel (Ni) film, followed by a rapid annealing at 1000 °C in N_2; similarly, the front electrode was obtained by Ni (100 nm thick) deposition, photolithographic patterning, selective metal etch and rapid thermal process at 700 °C in N_2 environment for the formation of the Ni_2Si layer (200 nm thick) and of the optimal Schottky electrode [18].

The M-SiC diodes have a square geometry with a total area of about 4 mm^2 and was polarised at 600 V in order to deplete all the 80 μm active region. The I-SiC have a total area of about 8 mm^2 and an area of about 6 mm^2 directly exposed to the impinging radiation (75% of fill factor being the single Ni_2Si strip width of about 3 μm and the distance between contiguous front metal stripes of about 10 μm). It was polarised at 20 V, bias enough to deplete the 4 μm active layer. More details of both detectors are given in [19, 20].

In this work we report measurements at low laser intensity, performed at INFN-Laboratori Nazionali del Sud in Catania and at high laser intensity, performed at the PALS facility in Prague. In the first case we employed a Nd:YAG laser operating at 1064 nm wavelength, with a 9 *ns* pulse duration, 200 mJ laser energy and 2×10^{10} W cm^{-2} pulse intensity. In the second case an Iodine laser operating in single pulse regime at 1315 nm wavelength, with 300 *ps* pulse duration, 600 J laser energy and 2 $\times10^{16}$ W cm^{-2} pulse intensity was used.

The detectors efficiency and resolution were checked by high energetic alpha particles using a radioactive source composed by three peaks, ^{239}Pm (5.15 MeV), ^{241}Am (5.48 MeV) and ^{241}Cm (5.80 MeV).

3 Results

The M-SiC and I-SiC detectors were used to monitor the radiation emitted by plasma generated by high and low intensity laser. Typically, high intensity laser generates high energy ions, and their detection needs very thick active region detector, as our M-SiC diode. In Fig.1 is shown a typical TOF spectrum acquired at PALS facility at 10^{16} W/cm^2 laser intensity by using an M-SiC device. The spectrum was obtained by irradiating a thin foil of Polyethylene (20 μm thick), and detecting radiation at 0° in forward direction with respect to laser incidence. The peak in the 0–10 ns region is related to hard X-ray radiation (from few keV up to few tens of keV) emitted during and after the laser–target interaction.

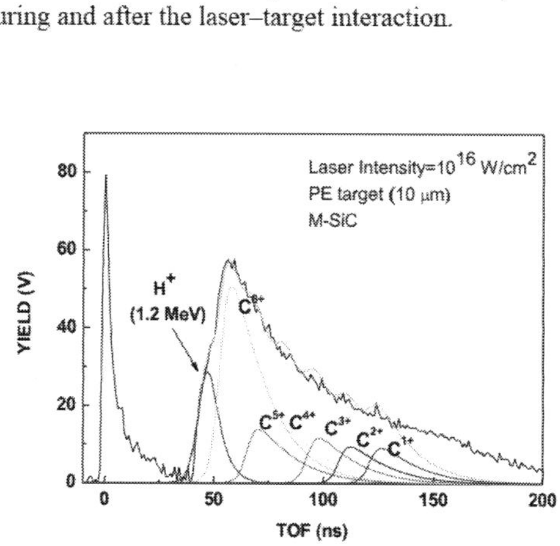

Fig.1. Time of flight spectrum of the plasma generated by high intensity pulsed laser irradiating a PE, 20 μm thick target, detected by M-SiC.

The spectrum structure between 10 and 30 ns is ascribable to XUV photons due to electron bremsstrahlung or to plasma electrons. The most interesting region of the spectrum is the one in the TOF range between 30 and 200 ns, due to the energetic ions of the plasma, from protons to carbon of different charge states and slower C_xH_y groups, detected at time of flight higher than 150 ns. The SiC detector allows separation of the photopeak from the ion component, observable as overlap of many different peaks. The fit of this spectrum with CBS (Coulomb-Boltzman-Shifted) model [21] allows the de-convolution of the ion peak with single components: the faster peak is related to the protons and it is followed by peaks related to C ions with different

charge states. The fit allows to evaluate some important plasma parameters, such as the equivalent temperature and the acceleration energy. We obtained an equivalent temperature of 54 keV, and an energy of 1.2 MeV for charge state, in good agreement with data reported in literature [22].
At low intensity laser, ions coming from laser-target interaction do not have enough energy to overcome the metallization layer of M-SiC detector, so the generated plasma can be monitored only by detector with active layer directly exposed to irradiation, that is the I-SiC. A typical example of TOF spectrum measured with an I-SiC detector by irradiating an Al target is shown in Fig.2. The spectrum of the Al target evidences a broad peak at 5.5 μs resulting in an energy of 1.2 keV, value typical of ions generated by low intensity lasers [23]. In this case, due to the low energy (200 mJ) of laser only the single charge state of Al is generated [24].

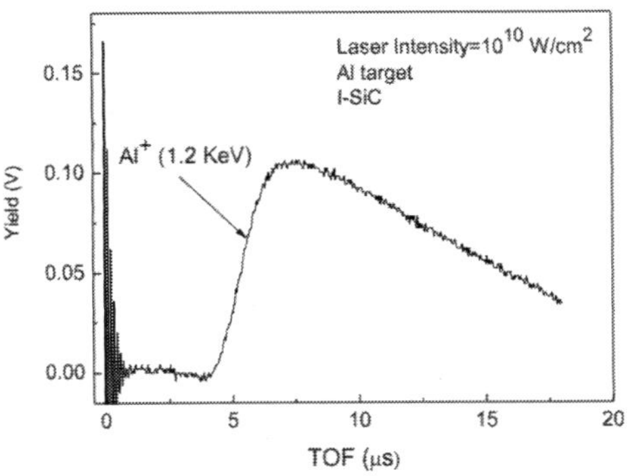

Fig. 2. Time of flight spectrum of the plasma generated by low intensity pulsed laser irradiating a Al target detected by I-SiC.

However, plasma exposure of SiC detectors generated some surface damage as evidenced in Fig.3, where are shown the photographs of I-SiC (Fig.3a) and M-SiC (Fig.3c) detector surface after laser irradiation. The I-SiC were exposed to low intensity laser in Catania to a number of about 100 shots, while the M-SiC were exposed to high intensity laser at PALS laboratory in Prague to a number of about 300 shots. Before irradiation, both detectors show a clean surface with a delineated geometry, while after plasma exposure the surface results covered by debris consisting of micrometric clusters. The cluster size distribution is quite large in both case, however for the I-SiC the main cluster size (Fig.3b) is 8 μm, while for the M-SiC (Fig.3d) is 18 μm. The increase of exposure time to the plasma clearly enhances the total number of such clusters, but their morphology and size depend on plasma parameters such as temperature, density, angular distribution of emitted particles and also on the detector position with respect to target as distance and angle. In particular, the wavelength and the intensity of laser influence the phothermal and photochemical ablation regime and at high intensity larger molecular or clusters are formed during plasma generation [25].

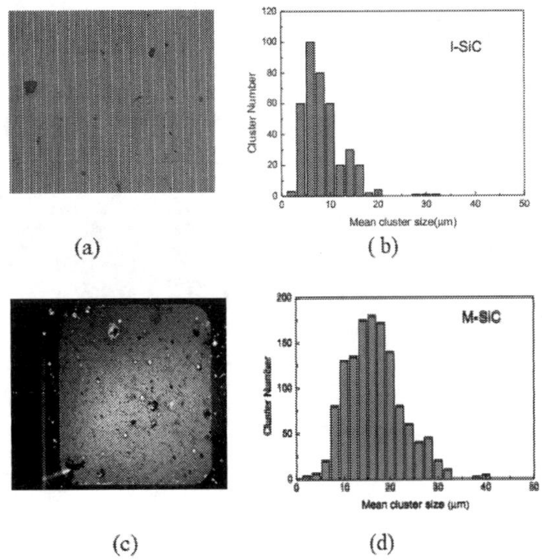

Fig. 3. Optical microscopy image (a) and cluster size distribution (b) of a portion of the I-SiC detector after exposure to low intensity laser. Optical microscopy image (c) and cluster size distribution (d) of M-SiC detector after exposure to high intensity laser.

The presence of damage has a considerable influence on the device electrical characteristics and in Fig.4 are shown the I-V measurements of I-SiC before and after laser exposure. The reverse current (Fig.4a) of detector before plasma exposure is very low, about $5x10^{-11}$ A, and increases of more than two order of magnitude after laser irradiation.
The forward characteristic (Fig.4b) of as prepared device evidences its excellent rectifying properties; it exhibits a region of linearity at low bias value (< 1.5 V) and a saturation at higher voltage, where the device behavior is dominated by the series resistances. The fit of this curve gives a value of 1.65 eV for the Schottky barrier height and of 1.05 for the ideality factor, close to literature values [26].
After laser exposure the linear region of the curves gets small and an increase of current at low bias occurs. In addition, a small decrease of current in the saturation region is observed related to the increase of series resistance of device [27].
Similar results are obtained in M-SiC detector irradiated at PALS whose I-V characteristics are reported in Fig.5. The reverse current after laser exposure (Fig.5a) increases of about four order of magnitude, while the forward characteristics shows a quite strange behavior: at low bias (< 0.3 V) it increases and then saturates at higher voltage.
The comparison between the forward I-V curves of I-SiC and M-SiC after laser exposure, shows that the M-SiC devices suffered a larger increase of current at low bias and a further current decrease at high bias. This difference is due both to the larger amount of damage and to the different type of damage produced in the M-SiC.

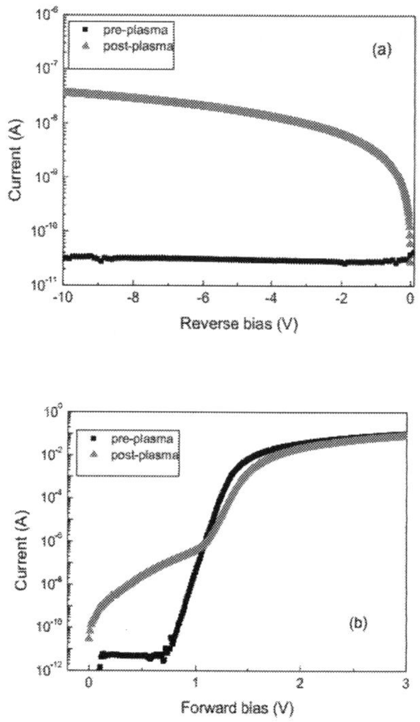

Fig. 4. Reverse (a) and forward (b) I-V characteristics of I-SiC detector before and after laser exposure.

In fact, we must underline that plasma exposure together to the formation of already described surface damage, induces the formation of defect localized into the active region of the device and due to penetrating ions. The energy loss of the ions produces the formation of point defects (interstitial, vacancies, antisite) and complex defects (dislocation, stacking faults, etc) localized into a depth close to the ion range. Because the laser produces ions with different charge states (i.e. energies), the defects are distributed into the entire irradiated depth. As already reported in literature [27] ion irradiation of SiC Schottky diodes is responsible of the increase of leakage current and of the decrease of forward current at high bias. In particular, point and complex defects induce the formation of deep levels in the forbidden gap which act as generation centers increasing the leakage current. At the same time, these defects imply a deactivation of dopant in the irradiated region, with a consequente increase of series resistance and decrease of forward current at high bias. This effect in the M-SiC is larger with respect to I-SiC, because of the different involved ion range, which is of the order of 40-50 μm in the M-SiC and of some hundred of Angstrom in I-SiC.

The increase of forward current at low bias can be, instead, ascribed to the surface damage as the presence of clusters and debris modifies the Schottky barrier, producing an inhomogenous barrier which is composed by a distribution of nanometric size "patches" with a low barrier height (< 1 eV) enclosed in a uniform high barrier background (1.65 eV barrier height in un-exposed diode). The formation of paths of low barrier height gives a strong contribution to the extra current measured at low bias [28].

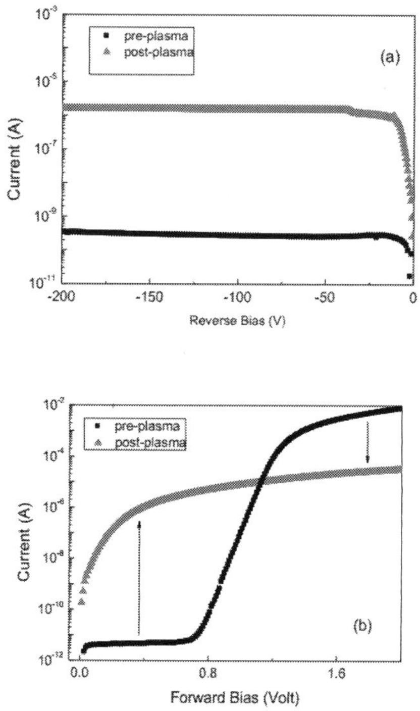

Fig. 5. Reverse (a) and forward (b) IV characteristics of M-SiC detector before and after laser exposure.

The effect of damage on the detection properties of the M-SiC detector was checked by a three peaks alpha source. The I-SiC cannot be used for these measurements because the range of alpha particles (18 μm) is higher than the active region thickness (4 μm).

The alpha spectra measured in as-prepared M-SiC diode and after two different number of laser exposures (300 and 400 shots), acquired during the same slot, are shown in Fig.6. The detected yield of all peaks decreases after exposition, and it reaches a value of about 30% after a number of laser shots of about 400.

This effect is mainly related to the coverage of surface by debris: the micrometric clusters induce a decrease of useful detector surface and a consequently reduction of the detection signal. In addition, a worsening of resolution from 0.9% to 1.5% is measured for the main peak (5.15 MeV), which is related both to bulk defects and to energy straggling due to surface micrometric clusters. In order to evaluate the fluence impinging on the detector surface during laser exposure, we can take into account that an average fluence of 10^{14} ion/cm^2 is produced by a single shot [17], then after 400 shots a total fluence of $4x10^{16}$ ion/cm^2 arrives on the detector surface. It is difficult to evaluate the absorbed dose

(MGy) due to the different ion energies arriving simultaneously on the detector, however this fluence value is rather high and evidences the radiation hardness of these devices.

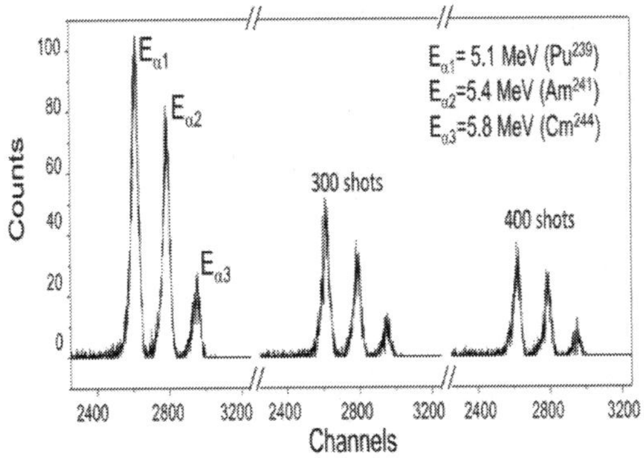

Fig.5: Three peaks alpha source spectra detected with M-SiC before and after two different numbers of laser shots.

4 Conclusions

In the present paper we report some aspects regarding the damage induced in SiC detectors when exposed for long time to laser generated plasma. Two different detectors were tested. The first detector is an interdigit front electrode device where a fraction of surface is directly exposed to irradiation and then it is able to monitor the low energy ions generated by low intensity laser. To detect the high energy ions generated by high intensity laser a continuous front electrode diode with a thick epitaxial layer was used: this second detector is able to detect proton and ion of different charge states and to extract important plasma parameters, as temperature and ion acceleration. Laser exposure of both detectors produces some damage consisting in two types of defects. The first type is localized in the detector surface and consists in ablated and evaporated target material producing micrometric clusters: the presence of these clusters induces the formation of an inhomogeneous barrier and reduces the effective detector area with a decrease of detector efficiency. The second type of defects are points or complex defects generated by the energy loss of the different plasma ions and are localized in the active detector region: these defects induces an increase of leakage current and a deactivation of semiconductor dopant.
However, we like to underline that these effects are observed after long exposure time, highlighting the high radiation hardness of SiC detectors.

Bibliography

1. E.V. Kalinina, A.M.Ivanov, N.B.Strokan, Tech. Phys.Lett. **34**, 210 (2008)
2. F.H. Ruddy, J.G.Seidel, Nucl.Instr.Meth. B **263**, 163 (2007)
3. G. Bertuccio, R.Casiraghi, A.Cetronio, C.Lanzieri and F.Nava, Nucl.Instr.Meth.A **522**,413 (2004)
4. F.Moscatelli, A.Scorzoni, A.Poggi, M.Bruzzi, S.Sciortino, S.Lagomarsino, G.Wagner, and R.Nipoti, IEEE Trans.on Nucl. Sci.**53**,1557 (2006)
5. S.Sciortino, F.Harties, S.Lagomarsino, F.Nava, M.Brianzi, V.Cindro, C.Lanzieri, and P.Vanni, Nucl.Instr.Meth. A **552**, 138 (2005)
6. F. Nava, G. Bertuccio, A. Cavallini and E. Vittone, Meas. Sci. Technol. **19**, 102011 (2008)
7. S.Sciortino, F.Harties, S.Lagomarsino, F.Nava, M.Brianzi, V.Cindro, C.Lanzieri, and P.Vanni, Nucl.Instr.Meth. A **552**, 138 (2005)
8. G.Raciti, M.De Napoli, F.Giacoppo, E.Rapisarda and C.Sfienti, Nucl. Phys. A **834**,784 (2010)
9. L.Torrisi, L.Calcagno, D.Giulietti, M.Cutroneo, M.Zimbone, J.Skala, Nucl.Instr.Meth., **355**, 221 (2015)
10. L.Calcagno, P.Musumeci, M.Zimbone and F.La Via, Rad. Effects and Defects in Solids **4**, 303 (2015)
11. M.Cutroneo, P.Musumeci, M.Zimbone, L.Torrisi, F.La Via, D.Margarone, A.Velyan, J.Ullschmied and L.Calcagno, J.Mat. Research **28**, 87 (2013)
12. L.Torrisi, L.Calcagno, D. Giulietti, M.Cutroneo, M.Zimbone and J.Skala, Nucl.Instr.Meth. B **355**, 221 (2015)
13. A.Sciuto, L.Torrisi, A.Cannavò, M.Mazzillo and L.Calcagno, J.Electronic Mater. (In Press)
14. L.Torrisi and A.Cannavò, IEEE Trans. Electron Dev. **63**, 4445 (2016)
15. A.Sciuto, F.Roccaforte, S.Di Franco, V.Ranieri, S.F.Liotta, S.Billotta, G.Bonanno and M.Belluso, Mat.Sci.Forum **556**, 945 (2007)
16. H.Shuoben, P.Hellström, C.M.Zetterling and M.Östling, IEEE Electr. Dev. Lett. **37**, 1594 (2016)
17. L.Torrisi and A.Cannavò Journal of Instr. **11**, 05009 (2016)
18. F.Roccaforte, F.La Via, V.Raineri, L.Calcagno, F.Mangano, J.App. Phys., **96**, 4313 (2004)
19. A.Sciuto, L.Torrisi, A.Cannavò, G.Ceccio, P.Musumeci, M.Mazzillo, L.Calcagno, Vacuum, **131**, 170 (2016)
20. P.Musumeci, M.Cutroneo, L.Torrisi, A.Velyan, M.Zimbone, L.Calcagno, Physica Scripta **T161**, 14021 (2014)
21. L. Torrisi, Radiat. Eff. Defect Solids **171**, 34 (2016)
22. L. Torrisi, S. Gammino, L. Andò, L. Laska, J. Krasa, K. Rohlena, J. Ullschmied, J. Wolowski, J. Badziak, and P. Parys, J. Appl. Phys. **99**, 083301 (2006)
23. L.Torrisi, S.Gammino, L.Andò and L.Laska, J.Appl.Phys. **91**, 4685 (2002)
24. L. Torrisi, F. Caridi, A. Picciotto, A. Borrielli, Nucl.Instr.Meth.B **252** (2006) 183
25. L.Torrisi, A.Borrielli and D.Margarone, Nucl.Instr.Meth. B **255** (2007) 373
26. F.Roccaforte, F.La Via, V.Raineri, R.Pierobon, E.Zanoni, J.Appl.Phys. **93** (2003) 9137
27. F.Roccaforte, S.Libertino, V.Raineri, A.Ruggiero, V.Massimino and L.Calcagno J.Appl.Phys. **99**, 13515 (2006)
28. L.Calcagno, A. Ruggiero, F.Roccaforte and F.La Via, J.Appl.Phys. **98**,23713 (2005)

Experimental and theoretical study on emission spectra of a nitrogen photoionized plasma induced by intense EUV pulses

Ismail Saber[1,2*], *Andrzej* Bartnik[1], *Wojciech* Skrzeczanowski[1], *Przemyslaw* Wachulak[1], *Roman* Jarocki[1], *Henryk* Fiedorowicz[1] and *Jiri* Limpouch[2]

[1]Institute of Optoelectronics, Military University of Technology, Kaliskiego 2, 00-908 Warsaw, Poland
[2]Faculty of Nuclear Sciences and Engineering Physics, Czech Technical University in Prague, Prague, Czech Republic

Abstract. Spectral lines of low-temperature nitrogen photoionized plasma were investigated. The photoionized plasma was created in the result of irradiation N_2 gas using laser plasma EUV radiation pulses. The source was based on a 10J/10ns Nd:YAG (λ = 1064 nm) laser system and a gas puff target. The EUV radiation pulses were collected and focused using a grazing incidence multifoil EUV collector. The emission spectra were measured in the ultraviolet and visible (UV/Vis) range. It was found that the plasma emission lines in the lower region of the UV range are relativley weak. Nonetheless, a part of the spectra contains strong molecular band in the 300 - 430 nm originated from second positive and first negative systems band transitions of nitrogen. These molecular band transitions were identified using a code for study the diatomic molecules, LIFBASE. The vibrational band of $\Delta v = 0$ and ± 1 transitions were significantly populated than of that with $\Delta v = \pm 2$ and 3 transitions. A comparison of the calculated and measured spectrum is presented. With an assumption of a local thermodynamic equilibrium (LTE), the vibrational temperature was determined from the integrated band intensities with the help of the Boltzmann plot method and compared to the temperature predicted by SPECAIR and LIFBASE simulations. A summary of the results and the variations in the vibrational temperatures was discussed.

1 Introduction

Photoionization of molecular nitrogen is performed and emission spectra in the result of the photoionization was measured. The plasma was produced by a laser-produced plasma (LPP) extreme ultraviolet (EUV) source. The experimental source was based on a Nd: YAG laser system delivering a 10 J, 10 ns pulse at the fundamental wavelength, λ =1064 nm. The laser power density in such regime can be easily achieved an order of 10^{12} W/cm^2 in the interaction region. Such intensity range can be enough to laser-produced xenon plasma and therefore the EUV radiation collected was used to ionize nitrogen gas up to first few ionization states. Mostly, spectral lines from molecular band transitions in the ultraviolet and optical wavelengths region were the dominant features in the spectra. The molecular emission spectra were found corresponding to the second positive (2PS) ($C^3\Pi_u$-$B^3\Pi_g$) and first negative (1NS) ($B^2\Sigma^+_u$-$X^2\Sigma^+_g$) systems of the molecular nitrogen. The identification of the molecular band transitions was performed using software for diatomic molecules simulation, LIFBASE [1]. The plasma parameters and its condition can be studied based on these molecular band transitions by evaluating vibrational temperature with an assumption of the local thermodynamic equilibrium. The vibrational temperatures provide insight to the molecular vibrational processes while the rotational temperature usually represents the gas temperature and it is responsible of reaction rates data of many molecular processes. In thermodynamic equilibrium condition, determination of the rotational and vibrational temperatures is often done by assuming the vibrational levels based on the Boltzmann distribution function. For nitrogen plasma, the band transitions from the upper electronic to the ground electronic states in 2PS and 1NS systems and their characteristics make possible to estimate the vibrational temperatures from emission intensities. Though, spectroscopic investigations on nitrogen plasmas or nitrogen-containing plasmas have been studied using the band spectra recorded in the various plasmas at different experimental conditions. Of these, include a compact helicon plasma source [2], nitrogen-containing plasmas [3] as well as in highly constricted plasma and applying the fitting methods [4]. Further, the plasma emission lines from 2PS are well-known and the best candidate used for estimation of the vibrational temperature in a nitrogen or nitrogen-containing plasma. Provided that if an assumption of the thermodynamic equilibrium between the ground and excited electronic states is achieved [5]. In this case, the required condition is the radiative rates are dominated by the collision rates to ensure the LTE condition in plasma. Knowledge of these population rates will provide more information

* Corresponding author: ismaailsaber@hotmail.com

about the plasma emissions processes and control conditions. However, if the plasma is far from LTE, determinations of vibrational temperature is more sophisticated unless the high-resolution spectra were used. In this case, the vibrational temperature inferred from the Boltzmann plot methods still give insight into the plasma condition. Though, the 2PS of nitrogen molecule was widely used to study the gas temperatures in nitrogen and nitrogen-containing plasmas [6-10].

In this work, the emission lines of nitrogen photoionized plasma created by laser plasma EUV radiation pulse is investigated. The electron vibrational temperature is determined using the integrated line intensities of the band transitions in the second positive system of nitrogen. This can be done with the help of the Boltzmann plot of vibrational and rotational levels. In this case, especially for the molecular spectroscopy, much attention must be paid to ensure the high-resolution of the molecular band spectrum which enables us to accurately determine the band head wavelength of a transition. On the other hand, the determination of the vibrational and rotational temperature can be predicted using software for the molecular spectral simulation, LIFBASE. In this case, the temperature was set manually in acquiring the simulated. The more accurate data on electron vibrational or rotational temperature can be utilized by direct fitting of the measured spectra [2]. The most applicable software used for this purpose is the commercial software SPECAIR [11].

2 Experimental setup

In the experiments, Nd: YAG laser system delivering the maximum energy of 10J/10 ns pulse and 10 Hz repetition rate was used. In the present measurements, energy about 5.6 J per pulse was employed. The laser pulse was focused onto a double stream gas-puff target, created synchronously with the laser pulse. The gas-puff target was created by pulsed injection of high-Z (argon or xenon) gas into a hollow stream of low-Z helium gas using a double nozzle set-up. The nozzle consists of a small central orifice surrounded by a ring-shaped outer nozzle and equipped with an electromagnetic valve system. More experimental detailed description of the double stream gas puff target can be found in [12,13]. Irradiation of nitrogen gas (4 bar) injected into the vacuum interaction chamber synchronously with the radiation pulses from laser plasma EUV source resulted in photoionized plasma. The photoionized plasma emissions and focusing conditions were adjusted in the way to obtain maximum power density in the interaction region. The maximum intensity is estimated around 10^8 W/cm^2 under EUV fluence of ~300mJ/cm^2. The EUV radiation was focused using a gold-plated grazing incidence multifoil EUV collector. The spectral analyzing instrument was an Echelle Spectra Analyzer ESA 4000 spectrograph, equipped with the ICCD Kodak KAF 1001 camera. The spectrometer system allowed for simultaneous measurements of complex spectra within the wide UV/Vis (200-780 nm) spectral range with considerably higher spectral resolution $\lambda/\Delta\lambda \approx 20000$. Schematic view of the EUV photoionization experimental setup is shown in Fig. 1.

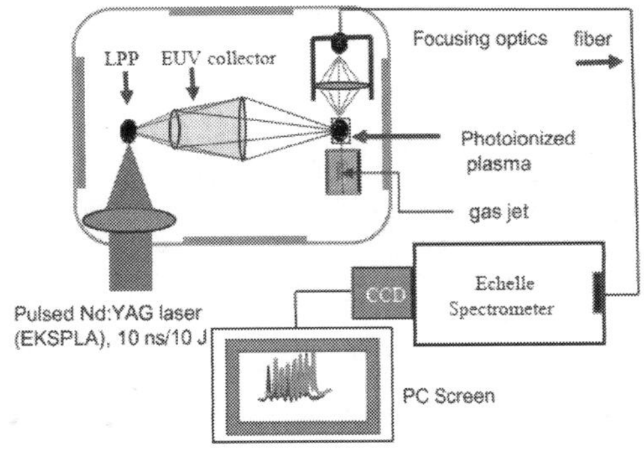

Fig. 1. Schematic view of the experimental setup for the EUV photoionization experiments.

2.1. Vibrational temperature

The spectrum of the generated photoionized plasma is dominated by the emission of second positive N_2 ($C^3\Pi_u$ - $B^3\Pi_g$) and first negative N^+_2 ($B^2\Sigma^+_u$ - $X^2\Sigma^+_g$) systems of molecular features of nitrogen. If the assumption of the upper populations having Boltzmann distributions with both vibrational and rotational temperatures T_v and T_r, respectively, the line intensity of the molecular band system between the upper v',J' and lower v'',J'' rovibrational levels can be expressed as following [14].

$$I(v',J' \to v'',J'') = \frac{CR_{SP}}{\lambda^4}\frac{q(v',v'')}{Q_{vib}}\frac{s(J',J'')}{Q_{rot}} \times \exp\left(\frac{G(v')}{k_B T_{vib}}\right)\exp\left(\frac{F(v',J')}{k_B T_{rot}}\right) \quad (1)$$

Here, C is the scaling factor related to the collection solid angle and the sample volume of the optical emission, R_{SP} is the spectral response function, $s(J',J'')$ is strength function and $F(v',J')$ rotational spectral term. Q_{vib}, Q_{rot} are the vibrational and rotational partition functions, respectively. $q(v',v'')$ is the Franck-Condon factor of the vibrational transition ($v' - v''$) [15].

For simplicity, we can obtain the T_v easily by from the Boltzmann plot using integrated band intensities $I(v',v'')$ as a function of the upper vibrational energies of different transition in the second positive system. In such case, as can be seen from equation (1), the total intensity is given for both vibrational and rotational levels. To deduce the vibrational temperature, we can use the vibrational dependent portion as follows [16,17]:

$$\ln\left(\frac{I(v',v'')\times\lambda(v',v'')}{A(v',v'')}\right) = \frac{G(v')}{k_B T_{vib}} + const. \quad (2)$$

Where $I(v',v'')$ is the intensity of vibrational band, $G(v')$ represents the vibrational energy of the upper level, T_v is the vibrational temperature, k_B is the Boltzmann constant and $A(v',v'')$ and $\lambda(v',v'')$ are respectively the transition probability and the wavelength of the band head.

3 Results and discussion

Identification of emission line in plasma has great importance and it is considered as the first step to specify the plasma species and their charge states. Fig.2 shows the measured and simulated emission spectra of molecular nitrogen. The dominant features in the spectrum are the molecular band transitions in 2PS and 1NS systems of N_2 in the ultraviolet (UV) wavelength range. Most of the emission lines were corresponding to the 2PS in the 300-400 nm range. The measured and SPECAIR simulation results were shown a good agreement with relatively different in their intensities, where the experimental emission lines are more strength. Although simulated with the T_v = 3358.9 K obtained from linearity of the Boltzmann plot, the spectra computed using LIFBASE also showed the strongest band transitions, (0-0) and (0-1) in 1NS (B-X).

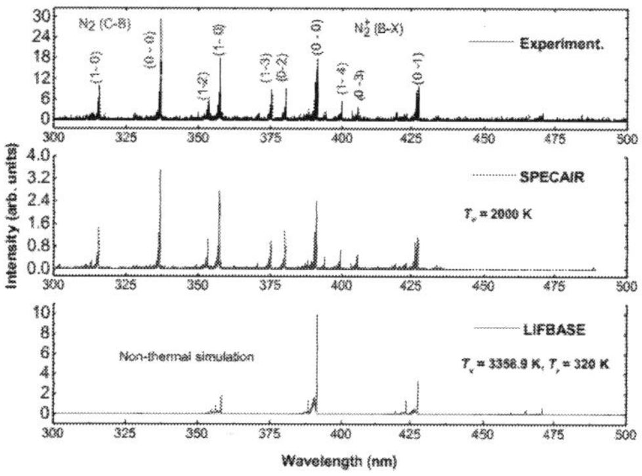

Fig. 2. Measured (top) and simulated UV/Vis emission spectra belong to the 1NS (B-X) and second positive 2PS (C-B) systems of nitrogen induced by the intense laser plasma EUV pulses.

It has been noticed that even for the T_v of 2000 K, the results do not show any significant difference in terms of molecular band transitions rather in the line intensity. The best T_v can be found by fitting the measured spectra with an appropriate fitting function.

In Fig.3, the experimental (top) and simulated (bottom) spectra corresponding to the (1-0) and (1-2) vibrational band transitions of the 2PS ($C^3\Pi_u$ - $B^3\Pi_g$) acquired at a pressure of 4 bar at a pulse repetition rate of 10 Hz with the energy of 5.6 J per pulse. A good agreement was found between the measured and simulated spectra for a temperature of 2000 K. This spectra is mainly part of the spectrum in Fig.2 and has been taken between 354 nm to 359 nm to clearly demonstrate the measured and calculated vibrational spectra.

To evaluate electron vibrational temperature, Fig.4 shows the Boltzmann plot of the integrated band intensities versus the upper vibrational energy levels corresponding to the transitions in 2PS. We have used six transitions in 2PS to infer the T_v because the spectra dominated mostly with molecular band transitions in the 2PS band system, Fig.2. The fitted line to the experimental points gives rise an approximate vibrational temperature of 3358.9 ± 665 K. This value is a bit higher compared to the best value of the T_v (2000 K) that used to obtain the best fit between the simulated and measured spectra with SPECAIR software under the present experimental conditions. Despite the temperature differences, the measured and calculated spectra using SPECAIR were found to be in good agreement. Such discrepancies are expected whether associated with experimental errors or attributed the use of a small number of vibrational transitions, as it is always recommended to include much more lines in constructing the Boltzmann plot. In the other hand, when the determined vibrational temperature 3358.9 K is used to simulate the measured spectrum with LIFBASE software, the computed spectra contain molecular bands mainly from 1NS (B-X), (0-0) and (0-1) emissions.

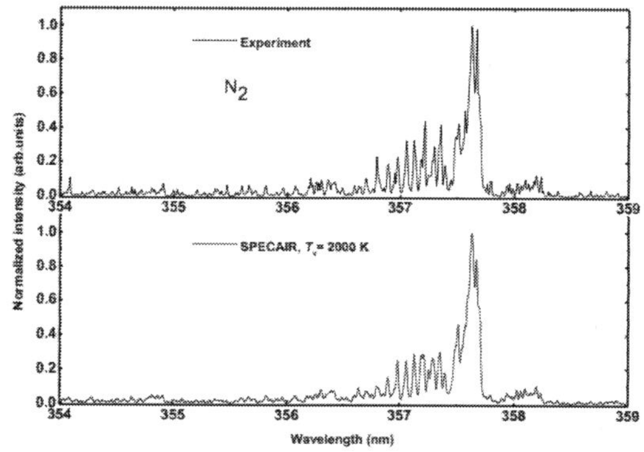

Fig.3. Comparison of measured (black) and calculated (red) emission spectra of a second positive ($C^3\Pi_u$-$B^3\Pi_g$) band transitions in plasma over a wavelength range from 354 nm to 359 nm

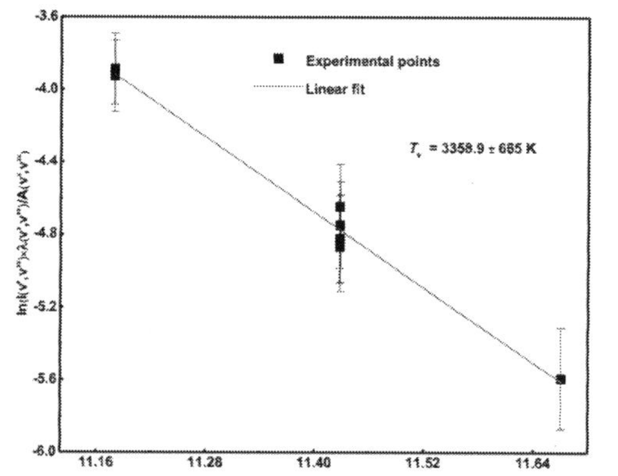

Fig.4. Vibrational temperature determination from the Boltzmann plot of the Δv = 0, ±1,-2, and -3 sequences in second positive 2PS band transitions.

In Fig.5 the vibrational population and its dependence on the T_v are examined using emission spectra in the 1NS. That calculated vibrational populations at a lower T_v, 2000 K, are found to be more vibrationally populated than that calculated at a higher T_v, 3358.9 K. This can be explained as at lower T_v, most of a gas molecule within molecule will be vibrated at their fixed vibrational levels. With Enhancing T_v the individual amplitude of each molecule increases before the molecules undergo the translational motion, this resulted in less populated states at higher vibrational temperatures. Therefore, the vibrational bands with the $\Delta v = 0$ and ± 1 transitions in $B^2\ \Sigma^+_u$ - $X^2\ \Sigma^+_g$ system are significantly populated than of that with $\Delta v = \pm 2$ and 3 transitions. As the plasma in LTE, the gas temperature can be considered as the rotational temperature T_r that measured in the upper states in a molecular transition. In contrast, for non-LTE plasmas, measurements of the gas temperature is possible via analyzing the rotational molecular emission provided that the rotational-translational relaxation is sufficiently fast to equilibrate the gas and rotational temperatures [18].

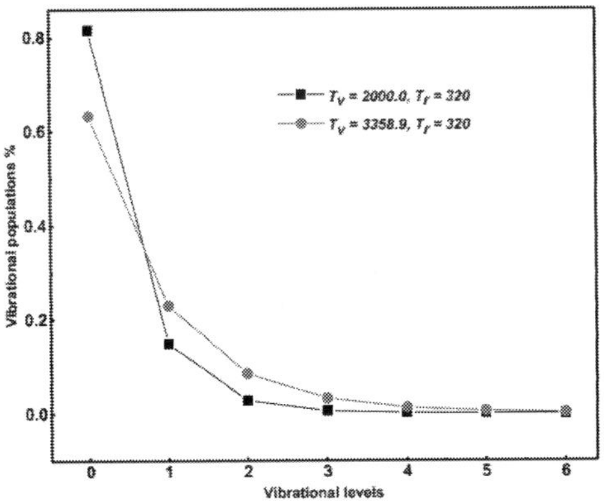

Fig. 5. Vibrational populations of non-thermal mode vibrational levels of 2PS, obtained at the end of the simulations for two different temperatures (T_v,T_r).

4 Conclusions

In summary, laser-plasma EUV source was used to create a low-temperature photoionized nitrogen plasma. High-resolution emission lines from photoionized plasma in the UV/Vis spectral range were recorded using echelle spectral analyzer ESA 4000. The dominant molecular band transitions in plasma were the electronic transitions from 2PS ($C^3\Pi_u$-$B^3\Pi_g$) and 1NS ($B^2\Sigma^+_u$ - $X^2\Sigma^+_g$) systems. The diatomic molecular simulation programs, LIFBASE and SPECAIR have been applied to reproduce the experimental spectra theoretically and interpretation. A comparison of the measured and calculated emission lines is given. With the assumption of the plasma in the LTE condition, the linearity of the Boltzmann plot is obtained using the integrated band intensities in 2PS vibrational transitions. A vibrational temperature of 3358.9 K is inferred from the used Δv = 0,±1,±2 and 3 transitions. The T_v calculated with the aid of the Boltzmann plot is comparable to the results obtained using the intensity ratio of two consecutive vibrational bands of the same sequence in 2PS from a helium microwave gas discharge [19].

The inconsistency between the simulated and measured vibrational temperatures was expected. The differences can be explained due to the departure of the plasma from the local thermodynamic equilibrium and low degree of ionization. The method of direct fitting the experimental emission spectra may yield reasonable results of vibrational temperature in plasma in the case where the Boltzmann plot is not accurate either for low intensity of the band emission or due to the weak radiative transition rates of a line.

The authors acknowledge the financial support from the EU FP7 Erasmus Mundus Joint Doctorate Programme EXTATIC under framework partnership agreement FPA-2012-0033 and partially with the support of the European Union's Horizon 2020 Programme (LASERLAB-EUROPE project grant agreement no. 654148. This work is also supported by the National Centre for Science; award number UMO-2016/23/B/ST7/00949.

References

1. J. Luque and D.R. Crosley. Lifbase: *Database and spectral simulation program* (version 2.1.1). SRI International Report MP 99-009, (1999)
2. C. Biloiu, X. Sun, Z. Harvey, and E. Scime, Rev. Scient. Inst. 77, 10F117 (2006)
3. E. Pawelec, Eur. Phys. J. Special Topics **144**, 227-231 (2007)
4. Q. Y. Zhang, D.Q. Shi, W. Xu, C.Y. Miao, C.Y. Ma, C. S. Ren, C. Zhang, and Z.Yi, AIP Advances **5**, 057158 (2015)
5. B. A. Cruden, M. V. V. S. Rao, S. P. Sharma, and M. Meyyappan, J. Appl. Phys. **91**, 8955 (2002)
6. V. M. Donnelly and M. V. Malyshev, Appl. Phys. Lett. **77**, 2467 (2000)
7. E. J. Tonnis and D. B. Graves, J. Vac. Sci. Technol. A **20**, 1787 (2002)
8. F. Iza and J. A. Hopwood, IEEE Trans. Plasma Sci. **32**, 498 (2004)
9. M. Tuszewski, J. Appl. Phys. **100**, 053301 (2006)
10. B. Bai, H. Sawin, and B. Cruden, J. Appl. Phys. **99**, 013308 (2006)
11. Laux et al, Plasma Sources Sci. Technol. **12**, 125-138 (2003)
12. A. Bartnik, H. Fiedorowicz, R. Jarocki, J. Kostecki, M. Szczurek, and P. W. Wachulak, Nucl. Inst. Methods Phys. Res. A **647**, 125-131 (2011)
13. A. Bartnik, W. Lisowski, J. Sobczak, P. Wachulak, B. Budner, B. Korczyc, and H. Fiedorowicz, Appl. Phys. A **109**, 39-43 (2012)
14. I. Kov'acs, *Rotational Structure in the Spectra of Diatomic Molecules* (American Elsevier Publishing Company, New York, 1969)
15. F. Roux, F. Michaud, and M. Vervloet, J. Mol. Spec. **158**, 270 (1993)
16. H. R. Griem, *Plasma Spectroscopy* (McGraw-Hill, New York, 1964)
17. M. Bashir, Julia M.Rees, S. Bashir, William B.Zimmerman, Phys. Lett. A **378**, 2395-2405 (2014)

18. G. Gardet, G. Moulard, M. Courbon, F. Rogemond and M. Druetta, Meas. Sci. Technol. **11**, 333-341 (2000)
19. G. L j. Majstorovic´, J. Jovovic´, N. M. Šišovic´, Contrib. Plasma Phys. **57**, 282-29 (2017)

Electro-optic analysis of the influence of target geometry on electromagnetic pulses generated by petawatt laser-matter interactions

Timothy Robinson[1,*], *Samuel* Giltrap[1], *Samuel* Eardley[1], *Fabrizio* Consoli[2], *Riccardo* De Angelis[2], *Francesco* Ingenito[2], *Nicholas* Stuart[1], *Claudio* Verona[3] and *Roland A* Smith[1]

1 Blackett Laboratory, Imperial College London, Prince Consort Road, London SW7 2AZ, United Kingdom
2 ENEA – C.R. Frascati, Dipartimento FSN, Via E. Fermi 45, 00044 Frascati, Italy
3 Dipartimento di Ingegneria Industriale, Università di Roma Tor Vergata, Via del Politecnico 1, I-00133 Roma, Italy

Abstract. We present an analysis of strong laser-driven electromagnetic pulses using novel electro-optic diagnostic techniques. A range of targets were considered, including thin plastic foils (20-550 nm) and mass-limited, optically-levitated micro-targets. Results from foils indicate a dependence of EMP on target thickness, with larger peak electric fields observed with thinner targets. Spectral analysis suggests high repeatability between shots, with identified spectral features consistently detected with <1 MHz standard deviations of the peak position. This deviation is reduced for shots taken on the same day, suggesting that local conditions, such as movement of metal objects within the target chamber, are more likely to lead to minor spectral modifications, highlighting the role of the local environment in determining the details of EMP production. Levitated targets are electrically isolated from their environment, hence these targets should be unable to draw a neutralization current from the earth following ejection of hot electrons from the plasma, in contrast to predictions for pin-mounted foils in the Poyé EMP generation model. With levitated targets, no EMP was measurable above the noise threshold of any diagnostic, despite observation of protons accelerated to >30 MeV energies, suggesting the discharge current contribution to EMP is dominant.

1 Introduction

Intense electromagnetic pulses (EMP) are often observed alongside high-intensity laser-matter interactions [1–9], such as those studied in petawatt-class laser facilities [10]. These transient radiofrequency (RF) fields can be problematic for experiments due to their coupling to electronic devices in the immediate vicinity, with the potential to damage valuable equipment or result in data losses from electronic diagnostics, such as CCD cameras. Despite these issues, EMP can be used as a complementary diagnostic, and as a source of intense fields for radiation-hardening tests or electromagnetic compatibility applications.

During intense laser-plasma interactions, hot electrons are ejected from the target, resulting in a net-positively charged plasma. EMP generation is typically attributed to two complementary models: firstly, the ejected electrons strike the metal walls of the target chamber, acting as a charge stimulus which excites the electromagnetic resonant cavity modes [10]. The second (Poyé) model relates to the dynamic target charging leading to a discharge current of cold electrons drawn through the target support assembly to maintain quasi-neutrality; this transient current results in antenna-like emission from the target holder [11,12]. However, target geometry's role in EMP generation has not been extensively studied. Poyé describes a dependence on the surface area of planar targets due to lowering of the potential barrier experienced by electrons as charge redistributes across the surface [11], but there are no published results using "unconventional" targets such as ultra-thin plastic foils or mass-limited isolated spheres. In this work carried out at Vulcan Petawatt and the Cerberus laser system at Imperial College (an OPCPA system operating at 1054 nm), we consider the effects on EMP of varying the thickness of plastic foil targets, and using levitated micro-targets trapped in-vacuo using an optical levitation trap developed by Price et al [13].

Electro-optic (EO) probing techniques are inherently low-noise, as signal transport is achieved via optical fiber-link to which ambient EMP fields do not couple. Furthermore, they provide *absolute* measurements of electric fields via the longitudinal Pockels effect in optical crystals, where polarization changes scale linearly with the electric field in a crystal [14]. Hence, no numerical integration is required to determine the fields, such that constant or low-frequency field components are recovered. The data presented in this article was obtained using the EO diagnostic described in detail in a separate publication [15], and a more compact version during Cerberus experiments, described in Section 2.

* Corresponding author: timothy.robinson10@imperial.ac.uk

2 Experimental Architecture

The diagnostic setups deployed for EMP measurement during the Vulcan and Cerberus experiments studying laser-interactions with novel targets are shown in Fig. 1.

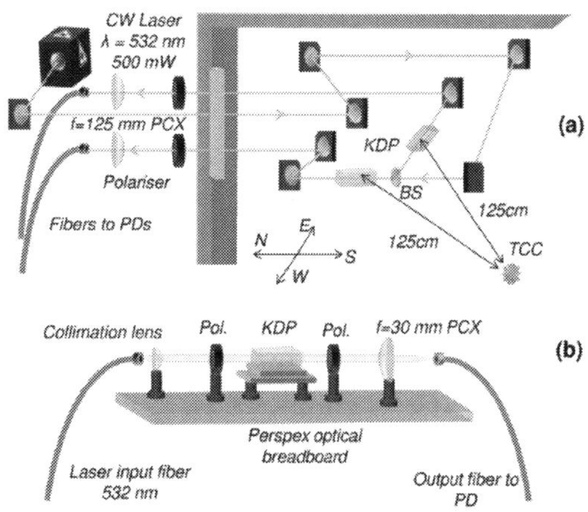

Fig. 1: The prototype and second generation electro-optic EMP diagnostics used for experiments at (a) Vulcan PW (adapted from [15]), using a free-space setup with the laser split between two KDP probes mounted on metal breadboards using a beam-splitter (BS), and (b) using the Cerberus laser.

At Vulcan, the probe KDP crystals were oriented to have their longitudinal axes aligned with the Vulcan target chamber North-South (N/S) and East-West (E/W) directions, and were located 1.25 m from target chamber centre (TCC), with no direct line of sight to the target; this setup is described in greater detail in [15]. For the Cerberus experiment, a more compact all-dielectric single channel diagnostic with fiber beam transport into and out of the chamber was used, such that no loss of optical alignment occurred upon moving the diagnostic or pumping down the chamber to low pressures.

As the Cerberus laser energies on-target were considerably less than at Vulcan (~1-2 J at ~650 fs instead of ~300-400 J, both focused to ~5 μm FWHM spots), the KDP was placed within the 1 m diameter steel spherical chamber, 0.25 m from the target, where the EMP fields radiated from the target assembly are expected to be stronger according to a $1/r^2$ scaling. The sensor crystals were shielded by 5 mm of plastic, with opaque tubes enclosing the beam path to minimize stray light collection by the output fiber-coupling optics.

3 Time-domain electric field behaviour

3.1 Foil targets

Key shot and target information relating to the ultra-thin plastic targets used during the Vulcan experiment can be found in Table 1. The Vulcan results give some indication that the peak EMP fields generated during the interaction lightly scale with the target thickness, with a trend towards peak fields increasing with decreasing thicknesses, as seen in Fig. 2. Only the dominant N/S mode results are shown, as the weaker E/W results suffer from signal-to-noise issues, making interpretation difficult. Conversely, the drive pulse energies on target do not seem to strongly affect the peak E-fields detected by the EO diagnostic, at least within the small energy range and target thicknesses considered, shown in Fig. 3.

Table 1. Shot and target information for shots with EMP data from Vulcan PW, where *T* is target thickness, U_{tar} is energy on target, I_0 is peak intensity and Δt is the FWHM pulse duration.

Shot #	Material	*T* [nm]	Δt [ps]	U_{tar} [J]	I_0 [10^{20} W cm^{-2}]
14	Parylene-N	550	1.00	372	14.3
18	Parylene-N	269	1.10	369	12.4
19	Formvar	20	1.30	341	9.7
20	Parylene-N	550	0.98	359	14.3
25	Parylene-N	550	1.00	319	10.9
29	Parylene-N	269	1.70	386	4.8

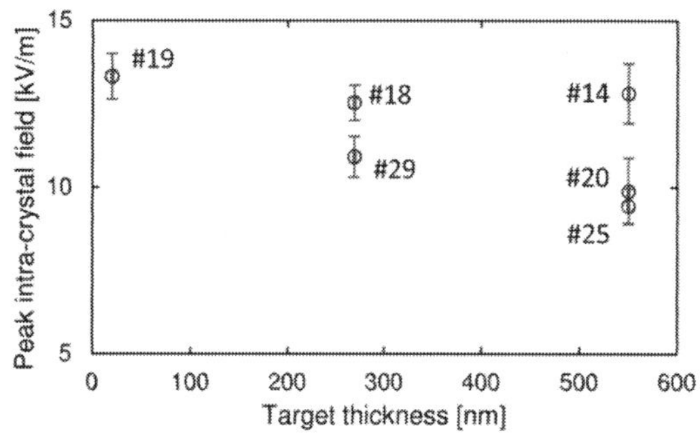

Fig. 2: Variation of the peak-intra crystal field detected by the longitudinal EO diagnostic along the Vulcan N/S direction with the target thicknesses. Shot numbers are given as annotations. The error bars are determined by the noise level on a given shot (annotated), where we have defined error bar lengths by the apparent E-field obtained by processing the reference noise channel data as if it were an E-field measured using EO probes.

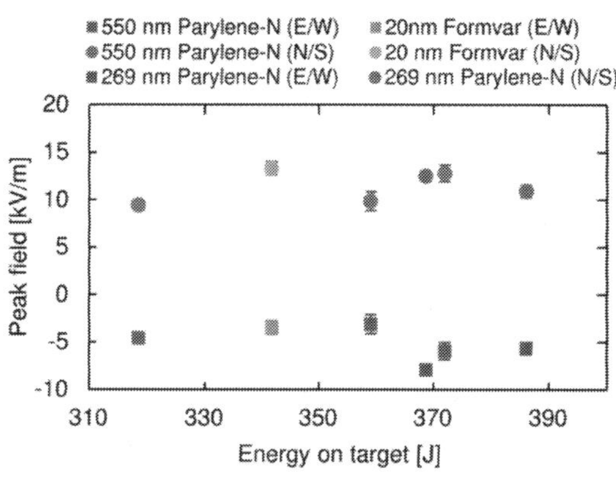

Fig. 3: Variation of the peak-intra crystal field detected by the EO diagnostic along the Vulcan N/S direction with the drive laser pulse energy on target. The error bars are found in the same way as in Fig. 2.

Although the dataset is small and thus care must be made when drawing conclusions, this data-set represents the first potential experimental observation of the phenomenon of enhanced EMP with thinner foil targets. This observation might be explained by a limited version of enhanced electron heating that occurring in the relativistic regime when the electron skin-depth (as seen by the drive laser) is comparable to the target thickness; even in a classically over-dense plasma, where fields do not propagate, the laser field still exerts a ponderomotive drive on the electrons within the target plasma, resulting in rapid and efficient heating of electrons. These effects have been observed for ultra-thin targets in both simulations and experiment by Yin and Albright, during investigations of ion acceleration in the relativistic transparency or laser break-out afterburner (BOA) regimes [16–18], although the pulse temporal contrast at Vulcan (~10^6 over 200 ps) is too low to properly access these regimes. No EMP was detected from Al-foil targets with Cerberus shots - only optical scatter or self-emission coupling into the output fiber was measured suggesting that much weaker EMP fields are generated by joule-level drive pulses.

During the experiment, the E-field detection threshold of the EO-EMP diagnostic was in the range of ≈400-700 V/m, depending on the base voltage level above noise on the photodiode for a given shot. This threshold is approximately half that achieved during the Vulcan experiment, due to the mean zero-field signal on the photodiodes being ≈50 mV instead of ≈20 mV. Hence, the Cerberus EO data can only tell us that the electric field component of the EMP from levitated target or solid target shots driven by few-TW laser pulses (focused to ≈ 5-10 μm spot diameters) is less than ≈500 V/m, despite the probe being placed 20 cm from TCC, compared to 1.25 m while at Vulcan PW. This highlights the relative insensitivity of EO diagnostics in comparison to conductive probes, although strategies for improving this are outlined in [15].

As the EO probe is mobile, measurements were attempted along and perpendicularly to the interaction chamber's radial axis, as well as at 180° to potentially reduce the amount of incident scattered light, in order to determine if peaks were self-emission or real fields. The distance of the probe crystal from TCC was kept fixed. This means that the radiated electric field strengths were less than ≈500 V/m in all probing directions, which is still a relatively large value. As an additional check, the Faraday cage containing the oscilloscope was left open for some shots to look for ambient electrical noise; none was detected on the oscilloscope. This suggests that there is very little EMP emission from the few-joule laser-matter interactions achievable with Cerberus.

3.2 Optically levitated micro-targets

According to the Poyé model, the return current drawn through the target assembly is a key contributor to EMP generated in a laser-plasma interaction. Hence, a target geometry which is completely electrically isolated from the interaction chamber and earth should remove the antenna-like RF radiation component from the target holder, and consequently reduce the EMP. Optically levitating targets is a practical method for achieving these conditions. At Vulcan, ≈10 μm silicone oil droplets were used due to their in-vacuo stability. Further development of the optical trapping technology by S. Giltrap has since enabled hollow spherical shell targets to be trapped, alongside oil droplets and glass spheres; these were also shot during the Cerberus campaign.

The EO diagnostic was unable to detect any EMP, above the minimum resolvable field-strength limit set by experimental noise, from interactions with levitated micro-targets using either Cerberus at the few J level, or more significantly, Vulcan PW at >300J, where accelerated protons of energies >30 MeV were observed from droplets with no accompanying EMP; the ion acceleration with EMP reduction results will be discussed separately in an upcoming publication led by S. Giltrap. Hence, any generated EMP fields were below the experimental electrical noise level, meaning any fields were less than ≈1 kV/m and ≈500 V/m during the Vulcan and Cerberus campaigns respectively, and are thus not a significant concern for electrical interference with laboratory equipment. Another upcoming article, led by F. Consoli, will separately cover the comparison between a suite of different EMP diagnostics.

4 Spectral analysis of EMP

EMP spectral information can be obtained by Fast Fourier Transform of the time-domain data; an example spectrum (shot #29) and the ambient EMP noise spectrum from an identical adjacent photodiode and fiber-coupling system are shown in Fig. 4, with the horizontal-plane characteristic modes of the Vulcan PW interaction chamber (without in-chamber metallic hardware), calculated by Mead et al. [10] is indicated by the dotted lines at 76, 101, 152, 202 and 228 MHz.

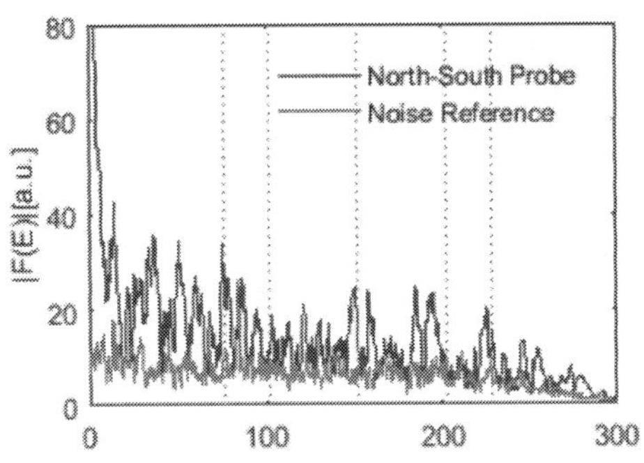

Fig. 4: An example EMP spectrum from shot #29 of the Vulcan PW experiment with shot and target parameters given in Table 1. The characteristic chamber cavity modes predicted by Mead for Vulcan PW [10] are indicated by the dotted lines. The reference channel spectrum, using an identical photodiode and fiber-coupling system, is shown to demonstrate the contribution of ambient EMP coupling to the read-out system.

From this, it is clear that the measured peaks are detected by the probe, with the exception of the 82 MHz feature, corresponding to the vertical fundamental mode of the Vulcan interaction chamber and the ≈140 MHz feature. The analysis in the following sections will examine the detailed shot-to-shot reproducibility and variation of spectral measurements with experimental parameters.

4.1 Reproducibility of spectral features

The reproducibility of the Vulcan EMP measurements shot-to-shot is important to quantify when deploying novel diagnostics. While there are indeed some differences in the spectra, numerous common spectral features, summarized in Table 2, are observable on all shots, irrespective of target or drive laser pulse parameters. Spectral components up to 200 MHz are shown, as above this frequency the signal-to-noise ratios were too small for meaningful conclusions to be made. Certain components, such as the chamber fundamental modes, are expected to be measured on every shot as the bulk in-chamber architecture (optical tables etc) remains fixed. As previously discussed, other variations in the EMP's spectral content can be attributed to the movement of conductive material such as optical mounts in the chamber between shots, which was not specifically controlled during the experiment. Within the diagnostic's bandwidth range (0-250 MHz), the spectral resolution is limited to ~0.21 MHz, in this case by the oscilloscope time-base settings and sampling frequency.

Table 2: Centre frequencies of the peaks measured on a given shot near the identified spectral features given in Column 1.

Peak [MHz]:	Observed frequency [MHz] at shot #					
	#14	#18	#19	#20	#25	#29
30	30.2	32.5	31.0	30.8	29.4	32.3
40	39.8	39.4	39.0	41.0	40	41.0
50	50	49.0	50	51.9	49.8	49.6
60	60	61.3	61.3	61.3	60.6	59.6
71	72	70.2	71.6	70.8	71.3	69.6
74	74.0	74.6	74.6	74.2	74.6	75.0
76	75.4	76.7	78	76.7	75.8	75.0
94	94.2	95.8	93.3	94.0	95	94.0
102	102.9	100.4	102.3	102.5	104.6	102.7
120	120.2	120.2	121.9	119.4	120.2	120.2
152	151.5	152.1	152.7	152.3	152.3	150.4
200	199.6	199.6	199.6	199.8	199.8	199.6

A higher degree of reproducibility between shots taken on the same day of the experiment can be seen in the spectra from shots #18, #19 and #20 of the Vulcan campaign. The other 3 shots, (#14, #25 and #29) all took place on different days during the experiment. The reproducibility is quantified by considering the standard deviation (SD) in the mean center-frequency $\sigma/\sqrt{n}$ of features measured for a given identified frequency value (Column 1 of Table 2), where we have divided by $\sqrt{n}$ to account for differing sample sizes. The SD for the set of shots restricted to the single day S_{rs}, was then subtracted from the SD all the shots, S_{all}, and the differences plotted in Fig. 5. A positive difference value indicates that the SD in the mean peak position for an identified spectral feature is smaller for the restricted dataset; a majority (8 of the 12) features, exhibit a value ≥ 0, suggesting a higher degree of spectral reproducibility between shots taken on the same day.

This chronological link, with no clear trend linked to target or shot parameters, suggests that minor spectral modifications are more likely to be linked to movement of conductive objects within the interaction chamber between shots on different days, which can affect the frequency modes supported within the chamber [3]. Although it was not monitored and thus cannot be properly confirmed, there is a greater likelihood that the positions of metal objects within the chamber were more similar for shots taken on the same day. While the dataset, and hence sample size, is small such that is difficult to draw reliable conclusions, this evidence suggests the internal chamber architecture and contents are a dominant factor in determining the spectral make-up of the EMP, rather than the thickness of the plastic foil targets used.

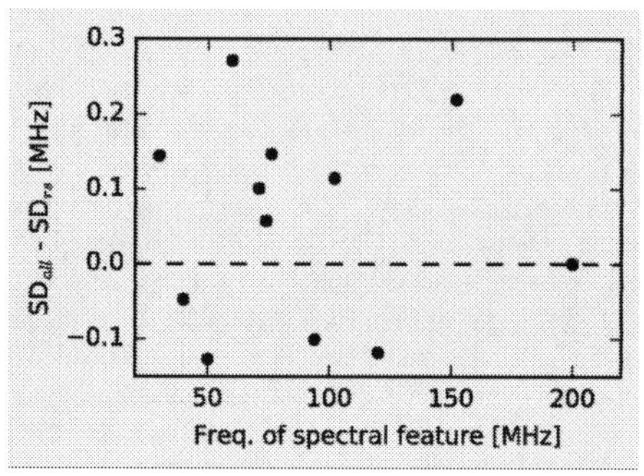

Fig. 5: A plot of the difference between the standard deviations (SD) S_{all} of the mean position of spectral features between all shots and the SD Sr_s calculated from a restricted population of shots taken on the same day (#18, #19, #20). The SD of the mean $\sigma/\sqrt{n}$, where n is the number of samples, has been used to account for the different sample sizes.

This does not imply that target geometry is unimportant, as the plastic foils mainly differed in thickness, and no EMP was observed with levitated targets. Furthermore, the target holder assembly was identical for foil shots; as this forms part of the radiating element according to the Poyé model, it follows that the frequency content of the initial radiated EMP will remain consistent. This lack of spectral dependence on the target thicknesses is in contrast to the peak E-fields measured, which are dependent on the ejection of hot electrons, and hence the plasma electron temperatures.

An interesting feature of the electro-optic EMP data is that some of the fundamental chamber resonant modes were observed in the measurement direction perpendicular to their expected direction; for example, the electric field directions of 76 and 152 MHz modes were predicted to be along the E/W chamber direction by Mead et al. [10], rather than N/S. In an electromagnetic parallelepiped-shaped hollow cavity, TE and TM cavity modes are allowed. Some of these modes are degenerate, so share resonance frequencies. Moreover, the presence of objects in the chamber can locally change the modal field distribution without changing the resonance frequency remarkably [3].

4.2 Time-frequency analysis

Short-time window Fourier analysis techniques can be used to infer information about the dynamical evolution of spectral modes emitted following the laser-plasma interaction. The spectrogram for electric fields measured along the N/S direction is shown in Fig. 6. From this, a late-time, secondary detection of EMP fields with frequencies matching the interaction chamber cavity modes was observed. The most likely explanation for this behaviour is that the two previously discussed mechanisms contributing to EMP generation are temporally offset; an electron bunch ejected from the target plasma will have a finite transit time to the interaction chamber walls, whereas a positively charged target can immediately draw a neutralization current through its supporting mechanical assembly.

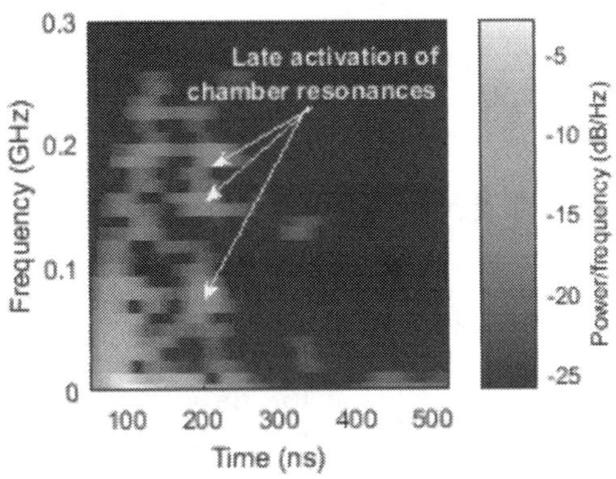

Fig. 6: Time-resolved spectrogram for EMP fields measured along the Vulcan N/S direction taken on shot #29 (see Table 1). There appears to be some late-time activation of the expected interaction chamber modes, likely due to the ejected electron bunch striking the chamber walls after a finite transit time, whereas EMP radiated due to the discharge current through the target holder can occur as soon as the target becomes positively charged. This plot has been adapted from [15].

This explanation can be tested against time-of-flight (TOF) measurements, sensitive to charged particles and X-rays. We were not able to obtain TOF diagnostic data for the shots with EO-EMP data, however the TOF data from one of the other shots in the series (#21) is still useful for providing some insight into the delayed activation of the chamber cavity modes. A synthetic monocrystalline CVD diamond detector was used as a TOF diagnostic. This was fabricated by Tor Vergata University laboratories with superficial interdigitalized electrodes [19,20] and operated by colleagues from ENEA-Frascati. It was placed in the Vulcan chamber's western wall, with a direct view of the target and 100 V biasing, at distance of ≈1 m from TCC.

A TOF trace for shot #21 is shown in Fig. 7; from this, we can observe the separate arrival of X-rays (at 390 ns) as well as "fast" and "slow" particles (above 490 ns) at the diamond detector. As the ambient EMP noise coupling proved problematic to shield, a 60 m length of RG58 cable was introduced between the detector and the oscilloscope, adding a delay of ≈311 ns and temporally separating the EMP background from the diamond signal.

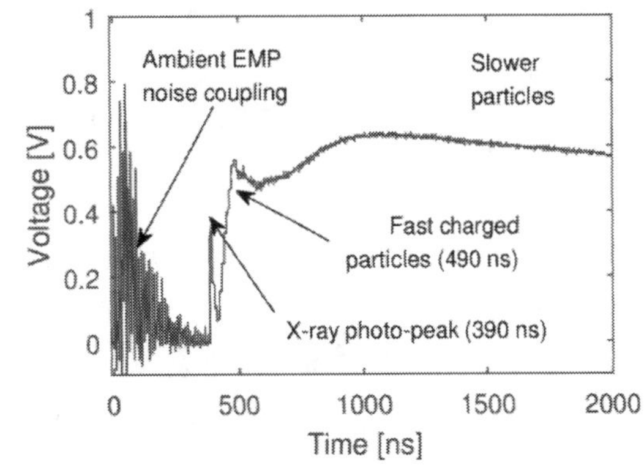

Fig. 7: A time-of-flight trace recorded using a biased CVD diamond detector. A 40 nm thick Formvar target was used for this shot, with a 1.6 ps FWHM, 357 J on target, with a peak intensity of 5.6×10^{20} W cm^{-2}. The temporal resolution of this measurement was 1 ns. The time axis has been shifted such that the initial laser strike $T_0 = 0$. A 60 m length of RG58 cable was used to temporally separate the diamond signal from the ambient EMP noise (beginning near $T_0 = 0$) coupling to the oscilloscope. The key contributions to the measured signal are labelled on the plot.

The X-ray photo-peak provides a reference for the shifted T_0, the time of the initial laser strike, relative to other features, as the time taken for prompt photons emitted from TCC to reach the diamond was calculated to be ≈3.3 ns. The arrival of fast particles (few-MeV fast ions, few-keV electrons) peaked ≈100 ns after the photo-peak, followed by a broad peak representing slower charged particles. These timescales are approximately consistent with the N/S mode EMP spectrogram for shot #29 (Fig. 6), where secondary activation of the fundamental chamber modes can be seen ≈110 ns after the initial EMP emission. Although we expect some shot-to-shot variation in the speeds to which ejected fast particles are accelerated, there should be some finite time between fast electrons striking the chamber walls and excitation of RF modes. This provides some evidence for the cavity modes being excited later in time by ejected particles striking the chamber walls, while the radiated fields emitted from the target assembly due to the neutralization current occur earlier in time, closer to T_0.

If true, this implies that the RF-emission from the target assembly is the larger contribution to EMP in experiments of this type, which could explain the lack of measurable EMP from shots with optically levitated micro-droplets, which have no physical discharge current path but still result in charged particle ejection.

5 Conclusions

Although the electro-optic EMP dataset obtained from Vulcan PW is limited to only 6 shots, there appear to be certain trends identifiable from the time- and frequency domain data. Firstly, the peak EMP E-fields from ultra-thin plastic foil targets appear to negatively correlate with the target thickness; this cannot be explained simply by energies on target being higher for certain shots, as there is no clear correlation between the input energy and the measured peak fields. Hence, at least within the restricted ranges of energy and target thicknesses considered, we suggest that EMP produced by a laser-plasma interaction scales with the target geometry, potentially due to enhanced electron heating when the thickness of targets is comparable to the electron skin depth. The lack of measurable EMP from electrically-isolated, optically-levitated, targets also indicates that removing the discharge current component of EMP is very significant in mitigating the EMP problem.

In contrast to the peak intra-crystal field values, the EMP spectra obtained do not appear to show any strong dependence on target thickness, with spectral features largely being reproduced in approximately the same positions on every shot. There appear to be minor shot-to-shot variations in the position of several identified spectral features, quantified by the standard deviations of the mean center frequency, which appear to be reduced for shots taken on the same day. This suggests that minor modifications to spectral features are more likely to be driven by movement of conductive objects within the chamber, as opposed to variation in target thicknesses.

Time-frequency analysis combined with time-of-flight measurements, which show a late secondary appearance of frequencies matching the resonant cavity modes $\approx$10 ns after the arrival of fast electrons at the chamber walls, also tentatively suggest that the cavity modes activate later in time than the initial EMP emission (which may include frequency components matching the cavity modes). We thus consider the early-time EMP contribution to be RF-emission from the target assembly, as in the Poyé discharge current model.

Additional shots at petawatt facilities are required to confirm these observations, however they provide a starting point which can be used to tailor experiments further studies into laser-driven EMP generation, in terms of experimental parameters such as target type and target holder design, particularly as ultra-high energy petawatt systems such as PETAL and Apollon 10 PW become active.

We would like to thank Margaret Notley, our link scientist, for her contributions to planning and running the Vulcan experiment. We acknowledge funding supporting the authors from EPSRC (Studentship Nos 1584149, 1367355, 1228252, 1738157 and Grant No. EP/K022415/1), Industrial CASE awards (30318116, 1103059, 30242504, 30353497) in partnership with AWE plc. The work of F.C and R.D.A. has been partially carried out within the framework of the EUROfusion Consortium and has received funding from the Euratom research and training programme 2014–2018 under grant agreement No. 633053. The views and opinions expressed herein do not necessarily reflect those of the European Commission.

References

1. C. G. Brown, A. Throop, D. Eder, et al., *Journal of Physics: Conference Series* (2008), **112**, p. 32025.
2. F. Consoli, R. De Angelis, P. Andreoli, et al., *2015 IEEE 15th International Conference on Environment and Electrical Engineering, EEEIC 2015 - Conference Proceedings* (2015), pp. 182–187.
3. F. Consoli, R. De Angelis, P. Andreoli, et al., *Physics Procedia* (2015), **62**, pp. 11–17.
4. M. De Marco, J. Krása, J. Cikhardt, et al., *J. Instrum.* **11**, C06004–C06004 (2016).
5. M. De Marco, J. Cikhardt, J. Krása, et al., *Nukleonika* **60**, 1–6 (2015).
6. J. L. Dubois, F. Lubrano-Lavaderci, D. Raffestin, et al., *Phys. Rev. E - Stat. Nonlinear, Soft Matter Phys.* **89**, 1–15 (2014).
7. D. C. Eder, R. W. Anderson, D. S. Bailey, et al., *Journal of Physics: Conference Series* (2010), **244**, p. 32018.
8. J. A. Miragliotta, B. Brawley, C. Sailor, et al., *SPIE Defense, Security, and Sensing* (2011), **8037**, p. 80370N1-8.
9. J. Miragliotta, J. Spicer, B. Brawley, et al., *Proc. SPIE 8381, Laser Technology for Defense and Security VIII, 83811N* (2012), **8381**, p. 83811N1-7.
10. M. J. Mead, D. Neely, J. Gauoin, et al., *Rev. Sci. Instrum.* **75**, 4225–4227 (2004).
11. A. Poyé, S. Hulin, M. Bailly-Grandvaux, et al., *Phys. Rev. E - Stat. Nonlinear, Soft Matter Phys.* **91**, 1–6 (2015).
12. A. Poyé, J. L. Dubois, F. Lubrano-Lavaderci, et al., *Phys. Rev. E - Stat. Nonlinear, Soft Matter Phys.* **92**, 1–17 (2015).
13. C. J. Price, T. D. Donnelly, S. Giltrap, et al., *Rev. Sci. Instrum.* **86**, (2015).
14. G. A. Massey, D. C. Erickson, and R. A. Kadlec, *Appl. Opt.* **14**, 2712–9 (1975).
15. T. S. Robinson, F. Consoli, S. Giltrap, et al., *Sci. Rep.* **7**, 983 (2017).
16. L. Yin, B. J. Albright, B. M. Hegelich, et al., *Laser Part. Beams* **24**, 291–298 (2006).
17. L. Yin, B. J. Albright, B. M. Hegelich, et al., *Phys. Plasmas* **14**, (2007).
18. B. J. Albright, L. Yin, K. J. Bowers, et al., *Phys. Plasmas* **14**, (2007).
19. M. Marinelli, E. Milani, G. Prestopino, et al., *Applied Surface Science* (2013), **272**, pp. 104–108.
20. R. De Angelis, F. Consoli, C. Verona, et al., *J. Instrum.* **11**, C12048–C12048 (2016).

Strong electromagnetic pulses generated in laser-matter interactions with 10TW-class fs laser

Piotr Rączka[1,*], *Marcin* Rosiński[1], *Agnieszka* Zaraś-Szydłowska[1], *Jerzy* Wołowski[1], and *Jan* Badziak[1]

[1]Institute of Plasma Physics and Laser Microfusion, 01-497 Warsaw, Poland

Abstract. The results of an experiment on the generation of electromagnetic pulses (EMP) in the interaction of 10TW fs pulses with thick (mm scale) and thin foil (μm scale) targets are described. Such pulses, with frequencies in the GHz range, may pose a threat to safe and reliable operation of high-power, high-intensity laser facilities. The main point of the experiment is to investigate the fine temporal structure of such pulses using an oscilloscope capable of measurements at very high sampling rate. It is found that the amazing reproducibility of such pulses is confirmed at this high sampling rate. Furthermore, the differences between the EMP signals generated from thick and thin foil targets are clearly seen, which indicates that besides electric polarization of the target and the target neutralization current there may be other factors essential for the EMP emission.

1 Introduction

The problem of strong electromagnetic pulses in the frequency range from tens of MHz to several GHz emitted as a result of laser-target interaction has a long history [1-6], but a complete quantitative understanding of this effect is still lacking. Such pulses strongly interfere with the data acquisition systems and may pose a threat to safe and reliable operation of high-power, high-intensity laser facilities, so it is no wonder that mechanisms for their generation continue to attract attention [7-14]. In particular, in a series of recent articles [7-9] attention was given to the phenomenon of electric polarization of the target as a result of laser-target interaction, and the effect of EMP generation via the target neutralization current. Investigations reported in [7,8] utilized a specific form of the target called "lollipop" - which facilitated measurement of the target neutralization current and determination of the target charge - and they focused on thick (mm scale) targets. This analysis was recently extended [15] to thin (μm scale) foils, which are commonly used in laser-ion acceleration experiments; a dedicated experiment devoted to this issue was performed at the Eclipse laser facility in CELIA.

In [15] two characteristic effects observed in an experiment were noticed: (1) the patterns of EMP emitted for various laser pulse energies and pulse durations appear to be reproducible to an astonishing degree, despite the appearance of EMP as a random and chaotic phenomenon; (2) the patterns of EMP registered for thick and thin targets were somewhat different; they were dominated by the 1 ns-scale pulsation of the neutralization current, but displayed a sub-ns structure, which is indicative of a deviation from the simple picture of the target acting as an antenna and EMP originating primarily as a result of electric polarization of the target and the neutralization current. However, the characteristic patterns observed at the sub-ns time scale were recorded at the limit of the oscilloscope sampling rate, so one could suspect that their appearance would change once the measurement of EMP would be performed with better temporal resolution. In order to explore this issue we looked into the data collected in a brief experimental campaign consisting of 25 shots on the 10 TW fs laser facility at IPPLM in the end of 2016, when we had the rare opportunity to use a very high bandwidth oscilloscope. In this note we report briefly on the results of this campaign.

2 The experimental setup

The high-power laser facility in IPPLM is a Ti:Sapphire laser delivering an s-polarized beam with 810 nm central wavelength [16]. The beam is focused by an off-axis parabolic mirror to a spot approximately 20 μm in diameter. The beam was incident on the target at the angle of approximately 5°. The pulse duration was kept at 50 fs, and the pulse energy was varied in the range of 140 mJ to 270 mJ. The ns contrast was better than 10^{-8}. Similarly as in the case of the Eclipse experiment [15] the targets we used were placed in a "lollipop" target holder which consisted of a brass ring 14.0 mm in diameter, supported on a thin brass wire stalk - which in our case extended 12.0 mm above the standard target mount, i.e a distance was much shorter than in the case of the Eclipse experiment [15]. Here we report on two types of targets: (a) a massive Cu "pill" 10.1 mm in diameter and 1.0 mm thick; (b) an Al foil 6.0 μm thick,

* Corresponding author: piotr.raczka@ifpilm.pl

https://doi.org/10.1051/epjconf/201816703008

with a 0.3 μm layer of polystyrene on the rear side, pasted on the rear side of a Cu pill with ten 1 mm holes for shots (denoted further as AlPs target). However, there was no arrangement to measure the target neutralization current, and there was no grounded plate to ensure the mirror charge and dipole character of the radiation pattern as in the experiments on the Eclipse laser [7,8,15].

In order to measure the magnetic field we used a Prodyn B24(R) B-dot probe with an average equivalent area of ~9×10^{-6} m^2, which was placed at a point located 7.3 cm above the target, 25.0 cm behind the target and 7.0 cm to the right relative to the incident beam direction. The probe was oriented so as to measure the tangential component (relative to the vertical axis defined by the target stalk) of the magnetic field. The probe was connected through a Prodyn BIB-100G balun[a] and 2.53 m long RG400 Pasternack coaxial cable to the Keysight MSOV334A oscilloscope with a 16 GHz bandwidth.

3 Experimental results

In order to determine the physical value of the tangential component of the magnetic field B_{tg} the following procedure was used: (a) the raw signal for dB_{tg}/dt was corrected for the 8 dB attenuation introduced by the balun; (b) a high pass Fourier transform (FT) filter was applied with the 180 MHz lower cutoff, determined empirically to reproduce null signal before the arrival of the main electromagnetic pulse; (c) a low pass FT filter was applied with a 10 GHz cutoff reflecting the location of the 3 dB point of the balun; (d) the correction was applied to take into account the frequency-dependent attenuation of the coaxial cable; (e) the inverse FT was performed to obtain the corrected and regularized data for dB_{tg}/dt; (f) the B_{tg} was obtained as a function of time by direct integration of dB_{tg}/dt. The overall B_{tg} signal has a well-known form a sequence of initial spikes followed by gradually decaying oscillations lasting approximately 200 ns. However, it is the first few ns of the signal that are most interesting, because that part originates directly from the target and is a direct reflection of the laser-target interaction; the signal at later times is a superposition of fields from various sources, including reflections of the original pulse from the chamber walls and equipment inside the chamber, and excitations of the chamber eigenmode oscillations. The B_{tg} obtained with the thick Cu target is illustrated in the Fig. 1, where two pulses recorded at different laser energies are compared as a representation of a broader sample. This figure is a good illustration for the astonishing reproducibility of the EMP patterns, which is fully confirmed in our study at a higher oscilloscope sampling rate.

In Fig. 2 we show the B_{tg} obtained with the thin AlPs target, where again two representative pulses recorded at different laser energies are displayed as an illustration for a broader sample. Also in the case of thin targets we find the pulse shape to be amazingly reproducible. Furthermore, we see that the pulses have a pronounced sub-ns structure which is different for thick and thin targets considered in our study. This is interesting, because our thin foil target support was designed so as to minimize electromagnetic differences between the thin foil and thick target setups. The case of a thin foil target may differ from a thick target by the fact that at the same laser pulse energy and duration one may expect somewhat larger electric charge to be generated on the target [15], but this is by itself is not sufficient to explain qualitative differences between EMP signal in those two cases. This indicates that in the setup we used the target charge and the neutralization current are not the only factors affecting the shape of the EMP signal.

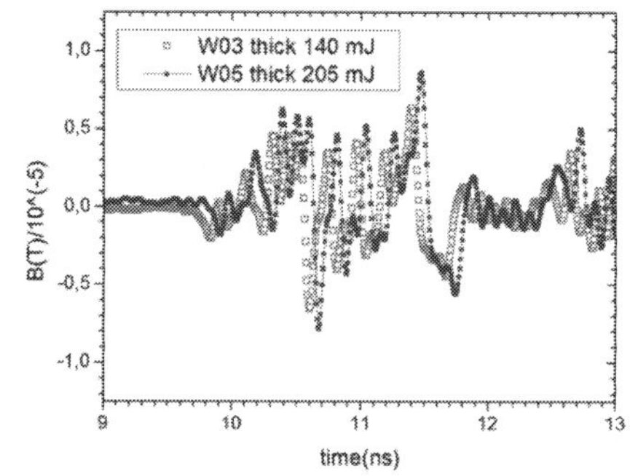

Fig. 1. The tangential component of the magnetic field as a function of time, as obtained in shots at the thick Cu target at two values of the laser pulse energy. Only the initial evolution of the pulse is shown, the total extent of the EMP signal is of the order of 200 ns.

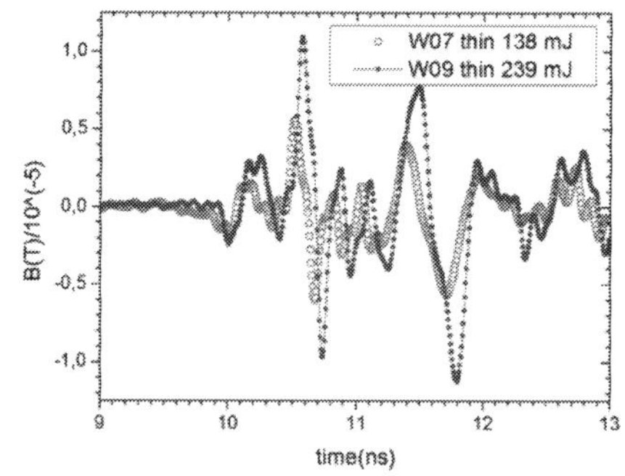

Fig. 2. The tangential component of the magnetic field, as obtained in shots at the thin foil AlPs target at two laser pulse energies.

For a further comparison of thin and thick targets we show in Fig. 3 a Fourier transform of a small part of the total signal, encompassing only the first 4 ns of the signal. We see that there are qualitative differences between these transforms: the short-time transform of the thin foil signal has a broad peak in the 1-2 GHz region and another one near 3.5 GHz, whereas the transform for the thick target is less pronounced in the 1-

[a] https://ppmtest.com/products/baluns/prodyn-baluns/bib-series/

2 GHz region, but shows a peak near 4.5 GHz. Both spectra show some structure at the top of the range.

4 Conclusions

On the basis of our short study we may say that the amazing reproducibility of the EMP signals from shot to shot is fully confirmed at high oscilloscope sampling rate. We may also say that the presence of a sub-ns/multi-GHz component in the EMP signals is unlikely to be an oscilloscope artefact. Furthermore, we see that despite general similarities between EMP signals generated from thin foil and thick targets there are also clear differences, related to the multi-GHz component, indicative of the presence of EMP generation mechanisms beyond simple target polarization and neutralization current flow. We hope to further clarify these points in forthcoming experiments. In particular, it would be important to perform EMP measurements using electromagnetic probes sensitive to frequencies above 10 GHz.

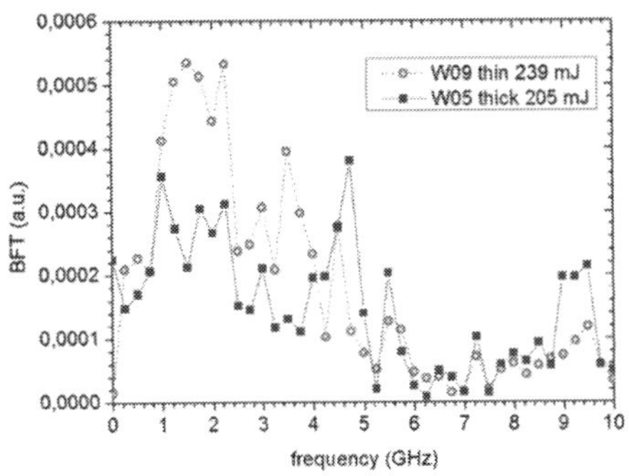

Fig. 3. The one-sided discrete Fourier transform spectrum of the first 4 ns of the EMP pulse generated from the thick Cu target (W05) and the thin AlPs target (W09).

The experiment on which we report here was closely modelled on the experiment devoted to EMP from thin targets at Eclipse facility [15], which was done in collaboration with J.-L. Dubois, S. Hulin, and V. Tikhonchuk (CELIA), to whom the present authors are greatly indebted. We are also grateful to D. Neely (RAL) for lending us the Prodyn B24 B-dot probe. We want to express our gratitude to the company AM Technologies (Warsaw) and in particular to their representative T. Szabłowski for giving as access to a 16 GHz oscilloscope. This research is supported by the Polish National Science Centre grant Harmonia 2014/14/M/ST7/00024. Our collaboration has profited greatly from the intellectual environment created by the COST Action MP1208.

References

1. J.S. Pearlman, G.H. Dahlbacka, Appl. Phys. Lett., **31**, 414 (1977).
2. M.J. Mead, D. Neely, J. Gauoin, R. Heathcote, P. Patel, Rev. Sci. Instrum. **75**, 4225 (2004).
3. J. Raimbourg, Rev. Sci. Instr. **75**, 4234 (2004).
4. C. Stoeckl et al., Rev. Sci. Instrum. **77**, 10F506 (2006).
5. J.L. Bourgade et al., Rev. Sci. Instrum. **79**, 10F301 (2008).
6. C.G. Brown Jr., A. Throop, D. Eder, J. Kimbrough, J. Phys. Conf. Ser. **112,** 032025 (2008).
7. J.-L. Dubois et al., Phys. Rev. **E 89**, 013102 (2014).
8. A. Poyé et al., Phys Rev. **E 91**, 043106 (2015).
9. A. Poyé et al., Phys Rev. **E 92**, 043107 (2015).
10. J. Krása et al., Laser Part. Beams **33**, 601 (2015).
11. F. Consoli et al., Sci. Rep. **6**, 27889 (2016).
12. T.S. Robinson et al., Sci. Rep. **7**, 983 (2017).
13. J. Krása et al., Plasma Phys. Control. Fusion **59**, 065007 (2017).
14. M. De Marco et al., Phys. Plasmas **24**, 083103 (2017).
15. P. Rączka et al., to appear in Laser Part. Beams (2017).
16. A. Zaraś-Szydłowska et al., Nukleonika **60**, 249 (2015).

https://doi.org/10.1051/epjconf/201816703009

Multi probes measurements at the PALS Facility Research Centre during high intense laser pulse interactions with various target materials

Massimo De Marco [1, 2,*], *Josef* Krása[1], *Jakub* Cikhardt[3,5], *Fabrizio* Consoli[4], *Riccardo* De Angelis[4], *Miroslav* Pfeifer[1,3], *Miroslav* Krůs[1,3], *Jan* Dostál[1,3], *Daniele* Margarone[1], *Antonino* Picciotto[6], *Andriy* Velyhan[1], *Daniel* Klír[3,5], *Roman* Dudžák[1,3], *Jiří* Limpouch[2] and *Georg* Korn[1]

[1]Institute of Physics CAS, v. v. i., Na Slovance 2, 182 21 Prague, Czech Republic
[2]Czech Technical University in Prague, FNSPE, Břehavá 7, 115 19 Prague 1, Czech Republic
[3]Institute of Plasma Physics CAS,Za Slovankou 3, 182 00 Prague 8, Czech Republic
[4]ENEA, Fusion and Nuclear Safety Department, C.R. Frascati, Via E. Fermi 45, 00044 Frascati, Italy
[5]Czech Technical University in Prague, Faculty of Electrical Engineering, Technická 2, 166 27 Prague 6, Czech Republic
[6]Micro-Nano Facility, Fondazione Bruno Kessler, 38123 Trento, Italy

Abstract. During the interaction of high intense laser pulse with solid target, a large amount of hot electrons is produced and a giant Electromagnetic Pulse (EMP) is generated due to the current flowing into the system target–target holder, as well as due to the escaping charged particles in vacuum. EMP production for different target materials is investigated inside and outside the target chamber, using monopole antenna, super wide-band microstrip antenna and Moebius antenna. The EMP consists in a fast transient magnetic field lasting hundreds of nanosecond with frequencies ranging from MHz to tens of GHz. Measurements of magnetic field and return target current in the range of kA were carried out by an inductive target probe (Cikhardt J. *et al.* Rev. Sci. Instrum. **85** (2014) 103507).

1 Introduction

When high intense laser pulse interacts with a solid target a strong electromagnetic pulse (EMP) is generated. The main sources of EMP are the return target current flowing through the target holder and the charged particles escaping into the vacuum; both these EMP sources are directly connected with hot electrons escaping the plasma [1-8]. To obtain a complex data set on this process, various probes are employed for observation of return current and EMP emission during 3TW PALS laser pulse-target interaction experiment. While the target current probe (T-Probe) allowed us to determine occurred current onto the target as well as magnetic field around the target holder, loop antennas and linear antennas were employed for measurements of electric field and magnetic field both inside and outside the interaction chamber.

The escaping electrons generate up to hundreds kV potential at the target and 10kA current flows from the target to the ground [9-12]. The return current, that compensates the net positive target charge, generates an electromagnetic pulse because the system target-target holder is in first approximation a linear antenna emitting toroidal radiation pattern in the far-field and more complex field pattern in the near field zone [13-20]. The charged particles moving inside the target chamber produce the Lenard-Wiechert potentials with radiative and non-radiative electric and magnetic fields depending on their both velocity and acceleration [21-24].

The generated EMP is a transient pulse lasting hundreds nanoseconds which is much longer than

*Corresponding author: massimo.demarco@eli-beams.eu

https://doi.org/10.1051/epjconf/201816703009

laser pulse target interaction [25-28]. The frequencies range from MHz to tens GHz.
EMP propagates in all directions and it resonates at the resonant frequencies of the target chamber that acts as not efficient resonant cavity with many leakages (e.g. big glass windows) [29]. The presence of the EMP inside the vacuum chamber as well as in the experimental hall can cause problems with sensitive diagnostics, sophisticated data acquisition systems and, in some case, even disruptions of electronic devices can occur.
The generation of this harmful EMP depends on laser pulse characteristics, target material and shape of the target chamber as well as accessories located within it.
This contribution reports measurement carried out at PALS facility and it is focused on characterizing the magnetic field using different antennas.

2 Method and results

The employed target current probe [20] makes it possible to measure magnetic field at 5 mm distance from the target holder, which is generated only by the neutralizing target current. Fig.1 shows the increase of the magnetic field associated with target-holder current with the enhancement of the laser energy.

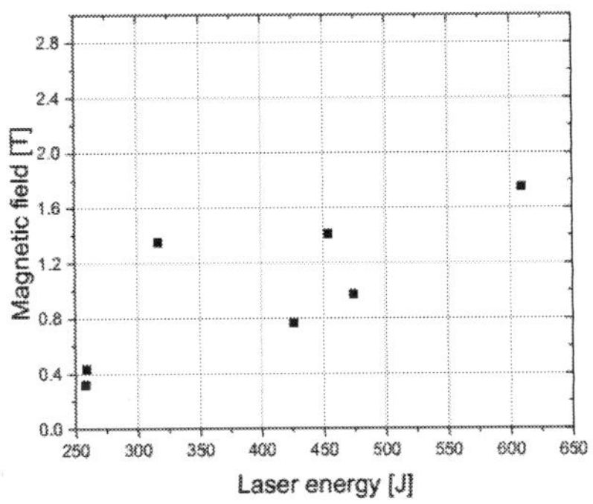

Fig.1 Laser pulse energy dependence of magnetic field produced by return target current which was detected by the inductive target probe at 5mm distance from the target holder. Target material was Tantalum.

The measurements of magnetic field generated by only the return current for different target materials, were carried out by using the target current probe [20] and they are presented in Fig.2(a). The magnetic field was normalized to the laser energy E_L, see Fig.2(b). The target material influence on the return target current is also affected by the amount of impurities chemisorbed on the irradiated target surfaces which significantly contribute to the emission of hot electrons [18]. For this reason, the study of the relation between target material and hot electron emission should be performed with perfectly cleaned targets and under ultra-high vacuum conditions.

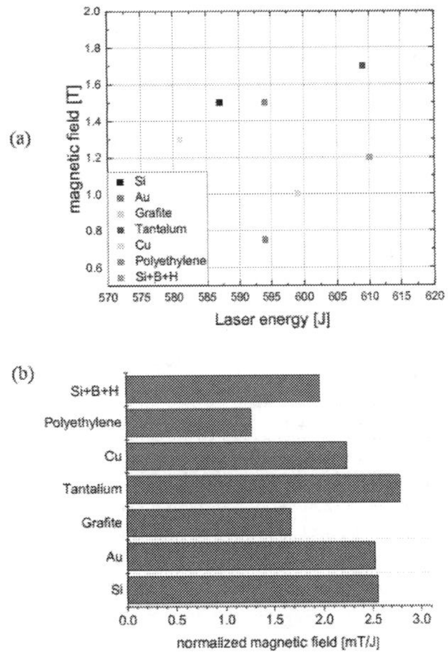

Fig.2 (a) Magnetic field at target holder, measured by the T-Probe, for different target materials. (b) Normalized magnetic field with E_L.

Moebius antennas were used outside the interaction chamber and placed at 3m distance from the PALS spherical chamber wall. The highest EMP signal was detected when the Moebius antenna was placed in front of the glass window (40 cm in diameter) serving as an input window into the chamber for the laser beam. The typical EMP signals and frequency spectra were collected for different target materials. The spectra have two dominant peaks, one between 500MHz and 1GHz and another between 1GHz and 1.5GHz. Although at the moment it is not clear the correlation of the EMP spectra with the target materials, one can presume the influence of the target surface impurities as hydrocarbons on the emission of hot electrons and consequently on the EMP intensity, as Fig.3 shows. The tantalum target exhibits the highest amplitude of the first peak (see Fig. 3a) probably denoting big amount of hydrocarbon impurities on its surface [9]. This clearly exceeds the EMP amplitudes registered for polyethylene and graphite targets.

https://doi.org/10.1051/epjconf/201816703009

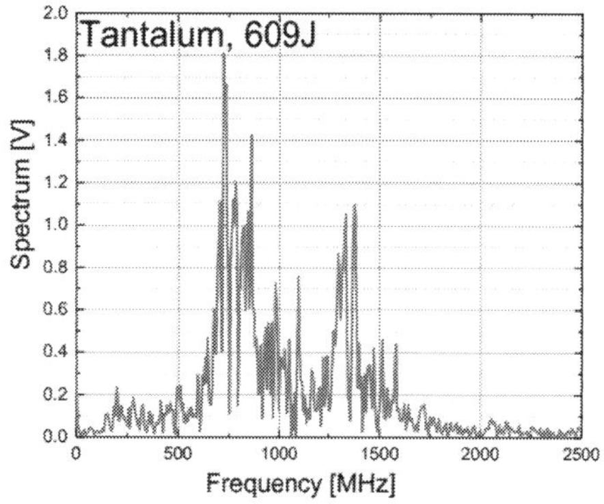

(a)

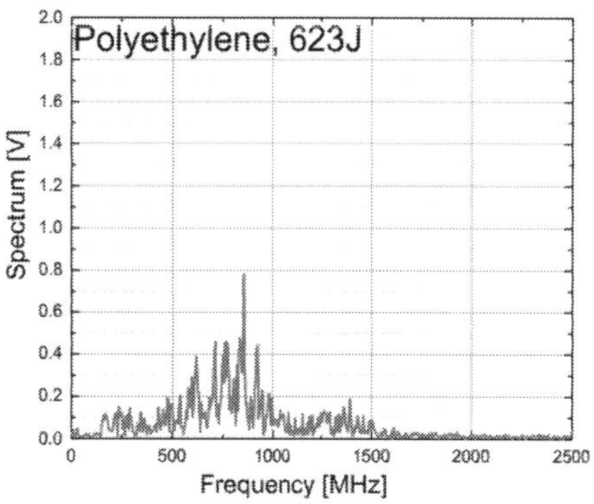

(b)

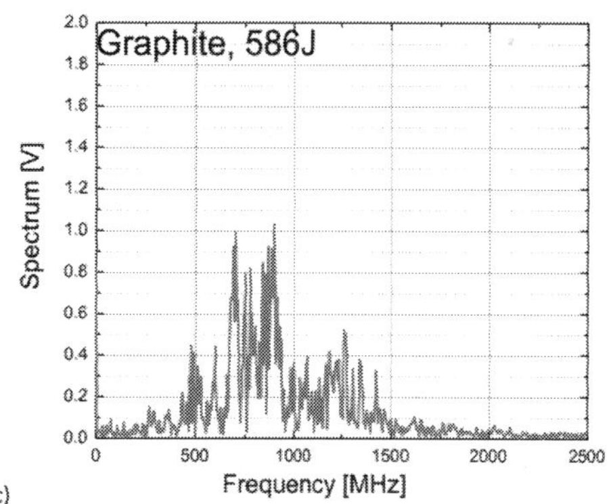

(c)

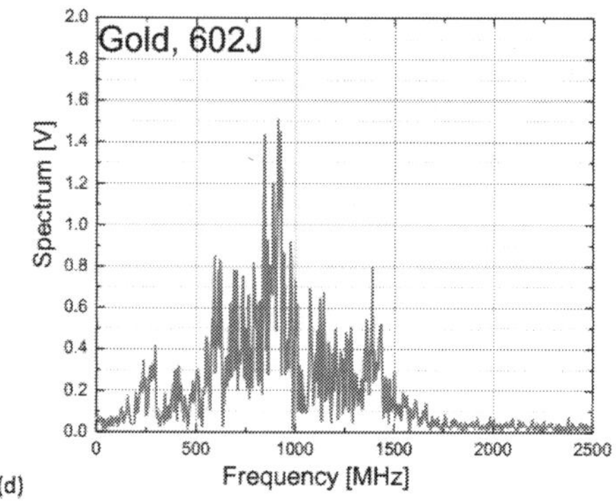

(d)

Fig.3 Frequency spectra of EMP emitted by interaction of laser pulse with various target under similar conditions (laser energy 586 – 623 J) detected by Moebius antenna outside the interaction chamber at 3m distance from the glass window. (a) Tantalum, (b) Polyethylene, (c) Graphite and (d) Gold.

Double exponential fits [14] were carried out for each detected EMP signal and both the time duration and rise time duration of the magnetic flux were determined; the mean values of rise time and FWHM time duration over 10 shots are collected in Fig.4.

The rise time is a parameter that reflects the rate of charge particles emission from target and it can be directly connected with both the atomic structure of the target material (e.g. atomic number) and the absorption of impurities on the target surface. In the case of Graphite target, the magnetic flux rise time reaches a very low value of only 8 ns while for the case of Tantalum and Si+B+H [30] the highest values are reached. In the case of Copper, Silicon, Gold and Polyethylene the rise time is very similar.

The time duration of the FWHM has a minimum for Gold at 81ns and a maximum for Tantalum, Polyethylene and Si+B+H [30] at about 110ns. The FWHM time durations of magnetic flux for different target materials are very similar and it is probably due to the same geometry of the used target chamber that plays an important role for the determination of the signals decays [19].

https://doi.org/10.1051/epjconf/201816703009

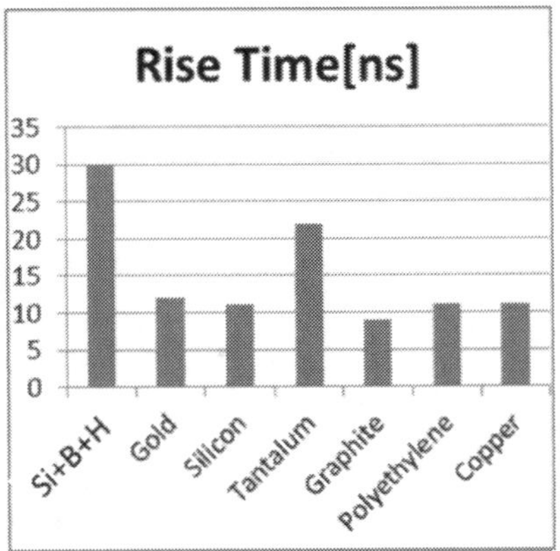

(a)

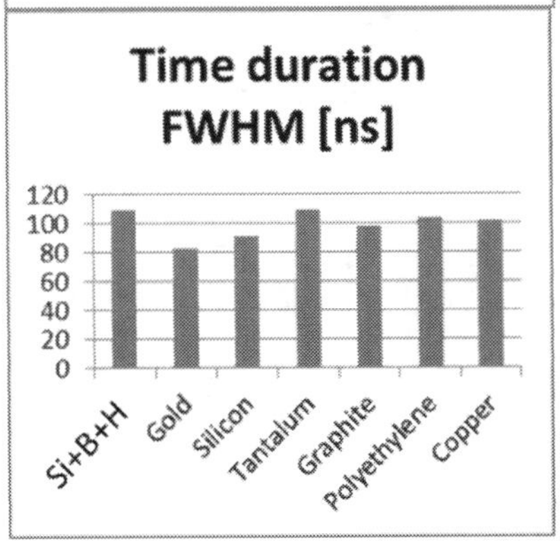

(b)

Fig.4 Mean rise time (a) and mean FWHM time duration (b) of EMP for different target materials.

The signal acquisitions carried out by linear monopole antenna [31] and super wide-band antenna [32] placed both inside the interaction chamber, above the target at 30cm from it, and protected by a thick stainless-steel slab, as from Fig.5(c), are shown in Fig.5(a) and in Fig.5(b), respectively. It is well known that it is very difficult to get reliable measurements of EMP by conductive probes when high intensity and energetic lasers are used [1-10, 12]. In fact, big fluctuations were present shot by shot on the measurements, due to background noise on the oscilloscopes caused by bad EMC-shielding of the scope and by spurious coupling of EMP with outside conductor of the coaxial cables. To resolve these problems, it was necessary to use a delay line to clean the signals, as described in references [33, 34]. Fig. 5 shows the laser energy dependence of the EMP amplitude for various target and antennas. For energies higher than 400 J, high values of EMP were obtained for gold target, and this is in good agreement with results from Fig. 2a achieved by T-probe. Unfortunately, we were not able to get reliable results from these probes for all the other materials studied, and for example for Tantalum. A clear and more complete comparison of absorbed EMP for other target materials, will be thus performed in future experimental campaigns [34].

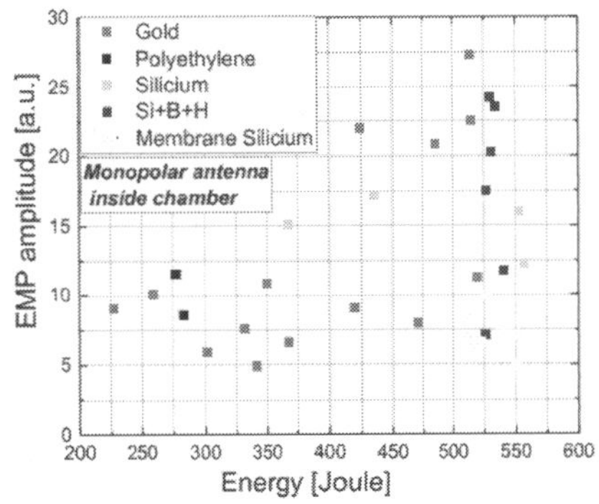

(a)

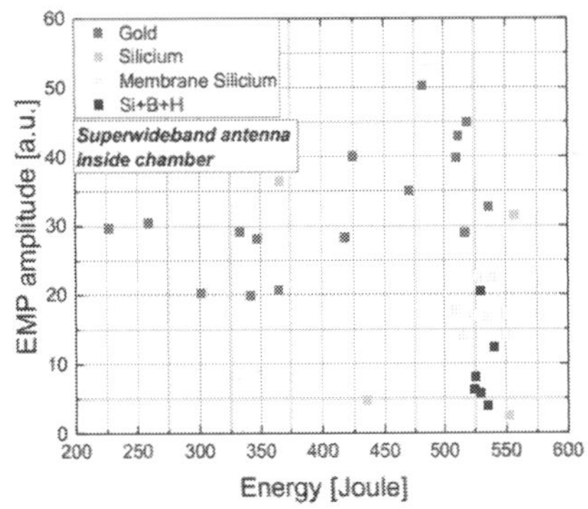

(b)

(c)

Fig.5 EMP amplitude for monopolar antenna (a) and SWB antenna (b) inside the experimental chamber; picture of the SWB magnetic antenna and linear antenna placed inside the target chamber at 30cm above the laser target interaction point (c).

3 Conclusions

In this work the results of EMP measurements are shown for different probes. The T-Probe allows us to measure the magnetic field produced only with the return target current. It was shown how the magnetic field increases with the laser pulse energy, as the target current scales with the laser intensity as $\sqrt[3]{I\lambda^2}$ [20]. The analysis of the collected data for various materials, show that EMP spectra have two dominant peaks which change in their ratios based on the target material; however their dependence is not yet understood. Nevertheless the influence of hydrogen or carbon impurities on the emission of hot electrons and, thus, on the EMP intensity seems to be evident. The EMP rise time and duration (FWHM) was determined for each of the calculated magnetic flux. Average over 10 shots shows that Graphite presents the lowest rise time while Tantalum and Si+B+H targets shows the highest ones. For different target materials, the time duration at FWHM is very similar and it is related to the common target chamber structure that induces similar decaying of the EMP pulses. Measurements performed by conductive antennas inside the chamber suffered of high background noise, but it was possible to find a suitable solution and reliable results were eventually achieved, which are in rough agreement with those from T-probe. More complete characterization and comparison of results for different target materials can be performed in future experimental campaigns.

The PALS staff and other researchers participating in the experimental campaign greatly appreciate the financial support from the Czech Ministry of Education within the project "Creating and probing dense plasma at the PALS facility" (CZ.02.1.01/0.0/0.0/16_013/0001552), and European Regional Development - the project ELI: Extreme Light Infrastructure (CZ.02.1.01/0.0/0.0/15_008/0000162). The research leading to these results has also received funding from the Czech Science Foundation (Grant No. 16-07036S). The support from the projects no. CZ.1.07/2.3.00/20.0087 and CZ.1.07/2.3.00/20.0279, which were co-financed by the European Social Fund and the state budget of the Czech Republic are also acknowledged. The work of F.C and R.D.A. has been partially carried out within the framework of the EUROfusion Consortium and has received funding from the Euratom research and training programme 2014–2018 under grant agreement No. 633053. The views and opinions expressed herein do not necessarily reflect those of the European Commission. The support by the Ministry of Education, Youth and Sports of the Czech Republic, Project No.: LM2015083 (PALS RI), is also gratefully acknowledged.

References

1. M. J. Mead et al., Rev. Sci. Instrum. 75, 4225 (2004).
2. M. De Marco et al., J. Phys.: Conf. Ser. **508** (2014) 012007.
3. M. De Marco et al., Physics of Plasmas **24**, 083103 (2017)
4. J.-L.Dubois et al., Phys. Rev. E 89, 013102 (2014).
5. R.F. Benjamin et al., Phys. Rev. Lett. **42** (1979) 890-893.
6. F. Consoli et al., Scientific Reports **6**, 27889 (2016).
7. C. G. Brown et al., Rev. Sci. Instrum. 83, 10D729 (2012).
8. C G Brown Jr. et al., J. Phys.: Conf. Series 112, 032025 (2008).
9. J. Krása et al., Laser Part. Beams 33, 601–605 (2015).
10. J.-L.Dubois et al., Phys. Rev. E 89, 013102 (2014).
11. S. Ramo et al., Fields and Waves in Communication Electronics (Wiley, New York, 1965), p.541.
12. A. Poyé et al.. Phys. Rev. E 91, 043106 (2015).
13. M. G. Drouet and H. Pépin. Appl. Phys. Lett. 28, 426-428 (1976).
14. Krása, et al., Plasma Phys. Control. Fusion 59 065007 (2017).
15. F. Consoli et al., Environment and Electrical Engineering (EEEIC), 2015 IEEE 15th International Conference on, Rome, 2015, pp. 182-187.
16. M. G. Drouet and R. Bolton. Phys. Rev. Lett. 36, 591-594 (1976).
17. C. J. Price et al., Rev. Sci. Instrum. 86, 033502 (2015).
18. A. Poyé et al., Phys. Rev. E 92, 043107 (2015).
19. J. Krása et al., Laser Part. Beams 32, 15–20 (2014).
20. M. De Marco et al., JINST 11, C06004 (2016).
21. J. Cikhardt et al., Rev. Sci. Instrum. 85, 103507 (2014).
22. C.G. Brown et al. Rev. Sci. Instrum. **83** (2012) 10D729.
23. C.G. Brown Jr. et al., J. Phys.: Conf. Ser. **244** (2010) 032001.
24. Zi-Yu Chen et al., Phys. Scr. **83**, 055503 (2011).

25. V. Nassisi and A. Luches, Rev. Sci. Instrum. **50** (1979) 900-902.
26. Hyoung Suk Kim. Electromagnetic Waves in Cavity Desig. Ed.: Ali Akdagli, ISBN: 978-953-307-302-6, InTech 2011, p. 77-100.
27. M. De Marco et al., Nukleonika (2015) 10.1515.
28. F. S. Felber. Appl. Phys. Lett. **86**, 231501 (2005)
29. F. Consoli et al Physics Procedia 62 (2015) 11 – 17
30. A. Picciotto et al, Phys. Rev. X, 4, 031030 (2014
31. SPE-12-8-046/E/SS, Technical Report, Taoglas, 2009
32. S. Barbarino, F. Consoli, IEEE Trans. Antennas Propag. 58 (2010) 4074
33. M. De Marco, et al, Journal of Physics: Conference Series 508 (2014) 012007
34. F. Consoli et al, submitted to the European Physical Journal Web Conferences

Emission of a low-power laser-induced vacuum discharge plasma in the EUV and SXR spectral ranges

Alexander Rupasov [1], *Igor* Romanov [1], *Andrey* Kologrivov [1], and *Viktor* Paperny [2]

[1] Lebedev Physical Institute, Russian Academy of Sciences, Moscow, Russia
[2] Irkutsk State University, Irkutsk, Russia

Abstract. X-ray spectral characteristics of a vacuum discharge plasma with the storage energy lower than 30 J initiated on an Al or a Fe cathode by a 10^{12} W/cm^2 neodymium laser were studied in the 30 – 300 Å wavelength range. It is shown that both the spectral composition and intensity of radiation of a micropinch plasma produced in the cathode jet of the discharge are determined by parameters of the discharge and laser pulse. These parameters were optimized to achieve a regime in which a considerable part of radiation energy was concentrated in the long-wavelength band of the quasi-continuum (230 - 270 Å and 160 - 200 Å for Al and Fe, respectively), which makes this discharge a source of narrowband X-ray radiation.

It is known that local regions of a hot micron-scale plasma produced in a high-current vacuum discharge plasma due pinching (micropinches) are sources of intense electromagnetic radiation in a broad spectral range. The emission spectra of micropinches in the hard X-ray range at wavelengths $\lambda < 2$ Å were studied in many works (see, for example, [1-3]). Soft X-ray (SXR) sources in the wavelength range above 20 Å attract considerable recent interest due to their possible scientific and technological applications including nanolithography, investigations of the inner structure of biological objects, astrophysical studies, etc. [4-7]. At present, as sources of such pulsed radiation, laser plasmas are mainly used. An important parameter for applications of radiation sources is the SXR conversion efficiency from a storage electric energy. From this point of view, it is interesting to use micropinch radiation sources. It was shown earlier that in vacuum discharge plasma with the storage energy about 10 J initiated on a cathode by a low-power laser pulse, micropinches emitting SXRs in the wavelength region > 40 Å are formed. The plasma pinching process stability and micropinch parameters were controlled by the laser pulse parameters [8, 9]. These results suggest the possibility of making a local source of intense SXR with a comparatively low energy deposition based on the laser-induced vacuum discharge.

The use of such a discharge as a SXR source requires the knowledge of the spectral composition of the radiation depending on the parameters of the discharge, laser pulse and cathode material. In turn, the maximum temperature of the micropinch plasma can be determined from shape of the emission spectrum of a metal plasma of some elements, which is important both for understanding the physics of processes involved and for improving parameters of the SXR source based on this discharge. However, the long-wavelength region of the SXR spectrum remains poorly studied, which is explained by the complexity of required measurements and their interpretation. Note that the mega-ampere Z-pinch plasma was studied in [10, 11] where only a few spectra were presented for fixed discharge parameters.

In this paper, we present the measurements of the emission spectrum of a micropinch plasma produced in the cathode jet of a low-power discharge in a broad spectral range from 30 to 300 Å for different laser pulse energies, different discharge voltages (currents), and different cathode materials.

Investigations were performed on a laboratory stand for studying X-ray and ion emission from discharge plasmas. The setup included a vacuum diode, a Q-switched 1.06 μm laser emitting ≤ 500 mJ, 6 ns pulses and also X-ray and corpuscular diagnostics. The scheme of experiments is shown in Fig. 1. The current in the discharge circuit was maintained with a storage 0.22 μF capacitor. The current was measured with a Rogowski coil placed in the cathode circuit of the diode. The semi-period of current oscillations in the shorting regime was 0.4 μs. Experiments were performed for discharge current amplitudes no more than 28 kA and storage voltages up to –16 kV. The maximum current rise rate was $5\cdot10^{11}$ A/s. The discharge was initiated on the cathode surface by a laser beam focused to a fine $0.4\cdot10^{-4}$ cm^2 spot. The laser pulse power density was varied between 10^{10} – 10^{12} W/cm^2. The angle of incidence of the laser beam on the cathode was 45^0. We used in experiments a ground steel cathode with a hole 6 mm in diameter in the middle (for ion emission experiments).

The interelectrode gap was 7 mm. The diode was placed in a chamber with oil-free pumping down to a residual pressure of 10^{-4} Pa. After each or several discharges, the electrode unit was displaced and the laser beam fell on a new spot on the cathode.

The discharge plasma radiation was studied in the integrated regime with a GIS-S grazing incidence spectrograph mounted perpendicular to the interelectrode gap at a distance of 65 cm from it. The entrance slit of the spectrograph was oriented along the discharge axis.

A concave 600 l/mm diffraction grating of the spectrograph with period d = 1.667 μm, the radius of curvature 1 m, and the grazing angle 4^0 has a W/Re coating. The non-Rowland configuration is used in which the spectrum is recorded in a plane perpendicular to diffracted beams. Thus, the accurate focusing is performed only for one wavelength corresponding to the point of intersection of the recording plane with the Rowland circle (accurate focusing point). For our spectrograph, the accurate focusing point corresponds to the wavelength λ = 140 Å. Because of a small angular aperture of the spectrograph, the beam defocusing in the non-Rowland configuration is small and therefore the spectrum can be recorded in a broad enough spectral range. The spectral resolution $\delta\lambda$ for this configuration and the 20 μm exit slit width of the spectrograph is 0.66 Å in the accurate focusing region (where $\lambda/\delta\lambda \approx 200$) and ~ 1 Å at the range boundaries. Note that the recording range is restricted in the short-wavelength region by the drastic decrease in the reflection coefficient of the grating at λ < 30 Å. Measurements of the 48.41 Å Ly-alpha line intensity for the B V ion in the first and second diffraction orders performed during spectrograph calibration showed that the second-order line intensity was ~ 15 times weaker than that in the first diffraction order [12]. Such a strong suppression of the second order allows us to neglect it in the processing of spectrograms, taking into account only the first order in the 30–90 Å region. For wavelengths above 90 Å, it is necessary to take into account the third order and above 120 Å – also the fourth order (radiation in the third and fourth diffraction orders is approximately two times attenuated compared to the first order). Thus, the wavelength range from 30 to 90 Å is "pure" from the overlap of diffraction orders, whereas at longer wavelengths the spectrum recovery procedure should be applied, which was not performed here.

Spectra were recorded with a Toshiba TCD 1304A linear CCD array coated with a P-46 luminophore converting SXRs to visible radiation. The array pixel size 8×200 μm provided the spectral resolution of the spectrograph indicated above. The total amplitude error of the detector was ~ 10%. Studies of similar CCD/luminophore detectors showed that the output signal amplitude linearly depended on the spectral radiation intensity. The output signal of the CCD array was fed to a PC. The device is described in more detail in [13].

We studied the influence of laser radiation on the plasma emission spectra by recording Al spectra for different laser pulse energies. Figure 2 shows plasma emission spectra in the range 30 - 120 Å for the fixed

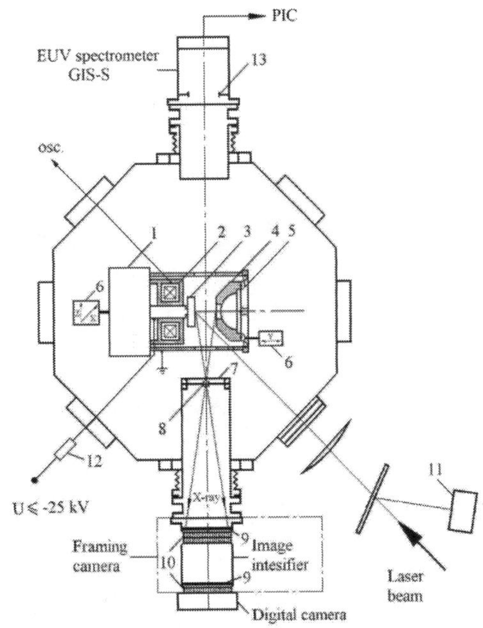

Fig.1. Experimental set-up. (1) capacitor; (2) Rogowski coil; (3) planar Al cathode; (4) anode; (5) flexible power input; (6) movable holders; (7) X-ray absorber; (8) pinholes; (9) MCP; (10) fiber optical interface; (11) calorimeter; (12) charging resistance of 100 MΩ; (13) entrance slit of EUV spectrometer.

storage voltage U = 6 kV (the discharge current amplitude is $I \approx 9$ kA) and increasing laser pulse energy J. To determine the contribution of laser plasma emission into the discharge spectrum for each energy J, we also recorded the laser plasma emission spectrum in the absence of the discharge.

The obtained spectra showed that for J < 80 mJ the discharge plasma emission in this spectral range lies below the detection threshold. The laser plasma emission in the absence of the discharge can be detected for $J \geq 150$ mJ. The obtained spectra showed that the increase in J from 80 to 110 mJ (Figs. 2a, b) resulted in the increase in the emission intensity and the shift of the $2p^4$-$2p^3$3s transition line of Al VI to the $2p^2$-2p3s line of Al IX, which suggest the increase in discharge plasma temperature T_e and/or ion population. This effect is probably explained by the increase in the cathode material amount evaporated by laser radiation, which leads to the increase in the plasma compression degree in a micropinch and its heating. The increase in J up to 150 mJ (Fig. 2c) results in the increase in the intensity of 2p –3d transition lines of the Al XI ion in the laser plasma and a drastic decrease in the mission lines of the discharge plasma. As J is further increased, this tendency is preserved, so that for J = 250 mJ and the corresponding laser radiation density on the cathode $P = 10^{12}$ W/cm^2 (Fig. 2d) and higher values of J, only the laser plasma emission is detected. This result shows that in the case of the high laser pulse energy and, therefore, the large mass of the plasma jet, the discharge current is not sufficient to obtain plasma pinching, in accordance with results [9, 14].

https://doi.org/10.1051/epjconf/201816703010

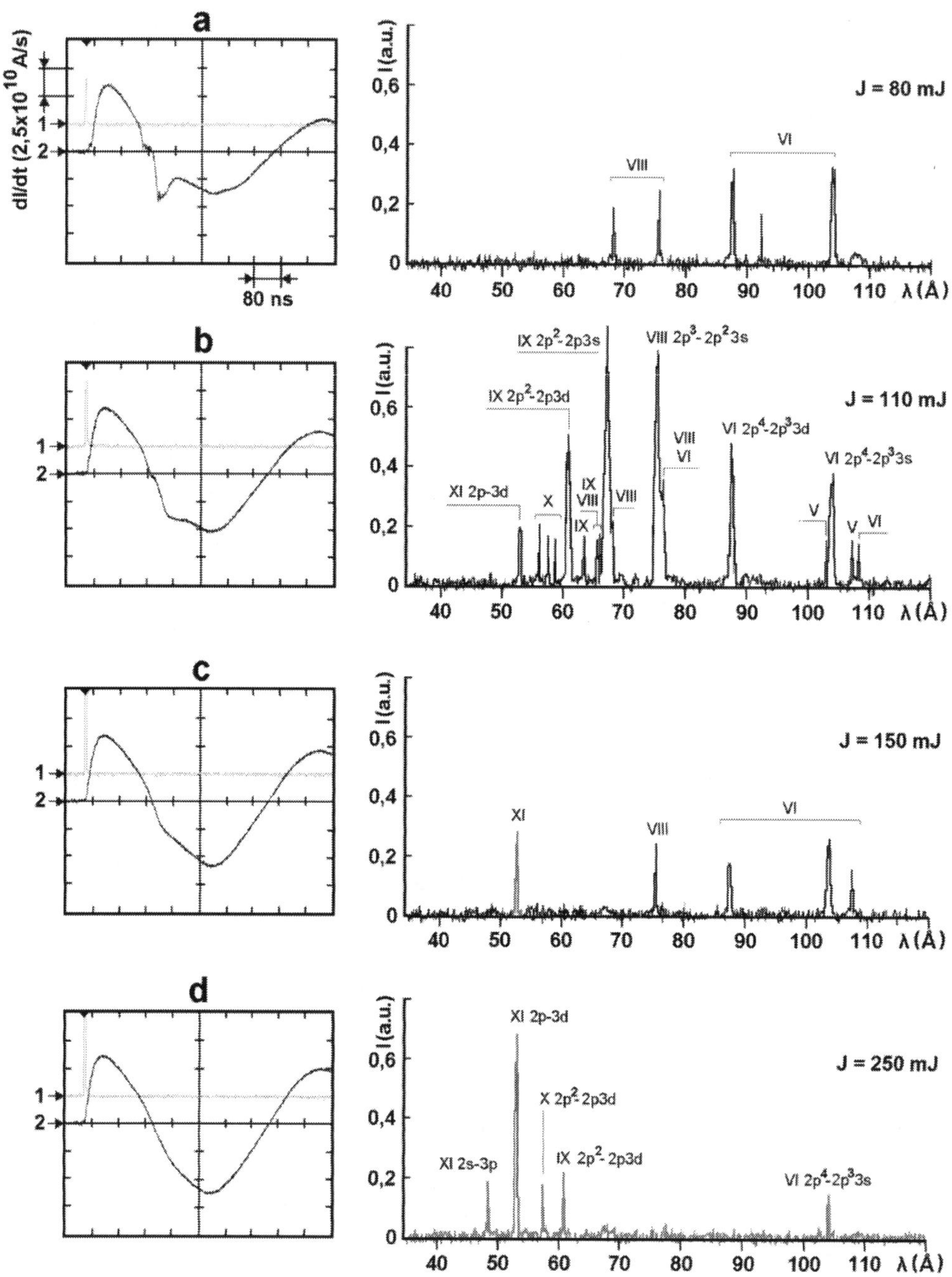

Fig. 2. Oscillograms of the discharge current derivative (at the left) and emission spectra of the Al plasma for U = 6 kV and different J (at the right). (1) laser pulse; (2) current derivative. The emission spectra of the discharge and laser plasma are shown by black and red curves, respectively.

Figure 2 shows that the amplitude of the singularity in the oscillogram of the discharge current derivative (Fig. 2a, left column) characterizing the restriction on the current in a plasma channel due to the neck formation does not correlate with conditions of the maximum plasma heating (Fig. 2b, right column). It was shown in previous experiments that pinching predominantly occurs in the laser plasma produced during the discharge ignition [9]. For a low laser pulse energy and, therefore, a low linear density N_0 of the laser plasma in the discharge gap, the singularity amplitude achieves the maximum. The increase in J leads to the "smoothing" of singularity. Thus, the initial value of N_0 plays a key role in the current flowing. For small N_0, the neck decay proceeds more efficiently, which can lead to the break of the current channel, the formation of a double dielectric layer and the anomalous acceleration of ions [8, 15]. However, such a density proves to be insufficient for efficient plasma pinching and achieving the density and temperature values providing the detected radiation level.

The emission spectrum of the discharge plasma depends on the discharge current (Figs, 3a, b). One can see that, for the fixed laser pulse energy 70 mJ, the increase in the voltage from 10 to 16 kV corresponding to the increase in the current amplitude from 17 to 28 kA results in the increase in the emission intensity by several times in the entire spectral range from 30 to 300 Å. The emission intensity predominantly increases (about 20 times) in the long-wavelength 230 - 270 Å region of the quasi-continuum (unresolved emission arrays (UTA)). This suggests that the mean discharge plasma temperature decreases probably due to the increase in the plasma volume with a relatively low temperature produced during the development of the emission surface of the cathode with increasing the discharge current. The increase of J from 70 to 200 mJ for U = 16 kV leads to the further increase in the emission intensity and its redistribution to the short-wavelength region down to 100 Å. This indicates to the increase in the plasma temperature, which is also confirmed by the appearance of the Al XII 2s – 3p lines of the He-like Al^{+11} ion (Fig. 3c). Our measurements showed that in this case the contribution of the laser plasma emission to the integrated spectrum does not exceed 5%. The further increase in J results in the decrease of temperature and emission intensity up to the complete cessation of plasma pinching. Note that emission spectra contain no traces of iron ions introduced to the total spectrum by the anode plasma emission, which means that the micropinch plasma temperature considerably exceeds the anode plasma temperature.

Unfortunately, because of the complex structure and specificity of VUV spectra of each element, at present there is no the universal procedure for determining the electron temperature T_e and plasma density N_e from them. These quantities are estimated using mathematical simulations. For some elements, in particular iron, a method is developed for determining the maximum temperature of electric discharge plasma by comparing its emission spectra with laser plasma spectra assuming the coronal equilibrium [10]. Experimental studies and mathematical simulations showed that the structure of VUV spectra of Fe ions in the laser plasma and the position of the maximum of the emission spectrum in the range 30 – 100 Å are very sensitive to a change in the electron temperature. As T_e increases, the ionization degree of Fe increases and the wavelength λ_{max} corresponding to the intensity maximum of the averaged spectrum for high-resolution spectra (or the intensity maximum for low-resolution spectrum) shifts to the blue. A comparison of these data obtained for laser plasma with data obtained for discharge emission spectra allows us to determine T_e for the micropinch plasma. In this connection, we obtained the emission spectra of the discharge plasma with an iron cathode.

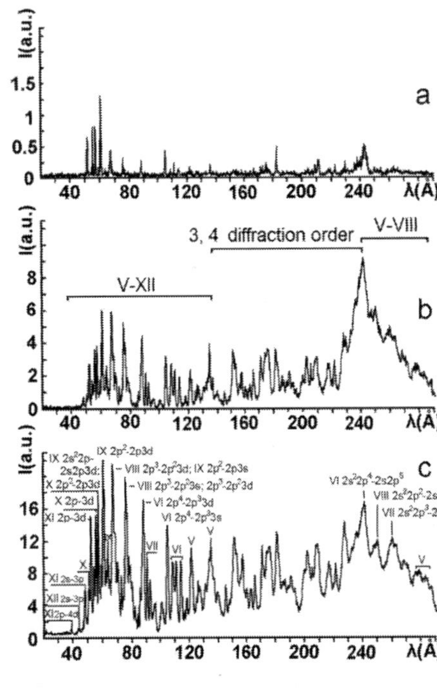

Fig. 3. Emission spectra of the Al discharge plasma for U = 10 kV, $I \approx$ 17 kA, J = 70 mJ (a); 16 kV, 28 kA, 70 mJ (b); and 16 kV, 28 kA, 200 mJ (c).

The behavior of emission spectra of Fe with changing the discharge current and laser pulse energy proved to be similar to that for Al emission spectra. The increase in the storage voltage (discharge current) for the laser pulse energy 80 mJ leads to the increase in plasma emission intensity (Figs. 4c, d). A detailed analysis of the structure of emission spectra in the short-wavelength almost did not change with increasing U from 8 to 14 kV, which means that the temperature of the hot plasma region did not increase. The emission intensity considerably increased in the 220 – 260 Å range, which means that the mean temperature of the emitting plasma volume decreased (this effect is more strongly manifested for J < 30 mJ). The increase in J up to 250 mJ (Figs. 4a – c) also leads to the increase in the emission intensity, which most strongly increases in the 30 – 150 Å region, in particular, in the 66.3 Å (3d – 4f transition of Fe XVI) and 67.2 Å (3d – 4p transition of Fe XVII) lines. However, any noticeably blue shift of the spectrum envelope maximum is absent, which means that the increase in T_e is negligible [10]. The observed decrease in the ratio of the UTA emission intensity in the 160 – 200 Å range to that in the short-wavelength region with increasing I and J corresponds to results [9] and is explained by the increase in the hot plasma volume.

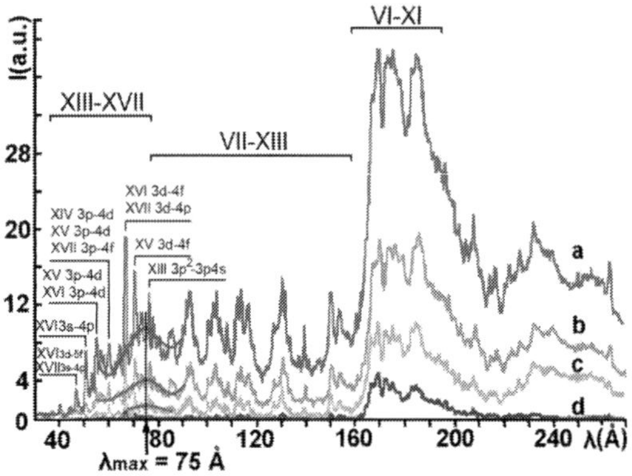

Fig. 4. Emission spectra of the Fe discharge plasma for U = 14 kV, J = 250 mJ (a); 14 kV, 130 mJ (b); 14 kV, 80 mJ (c); and 8 kV, 80 mJ (d). The arrow shows the approximate position $\lambda_{max} \approx 75$ Å of the maximum of the averaged spectrum in the 30–90 Å range.

Based on these data, we will estimate the maximum temperature of the discharge plasma by comparing the spectrum in the 30–90 Å region (Fig. 4a), where the overlap of diffraction orders is absent, with the VUV spectra of Fe ions in the laser plasma observed in experiments and synthesized in [10]. The comparison shows that the obtained spectrum is similar in its ion composition to the emission spectrum of the laser plasma. The analysis of the structure of the spectrum shows that the discharge plasma temperature lies in the range from 100 to 200 eV (see Figs. 6 – 8 in [10]). To estimate the temperature more accurately, we will use empirical formula (4) in [10]:
$ln(\lambda_{max}, Å) = A \times (1/T_e, keV) + B$, where A = 0.082 and B = 3.82. The estimate of T_e by this formula gives $T_e \approx 165$ eV for $\lambda_{max} \approx 75$ Å.

Thus, our studies have shown that:

(i) The SXR generation efficiency in a laser-triggered vacuum discharge depends on the relation between the laser pulse energy determining under these experimental conditions the mass of the evaporated plasma-producing cathode material, and the discharge current determining the plasma pinching rate. If the material mass is small, the amount of the produced plasma, its compression and heating degrees in a micropinch are insufficient for generating an intense SXR flow. On the contrary, in case of the mass excess, the current value is insufficient for the efficient plasma jet compression and heating.

(ii) The intense UTA bands in the 230–270 Å and 160–200 Å ranges for Al and Fe, respectively, are inherent in the emission spectra of the discharge plasma. They suggest the formation during pinching, along with the hot plasma nucleus with temperature $T_e > 150$ eV, also a relatively cold shell. There exists the range of laser pulse and discharge current parameters for which a considerable fraction of the radiation energy is concentrated in UTA.

(iii) The laser-induced vacuum discharge with the storage energy up to 30 J can be a promising for various applications as a SXR source with a broad line emission spectrum with parameters controlled by the laser pulse characteristics and the discharge current.

This work was supported by the Russian Foundation for Basic Research (grants nos. 15-02-03757, 16-02-01-140, 17-02-00572).

References

1. C.R. Negus, N.J. Peacock, J. Phys. D: Appl.Phys., **12,** 91 (1979)
2. A. Schulz, R. Burhenn, F.B. Rosmej, H.J. Kunze, J. Phys. D: Appl. Phys., **22,** 659 (1989)
3. O.A. Bashutin, M.A. Alkhimova, E.D. Vovchenko, et al, Plasma Physics Reports, **39,** 11, 900 (2013)
4. .Y. Vinokhodov, V.N. Krivtsun, A.A. Lash, et al, Quantrum Electron., **46,** 1, 81 (2016)
5. D. Thanh-Hung, S. Yuhei, A. Goki, et al, Appl. Phys. Lett., **107,** 12, 121101 (2015)
6. A. Sakdinawat, D. Attwood, Nature Photonics, **4,** 840 (2010)
7. P.T. Springer, et al, J. of Quantitative Spectroscopy and Radiation Transfer, **58,** 4-6, 927 (1997)
8. I.V. Romanov, V.L. Paperny, Yu.V. Korobkin, et al, Tech. Phys. Lett., **39,** 388 (2013)
9. I.V. Romanov, V.L. Paperny, Yu.V. Korobkin, et al, Physics of Plasmas, **23,** 023112 (2016)
10. A.P. Shevelko, D.E. Bliss, E.D. Kazakov, et al, Plasma Phys.Reports, **34,** 11, 944 (2008)
11. V.V. Aleksandrov, V.A. Gasilov, E.V. Grabovskiy, et al, Plasma Phys.Reports, **40,** 12, 939 (2014)
12. G.A. Vergunova, A.S. Grushin, A.A. Kologrivov, et al, Plasma Phys.Reports, **41,** 5, 408 (2015)
13. A.P. Shevelko, O.F. Yakushev, Poverkhnost, **2,** 51 (2003)
14. I.P. Tsygvintsev, A.Yu. Krykovskiy, V.A. Gasilov, et al, Mathematical Models and Computer Simulation, **8,** 5, 595 (2016)
15. I.V. Romanov, A.A. Rupasov, A.S. Shikanov, et al, J. Phys. D: Appl. Phys., **43,** 465202 (2010)

Magnetic and electric deflector spectrometers for ion emission analysis from laser generated plasma

Lorenzo Torrisi[1,*], *Giuseppe* Costa[1], *Giovanni* Ceccio[1], *Antonino* Cannavò[1], *Nancy* Restuccia[1], *Mariapompea* Cutroneo[2]

[1]Dipartimento di Scienze Fisiche, MIFT, V.le F.S. d'Alcontres 31, 98166 S. Agata, Messina, Italy
[2]Nuclear Physics, Institute, CAS 25086, Rez, Czech Republic

Abstract. The pulsed laser-generated plasma in vacuum and at low and high intensities can be characterized using different physical diagnostics. The charge particles emission can be characterized using magnetic, electric and magnet-electrical spectrometers. Such on-line techniques are often based on time-of-flight (TOF) measurements. A 90° electric deflection system is employed as ion energy analyzer (IEA) acting as a filter of the mass-to-charge ratio of emitted ions towards a secondary electron multiplier. It determines the ion energy and charge state distributions. The measure of the ion and electron currents as a function of the mass-to-charge ratio can be also determined by a magnetic deflector spectrometer, using a magnetic field of the order of 0.35 T, orthogonal to the ion incident direction, and an array of little ion collectors (IC) at different angles. A Thomson parabola spectrometer, employing gaf-chromix as detector, permits to be employed for ion mass, energy and charge state recognition. Mass quadrupole spectrometry, based on radiofrequency electric field oscillations, can be employed to characterize the plasma ion emission. Measurements performed on plasma produced by different lasers, irradiation conditions and targets are presented and discussed. Complementary measurements, based on mass and optical spectroscopy, semiconductor detectors, fast CCD camera and Langmuir probes are also employed for the full plasma characterization. Simulation programs, such as SRIM, SREM, and COMSOL are employed for the charge particle recognition.

1 Introduction

The class of charge particle spectrometers based on electrical or magnetic deflections and both on electrical and magnetic deflection systems is large. It contains devices to select the mass-to-charge ratio, to the ion or electron recognition, to evaluate the particle velocity, energy, charge state and directionality.
When the deflection is controlled by an electrical field generated between two electrodes placed transversally to the path of the particles, the technique is called electrostatic deflection because the strength and direction of the applied field changes slowly with respect to the time it takes for the particles to cross the field. Thus the deflection introduces a change in the charge particle velocity directed transversally to the incident direction, which component velocity vT depends on the following parameters:

$$v_T = \sqrt{\frac{2ezV}{m}} \qquad (1)$$

where e is the electron charge, z is the charge state, V is the applied static potential and m is the particle mass. Inside the electric field the trajectory of the particle is parabolic and, if d is the distance between the two electrodes and L the length of the electric field, the particle deviation is calculable from the relation [1, 2]:

$$tg\alpha = \frac{ez(V/d)}{mv_0^2} L \qquad (2)$$

where α is the angle that the velocity vector crossing the electric field forms with the incident direction and v_0 represents the velocity of the particle incoming orthogonally to the electric field direction. It is possible to observe that the deviation depends on the charge-to-mass ratio and on the initial particle velocity.
When the deflection is controlled by a constant magnetic field B placed orthogonally to the incident particle direction, the Lorenz force acts and a rotation on the particle is induced according to the Larmor radius:

$$r = \frac{m}{ez}\frac{v_0}{B} \qquad (3)$$

Also from this relation it is possible to observe that the particle trajectory depends on the mass-to charge ratio and initial velocity of the particle.
The non perfect orthogonality of the incident particle with respect to the magnetic field direction will produce a helicoidally trajectory which radius depends on the component orthogonal and the pitch on the component parallel.
The use of both electrical and magnetic fields, in stationary regime, permits to discriminate very well the trajectory for the mass-to-charge ratio also for high ion

* Corresponding author: lorenzo.torrisi@unime.it

velocities, as occurs in the Thomson parabola spectrometer [3, 4]. In this case, assuming the particles moving along the z direction toward a magnetic (input) and electric (output) fields, both parallel to x direction and orthogonal to the particle incoming direction z, the magnetic field will produce a deflection on the plane z-y and the electric field a deflection on the plane z-x. The particle projection on a final shield, orthogonal to the incident direction, where are placed the detectors, will show an electrostatic deflection X given by:

$$X = k\frac{ezV}{\varepsilon} \tag{4}$$

where $k = \left(\frac{L_E}{2} + D\right) \cdot \frac{L_E}{2d_E}$, L_E represents the length of the electric field plates, V the applied voltage, D the final shield distance from the electric plates, and ε is the kinetic energy of the incident particle. The Y magnetic deflection, instead depends on the relation:

$$Y = k'\frac{ezB}{\sqrt{m\varepsilon}} \tag{5}$$

where $k' = (\frac{L_B}{2} + l + L_E + L) \cdot \frac{L_B}{\sqrt{2}}$, L_B represents the length of the magnetic field plates, l the distance between the magnet and the electric plates.

Thus both deflections will produce a parabolic trajectory according to the final relation:

$$X = \frac{kmV}{k'^2\, ezB^2} \cdot Y^2 \tag{6}$$

Many spectrometers are based on such relations but often their plates are not parallel and the electric and magnetic fields can vary up to MHz in order to obtain compact instrument and high sensitivity. One example is represented by the mass quadrupole spectrometer, which uses a quadrupole consisting in four cylindrical rods, set parallel to each other [5]. Ions are separated in base on the stability of their trajectories in the radio-frequency oscillating electric fields that are applied to the rods.

For the different species of charge particle spectrometers it is possible to use different kind of detectors, of which the most diffused are based on Faraday cups and secondary electron multipliers.

In this paper will be presented four kind of spectrometers employed mainly for analysis of the particles emitted from high-intensity lasers generating plasma in high vacuum.

2 Experimental section

Generally the presented spectrometers are employed to analyze the plasma produced by pulsed laser irradiating solid target in high vacuum. To this are employed a Nd.YAG laser operating at 1064 nm wavelength, 3 ns pulse and an intensity up to 10^{11} W/cm^2 at Messina University and a Iodine laser operating at 1315 nm wavelength, 300 ps and an intensity up to 10^{16} W/cm^2 operating at PALS in Prague (Czech Republic).

The first ion mass spectrometer used in diagnostics of laser-generated plasma is known as ion energy analyzer (IEA). It consists in an electrostatic deflector of ions at 90° with respect to the incident direction based on two curved metallic plates from which is applied an high electric field. The system acts such as a filter permitting the deflection at 90°, towards the detector, only of the ions having the same energy-to-charge ratio (E/z), according to the equation [6]:

$$\frac{E}{z} = KeV \tag{7}$$

being V the voltage applied to the two electrodes and K a factor depending on the geometry of the spectrometer. Often the system is realized using an electric field of the order of 1-10 kV/cm, and maintain the *k* parameter to a value of 10. Thus, for example, if is E/z=10 keV it means that the system are using an applied voltage of 1 kV and that the electrostatic filter can deflect 10 keV ions with a single charged, 20 keV with a double charged, 30 keV with a triple charged and so on. Limitations are due to the use of the high voltage applied to the electrode in high vacuum. Up to 30 kV have been applied without discharge problems.

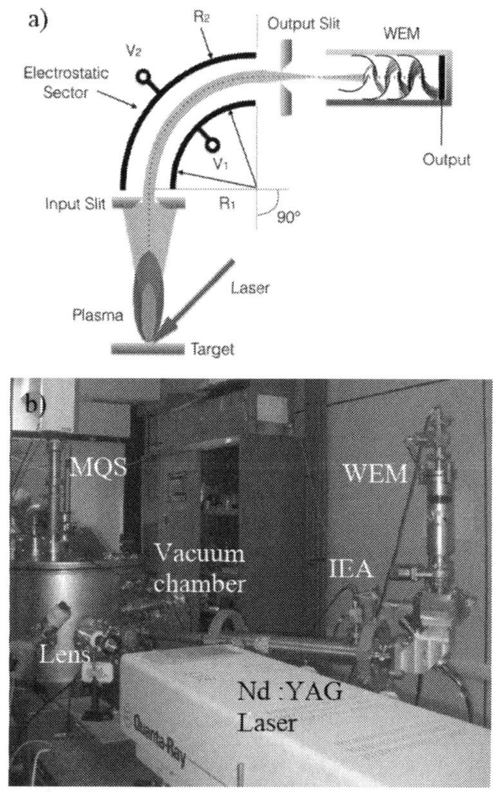

Fig. 1. Scheme of the IEA spectrometer (a) and photo of the experimental set up (b).

Fig. 1a shows a typical scheme of IEA spectrometer. In order to increase the sensitivity of the instrument

https://doi.org/10.1051/epjconf/201816703011

generally as detector is employed an electron multiplier and the first dynode is opened to vacuum in order to be hit by the ions. The input collimation need to determine very well the incident direction of the detected particles to be deflected. Fig. 1b shows a photo of the detector at the laboratory of plasma laser in Messina University. The system resolves the different mass-to charge using a time-of-flight (TOF) technique separating in time the faster ions from the slower ones from the time of arrival to the detector. Changing the bias voltage V it is possible to change the E/z ratio and to have information about the ion energy distributions. For example Boltzmann ion distributions emitted from hot plasma were detected according to previous papers [7, 8].

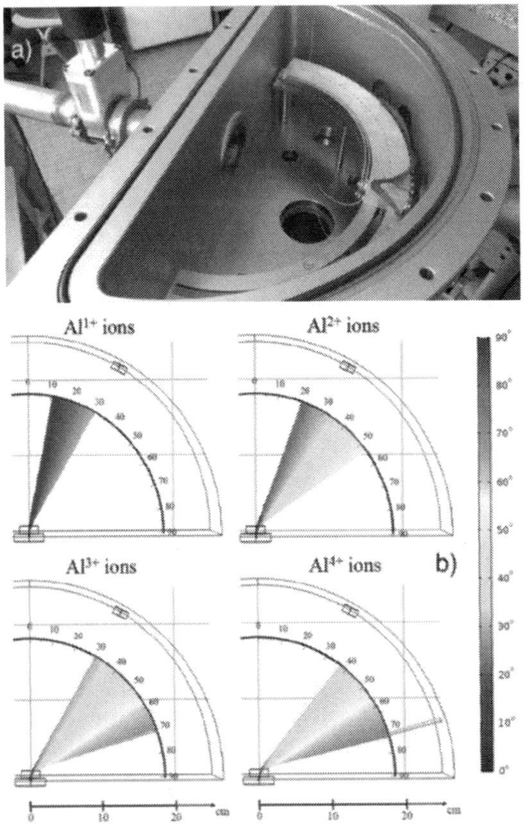

Fig. 2. Experimental set-up of the magnetic spectrometer (a) and simulations of the Bolzmann distributed Al ions (0.4 keV mean energy per charge state) deflected by 0.3 T vs. their charge state (b).

An example of magnetic deflector is represented by the system used in our laboratory to separate the different mass-to-charge ratio using the horizontal ion deflection produced by a magnetic field of the order of 0.3 T with a length of 1 cm orthogonal to the input direction of ions accelerated from the laser generated plasma. A photo of the experimental apparatus using 32 different ion collectors (IC) placed at 2.9° for each other for a total angular distribution of 90° is reported in Fig. 2a. A COMSOL Multiphysics code simulation [9] data relative to the first four charge states of the Al ions angular separation accelerated from laser-generated plasma (0.4 keV mean energy per charge state) is reported in Fig. 2b. Limitations are due to the maximum magnetic field applicable using a permanent magnet up to 0.3 T permitting a deflection of protons of 6.3° at 10 keV energy and of 1.7° at 100 keV, thus not usable for protons at energy higher than about 100 keV for the lower angular deflection.

A Thomson parabola spectrometer (TPS) used for low energetic ions, of the order of 100 keV per charge state is employed in our laboratory. In this case the ions are produced by laser-generated plasma, two pin-holes are employed to fix the incident direction of the ions, a permanent magnetic field B of 0.2 T is employed with a space of 1 cm (vertical) x 2.5 cm (horizontal) input cross-section and 2.5 cm length, followed, by an electric field E which applied voltage is adjustable up to 10 kV having and aperture cross section of 3 cm (horizontal) x 2.5 cm (vertical) and 2.5 cm length. A photo of the TPS is reported in Fig. 3a while Fig. 3b reports a scheme of the real parabola system realized at Messina University. The presented model can be employed both using a gaf-chromix film, or a micro-channel plates (MCP) coupled to a phosphorus screen and to a fast CCD camera, as detector.

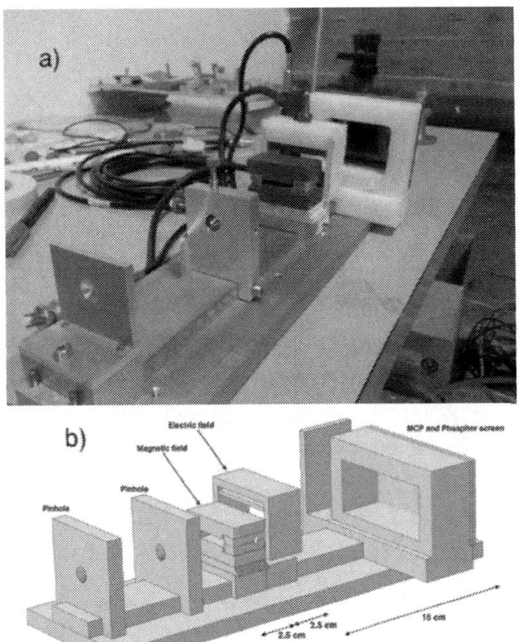

Fig. 3. Photo (a) and scheme (b) of the little TPS employed for the ion recognition using post ion acceleration system.

For higher energetic ions, of the order of 1MeV per charge state, higher and adjustable electric and magnetic fields are employed, such as the TPS used in our experiment at PALS laboratory in Prague (Czech Republic) at which also protons up to 10 MeV can be measured [10].

Finally we should also mention our mass quarupole spectrometer (MQS), a Prisma-Pfeifer from Mod.100-300, permitting the atom mass recognition from 1 up to 300 amu using an electron multiplier (SEM) as detector with a sensitivity lower than 1 ppm (part per million), as presented in our previous papers [11].
Faraday cup, ion collectors with electron suppression grid, fast storage oscilloscope, semiconductor detectors, plastic scintillators, electron multipliers, MCP, gaf-chromix and track detectors are employed for the particle detection.
Our measurements are facilitates using simulation programs permitting the comparison between the experimental data and the simulation ones. To this, programs such as SRIM, SREM, and COMSOL are employed [12]. NIST data base are also used for X-rays absorption, electron and ion stopping powers, and for natural isotope abundances [13].

3 Results

Fig. 4a reports a typical TOF spectrum of Ta ions detected using the IEA spectrometer. Ions are generated by the Nd:YAG laser interacting with a solid Ta target and detected using an electron multiplier which output signal is acquired by a fast storage oscilloscope permitting TOF measurement from the start laser shot to the stop ion detection.

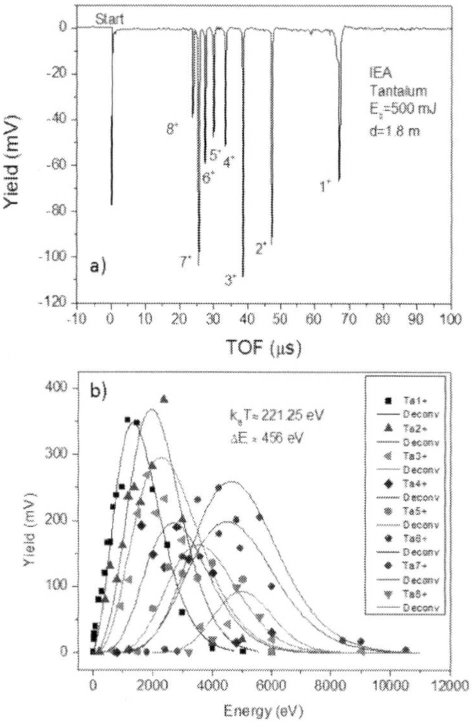

Fig. 4. Typical IEA spectrum recording Ti ions by plasma in TOF regime (a) and plot of the ion energy distribution obtained by varying the E/z IEA filter (b).

The flight distance from the target up to the first dynode where ions are impacted is 1.8 m, thus the single charged Ta ions detected at a TOF time of 68 μs has a mean kinetic energy of about 660 keV. By varying the E/z ratio through the voltage applied to the curved plates it is possible to measure the yield of ions at different kinetic energies filtered through IEA spectrometer. The plot of such result permits to evince that the ion energy follows the Coulomb-Boltzmann shifted (CBS) energy distribution [14], as reported in the fit of experimental point reported in Fig. 4b. Ions are shifted toward higher energy increasing their charge state. From the Boltzmann curve widths and from the regular shift of the distributions with the charge state it is possible to evaluate a plasma temperature of 220 eV and a mean plasma acceleration voltage of 456 V, respectively.

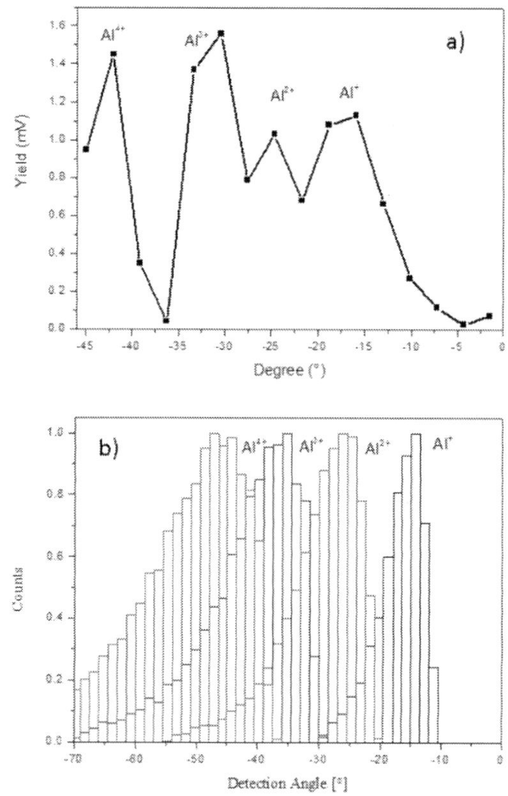

Fig. 5. Typical plot of the experimental angular distribution vs. charge state of the Al ions laser-generated analyzed with the magnetic defector (a) and COMSOL simulation of the expected ion angular distribution (b).

Fig. 5a reports a typical spectrum obtained analyzing with a magnetic deflection the Al charge states and neutrals coming from a laser-generated plasma at 10^{10} W/cm^2 intensity. In this case the IC at different angles was employed as detectors, connected to a fast storage oscilloscope. Fig. 5b shows, for comparison, the COMSOL normalized simulation data, indicating the mean angular deflections of the ions Al^+, Al^{2+}, Al^{3+}, Al^{4+}, which are 15°, 25°, 35° and 45°, respectively, confirming the obtained experimental results.

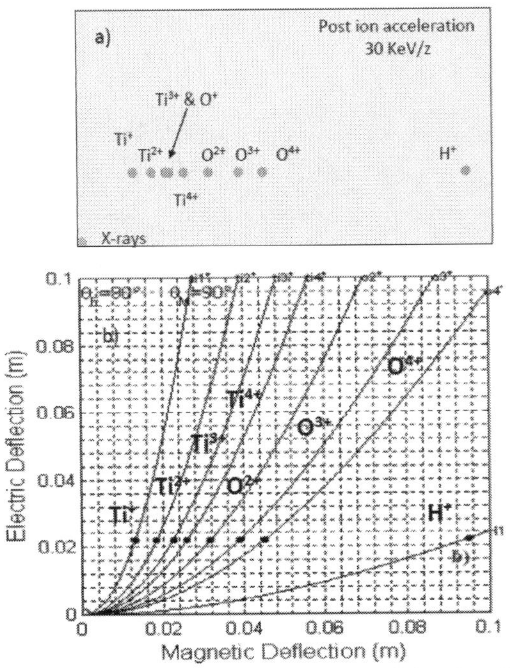

Fig. 6. Typical spectrum of ions recorded in gaf-chromix with the little TPS of our laboratory during post-acceleration using 30 kV (a) and comparison with the COMSOL simulation for the ion recognition (b).

Measurements of ions emitted from laser-generated plasma were performed using both our TPS instrument reported in Fig. 3 and the TPS of PALS Laboratory facility in Prague. In the first case a suitable gaf-chromix sheet is employed to detect ions, while in the second case a MCP with coupled red phosphorus plate was used as on line monitor. Fig. 6a reports the measurements of the ions obtained by irradiating a Ti target at 10^{10} W/cm^2 laser intensity, using a post ion acceleration of 30 kV (it means that ions are accelerated at 30 keV per charge state) and a gaf-chromix shield for the ion detection. The measurements indicate that plasma permits to obtain the ionization of Ti up to four charge states and that the Ti target contains also oxygen, which is detected with a maximum ionization up to four, and hydrogen contamination. Fig. 6b shows the simulated ion parabolas obtained using COMSOL code, reporting as electric deflection vs. magnetic deflection of the different ion charge states, that are in agreement with the ions detected using the gaf-chromix detector. The maximum Ti charge state is four; this indicates that the plasma electron temperature ranges between the fourth and fifth ionization potentials of Ti which are of 43 eV and 99 eV, respectively. Ions are extracted by the ablation chamber and submitted to a post acceleration of 30 kV, as reported in a previous paper [15], thus a multi energetic ion beams is generated where, for example, protons are accelerated monochromatically at 30 keV and Ti at the energies of 30 keV, 60 keV, 90 keV and 120 keV, proportionally to their charge state, respectively.

The PALS-TPS uses a magnetic field of 0.1 T, an electric field of 2.0 kV/cm and has the following parameters: $L_B = L_E$= 8 cm, l=0, D=16.5cm, which geometry is reported in literature. Using such parameters the COMSOL simulation permits the recognitions of the ions accelerated from laser-generated plasma. One example of such application is obtained for a thin target of CD_2 irradiated at 500 J pulse energy, with a focal distance of -160 μm (in front of the target), and recording the ion emitted in forward direction in TNSA conditions. A typical spectrum for this experiment is reported in Fig. 7a, as photo of the CCD camera coupled to the MCP detector, while Fig. 7b reports the simulations of the ion recognition using COMSOL for the electric vs. magnetic deflections on the base of the TPS geometry.

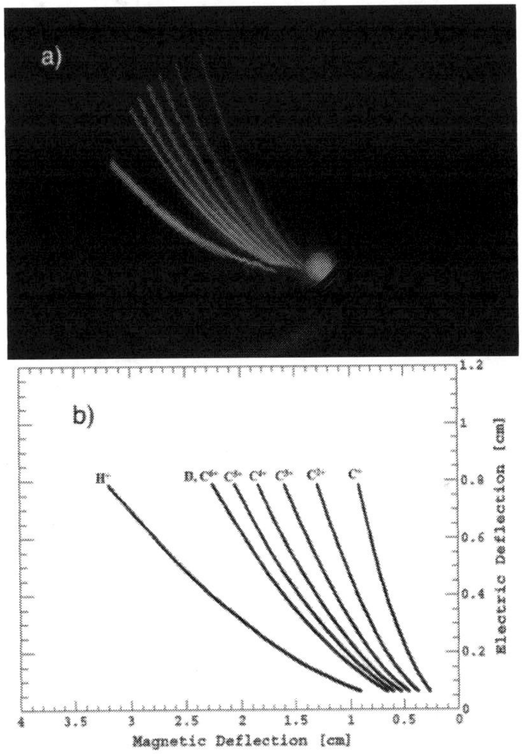

Fig. 7. Typical spectrum of ions recorded in MCP/Phosphorus screen with the TPS of PALS facility irradiating a CD_2 target (a) and comparison with the COMSOL simulation for the ion recognition (b).

The maximum acceleration is of about 3.0 MeV per charge state, as measurable from the nearest parabola points to the zero circle generated by the X-rays on the image. Thus protons have a maximum kinetic energy of 3.0 MeV with a Boltzmann distribution, carbon ions have a maximum energy of 18 MeV relative to the C^{6+}. The parabolas due to C^{6+} and D^+ ions are indistinguishable due to their same mass to charge ratio. In this case the plasma temperature and density is higher with respect to the case employed in our laboratory due to the intensity higher five order of magnitude. The temperature is higher than the

ionization potential of C^{6+} which is of 490 eV, and the plasma density more than three orders of magnitude higher just after the laser pulse. Further details on the production of CD_2 plasma at PALS, its characterization and application, are presented in our previous article [16].
Further techniques of ion deflection can be employed for plasma characterization and other applications, using electric and/or magnetic deflections static and/or dynamic. One of these is represented by the MQS which is based on the ion mass-to-charge ratio (m/z) selection on the base of the ion trajectory stability under radiofrequency oscillating electric fields applied to the rods. Such spectrometer can be used to analyze the neutral atoms present in plasma, to measure the m/z ratio, to detect particular isotopes, to evaluate the ion yield emitted during the laser irradiation of a solid target, to measure the ion energy distributions and to other scopes. Fig. 8 shows a typical MQS spectrum relative to the analysis of Cu species emitted in vacuum during the laser irradiation of a Cu target. The spectrum is acquiring fixing specific masses of interest and reporting their yield versus time. At the laser ON the laser hit the target, at the laser OFF the laser is switched off. The spectrum shows the two Cu stable isotopes and some Cu compounds produced in the laser generated plasma at a laser intensity of about 10^{10} W/cm^2.

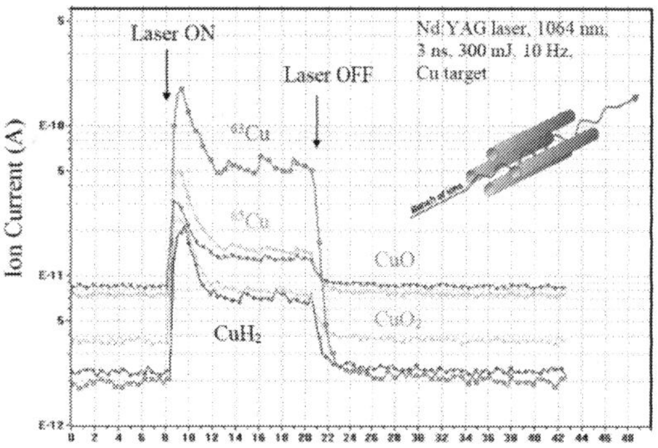

Fig. 8. Typical MQS spectrum obtained fixing the masses of interest vs. the time for a Cu target laser ablated.

Complementary measurements, based on mass and optical spectroscopy, semiconductor detectors, fast CCD camera and Langmuir probes are also employed for the full plasma characterization.

4 Discussion and Conclusions

It has been demonstrated that the laser-generated plasma characterization can be performed using suitable electrical and/or magnetical spectrometry both using static and/or dynamical fields.
In particular the fast and hot plasma emitting ions from keV up to hundred MeV can be analyzed using TPS spectrometer giving information on the mass-to-charge ratio an on the ion energy distribution. TPS can be employed both at low and high laser intensities, from about 10^{10} W/cm^2 up to 10^{16} W/cm^2 and higher. It is fast and generates images of the ions as a dependence of their m/z ratio. The on-line spectrometer uses MCP coupled to phosphorus screen and CCD camera, as that used at PALS facility.
IEA spectroscopy can be employed for ions up to energies of the order of 100 keV because it is limited by the high electric fields necessary for their 90° deflection in high vacuum. Thus IEA is useful to study mainly ion emission form thick target irradiated at relatively low laser intensities and it is often coupled to TOF analyses for the ion recognition.
Magnetic deflection permits not only the ion focalization using the axial geometry and the ion acceleration due to the generation of electron traps in front of the target, as reported in a recent our article [17], but also the ion selection in transversal application mode. The limitations for the magnetic field analysis are due to the application of high magnetic fields in the case of high energetic ions of the order of 100 keV or more.
MQS spectrometer is adapts to analyze the neutral component emitted from laser ablation and plasma emission and permits the recognition of the isotopes ad of the chemical species generated in the plasma. For low plasma temperature it permits the evaluation of the fractional ionization of the plasma and measurements of the ion energy distributions [18].
Of course the plasma characterization not only uses the electro-magnetic deflection of ion but also many other complementary analysis to be investigated in detail. Optical spectroscopy, streak camera imaging, X-ray detectors, Langmuir probes, semiconductor detectors and electron detectors, for example, represent a lot of instrumentation necessary to have information on the plasma nature, temperature and density dynamics, expansion and charge separation effects, plasma frequency, and other parameters characterizing its expansion and angular distribution.

This work was supported by the "Research and Mobility" project of Messina University No. 74893496, scientifically coordinated by Professor L. Torrisi and by CANAM in Rez (Czech Republic) Project No. CZ.02.1.01/0.0/16_013/0001812.

References

1. M. Colucci, Elettronica Generale, Cedam Ed., Padova (1975).
2. LD Didactic GmbH, Investigating the deflection of electrons in electrical and magnetic fields. Actual website (2017): https://www.ld-didactic.de/literatur/hb/e/p3/p3851_e.pdf
3. L. Torrisi, M. Cutroneo, L. Andò, and J. Ullschmied, Phys. Plasmas **20**, 023106 (2013)
4. M. Cutroneo, L. Torrisi, S. Cavallaro, L. Ando' and A. Velyhan, Journal of Physics: Conference Series 508, 012020 (2014).
5. P.H. Dawson, Quadrupole mass spectrometry and its applications, Elsevier, New York (1976).
6. E. Woryna, P. Parys, J. Wołowski and W. Mróz, Laser and Particle Beams **14**, 293-321 (1996).

7. L. Laska, J. Badziak, F.P. Boody, S. Gammino, K. Jungwirth, J. Krasa, M. Pfeifer, K. Rohlena, J. Ullschmied, P. Parys, J. Wołowski, E. Woryna, and L. Torrisi, Brazilian Journal of Physics **34**(4B), 1615-1620 (2004).
8. L. Torrisi, S. Gammino, L. Andò and L. Laska, J. Appl. Phys. **91**(5), 4685-4692 (2002).
9. COMSOL Multiphysics code, actual website 2017: https://www.comsol.com/blogs/computational-electromagnetics-modeling-which-module-to-use/
10. L. Torrisi, Physics of Plasma **24**, 023111 (2017).
11. L. Torrisi, A. Italiano and A. Torrisi, Appl. Surf. Sci. **387**, 529–538 (2016).
12. J. Ziegler, SRIM, The stopping and range of ions in matter, actual website 2017: http://www.srim.org/
13. NIST, Stopping-Power & Range Tables for Electrons, Protons, and Helium Ions, actual website 2017: https://www.nist.gov/pml/stopping-power-range-tables-electrons-protons-and-helium-ions
14. L. Torrisi, Rad. Eff. and defects in Solids **171**(1-2), 34-44 (2016).
15. L. Torrisi, L. Giuffrida, M. Rosinski and C. Schallhorn, Nucl. Instr. and Methods in Phys. Res. A **268**, 2808-2814 (2010).
16. L. Torrisi, S. Cavallaro, M. Cutroneo, L. Giuffrida, J. Krasa, D. Margarone, A. Velyhan, J. Kravarik, J. Ullschmied, J. Wolowski, A. Szydlowski and M. Rosinski, Review of Scientific Instruments **83**, 02B11, 1-3 (2012).
17. L. Torrisi and G. Costa, Laser and particle Beams **35**(2), 202-209 (2017).
18. L. Torrisi, F. Caridi, D. Margarone and A. Borrielli, Nucl. Instr. and Methods in Physics Res. B **266**, 308-315 (2008).

Plasma Mirrors for Cleaning Laser Pulses from the Infrared to the Ultraviolet

István B Földes[1,*], *Barnabás* Gilicze[2], *Zsolt* Kovács[1] and *Sándor* Szatmári[2]

[1]Plasma Physics Department, Wigner Research Centre for Physics of the Hungarian Academy of Sciences, H-1525 Budapest P.O.B. 49, Hungary
[2]Department of Experimental Physics, University of Szeged, Dóm tér 9, H-6720 Szeged, Hungary

Abstract. Ultrashort laser pulses are generally preceded by prepulses which – in case of high main pulse intensities - may generate preplasmas on solid surfaces, thus making the initial conditions for the interactions ambiguous. Infrared laser systems applied successfully, with high efficiency self-induced plasma mirrors for improving the contrast of the beam. Short wavelength laser beams however have a larger critical density in the plasma, and due to their deeper penetration the absorption is higher, the reflectivity, and the corresponding plasma mirror efficiency is lower. We show herewith that with carefully planned boundary conditions plasma mirrors can reach up to 70% efficiency even for KrF laser radiation. Our observations can be qualitatively explained by the classical Drude model. The high reflectivity allows the use of plasma mirrors even after the final amplification or before the last amplifier. Different arrangement proposals for its integration to our high power KrF laser system are given as well.

1 Introduction

One can speak about ultrashort laser pulses when the duration of the pulse is less than a picosecond (10^{-12}s). Laser pulses with durations as short as 4-5 fs (1 fs=10^{-15} s), i.e. less than 2 cycles of the electromagnetic pulse can be generated. They are however generally not purely Gaussian pulses, as the short pulse generally follows an amplified spontaneous emission (ASE) pulse with nanosecond duration from the amplifier, and/or a shoulder of a few picosecond due to the incomplete compression after the chirped pulse amplification (CPA) [1]. CPA systems avoid the damage and nonlinearities of the laser amplifier, as the amplification occurs after spectrally selective stretching of the laser pulse. The amplified pulse is then recompressed after the amplifier stage. Most of the solid state laser systems operate with CPA amplification, providing short pulses in the infrared. Excimer lasers however allow ultrashort pulse generation as well, and - as they are pumped by gas discharges - the active gas material allows direct amplification as well, i.e. they are non-CPA system [2].

In case of high-power laser-matter interactions ultrashort pulses allow the generation of intensities exceeding the relativistic limit of $>10^{18}$ W/cm^2 at which the free electrons oscillate with velocities almost the speed of light. Clearly, even a low intensity pedestal of the pulse is sufficient to generate preplasma on the surface of solid targets, thus preventing one to obtain clean boundary conditions. Preplasmas can be detrimental for phenomena as for example high order harmonic generation from the surfaces of solid targets [3].

Plasma mirror or self-induced plasma shuttering [4] is one of the most efficient methods to remove prepulses. In this case the laser beam is focused onto an antireflection-coated transparent glass or quartz target inside a vacuum chamber. The low intensity part of the beam arriving before the main pulse is transmitted through the target. Once the ionization threshold is reached, a high density overdense plasma is generated which reflects the main pulse of the beam with an improved temporal contrast. Clearly more than 2 orders of magnitude improvement can be obtained in a single stage. This has been well applied for the infrared laser systems, in which case reflectivity was obtained up to 80% [5]. Plasma mirror is probably the ultimate method to remove prepulses. It is generally applied after the last amplifier, in the already compressed beam, additionally to the cross polarized wave generation (XPW) method [6], which is - in general - applied between two CPA amplifier stages.

Due to the availability of high reflectivity in the infrared, double plasma mirrors were introduced [7] with 50% efficiency, resulting in 4 orders of magnitude contrast improvement. It must be noted that in case of the CPA systems not only the ASE pedestals are eliminated but the shoulders of a few picosecond duration (caused by the imperfect compression and uncompensated high-order dispersion) are suppressed as well.

Different is the case for the UV lasers. Although KrF systems operate with direct amplification

* Corresponding author: foldes.istvan@wigner.mta.hu

and prepulses from incomplete compression do not appear, the high photon energy is able to modify laser-matter interactions even at lower intensities. The 5 eV photons of the ASE of the KrF amplifiers may generate ions due to photoablation and photoionization of the solid matter even if the prepulse of nanosecond duration has intensity as low as 10^7 W/cm^2 [8]. It means that altogether more than 10-13 orders of magnitude intensity contrast is needed for high intensity interactions, i.e. for 10^{17}-10^{20} W/cm^2. On the other hand, until recently no plasma mirrors were used for short-pulse UV lasers. The reason of it is that the UV radiation of the KrF laser penetrates deeply into the plasma generated on solid surfaces, namely to a critical density which is 16 times higher than for an infrared laser beam with wavelength of 1 μm. If the penetration depth is larger, then the reflectivity is lower because of the collisional absorption in the underdense plasma. A high reflectivity, i.e. 70% efficiency was just recently obtained for laser pulses of 248 nm wavelength [9] which opens the possibility to its application in UV laser systems.

In the case of IR lasers detailed studies of Ziener et al [5] show that plasma mirrors can work well up to quite high laser intensity on the plasma mirror target until nonlinear interactions set in. The maximum applicable intensity depends on the main pulse duration [10], generally it is less than 10^{17} W/cm^2. Therefore the laser beam cannot be tightly focused onto the target, which is positioned in most cases behind the focal plane [11]. Rödel et al [11] obtained three orders of magnitude contrast improvement until 3 ps before the main pulse with a single stage plasma mirror by precise choice of the target coating and the experimental geometry. As the high beam quality on the plasma mirror target is a critical issue, long focal length focusing is often used in order to keep the intensity low. In the case of the double plasma mirror [7] the two plasma mirror targets are inside an f=10 m telescope.

For UV lasers however the ASE prepulse can be relatively strong even if it is minimized by off-axis amplification [2]. Note that for the 248 nm wavelength of the KrF systems the possibilities of filtering the prepulses by nonlinear crystals (Pockels cell or XPW) is strongly limited. It was already shown by the early works of Fedosejevs et al [12] that even the ASE prepulse must be kept low in order to obtain high reflectivity. In the following the methods of optimizing the plasma mirror for the short pulse KrF laser is described in detail which resulted in nearly 70% reflectivity either for 600 fs or for 200 fs ultrashort pulse duration [9].

2 Experimental arrangement

A twin-tube excimer laser system was the basis of the experiments [2]. The first laser tube is an ordinary XeCl laser. It serves as a pumping of the dye-laser chain, which finally gives an ultrashort laser pulse of 500 fs duration on the 496 nm wavelength. After frequency conversion in a BBO crystal it serves as a seed pulse for the second, KrF amplifier. It was used either in a 2-times or in a 3-times off-axis pass configuration providing 10 and 15 mJ pulses of ~500 fs duration, respectively. A further power amplifier has been recently added, in which after 2-passes off-axis amplification up to 80 mJ energy could be obtained (Fig. 1).

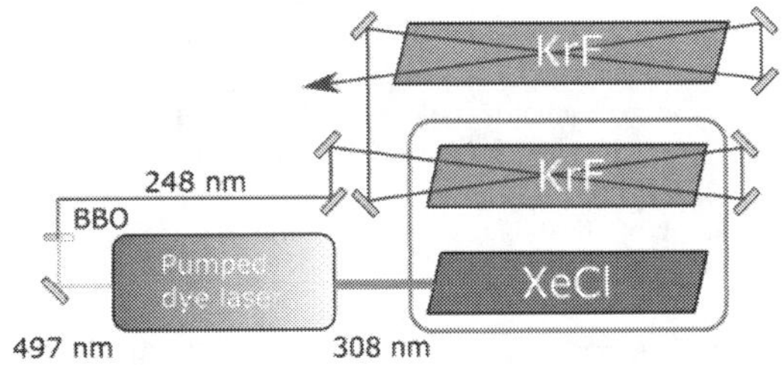

Fig. 1. The scheme of the KrF laser system.

As mentioned above, the only source of prepulses is the ASE of the amplifiers. Off-axis amplification is advantageous in this respect because most of the ASE propagates along the axis of the amplifier [13]. A part of the ASE can be excluded as well, if the pulse to be amplified is coming in the first half of the (discharge) pumping and the corresponding ASE duration. Therefore the first amplifier was used only in a 2-pass configuration, when the energy contrast was better than 100. Taking into account the duration time of the main pulse and that of the ASE this energy contrast corresponds to more than 6 orders of magnitude in the power contrast and up to 10 orders of magnitude intensity contrast in case of tight focusing because the focusability of ASE is much worse than that of the main pulse.

In the case of using the final amplifier a spatial filter is situated between the two amplifier stages. The result is that the energy contrast remains very high (better than 100) but the ASE transmitted by the spatial filter will be the focusable part, therefore the intensity contrast will be not better than 8-9 orders of magnitude, corresponding to ~10^8–10^9 W/cm^2 ASE intensity in the focal plane resulting in a preplasma caused by photoablation and photoionization. Plasma mirror experiments were carried out after the first amplifier, in which case a modest f/25 or similar focusing was sufficient to investigate and optimize the plasma mirror, and in which case the ASE intensity was less than 10^6 W/cm^2, thus below the threshold of photoablation [8,14].

Part of the experiments were carried out after compressing the laser pulse to 220 fs pulse duration, in which case the laser energy was reduced to 0.5-1 mJ.

The target was an antireflection-coated quartz plate in a vacuum chamber. It was antireflection-coated which was moved by stepper motors to provide fresh surface area for each laser shot. The laser beam was of 15 mm diameter in front of the chamber and it was focused by a 40 cm lens onto the target surface by 12° angle of incidence. At this nearly perpendicular angle of incidence the reflectivity does not depend strongly on the polarization of the laser beam, nevertheless the experiments were carried out both by p- and s-polarized incident radiation.

https://doi.org/10.1051/epjconf/201816704001

Care was taken to avoid any beam inhomogeneities on the target, especially that of astigmatism. Therefore the beam profile was monitored by imaging the focal spot to a UV-sensitive Hamamatsu T7040 CCD detector with a 12 times magnification. This imaging was carried out after refocusing the beam reflected by the plasma mirror, too, in order to confirm the applicability of the used configuration.

The incoming and the reflected energy was monitored by fiber coupled energy-meters using peak-hold detection [15]. Note that in the experiments using discharge-pumped lasers fiber-coupling is essential to reduce electrical noises. Energy monitoring was carried out by each shot by splitting out a small part of the beam using a quartz plate. The energy monitor was then calibrated using a Gentec Joulemeter QE50LP. The energy monitors were equipped with a microprocessor, thus the ratio of the incoming and that of the reflected energy could be immediately obtained by the computer.

3 Experimental results

a)

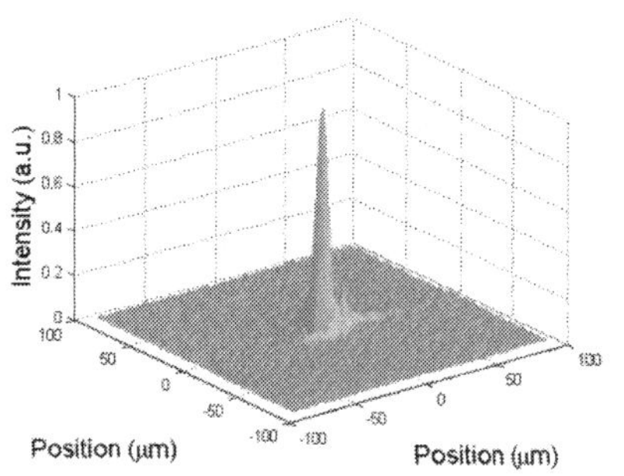

b)

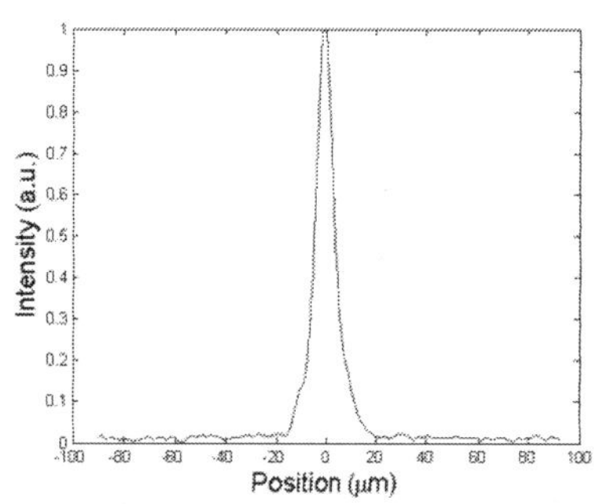

Fig. 2. Image of the 12 times magnified focal spot of the laser (a) and a cross-sectional view (b).

Fig. 2 shows the 2-dimensional focal distribution of the incoming beam and a typical cross-section. The beam is nearly Gaussian, and it corresponds to the focal spot of a 1.5 times diffraction limited rectangular beam. In this case approximately 72% of the laser energy is in the central spot of the Airy-pattern.

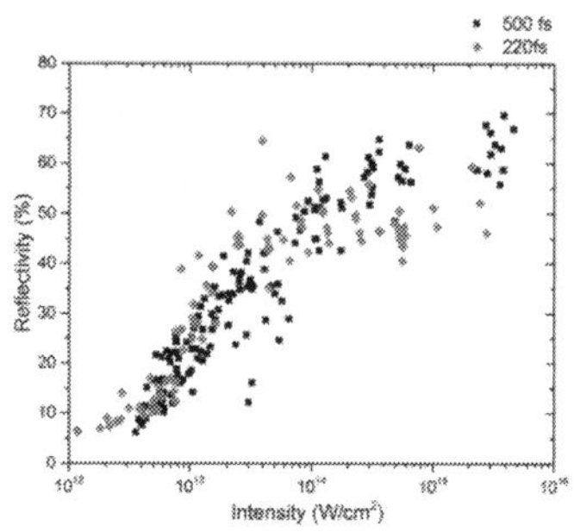

Fig. 3. Intensity dependence of plasma mirror reflectivity. The reflectivity for the s-polarized KrF laser beam is given for 500 fs (black squares) and for 220 fs (red diamonds) pulse duration.

High reflectivity, i.e. more than 60%, in some shots up to 70% was measured both with 500 fs and with 220 fs pulse duration. In the results shown in Fig. 3. s-polarized incoming radiation was used. It must be mentioned that changing the direction of the linear polarization to p-polarization did not affect the results strongly due to the nearly perpendicular angle of incidence. The only difference is that at higher intensities the scatter of data is larger due to the increasing importance of nonlinearities for p-polarized laser beam. It can be seen that below 10^{12} W/cm^2 short-pulse intensity the target remains practically transparent, the reflectivity remains in the noise level. It starts to increase if the intensity reaches 10^{13} W/cm^2 (i.e. above the plasma threshold) and saturates above 10^{14} W/cm^2 remaining approximately constant up 10^{16} W/cm^2. The observed high reflectivity [9] which allows the direct applicability of the plasma mirror for KrF systems can be attributed mainly to the already good initial contrast. In the present case the intensity of the ASE pedestal of 15 ns duration on the target was less than 10^5 W/cm^2. If the prepulse level is higher (e.g. 10^8 W/cm^2 after 3 passes in the amplifier) the reflectivity drops below 50% [16] in agreement with the early observations of Fedosejevs et al. [14]. The nearly perpendicular angle of incidence is needed as well, because in the first plasma-mirror experiments with the KrF laser [17] the angle of incidence was 45°and the reflectivity was as low as 30%. The shot-to-shot variation of reflectivity is attributed to the beam quality nonuniformities due to the shot-to-shot variation of the discharge-pump of the laser. In case the beam quality was lower than illustrated in Fig. 2, then the observable reflectivity could be significantly lower, too. It must be mentioned that the maximum reflectivity is similar in the cases of the two applied pulse durations, especially when taking into account that in the case of the shorter pulse the maximum incident intensity was limited by the size of the compressor optics, therefore the full saturation was not reached.

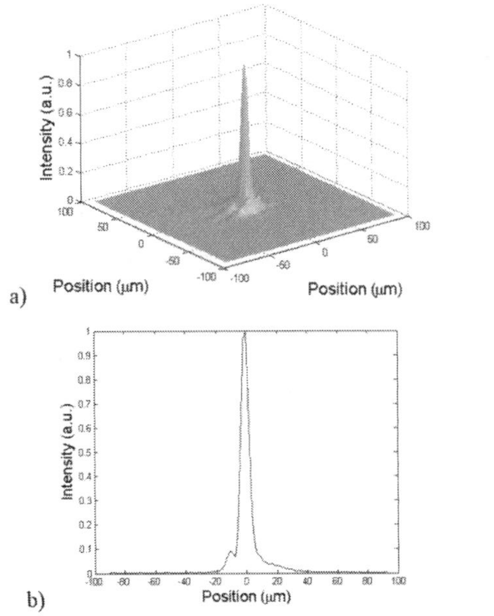

Fig. 4. Image of the 12 times magnified focal spot of the beam after the plasma mirror (a) and a cross-sectional view (b).

The beam quality of the reflected beam is illustrated in Fig. 4. It is similar to that of the incoming beam, as it is well focusable to a 1.75 times Gaussian spot. The energy contained in the central spot is 70%.

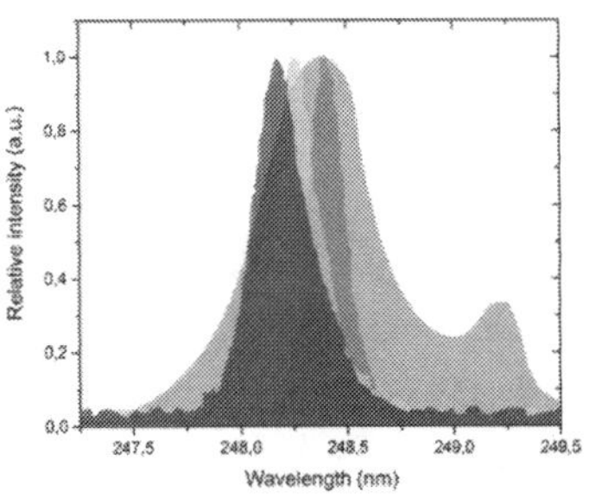

Fig. 5. Laser spectra. The incoming laser beam (red) as compared with the approximate gain bandwidth (salmon) and the reflected pulses at intensities of 2×10^{14} W/cm^2 (green) and 8×10^{15} W/cm^2 (blue).

In case one wants to further amplify the laser beam cleaned by a plasma mirror, it is essential that the reflected and therefore Doppler-shifted beam should be within the gain bandwidth of the amplifier. It is more critical in our case of KrF amplifiers than for the Ti:sapphire laser systems which have much broader bandwidth. Figure 5 shows the laser spectra as compared with the gain spectrum [18]. Clearly the Doppler-shift is increasing with increasing laser intensity [19] but it remains within the spectral range of the gain bandwidth up to the applied maximum intensity of ~10^{16} W/cm^2.

4 Discussion

Plasma reflectivity is generally calculated by applying the Drude model [14] and it gives a rough agreement with our results in case of assuming an average plasma scalelength of L/λ~0.1 during the interactions which gives a reflectivity of 70% corresponding to our experimental observations. Even recent plasma mirror experiments [10] used similarly the Drude model to explain their findings. It must be mentioned however that this model cannot describe the phenomena appropriately.

In the Drude model Fresnel reflection is calculated on the border of the plasma medium which has an index of refraction

$$n_r^2 = 1 - \frac{n_e / n_c}{1 + i v_0 / \omega}. \qquad (1)$$

Here n_e is the electron density in the high density (overdense) region, $n_c = m_e\omega^2/(4\pi e^2)$ is the critical density, ω is the laser frequency, with m_e electron mass and e charge. The electron-ion collision frequency in the dense region is assigned by υ_0. The Fresnel reflectivity for an angle of incidence of θ_{inc} in the case of s-polarized radiation is

$$R_s = \left| \frac{\sin(\theta_{inc} - \theta_{\text{int}})}{\sin(\theta_{inc} + \theta_{\text{int}})} \right|^2. \qquad (2)$$

Here θ_{int} is the complex angle of propagation inside the dense plasma [14] which can be obtained from the generalized Snell law as $\theta_{int} = arcsin[\theta_{inc}/n_r]$. The imaginary part of the index of refraction, and thus the absorption strongly depends on the electron-ion collision frequency. Fedosejevs et al [14] made calculations for different υ_0 values in case of step function vacuum-plasma interface and they obtained high reflectivity when υ_0 was significantly less, typically some tenth of the laser frequency. The electron-ion collision frequency in a plasma is given [20] by

$$v_0 = \frac{4(2\pi)^{1/2}}{3} \frac{Z n_e e^4 \ln\Lambda}{m_e^{1/2}(k_B T)^{3/2}} \qquad (3)$$

with Z ion charge, T temperature and k_B Boltzmann-constant. It also depends on the $ln\Lambda$ Coulomb logarithm. Clearly, the collisional frequency is strongly dependent on the temperature and the density of the plasma. A special problem arises with the Coulomb logarithm which is derived from the integration limits of the Rutherford scattering of an electron on ions. In the case of very high densities and low temperatures corresponding to the state of the solid matter the classical Coulomb logarithm becomes less than or equal to zero, and then the classical [20] (Spitzer) expression of the Coulomb logarithm cannot be applied any more. In that case namely the plasma is strongly coupled, the potential energy in the electric field is comparable with the kinetic energy of the particles. For this case more accurate quantum mechanical calculations [21] avoid this divergence but further increasing the density the plasma is becoming Fermi degenerate. Due to these problems most textbooks [22, 23] suggest to take the Coulomb logarithm simply equal to 2, a rough estimation. Fedosejevs et al [14] avoided this problem by choosing only some finite discrete values of the electron-ion collision frequency by taking e.g. a typical $\upsilon_0/\omega \sim 0.1$. We also carried out some calculations for steep plasma gradients of our case of 12°angle of incidence. Clearly more than 60% reflectivity was obtained (depending on the temperature) in case of using lnΛ=2 and it can be even higher when using the Coulomb logarithm of Mulser et al [21]. When using

$$\ln \Lambda_0 = \ln \frac{\lambda_D}{\lambda_B} + 0.2 + \frac{\lambda_B^2}{4\lambda_D^2} \qquad (4)$$

with λ_D Debye length and λ_B de Broglie length, nearly 80% reflectivity was obtained in case of 4eV plasma temperature of the plasma step having a 10 times critical density. Thus our experimental finding can roughly be explained by the theory but in order to a full understanding of the phenomena the exact equation of state and the plasma density and temperature profile is needed at the arrival and during the interaction with the main ultrashort laser pulse.

5 Conclusion

We could demonstrate the high, ~70% efficiency even for the KrF laser system, although its 248 nm wavelength in the ultraviolet causes a deep penetration to the plasma, and therefore generally lower absorption than in the infrared. The method for obtaining it is that we could shorten the interaction length by choosing appropriate initial conditions with shorter scalelength plasmas. Basically there are 2 possible method to integrate it into the laser system In our previous paper [9] we suggested an arrangement in which the plasma mirror is applied in front of the final amplifier. It became possible because although the reflected beam suffers a spectral Doppler shift, the spectrum remains within the gain bandwidth of the final amplifier. In this case the final energy of the laser system will not decrease significantly because the amplifier operates in a saturation regime.

The observed high efficiency of the plasma mirror allows its direct use after the first amplifier, as even in this case more than 60% of the energy can be coupled to the target. We mentioned however that the high reflectivity requires high quality beam, i.e. best is to put the plasma mirror target in focus. In order to avoid nonlinearities the intensity must be limited to the values given above. Therefore in our case of 80 mJ /600 fs laser beam a loose focusing with at least f=5m focusing lens seems to be necessary. Its problem lies in the final size of the laboratory and the long vacuum beamlines. There is however another possibility. As the saturation intensity of the KrF laser is significantly lower than the damage threshold of the optics, it is possible first to reduce the diameter of the beam a telescope, and then to focus it with a shorter focal length focusing optics. Alternatively a galilei telescope may be directly applied to provide a larger focal spot in shorter distance and the near optimal 10^{15} W/cm^2 intensity on the plasma mirror.

We can thus summarize that the observed high reflectivity of plasma mirrors opens the possibility of their application for improving the contrast of lasers even in the ultraviolet.

This work has been carried out within the framework of the EUROfusion Consortium and has received funding from the Euratom research and training programme 2014-2018 under grant agreement 633053. The views and opinions expressed herein do not necessarily reflect those of the European Commission. The authors also acknowledge the support of the Hungarian Scientific Research Fund - OTKA 113222 and the COST MP1208 activity. The technical assistance of Rita Dajka is acknowledged this way, too.

References

1. D. Strickland and G. Mourou, Opt. Commun. **56**, 219 (1985)
2. S. Szatmári, Appl. Phys. B **58**, 211 (1994)
3. C. Thaury and F. Quéré, J. Phys. B **43**, 213001 (2019)
4. H. Kapteyn, M. Murnane, A. Szoke, R. Falcone, Opt. Lett. **16**, 491 (1991)
5. CH. Ziener et al., J. Appl. Phys. **93**, 768 (2003)
6. A. Jullien et al., Appl. Phys. B **84**, 409 (2006)
7. T. Wittmann et al., Rev. Sci. Instrum. **77**, 083109 (2006)
8. I.B. Földes et al., Laser Physics **10**, 264 (2000)
9. B. Gilicze, A. Barna, Zs. Kovács, S. Szatmári, I.B. Földes, Rev. Sci. Instrum. **87**, 083101 (2016)
10. G. Doumy et al., Phys. Rev. E **69**, 026402 (2004)
11. C. Rödel et al., Appl. Phys. B **103**, 295 (2011)
12. R. Fedosejevs, R. Ottmann, R. Sigel, G. Kühnle, S. Szatmári, F.P. Schäfer, Phys. Rev. Lett. **64**, 1250 (1990)
13. G. Almási, S. Szatmári, P. Simon, Opt. Commun. **88**, 231 (1992)
14. R. Fedosejevs, R. Ottmann, R. Sigel, G. Kühnle, S. Szatmári, F.P. Schäfer, Appl. Phys. B **50**, 79 (1990)
15. A. Barna, I.B. Földes, Z. Gingl, R. Mingesz, Metrol. Meas. Syst. **20**, 183 (2013)
16. I.B. Földes, A. Barna, D. Csáti, F.L. Szűcs, S. Szatmári, J. Phys. Conf. Ser. **244**, 032004 (2010)
17. I.B. Földes, D. Csáti, F.L. Szűcs, S. Szatmári, Rad. Effects & Defects in Solids, **165**, 429 (2010)
18. W.T. Silfvast, *Laser Fundamentals* (Cambridge University Press, 2004) ISBN 0-521-83345-0
19. R. Sauerbrey, Phys. Plasmas, **3**, 4712 (1996)
20. S. Atzeni, J. Meyer-ter-Vehn, *The Physics of Inertial Fusion* (Oxford Science Publications, Clarendon Press-Oxford 2004) ISBN 0-198562640
21. P. Mulser, G. Alber, M. Murakami, Phys. Plasmas **21**, 042103 (2014)

https://doi.org/10.1051/epjconf/201816704001

22. S. Eliezer, *The interaction of high-power lasers with plasmas* (IOP Bristol and Philadelphia, 2002) ISBN 0 7503 0747 1

23. S. Pfalzner, *An introduction to inertial confinement fusion* (CRC Press Tayleo & Francis New York, London 2006) ISBN 0-7503-0701-3

 https://doi.org/10.1051/epjconf/201816704002

Prediction of strontium bromide laser efficiency using cluster and decision tree analysis

Iliycho Iliev[1,*], *Snezhana* Gocheva-Ilieva[2], and *Chavdar* Kulin[2]

[1]Technical University of Sofia – branch Plovdiv, Department of Physics, 25 Tzanko Dujsstabanov str., 4000 Plovdiv, Bulgaria
[2]University of Plovdiv Paisii Hilendarski, Faculty of Mathematics and Informatics, 24 Tsar Asen str., 4000, Plovdiv, Bulgaria

Abstract. Subject of investigation is a new high-powered strontium bromide ($SrBr_2$) vapor laser emitting in multiline region of wavelengths. The laser is an alternative to the atom strontium lasers and electron free lasers, especially at the line 6.45 μm which line is used in surgery for medical processing of biological tissues and bones with minimal damage. In this paper the experimental data from measurements of operational and output characteristics of the laser are statistically processed by means of cluster analysis and tree-based regression techniques. The aim is to extract the more important relationships and dependences from the available data which influence the increase of the overall laser efficiency. There are constructed and analyzed a set of cluster models. It is shown by using different cluster methods that the seven investigated operational characteristics (laser tube diameter, length, supplied electrical power, and others) and laser efficiency are combined in 2 clusters. By the built regression tree models using Classification and Regression Trees (CART) technique there are obtained dependences to predict the values of efficiency, and especially the maximum efficiency with over 95% accuracy.

1 Introduction

A key element in the development of laser systems is the improvement of their output characteristics – average output power, laser efficiency, etc. With the accumulation of experiment data from the operation of the lasers, there is an opportunity to gather significant new information on the relationships and dependencies between the measured operational parameters for the specific device.

The subject of investigation in this study is the new high-powered strontium bromide ($SrBr_2$) vapor laser emitting in a longitudinal pulse helium discharge [1]. The laser is characterized by a multiline radiation at several wavelengths in the range from 2.06 μm to 6.45 μm. It shows the greatest promise for practical medical applications such as the processing of biological tissues and bones with minimal damage at a wavelength of 6.45 μm [2].

In order to extract relationships from the available experiment data, in this investigation, two statistical data mining techniques are applied – exploratory cluster analysis and classification and regression trees (CART, [3, 4]). With their help, seven operational parameters are used as input independent variables - laser tube length and diameter, supplied electric power, pulse frequency, and helium buffer gas pressure, as well as the overall output laser efficiency. Through cluster analysis, on linear type classification, it is found that the parameters can be grouped in two clusters and that for efficiency the geometric parameters are the determining factors. Using the CART technique, models are obtained where the experiments are classified according to the measured laser efficiency values. Stable CART models are built, the relative importance of each operating characteristic for laser efficiency is determined. The models are validated through machine learning approach. The values for the efficiency, predicted by the models describe those measured including the maximum values with an accuracy of over 95-96%. The results of the statistical analysis can be applied to guide experimental investigations for the increase of the efficiency of the considered type of $SrBr_2$ lasers. The approach is also fully applicable for other types of laser systems.

All statistical analyses are performed by using IBM SPSS statistical software [5].

2 Description of strontium bromide laser and experimental data

Until recently, the only known source of laser generation with a wavelength of 6.45 μm were free electron lasers. But their use is difficult because of the high price and service costs, which is not acceptable for regular medical practice. A new alternative to the free electron lasers are strontium vapour lasers (atom and ion transfers) with generation at several wavelengths in the infrared spectrum: 2.06, 2.20, 2.69, 2.92, 3.01, 3.07, and 6.45 μm [6, 7]. These lasers have acceptable price, compact size, easy operation, and lower service costs than the free electron lasers. In this study we investigate the new

* Corresponding author: iliev55@abv.bg

$SrBr_2$ lasers, developed and patented at the Institute of Solid State Physics "Academician Georgi Nadjakov" of the Bulgarian Academy of Sciences [1].

The current stage of the development of $SrBr_2$ laser and its physical characteristics are presented in papers [8, 9], where a laser based on $SrBr_2$ with laser output power of 2.4 W is reported. In [1, 10] the laser output power reaches 4.27 W, with 90% of the laser generation at the 6.45 μm line. The obtained values of laser output power are fully comparable with these for metal strontium lasers.

The laser tube of $SrBr_2$ laser is entirely designed of quartz. Because of the high internal temperature in the active laser volume (more than 1000°C), a ceramic tube insert made out of Al_2O_3 is used to assure the thermo-chemical stability of the discharge. Principal scheme of the laser tube of $SrBr_2$ laser with its components is shown in Fig. 1.

The operation of the strontium bromide ($SrBr_2$) laser depends on the following main characteristics, or input variables: *D*1 (mm) – the internal diameter of the outer (quartz) tube; *D*2 (mm) – the internal diameter of the ceramic tube insert; *La* (cm) – distance between the electrodes (active volume); *Ceq* (pF) – equivalent capacity of the capacitor battery; *Pin* (kW) – supplied electric power in the active laser volume; *PRF* (kHz) frequency impulse repetition rate; *PHe* (Torr) – pressure of the buffer gas helium. To account for 50% loses, during the exploitation of the tube, in all numerical procedures the values of *Pin* are reduced twice, and the new variable is noted by *PIN*2.

We carry out statistical studies by using data from 167 available experiments. The dependent variable is the laser efficiency *Eff* (%). All data are collected from publications [1, 8-10].

The descriptive statistics of the processed data is given in Table 1. It could be mentioned that the distribution differs from the normal distribution and standard statistical analyses as regression are not appropriate.

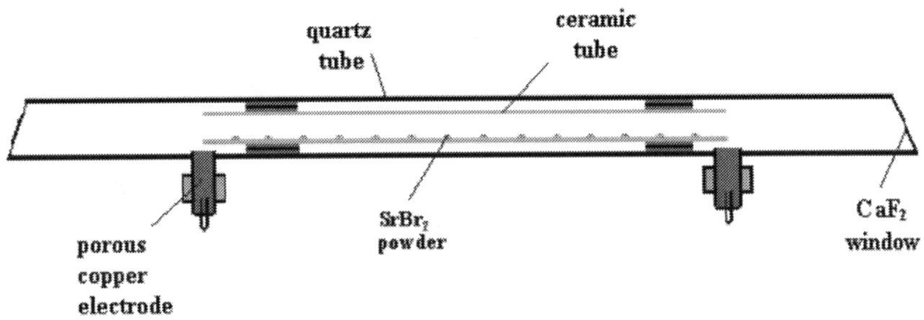

Fig. 1. Longitudinal principal scheme of strontium bromide laser.

Table 1. Descriptive statistics of experiment data[1].

	D1, mm	*D2*, mm	*La*, cm	*Ceq*, pF	*PIN2*, kW	*PRF_* KHz	*PHE*, Torr	*Eff*, %
Mean	46.548	19.14	87.955	577.795	0.928	19.231	40.954	1.61
Median	46.000	19.80	98.000	632.300	0.962	19.000	41.600	1.78
Mode	46.0	20	98.0	632.3	1.05	19.0	38.00	1.16
Std. Deviation	0.724	0.869	13.281	142.422	0.116	0.817	5.894	0.400
Variance	0.525	0.756	176.391	20284.079	0.014	0.668	34.734	0.160
Skewness	0.565	-0.565	-0.565	0.181	-0.557	-1.559	-0.544	-0.412
Kurtosis	-1.702	-1.702	-1.702	-1.373	-1.233	8.556	-0.404	-1.552
Minimum	46.0	18	70.5	398.0	0.70	15.0	25.00	0.83
Maximum	47.5	20	98.0	793.1	1.05	22.0	50.00	2.08

[1] Std. Error of Skewness for all variables is 0.188 and Std. Error of Kurtosis is 0.374.

3 Statistical data mining techniques

In this study we apply two types of data mining techniques to process the experimental data for strontium bromide laser. Cluster analysis is used for grouping investigated laser variables in relatively homogeneous groups (clusters) and classification and regression tree (CART) method for determining the dependence of laser efficiency from the operational variables. This makes it possible on the basis of real measurements to find new relationships in description of laser operation.

3.1. Cluster analysis

Cluster analysis or clustering is well known statistical technique aiming to find an optimal grouping of observations or variables, for which the elements of a given cluster are similar (homogeneous) but the clusters are clearly distinguishable from each other. Cluster analysis is not well formalized unique method, but is a set of procedures for identifying the best cluster model and the appropriate number of clusters. In this paper we apply the agglomerative hierarchical cluster analysis, which is recommended for relatively small data samples. The similarity of variables is measured by the usual Euclidean distance $d(\mathbf{a},\mathbf{b})$ between two vectors (points) $\mathbf{a}=(a_1,a_2,\dots,a_n)$ and $\mathbf{b}=(b_1,b_2,\dots,b_n)$, defined by $d(\mathbf{a},\mathbf{b})=\sqrt{\sum_{k=1}^{n}(a_k-b_k)^2}$ or squared Euclidean distance. Two clusters are considered to be similar and are grouping when the selected distance between them is small. To avoid different physical measures, all variables are preliminary transformed into dimensionless.

In the beginning of CA all objects (i.e. variables) are considered as initial clusters. The hierarchical procedure is a step by step process for consecutive formation of a new cluster at each step, by merging the two clusters with the smallest distance between them. Different cluster methods with specific distance between clusters could be applied. We will use the following four cluster methods, provided in SPSS: Average linkage (Within-groups), Average linkage (Between-groups), Nearest neighbor (Single linkage) and Ward's method. [5, 11, 12]. As an example, in Average linkage (Between groups), the distance $D(A,B)$ between two clusters A and B is defined using all n_A points of A and all n_B points of B by the expression

$$D(A,B)=\frac{\sum_{i,j}d(x_i,x_j)}{(n_A+n_B)(n_A+n_B-1)} \qquad (1)$$

The optimal number of clusters is achieved when the distance between the formed clusters is relatively large. Usually it can be equal or bigger than 20% of the maximum of all initial distances d. For convenience, the distances are graphically presented in rescaled units, where the maximum is 25 units.

3.2. CART method

Classification and regression tree (CART) method, known also as "decision trees" is a powerful data mining technique, proposed by Breiman et al. in 1984 [3]. This is a supervised learning algorithm used both for classification and regression purposes. If the response variable Y is binary-valued the CART technique is used for classification problems, whereas for regression problems Y could be a continuous variable. The algorithm of the method is capable to process very large or small datasets of arbitrary type of variables $X_1,X_2,\dots,X_p$. CART tree can handle linear, nonlinear effects and interaction terms in a rule sequence and build the easily interpreted models. CART is characterized as a multivariable nonparametric statistical method [12].

The CART methodology is based on a recursive type partitioning of the initial data of $\mathbf{R}^p$ into non-overlapping multidimensional sub-regions (classes). The structure of a CART tree consists of a root node, internal nodes and terminal nodes, each of them representing a subset of cases falling within a sub-region. At any step the cases in a current node τ could be or not be splitted into two daughter nodes by asking the unique question of the type $x_{i,j}\le\theta_j$, where $x_{i,j}$ is the i-th value of the variable X_j in the current node, and θ_j is a threshold value. The CART algorithm selects the values of i and j from all possible variables $X_j, j=1,2,\dots,p$ and θ_j from cases in the current node which gives the minimum error in predicting the dependent variable Y. This way, beginning from the root node, each node identifies a specific decision sequence of rules. In regression problems the predicted value $\hat{Y}_\tau$ for cases in a node τ is simply equal to the mean value of their corresponding responses. The usual way to calculate the overall error is the least square error $\sum_{\tau=1}^{M}\left(Y_\tau-\hat{Y}_\tau\right)^2$. The terminal nodes give the final decision about the distribution of dataset and model prediction of existing or new observations.

In machine learning the model performance is estimated not only for its overall accuracy, but rather than from the prediction accuracy for a random test subsample, which has not been used in the modelling procedure. This procedure is called cross-validation [3].

4 Results

We applied the two described techniques, which have to be considered as independent each to other. Cluster analysis is used to classify the variables by similarity principle and to determine the place of the efficiency Eff between the operational variables. CART method is used to establish the relationships in order to construct appropriate regression model, capable to predict the efficiency values and this way to guide the experiment.

4.1 Results from cluster analysis

We used four different cluster methods with Euclidean and squared Euclidean distance. In all analyses the optimal cluster solutions are very similar. They consist of two clusters. As a presentative cluster model, we selected the model obtained by the average linkage (between groups) cluster method with squared Euclidean distance. To note that all variables were preliminary transformed into dimensionless form using the zscores transformation of an variable x, given by the expression $zx_i = (x_i - \bar{x})/s,\ i = 1,...,n$, where $\bar{x}$ is the mean value and s is standard deviation of the variable, n is the volume of the sample. The dendrogram of all clusters is shown in Fig. 2.

From Fig. 2 it can be seen that the solution with two clusters: $C1 = \{La, D2, PIN2, PHe, Ceq, Eff\}$ and $C2 = \{D1, Prf\}$ has the largest rescaled distance between all clusters.

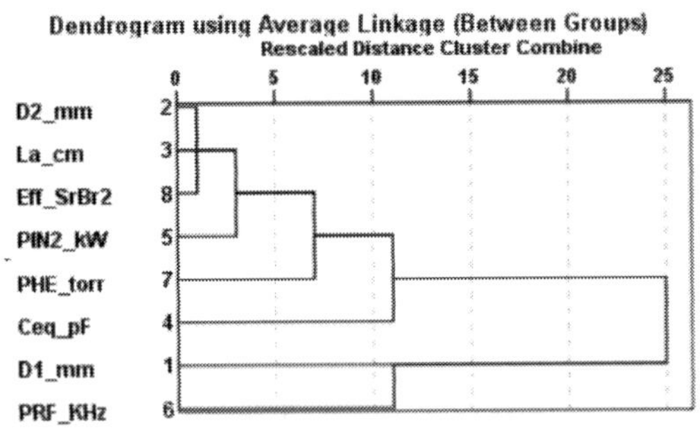

Fig. 2. Dendrogram of all seven input variables and laser efficiency *Eff*, obtained by the Average linkage (Between groups) cluster method with squared Euclidean distance.

The gap from the two clusters *C*1 and *C*2 shows that they are divided by 14 rescaled units one from another. This corresponds to the bigger distance from 224.138 to 512.718 obtained at stage 7 for the two-cluster solution. From Fig. 2 it is also observed that laser efficiency *Eff* (number 8) is grouped at stage 2 with the laser tube length *La* (number 3) and internal diameter of the ceramic tube *D*2 (2). This result could be interpreted, that these two variables are closely related to the efficiency. Then to this cluster is added *PIN*2 (5), etc. to obtain cluster *C*1.

Very similar results are obtained by other cluster methods. For the same type of distance the different cluster methods give the biggest rescaled distance for 2 clusters as follows: Average linkage (Within groups) – 7 rescaled units, Nearest neighbour (Single linkage) – 10 rescaled units, Ward's method – 18 rescaled units. In all models the stages of grouping clusters are the same. We can conclude that the two-cluster solution, consisting of

$$C1 = \{La, D2, PIN2, PHe, Ceq, Eff\} \qquad (2)$$
$$C2 = \{D1, Prf\}$$

is very stable and laser efficiency *Eff* for the examined dataset of experiments is closely related to *La*, *D*2, and *PIN*2.

4.2 Results from CART

There were carried out different CART analyses, by experimenting with the so called control parameters of the method and removing the correlated variables. As the number of experiments N=167, it is taken a minimum from 10 to 20 cases in the parent node of the tree and a minimum of 5 to 10 cases in the child node, respectively. We obtained similar solutions by changing the minimum cases in parent and child nodes and removing some of the variables. All models are tested by the usual 10-fold cross-validation procedure against overfitting [3, 12].

We present in this paper one of the best from the obtained CART models. It was established that variable Seq has no significant influence on the efficiency, and *D*1 correlate strongly with *D*2, so that these variables were removed from the CART procedure. The plot of the regression tree until depth three is shown in Fig. 3. The maximum predicted value of *Eff* is calculated in the node

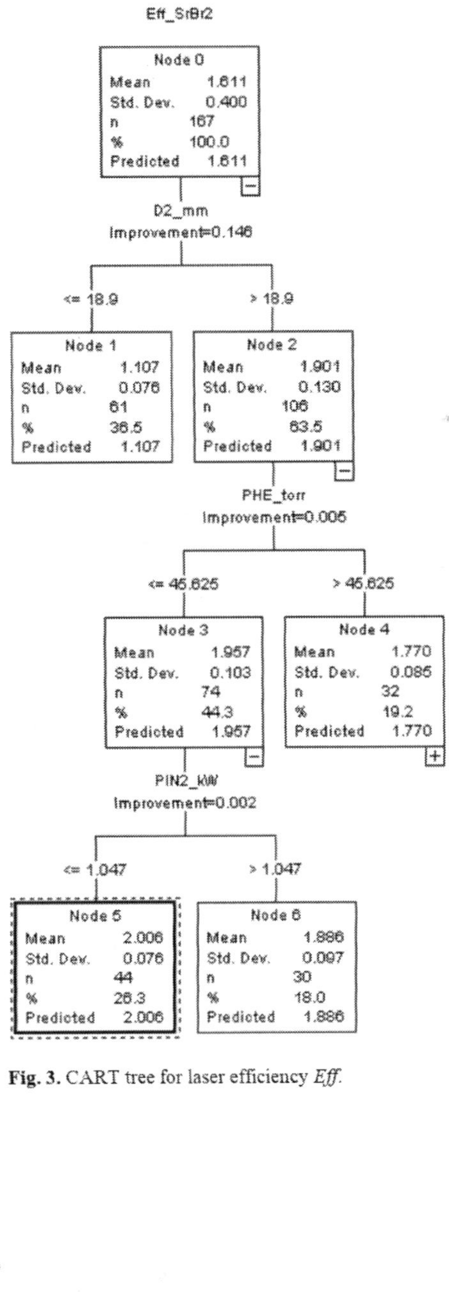

Fig. 3. CART tree for laser efficiency *Eff*.

5 and is equal to 2.006%. This value gives 96.4% accuracy in comparison with the maximum experiment value of *Eff*, which is 2.08% (see also Table 1). This node contains 44 experiments, or 26.3% of all data. To note that the branch from node 4 is truncated and not included in this figure.

CART model is very intuitive and could be easily used for implementation. For example, by following the tree rules in Fig. 3 it is seen that every experiment with $D2$>18.9 mm falls into node 2. Then if $PHe \leq 46.625$ Torr it goes to node 3. In the case of supplied power $PIN2 \leq 1.047$ kW, it will belong to node 5 and one can expect a maximum laser efficiency *Eff*.

The relative variable importance in the model is as follows: *La* and *D2* - 100%, *PIN2* – 86.6%, *PRF* – 76.3% and *PHe* – 70.9%. Thus, from the observed experiment data *D2* and *La* have the highest influence on efficiency *Eff*. This corresponds to the results from cluster analyses. Some difference with respect to cluster model could be explained by the nonlinear regression dependence between *Eff* and all variables in the CART model.

The overall fitting of the selected CART model to the entire (learn) dataset of experiments is illustrated in Fig. 4. The simple comparison between the experiment and predicted values of efficiency *Eff* shows that the coefficient of determination $R^2 = 96.0\%$, i.e. the model explains 96% of the dataset.

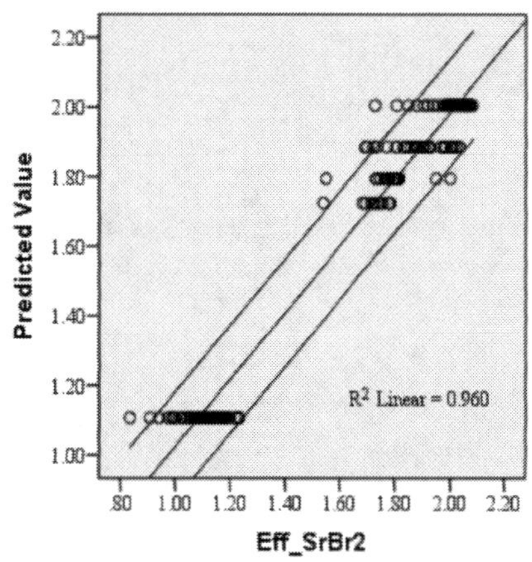

Fig. 4. Experimental versus predicted values of laser efficiency *Eff*, using the selected CART model with 5% confidential interval.

5 Conclusions

By applying two data mining statistical techniques the relations between laser efficiency and seven operational characteristics of $SrBr_2$ laser are obtained based on available experiment data. It was obtained a cluster model for linear type classification of variables in two cluster solution. The more precise results are obtained by CART modelling. The obtained CART model determined the nonlinear type of influence of the operational characteristics on laser efficiency. The predicted values reached over 95% from the maximum measured laser efficiency. There are extracted the main local variation of the investigated laser operational characteristics in order to achieve higher efficiency. The proposed approach can be used for guiding future experiments.

This study was supported by the Scientific and Research Sector of Technical University of Sofia, and the Scientific and Research Department (NPD) of University of Plovdiv Paisii Hilendarski, grant MU17-FMI-003.

References

1. A. Temelkov, N. Vuchkov, I. Freijo-Martin, A. Lema, L. Lyutov, N. Sabotinov, J. Phys. D: Appl. Phys. **42**(11), 115105 (6 pp), (2009)
2. G. Edwards, R. Logan, M. Copeland et al., Nature, **371**.6496, 416-419 (1994)
3. L. Breiman, J.H. Friedman, R.A. Olshen, C.J. Stone, *Classification and regression Trees* (Wadsworth Advanced Books and Software, Belmont, Canada, 1984)
4. X. Wu, V. Kumar (eds.), *The Top ten algorithms in data mining* (Chapman & Hall/CRC, Boca Raton, 2009)
5. IBM SPSS Statistics, http://www-01.ibm.com/software/analytics/spss/products/statistics/
6. G.D. Chebotarev, E.L. Latush, O.O. Prutsakov, A.A. Fesenko, Quantum Electron. 38(4), 299-308 (2008)
7. G.D. Chebotarev, E.L. Latush, A.A. Fesenko, Quantum Electron. 38(4), 309-318 (2008)
8. K.A. Temelkov, N.K. Vuchkov, B.L. Pan, N.V. Sabotinov, B.Ivanov, L.Lyutov, J. Phys. D: Appl. Phys. **39**, 3769-3772 (2006)
9. K.A. Temelkov, N.K. Vuchkov, B.L. Pan, N.V. Sabotinov, B.Ivanov, L.Lyutov, Proc. of SPIE, Bellingham, WA, **6604**, 660410, 1-5 (2007)
10. K.A. Temelkov, N.K. Vuchkov, B. Mao, E.P. Atanasov, L. Lyutov, N.V. Sabotinov. IEEE J. Quant. Electron. **45**(3), 278-281 (2009)
11. S. Gocheva-Ilieva, I. Iliev, *Statistical models of characteristics of metal vapor lasers* (Nova Science, New York, 2011)
12. A.J. Izenman, *Modern multivariate statistical techniques: regression, classification, and manifold learning* (Springer, New York, 2008

Plasma-laser ion discrimination by TOF technique applied to coupled SiC detectors.

Salvatore Cavallaro[1,2,3]

[1]Department of Physics and Astronomy, University of Catania, via S.Sofia,64 95123 Catania, Italy
[2]LNS-INFN via S. Sofia, 62 95123 Catania, Italy
[3]CSFN&SM via S.Sofia,64 95123 Catania, Italy

Abstract. The rate estimation of nuclear reactions induced in high intensity laser-target interaction ($\geq 10^{16}$ W/cm^2), is strongly depending on the neutron detection efficiency and ion charge discrimination, according to particles involved in exit open-channels. Ion discrimination is basically performed by means of analysis of pits observed on track detector, which is critically dependent on calibration and/or fast TOF devices based on SiC and diamond detectors. Last setup is used to determine the ion energy and to obtain a rough estimation of yields. However, for each TOF interval, the dependence of yield from the energy deposited in the detector sensitive region, introduces a distortion in the ion spectra. Moreover, if two ion species are present in the same spectrum, the discrimination of their contribution is not attainable. In this paper a new method is described which allows to discriminate the contribution of two ion species in the wide energy range of nuclear reactions induced in laser-target interactions. The method is based on charge response of two TOF-SiC detectors, of suitable thicknesses, placed in adjacent positions. In presence of two ion species, the response of the detectors, associated with different energy losses, can determine the ion specific contribution to each TOF interval.

1 Introduction

With the availability of high intensities ($\geq 10^{16}$ W/cm^2) laser-target interaction has evidenced nuclear events characterized by fast neutrons and/or recoiling ions involved in the induced specific reactions [1-8]. Estimated reaction rates and, consequently potential applications, are strongly depending on the neutron efficiency values and ion discrimination. Efficiencies of fast neutron detection are rather small, 10^{-4} to 10^{-3}, depending on neutron energies [9-11]. Ion discrimination is performed with solid state nuclear track detectors (SSNTD), as CR-39 or PM-355, by analyzing the distribution of diameter etched pit observed into sample field of optical microscope. However the discrimination effectiveness is critically dependent on calibration (especially etching time), often coupled to thin foil degraders, or restricted to short energy range. [7],[12,13]. Furthermore atomic ion flux due to target impurities, which is expected much larger than nuclear one, can contribute to produce a false signal or, at least, an high statistical error. Thompson Spectrometer (TP), which could give almost complete ion discrimination [14,15], presents the disadvantage of very small acceptance because of needed entrance-pinhole. On the other side, many attempts have been introduced in the past, to get more efficient CR-39 discrimination by coupling the diameter with other parameters (length [16,17]), (grey levels [18]), with scarce experimental pursuance. TOF devices based on Silicon Carbide (SiC) and diamond detectors, almost insensitive to visible and near UV light, have been used to determine the possible energy of ions and to obtain a rough estimation of yields. An intrinsic problem can arise because of dependence of yields from energy deposited in the depleted region of detector subject to drifting electric field (so-called *active sample*). In fact, yield for each time interval (named also *time bin*), associated with sampling unit of TOF oscilloscope, depends on both number of ions arriving within and on energy deposited by each ion into active sample and converted in a corresponding number of electron-hole. The overall mechanisms and characteristics of charge drifting along the active electric field to the electrodes and finally converted in measured current-yield at readout circuit, are well described in many papers, as, for instance, in refs. [19,20]. Here it must be noticed that, because of energy dependence, the interpretation of measured yield as histogram height, is basically wrong. Moreover the active sample can represent full energy absorption for some ion and partial absorption (in punching through case) to others. In this paper, investigating on corrections needed to measured yields to obtain the effective distributions, a new method has been developed which, using correlated SiC

Corresponding author: cavallaro@lns.infn.it

detectors allows to discriminate the contribution of two ions in the wide energy range of nuclear reactions involved in laser-target interactions.

2. Method description

The method is based on the response of two TOF-SiC devices of appropriate depletion layers placed in adjacent positions and subjected to the simultaneous action of two species of ions produced by a plasma-induced nuclear reaction. In the absence of experimental data obtained in such conditions, the following procedure was referring to a nuclear reaction whose experimental data, obtained with a standard accelerator and ion beam hitting target, was as complete as possible. *Thus, the available data was processed in such a way as to have the required characteristics.*

To make the description of the method clear and effective, it is assumed an experimental setup similar to those commonly used today. The iodine laser system of PALS (Prague Laboratory), operating at single impulse at 1315nm, duration of 300ps, is considered. Laser is supposed to irradiate in normal incidence a solid target where the generated plasma induces nuclear reactions. Several detectors are surrounding the target to monitoring the experiment in suitable pressure conditions (10^{-5} to 10^{-6}mbar). Two contiguous SiC-TOF devices are supposed backward at $\theta = 60°$ with respect to target normal. The TOF flight distance is d = 1.25m. SiC devices are constituted by 2 SiC diodes $1mm^2$ in active surface and in adjacent position (1mm far), depletion layer of 5μm and 80μm, respectively.

The SiC detectors are generally built on a 4H-SiC epitaxial layer, grown with the chemical vapor deposition technique in HCl atmosphere, which allows one to obtain a very thick depletion layer (~100μm) with low dopant concentration and with very low defect concentration [19]. The Schottky diodes, reverse polarized and suitably equipped with an output-signal circuit, are coupled with a 50 Ohm input resistor to a fast digital oscilloscope (2GHz,20GS/s). Typical active areas of detectors are $1mm^2$ to $5mm^2$, and depletion layer of 5μm to ~100μm TOF time resolution of order of 1ns can be easy achieved [20]. In the passing through the active layer, an ion spends almost fully the energy loss in pair electron-hole production and output signal formation. (e-hole pair E≅7.8 eV).When a flux of ions goes into the active area of detector simultaneously (i.e., into the time bin selected by TOF setup device), the associated charge depends on both the number of ions and the energy loss of ion into the active thickness of detector; which in turn depends on thickness and ion specie and energy.

2.1 Ion discrimination in plasma-laser nuclear reactions

Nuclear reactions produced in high intensity laser-target interaction involve, depending on atomic composition of the target, the simultaneous presence of different ion species, as proton, deuterons, ^{3}He, ^{4}He, etc. Protons, in particular, are also present due to superficial impurities contained in targets. Typically, nuclear reactions, as d + d, d+^{3}He, p+^{7}Li, d+^{6}Li , d+^{7}Li, p+^{11}B, etc, produce output ion energies of several MeV which can be mixed its self or mixed with impurity-protons. In presence of two ion species of M_1 and M_2 masses respectively, the TOF technique, for each time bin selects two mean energies, E_1, E_2 which are proportional to the ratio of masses M_1/M_2. So, for p and α, it will be $E_\alpha = 4\times E_p$. For example, in the p+^{11}B reaction, 1 MeV of proton can compete with 4 MeV-α which is rather a central energy value for last reaction (see later). In this and similar cases, the discrimination is needed.

2.2 Simulation of TOF-SiC response and results

Let one to consider, as a physical case, the reaction p+^{11}B, which has been object of many investigations in plasma-physics [1,3,5,6,8].

At proton low energy (E_p < 3MeV) The exit open channels are:

1) Be^8(GS)+ α_0 , (Q_0=8.59MeV) with ^{8}Be(GS) -> 2α by sequential decay, Q-value=92keV; and cross section ranging from 0.2 to 6 mbarn as proton energy varies from Ep = 0.163MeV to 1.388MeV.

2) Be^{8*}(excited state, E*=2.9MeV) + α_1 (Q_1=5.69MeV) with cross section of 10-600-150 mbarn in the energy range Ep =0.165-0.600-1.388 MeV with a resonance at 0.600 MeV.

3) ^{12}C+γ with cross section negligible with respect to previous channels.

4) ^{11}B+p elastic scattering with cross section decreasing from 500 to 200 barns with increasing energy of 0.3MeV to 1.5MeV.

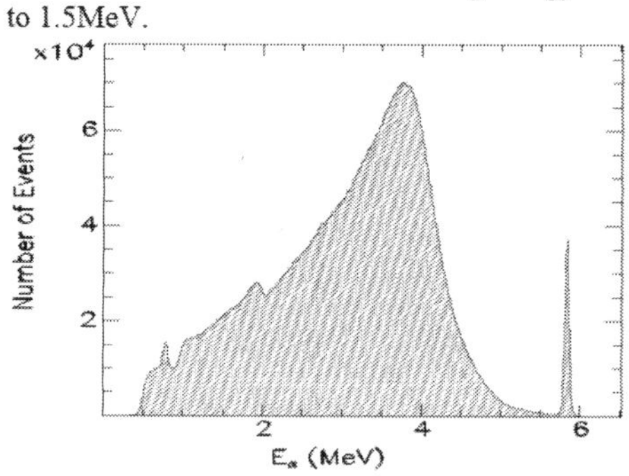

Fig. 1. Single-alpha Energy spectrum reproduced from ref. [21].

The overall contribution of channels 1 and 2 producing alpha particles, gives the energy single α-spectrum shown in Figure 1. As reported in ref.[21], this spectrum has been measured with large area, two faced silicon detectors, equipped to measure α-α coincidences. So, it can be considered as representative shaping of alpha energy spectrum. Taking into account of cross sections of channels 1 and 2, a rough estimation of alpha cross section can be of dσ/dω ≅ 230mbarn/sr. Elastic scattering contribution, as expected from Rutherford approximation, should be orders of magnitude higher.

However, to make the basic conditions simpler (and clear the potentiality of the method), its contribution has been fixed to the same value assumed for alpha. To convert in TOF domain the energy spectra of proton and α-particles, energy bins of 10keV and 0.10ns time bins have been used, respectively. Figure 2 reports the proton elastic scattering, the α particles and the total p+α contributions, as function of Energy (top part) and as function of TOF (bottom part), respectively. The proton and α yields are normalized to the integrated contributions. in energy ranges, by the expression:

$$\sum_{Epm}^{Epx} Yp(Ep)\,DEp = \sum_{E\alpha m}^{E\alpha x} Y\alpha(E\alpha)\,DE\alpha = 230 \text{ mbarn} \quad (1)$$

where Epm = 0.28MeV and Epx =1.5MeV are lower and higher limits of proton energy, Eαm=0.30MeV, Eαx=5.99MeV the corresponding values for α–particle and DEp = DEα = 10keV the energy bins.

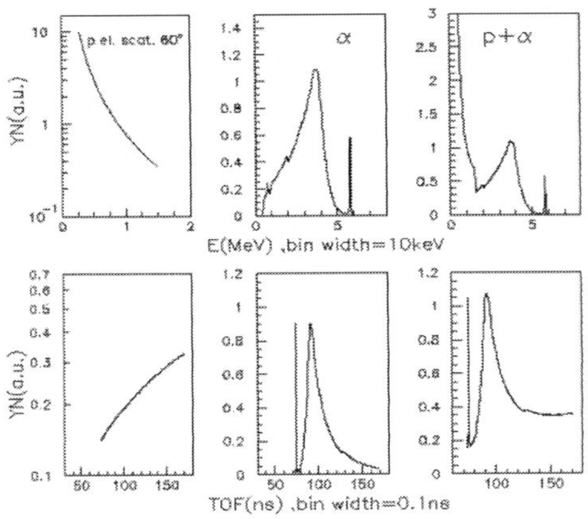

Fig.2. p+[11]B reaction contributions for p, α and total p+α, to Energy (top) and TOF spectra (bottom), respectively. TOF spectra have been calculated in the interval 73.3-170ns with time bin of 0.10ns.

Now, by assuming these contributions as representative of data obtainable with *discriminative* detectors, we regard to the signal amplification factors related to energy losses in the depletion layer of detectors.

Taking into account these factors, the TOF spectra of Figure 2 are modified as shown in Figure 3, for devices SiC 5μm (top) and SiC 80μm (bottom). As it will be shown, last (p+α) spectra can be considered as simulated response of SiC 5μm (top) and SiC 80μm (bottom), respectively.

Let us consider a normalized yield ΔY_t, at time bin ΔT centered at time Ti were contributions of both protons and alpha particles are present:

We have: $$\Delta Y_t = \Delta Y_p + \Delta Y_\alpha \quad (2)$$

with $\Delta Y_p = Y_p(Ti)\Delta T$ and $\Delta Y_\alpha = Y_\alpha(Ti)\Delta T$ (3)

To Ti correspond Energy values E_p and E_α as the ratios

$E_p/E_\alpha = M_p/M_\alpha$ and energy losses in the active region of detector ΔE_p, ΔE_α.

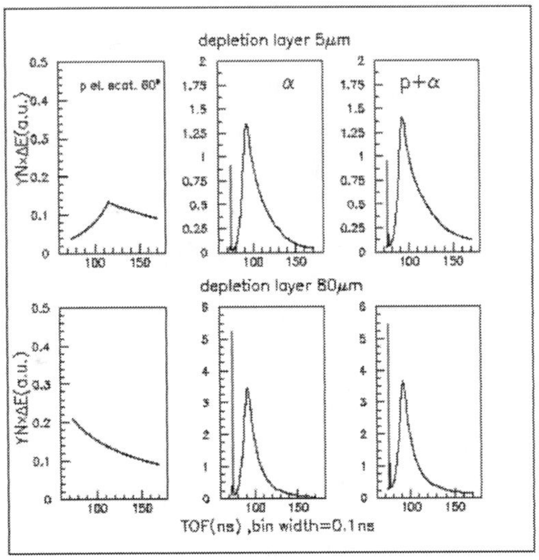

Fig. 3. Spectra of p, α, (p+α) as modified by energy loss factors, for TOF with: SiC 5μm (top), SiC 80μm (bottom), respectively.

As first approximation, it is possible to assume these energy losses to be fully converted in e-hole pairs production. So, for each p and α particle, the associated charge signal would be proportional to ΔE_p, ΔE_α. Then the charge signals (yields) in each time bin are:

$$\Delta Y_p = c\, N_p\, \Delta E_p \quad \text{and} \quad \Delta Y_\alpha = c\, N_\alpha\, \Delta E_\alpha \quad (4)$$

where c is a constant depending on readout circuit and, N_p, N_α are the number of ions arriving into detector in time bin ΔT, as proton and alpha respectively.

Spectra (p+α) of figure 3 have been obtained by means of two equations (4), with c = 1, and represent the simulated responses of two SiC detectors.

We observe as the expected yields depend very strongly on ion energy losses in the active layer of detector so being possible be very different from the true values.

The method is based on the assumptions of:

(a) The active area of detectors to be equal/scaled because of collimating diaphragms and/or preventive calibration. The readout circuits of both detectors to have the same characteristics of gain and time resolution, (c=1).The contiguity, due to very small areal dimensions, and the long distance path of flight, to produce a very small change in the observation direction, assuring (within the statistical fluctuations) the same number of ions, N_{ion}, to impinge on two SiC detectors.

(b) The charge measured by SiC device be proportional to product $N_{ion} \times \Delta E_{ion}$.

Under the validity of **(a)** assumptions, it is expected, for each time Ti ,within the time bin, the number of ions arriving into two detectors to be the same.

Then: $N_{p1} = N_{p2} = N_p(Ti)$, $N_{\alpha 1} = N_{\alpha 2} = N_\alpha(Ti)$ (5)

for each time bin centered at Ti:

$$\text{det. 1} \quad (\Delta Y_{Ti})_1 = N_p \times \Delta E_{p1} + N_\alpha \times \Delta E_{\alpha 1} \qquad (6)$$

$$\text{det. 2} \quad (\Delta Y_{Ti})_2 = N_p \times \Delta E_{p2} + N_\alpha \times \Delta E_{\alpha 2} \qquad (7)$$

$(\Delta Y_{Ti})_1$ and $(\Delta Y_{Ti})_2$ are the measured total charges for each bin centered to Ti, into detector 1 and detector 2, respectively.

As it is shown in eqs. (6) and (7), first terms $(\Delta Y_{Ti})_1$ and $(\Delta Y_{Ti})_2$ can be measured; the right sides of two equations contain the energy losses of proton and alpha into detectors, which can be calculated by SRIM2008 code [22], so that, for each time Ti, the true N_p and N_α can be determined.

Figure 4 shows the reconstructed spectra of proton, α-particles and sum contributions.

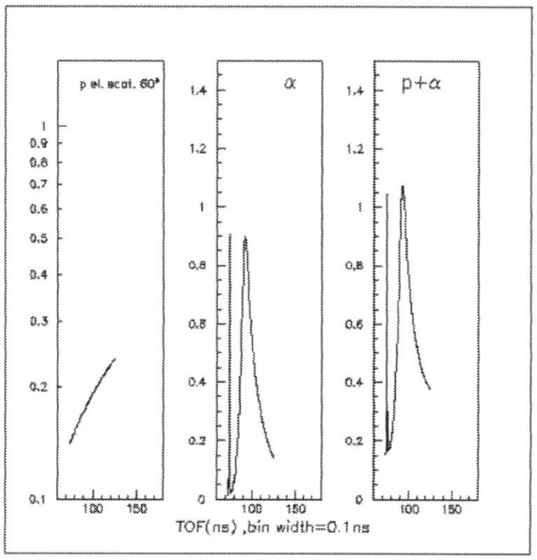

Fig.4. Reconstructed TOF spectra of proton, α-particles and sum contributions.

3. Discussion and summary

From compared inspection of Figure 2 and Figure 4 it can be noticed how the described method works. In fact, Figure 2 (bottom) reports the true expected TOF-spectra for proton and α-particles, whereas Figure 4 shows the same spectra *reconstructed* by means of eq.(6) and eq.(7) joint to simulated responses (p+α) of SiC-5μm and SiC-80μm. The obtainable proton and alpha spectra are limited to TOF range where the corresponding energy losses for p and/or α–particles are different in one, at least, detector.

Assumption **(a)** can be verified directly by setting in contiguous positions two SiC detectors of same depletion layer and readout characteristics.

To verify assumption **(b)**, some dedicate experiment (where only one of ions to discriminate could be present) can be performed by using coupled detectors. For instance, SiC detector contiguous to an ion collector (IC) or SiC detector contiguous to pin-hole of a Thomson spectrometer equipped with CR-39 strips on focal plane. In that cases, SiC detector yield could give information on product $N_{ion} \times N_{pair}$, per each bin, whereas IC or CR-39 detectors, on true yield (N_{ion}). The discrimination has been treated only for proton and alpha particles but it can be extended to other cases as (p,d), (d,^{3}He), etc and different depletion layer SiC detectors.

To summarize, in this paper the problem of Yield-dependence on energy loss of ions in TOF-SiC technique, has been analyzed, showing the need to take into account for. On the assumption of few conditions, it has been found how the using of two contiguous SiC detectors of suitable characteristics and depleted layers, can allow to discriminate, in a large energy range, the contribution of competing light ions involved in plasma-laser induced nuclear reactions. However it has been shown the basic applicability of the method, without taking into account of limitations due to experimental sensitivities and statistical errors. Work is in progress in this perspective as well as to compare expected previsions with data of a real experiment.

References

1. V. S. Belyaev, A. P. Matafonov, and V. I. Vinogradov P. Krainov, S. Lisitsa, A. S. Roussetski, G. N. Ignatyev and V. P. Andrianov, Phys. Rev. **E 72**, 026406 (2005).

2. W. Bang, M. Barbui, A. Bonasera, H. J. Quevedo, G. Dyer, A. C. Bernstein, K. Hagel, K. Schmidt, E. Gaul, M. E. Donovan, F. Consoli, R. De Angelis, P. Andreoli, M. Barbarino, S. Kimura, M. Mazzocco, J. B. Natowitz, and T. Ditmire, Phys. Rev. **E 88**, 033108 (2013).

3. C. Labaune, C. Baccou, S. Depierreux, C. Goyon, G. Loisel, V. Yahia J. Rafelski, Nature Comms **4**:2506 doi:10.1038/ncomms3506 (2013).

4. S Cavallaro, L Torrisi, M Cutroneo, J Krása and J Ullschmied, IOP Publishing Journal of Physics: Conference Series 508 (2014) 012023 doi:10.1088/ 17426596/508/1/012023.

5. D Margarone, A Picciotto, A Velyhan, J Krasa, M Kucharik, A Mangione, A Szydlowsky, A Malinowska, G Bertuccio, Y Shi, M Crivellari, J Ullschmied, P Bellutti and G Korn, Plasma Phys. and Contr. Fusion, Volume **57**, Number 1 0741-3335/57/1/014030 . Published 28 November 2014 • © 2015 IOP Publishing Ltd.

6. A. Picciotto, D. Margarone, A. Velyhan, P. Bellutti, J. Krasa, A. Szydlowsky, G. Bertuccio,Y. Shi,

A. Mangione, J. Prokupek, A. Malinowska,E. Krousky, J. Ullschmied, L. Laska, M. Kucharik, and G. Korn, Phys. Rev. **X 4**, 031030 (2014).

7. J. Krása, D. Klír, A. Velyhan, E. Krouský, M. Pfeifer, K. Řezáč, J. Cikhardt, K. Turek, J. Ullschmied and K. Jungwirth, (2014). High Power Laser Science and Engineering, **2**, e19. doi:10.1017/hpl.2014.25.

8. D. Giulietti, P. Andreoli, D. Batani, A. Bonasera, D.G. Boutoux, F. Burgy, M. Cipriani, F. Consoli, G. Cristofari, R. De Angelis, G. Di Giorgio, J.E. Ducret, F. Ingenito, K. Jakubowska, C. Verona, G. Verona-Rinati, Nucl. Instr. Methods **B** March 2017 DOI: 10.1016/j.nimb. 2017.03.076.

9. W. Bang, H. J. Quevedo, G. Dyer, J. Rougk, I. Kim, M. McCormick, A. C. Bernstein, and T. Ditmire, Rev. Sci. Instrum. **83**, 063504 (2012).

10. S. Cavallaro , L. Torrisi, 41st EPS Conference on Plasma Physics Berlin 23-27 June2014P2.115. http://ocs.ciemat.es/EPS2014PAP/html/.

11. S. Cavallaro, Rev. Sci. Instrum. **86**(3):036103·March 2015

12. A. Szydlowski, Rad. Meas. **36** (2003) 35–42.

13. C. Baccou,V. Yahia, S. Depierreux, C. Neuville, C. Goyon, F. Consoli, R. De Angelis, J. E. Ducret, G. Boutoux J. Rafelski, and C. Labaune, Rev. Sci. Instrum. **86**, 083307 (2015).

14. A. Alejo, S. Kar, H. Ahmed, A. G. Krygier, D. Doria, R. Clarke, J. Fernandez, R. R. Freeman, J. Fuchs, A. Green, J. S. Green, D. Jung, A. Kleinschmidt, C. L. S. Lewis, J. T. Morrison, Z. Najmudin, H. Nakamura, G. Nersisyan, P. Norreys, M. Notley, M. Oliver, M. Roth J. A. Ruiz, L. Vassura, M. Zepf, and M. Borghesi, Rev. Sci. Instrum. **85**, 093303 (2014).

15. S. Cavallaro, L. Torrisi ,A. Amato ,L. Andò, M.Cutroneo, 39th EPS Conference & 16th Int. Congress on Plasma Physics. http://ocs.ciemat.es/ epsicpp2012/ pap/P4.121.pdf.

16. V. Togo, S. Balestra, S. Cecchini D. Di Ferdinando, M. Frutti, G. Giacomelli, M. Giorgini, A. Kuma, G. Mandrio, S. Manzoor, A. Margiotta, E. Medinaceli, L. Patrizi, V. Popa, M. Spurio, Nucl. Instr. Methods **A 580** (2007) 58–61.

17. S. Cavallaro, D. Margarone, L.Torrisi 35th EPS Conference on Plasma Phys. Hersonissos, 9-13 June 2008 ECA Vol.**32D**, P-4.154 (2008).

18. A. Malinowska, M. Jaskóła, A. Korman, A. Szydłowski, and M. Kuk, Rev. Sci. Instrum. **85**, 123505 (2014); doi: 10.1063/1.4903755.

19. M. Cutroneo, P. Musumeci, M. Zimbarnone, L. Torrisi, F. La Via, D. Margarone, A. Velyhan, J. Ullschmied,L. Calcagno, J. Mater. Res. 2012 DOI: 10.1557/jmr. 2012.211.

20. Giuseppe Bertuccio, Donatella Puglisi, Lorenzo Torrisi, Claudio Lanzieri, Appl. Surf. Sci. **272** (2013) 128–131 **381** (2012) 012116 doi:10.1088/1742-6596/381/1/012116.

21. C A Diget, H O U Fynbo, K L Laursen, K O Madsbøl, LJørgensen, O S Kirsebom,, and K Riisager Rutherford Centennial Conference on Nuclear Physics IOP Publishing Journal of Physics: Conference Series **381** (2012) 012116 doi:10.1088/17426596/381/1/012116 .093303 (2012).

22. J. F. Ziegler, J. P. Biersack, SRIM-2008 code, NEA-0919/06.

On a 'time' reparametrization in relativistic electrodynamics with travelling waves

Gaetano Fiore[1,2,⋆]

[1] *Dip. di Matematica e Applicazioni, Università di Napoli "Federico II", Complesso Universitario MSA, Via Cintia, Napoli, Italy*
[2] *INFN, Sezione di Napoli, Complesso Universitario M. S. Angelo, Via Cintia, Napoli, Italy*

Abstract. We briefly report on our method [23] of simplifying the equations of motion of charged particles in an electromagnetic (EM) field that is the sum of a plane travelling wave and a static part; it is based on changes of the dependent variables and the independent one (light-like coordinate ξ instead of time t). We sketch its application to a few cases of extreme laser-induced accelerations, both in vacuum and in plane problems at the vacuum-plasma interface, where we are able to reduce the system of the (Lorentz-Maxwell and continuity) partial differential equations into a family of decoupled systems of Hamilton equations in 1 dimension. Since Fourier analysis plays no role, the method can be applied to all kind of travelling waves, ranging from almost monochromatic to socalled "impulses".

1 Introduction and set-up

The equation of motion of a particle with charge q in external electric and magnetic fields $\mathbf{E}(\mathrm{x}), \mathbf{B}(\mathrm{x})$ $[\mathrm{x} \equiv (ct, \mathbf{x})]$ in its general form is non-autonomous and highly nonlinear:

$$\begin{aligned} \dot{\mathbf{p}}(t) &= q\mathbf{E}[ct, \mathbf{x}(t)] + q\boldsymbol{\beta}(t) \wedge \mathbf{B}[ct, \mathbf{x}(t)], \\ \dot{\mathbf{x}}(t) &= \frac{c\mathbf{p}(t)}{\sqrt{m^2c^2 + \mathbf{p}^2(t)}}; \end{aligned} \tag{1}$$

here $\boldsymbol{\beta} \equiv \mathbf{v}/c$, $\mathbf{p} \equiv m\mathbf{v}/\sqrt{1-\boldsymbol{\beta}^2}$ is its relativistic momentum.

Usually, (1) is simplified assuming:

1. $\mathbf{E}, \mathbf{B}$ are constant or vary "slowly" in space/time; or
2. $\mathbf{E}, \mathbf{B}$ are "small" (so that nonlinear effects in $\mathbf{E}, \mathbf{B}$ are negligible); or
3. $\mathbf{E}, \mathbf{B}$ are monochromatic waves, or slow modulations of; or
4. the motion of the particle keeps non-relativistic.

The astonishing developments of Laser technologies (especially *Chirped Pulse Amplification* [2, 3]) today allow the construction of compact sources of extremely intense (up to 10^{23} W/cm^2) coherent EM waves, possibly concentrated in very short laser pulses ($\gtrsim$ fs). Even more intense/short (or cheaper) laser pulses by new technologies (thin film compression [4], etc.) will be soon available. In particular, these lasers can be used for making small particle-accelerators based on Laser Wake Field Acceleration (LWFA) [5] in plasmas. Extreme conditions are present also in several violent astrophysical processes (see e.g. [6] and references therein). In either case the effects are so fast, huge, highly nonlinear, ultra-relativistic that conditions 1-4 are not fulfilled. Alternative simplifying approaches are therefore desirable.

Here we summarize a new approach [23] that is especially fruitful if in the spacetime region Ω of interest (i.e., where we wish to follow the charged particles' worldlines) $\mathbf{E}, \mathbf{B}$ can be decomposed into a static part and a plane transverse travelling wave propagating in the z direction:

$$\begin{aligned} \mathbf{E}(\mathrm{x}) &= \underbrace{\boldsymbol{\epsilon}^{\perp}(ct-z)}_{pump=travelling\ wave} + \underbrace{\mathbf{E}_s(\mathbf{x})}_{static}, \\ \mathbf{B}(\mathrm{x}) &= \underbrace{\mathbf{k} \wedge \boldsymbol{\epsilon}^{\perp}(ct-z)}_{travelling\ wave} + \underbrace{\mathbf{B}_s(\mathbf{x})}_{static}, \end{aligned} \tag{2}$$

$\mathbf{x} = x\mathbf{i}+y\mathbf{j}+z\mathbf{k}$, $\boldsymbol{\epsilon}^{\perp} \perp \mathbf{k}$. We decompose vectors as $\mathbf{u} = \mathbf{u}^{\perp} + u^z\mathbf{k}$. We assume *only* that $\boldsymbol{\epsilon}^{\perp}(\xi)$ is piecewise continuous and

$$\begin{aligned} &\mathbf{a)} \quad \boldsymbol{\epsilon}^{\perp} \text{ has a compact support } [0, l], \\ \text{or} \quad &\mathbf{a')} \quad \boldsymbol{\epsilon}^{\perp} \in L^1(\mathbb{R}), \end{aligned} \tag{3}$$

$$\Rightarrow \quad \boldsymbol{\alpha}^{\perp}(\xi) \equiv -\int_{-\infty}^{\xi} dy\, \boldsymbol{\epsilon}^{\perp}(y) \to 0 \quad \text{as } \xi \to -\infty; \tag{4}$$

$\boldsymbol{\alpha}^{\perp}$ is the travelling-wave part of the transverse EM potential $\mathbf{A}^{\perp}$. **a)** $\Rightarrow \boldsymbol{\alpha}^{\perp}(\xi) = 0$ if $\xi \leq 0$, $\boldsymbol{\alpha}^{\perp}(\xi) = \boldsymbol{\alpha}^{\perp}(l)$ if $\xi \geq l$. We can treat on the same footing all such $\boldsymbol{\epsilon}^{\perp}$, in particular:

1. A modulated monochromatic wave fulfilling (3):

$$\boldsymbol{\epsilon}^{\perp}(\xi) = \underbrace{\epsilon(\xi)}_{\text{modul.}} \underbrace{[\mathbf{i}a_1 \cos(k\xi+\varphi) + \mathbf{j}a_2 \sin(k\xi)]}_{\text{carrier wave } \boldsymbol{\epsilon}_o^{\perp}(\xi)} \tag{5}$$

$$\Rightarrow \quad -\boldsymbol{\alpha}^{\perp}(\xi) = \frac{\epsilon(\xi)}{k^2}\, \boldsymbol{\epsilon}_o^{\perp\prime}(\xi) + O\!\left(\frac{1}{k^2}\right) \simeq \frac{\epsilon(\xi)}{k^2}\, \boldsymbol{\epsilon}_o^{\perp\prime}(\xi); \tag{6}$$

$\simeq$ holds if $|\epsilon'| \ll |k\epsilon|$ (slow modulation).

⋆e-mail: gaetano.fiore@na.infn.it

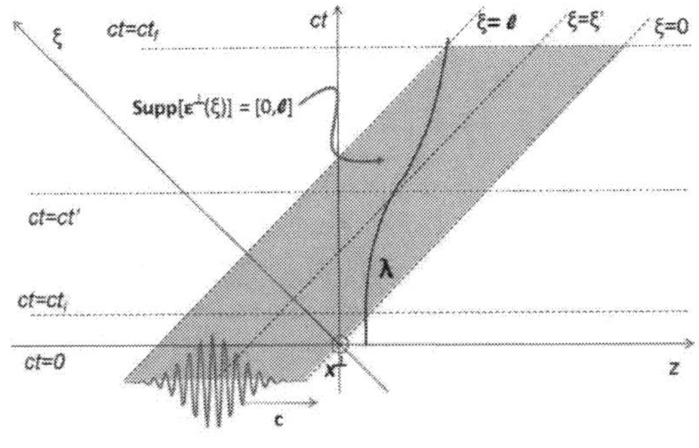

Figure 1. Every worldline λ and hyperplane ξ=C intersect once.

2. A superposition of waves of type 1.

3. An 'impulse' (few cycles, or even a fraction of).

The idea is: as no particle can reach the speed of light c, $\tilde{\xi}(t)=ct-z(t)$ is strictly growing, and we can adopt $\xi=ct-z$ as a parameter on the worldline λ (see fig. 1) and in the action functional of the particle:

$$S(\lambda)=-\int_\lambda mc^2 d\tau+qA(x)$$
$$=-\int_{t_0}^{t_1}dt\,\underbrace{\frac{mc^2+qu^\mu A_\mu}{\gamma}}_{L[\mathbf{x},\dot{\mathbf{x}},t]}=-\int_{\xi_0}^{\xi_1}\frac{d\xi}{c}\,\underbrace{\frac{mc^2+q\hat{u}^\mu\hat{A}_\mu}{\hat{s}}}_{\mathcal{L}[\hat{\mathbf{x}},\hat{\mathbf{x}}',\xi]};$$

$(u^\mu)=(u^0,\mathbf{u})\equiv(\gamma,\gamma\boldsymbol{\beta})$ is the 4-velocity, i.e. the dimensionless version of the 4-momentum. $A(x)=A_\mu(x)dx^\mu=A^0(x)cdt-\mathbf{A}(x)\cdot d\mathbf{x}$ is the EM potential 1-form, $\mathbf{E}=-\partial_t\mathbf{A}/c-\nabla A^0$, $\mathbf{B}=\nabla\wedge\mathbf{A}$ (we use Gauss CGS units), and we denote $\hat{\mathbf{x}}(\xi)=\mathbf{x}(t)$, $\hat{f}(\xi,\hat{\mathbf{x}})\equiv f(ct,\mathbf{x})$, $\hat{f}'\equiv d\hat{f}/d\xi$ for all functions $f(ct,\mathbf{x})$. Applying Hamilton's principle and the Legendre transform we find simplified Lagrange, and Hamilton equations where the argument of $\boldsymbol{\epsilon}^\perp$ is the independent variable ξ, rather than the unknown $ct-z(t)$, and the new kinetic momenta are the dependent variables.

2 General results for one particle

To parametrize λ by ξ we have to replace $d\tau/dt=1/\gamma=\sqrt{1-\dot{\mathbf{x}}^2/c^2}$ (τ is the particle proper time) by

$$\frac{1}{\hat{s}}\equiv\frac{d(c\tau)}{d\xi}=\sqrt{1+2\hat{z}'-\hat{\mathbf{x}}^{\perp\prime 2}}>0. \tag{7}$$

From $\mathbf{p}=md\mathbf{x}/d\tau$, $\gamma=dt/d\tau$ we find that the *s-factor* $\hat{s}$ is the light-like component $\hat{u}^-=\hat{\gamma}-\hat{u}^z$ of the 4-velocity $u=(u^0,\mathbf{u})\equiv(\gamma,\gamma\boldsymbol{\beta})=\left(\frac{p^0}{mc^2},\frac{\mathbf{p}}{mc}\right)$ (all these are dimensionless), and $\hat{\mathbf{u}}=\hat{s}\hat{\mathbf{x}}'$. $\hat{\gamma},\hat{u}^z,\hat{\boldsymbol{\beta}},\hat{\mathbf{x}}'$ can be expressed as *rational functions* of $\hat{\mathbf{u}}^\perp,\hat{s}$:

$$\hat{\gamma}=\frac{1+\hat{\mathbf{u}}^{\perp 2}+\hat{s}^2}{2\hat{s}},\quad \hat{u}^z=\hat{\gamma}-\hat{s},\quad \hat{\boldsymbol{\beta}}=\frac{\hat{\mathbf{u}}}{\hat{\gamma}}, \tag{8}$$

$$\hat{\mathbf{x}}^{\perp\prime}=\frac{\hat{\mathbf{u}}^\perp}{\hat{s}},\qquad \hat{z}'=\frac{1+\hat{\mathbf{u}}^{\perp 2}}{2\hat{s}^2}-\frac{1}{2} \tag{9}$$

By Hamilton's principle, any extremum λ of S is the worldline of a possible motion of the particle with initial position $\mathbf{x}_0$ at time t_0 and final position $\mathbf{x}_1$ at time t_1. Hence it fulfills Euler-Lagrange equations in both forms $\frac{d}{dt}\frac{\partial L}{\partial\dot{\mathbf{x}}}=\frac{\partial L}{\partial\mathbf{x}}$ and $\frac{d}{d\xi}\frac{\partial\mathcal{L}}{\partial\hat{\mathbf{x}}'}=\frac{\partial\mathcal{L}}{\partial\hat{\mathbf{x}}}$, equivalent to (1). The Legendre transform yields the Hamiltonians $H\equiv\dot{\mathbf{x}}\cdot\frac{\partial L}{\partial\dot{\mathbf{x}}}-L=\gamma mc^2+qA^0$ and $\hat{H}\equiv\hat{\mathbf{x}}'\cdot\frac{\partial\mathcal{L}}{\partial\hat{\mathbf{x}}'}-\mathcal{L}=\hat{\gamma}mc^2+q\hat{A}^0$. $\hat{H}$ is a *rational* function of $\hat{\mathbf{x}},\hat{\boldsymbol{\Pi}}\equiv\frac{\partial\mathcal{L}}{\partial\hat{\mathbf{x}}'}$, or, equivalently, of $\hat{s},\hat{\mathbf{u}}^\perp$:

$$\hat{H}(\hat{\mathbf{x}},\hat{\boldsymbol{\Pi}};\xi)=mc^2\frac{1+\hat{s}^2+\hat{\mathbf{u}}^{\perp 2}}{2\hat{s}}+q\hat{A}^0(\xi,\hat{\mathbf{x}}), \tag{10}$$

$$\text{where}\quad\left\{\begin{array}{l} mc^2\hat{\mathbf{u}}^\perp=\hat{\boldsymbol{\Pi}}^\perp-q\hat{\mathbf{A}}^\perp(\xi,\hat{\mathbf{x}}),\\ mc^2\hat{s}=-\hat{\Pi}^z-q\hat{A}^-(\xi,\hat{\mathbf{x}}),\end{array}\right. \tag{11}$$

while $H(\mathbf{x},\mathbf{P},t)=\sqrt{m^2c^4+(c\mathbf{P}-q\mathbf{A})^2}+qA^0$ ($\mathbf{P}\equiv\frac{\partial L}{\partial\dot{\mathbf{x}}}=\mathbf{p}+\frac{q}{c}\mathbf{A}$) is not. Eq. (1) are also equivalent to the Hamilton equations $\hat{\mathbf{x}}'=\frac{\partial\hat{H}}{\partial\hat{\boldsymbol{\Pi}}}$, $\hat{\boldsymbol{\Pi}}'=-\frac{\partial\hat{H}}{\partial\hat{\mathbf{x}}}$. All the new equations (in particular these ones) can be also obtained more directly from the old ones by putting a caret on all dynamical variables and replacing d/dt by $(c\hat{s}/\hat{\gamma})d/d\xi$. Along the solutions $\hat{H}$ gives the particle energy as a function of ξ. Under the EM field (2) eqs (1) amount to (9) and

$$\begin{aligned}\hat{\mathbf{u}}^{\perp\prime}&=\frac{q}{mc^2}\left[(1+\hat{z}')\hat{\mathbf{E}}_s^\perp+(\hat{\mathbf{x}}'\wedge\hat{\mathbf{B}}_s)^\perp+\boldsymbol{\epsilon}^\perp(\xi)\right],\\ \hat{s}'&=\frac{-q}{mc^2}\left[\hat{E}_s^z-\hat{\mathbf{x}}^{\perp\prime}\cdot\hat{\mathbf{E}}_s^\perp+(\hat{\mathbf{x}}^{\perp\prime}\wedge\hat{\mathbf{B}}_s^\perp)^z\right],\end{aligned} \tag{12}$$

while the *energy gain* (normalized to mc^2) is

$$\mathcal{E}\equiv\frac{\hat{H}(\xi_1)-\hat{H}(\xi_0)}{mc^2}=\int_{\xi_0}^{\xi_1}d\xi\,q\boldsymbol{\epsilon}^\perp\cdot\frac{\hat{\mathbf{u}}^\perp}{\hat{s}} \tag{13}$$

in the interval $[\xi_0,\xi_1]$. Once solved (9-12), analytically or numerically, to obtain the solution as a function of t we just need to invert $\hat{t}(\xi)=\xi+\hat{z}(\xi)$ and set $\mathbf{x}(t)=\hat{\mathbf{x}}[\xi(t)]$. If $\mathbf{E}_s,\mathbf{B}_s$=const then eq. (12) are solved by

$$\begin{aligned}\hat{\mathbf{u}}^\perp&=\frac{q}{mc^2}\left[\mathbf{K}-\boldsymbol{\alpha}(\xi)+(\xi+\hat{z})\mathbf{E}_s+\hat{\mathbf{x}}\wedge\mathbf{B}_s\right]^\perp,\\ \hat{s}&=\frac{-q}{mc^2}\left[K^z+\xi E_s^z-\hat{\mathbf{x}}^\perp\cdot\mathbf{E}_s^\perp+(\hat{\mathbf{x}}^\perp\wedge\mathbf{B}_s)^z\right]\end{aligned} \tag{14}$$

(K^j are integration constants) whereby (9) become three 1st order ordinary differential equations (ODE) *rational* in the unknown $\hat{\mathbf{x}}(\xi)$.

Contrary to (9-12), (1) is a transcendental system, and the unknown $z(t)$ appears in the argument of the rapidly varying functions $\boldsymbol{\epsilon}^\perp,\boldsymbol{\alpha}^\perp$ in $(1)_1$, which now reads:

$$\tfrac{1}{q}\dot{\mathbf{p}}(t)=\mathbf{E}_s+\boldsymbol{\beta}\wedge\mathbf{B}_s+\boldsymbol{\epsilon}^\perp[ct-z(t)]\,(\boldsymbol{\beta}\mathbf{k}+1-\beta^z).$$

Also determining $\mathcal{E}(t)$ is more complicated.

2.1 Dynamics under $A^\mu=A^\mu(t,z)$

This applies in particular to (2) if $\mathbf{E}_s=E_s^z(z)\mathbf{k}$, $\mathbf{B}_s=\mathbf{B}_s^\perp(z)$, choosing e.g. $A^0=-\int^z d\zeta E_s^z(\zeta)$, $\mathbf{A}^\perp=\boldsymbol{\alpha}^\perp-\mathbf{k}\wedge\int^z d\zeta\mathbf{B}^\perp(\zeta)$, $A^z\equiv 0$. As $\partial\hat{H}/\partial\hat{\mathbf{x}}^\perp=0$, we find $\hat{\boldsymbol{\Pi}}^\perp=q\mathbf{K}^\perp=$ const, i.e. the known result $\frac{mc^2}{q}\hat{\mathbf{u}}^\perp=\mathbf{K}^\perp-\hat{\mathbf{A}}^\perp(\xi,\hat{z})$. Setting $v:=\hat{\mathbf{u}}^{\perp 2}$ and replacing in $(9)_2$, $(12)_2$ we obtain

$$\hat{z}'=\frac{1+\hat{v}}{2\hat{s}^2}-\frac{1}{2},\quad \hat{s}'=\frac{-q}{mc^2}E_s^z(\hat{z})-\frac{1}{2\hat{s}}\frac{\partial\hat{v}}{\partial\hat{z}}. \tag{15}$$

Once solved the system (15) in the unknowns $\hat{z}(\xi)$, $\hat{s}(\xi)$, the other unknowns are obtained from

$$\hat{\mathbf{x}}(\xi) = \mathbf{x}_0 + \hat{\mathbf{Y}}(\xi), \qquad \hat{\mathbf{Y}}(\xi) \equiv \int_{\xi_0}^{\xi} dy\, \frac{\hat{\mathbf{u}}(y)}{\hat{s}(y)}. \tag{16}$$

If in addition $\mathbf{B}_s \equiv 0$, then $\mathbf{A}_s \equiv 0$ (in the Coulomb gauge), $\hat{\mathbf{u}}^{\perp}(\xi) = \frac{q}{mc^2}[\mathbf{K}^{\perp} - \boldsymbol{\alpha}^{\perp}(\xi)]$ and $\hat{v} = \hat{\mathbf{u}}^{\perp 2}$ are already known. The system (15) to be solved simplifies to

$$\hat{z}' = \frac{1+\hat{v}}{2\hat{s}^2} - \frac{1}{2}, \qquad \hat{s}' = \frac{-q}{mc^2} E_s^z(\hat{z}). \tag{17}$$

Some remarkables properties of the solutions are [23]:

1. Where $\boldsymbol{\epsilon}^{\perp}(\xi) = 0$ then $\hat{v}(\xi) = v_c$ =const, $\hat{H}$ is conserved, (17) is solved by quadrature.
2. The final transverse momentum is $mc\mathbf{u}^{\perp}(\xi_f)$. If ϵ of (5) varies slowly and $\mathbf{u}^{\perp}(0) = \mathbf{0}$, then $\mathbf{u}^{\perp}(\xi_f) \simeq 0$.
3. $\hat{s}(\xi)$ is insensitive to fast oscillations of $\boldsymbol{\epsilon}^{\perp}$, contrary to $\mathbf{u}, \gamma, \boldsymbol{\beta}$, which can be reobtained via (8).

3 Some exact solutions for $\mathbf{B}_s$, $\mathbf{E}_s$ =const

Let $\mathbf{b}^{\perp} + b\mathbf{k} \equiv q\mathbf{B}_s/mc^2$, $\mathbf{e}^{\perp} \equiv q\mathbf{E}_s^{\perp}/mc^2$ (constants), $\boldsymbol{w}(\xi) \equiv q[\mathbf{K} - \boldsymbol{\alpha}^{\perp}(\xi) + \xi\mathbf{E}_s]/mc^2$ (all dimensionless); (14) take the more explicit form

$$\begin{aligned} \hat{u}^x &= (e^x - b^y)\hat{z} + b\hat{y} + w^x(\xi), \\ \hat{u}^y &= (e^y + b^x)\hat{z} - b\hat{x} + w^y(\xi), \\ \hat{s} &= (e^x - b^y)\hat{x} + (e^y + b^x)\hat{y} - w^z(\xi), \end{aligned} \tag{18}$$

For any $E_s^z, B_s^z, \mathbf{E}_s^{\perp}$, if $\mathbf{B}_s^{\perp} = \mathbf{k} \wedge \mathbf{E}_s^{\perp}$, setting $\kappa \equiv \frac{qE_s^z}{mc^2}$ we find the following exact solutions (part of them are new):

$$\begin{aligned} (\hat{x} + i\hat{y})(\xi) &= (1-\kappa\xi)^{ib/\kappa} \int_0^{\xi} d\zeta\, \frac{(w^x + iw^y)(\zeta)}{(1-\kappa\zeta)^{1+ib/\kappa}}, \\ \hat{z}(\xi) &= \int_0^{\xi} \frac{d\zeta}{2}\left[\frac{1}{(1-\kappa\zeta)^2} + \hat{\mathbf{x}}^{\perp\prime 2}(\zeta) - 1\right], \quad \hat{s}(\xi) = 1 - \kappa\xi, \\ \hat{\mathbf{u}}^{\perp}(\xi) &= (1-\kappa\xi)\,\hat{\mathbf{x}}^{\perp\prime}(\xi), \quad \hat{\gamma}(\xi) = 1 - \kappa\xi + \hat{u}^z(\xi) \end{aligned} \tag{19}$$

$$\hat{u}^z(\xi) = \frac{1}{2(1-\kappa\xi)} + (1-\kappa\xi)\,\frac{\hat{\mathbf{x}}^{\perp\prime 2}(\xi) - 1}{2};$$

here we have adopted the initial conditions $\mathbf{x}(0) = \mathbf{0} = \mathbf{u}(0)$. We next analyze a few special cases.

3.1 Case $\mathbf{E}_s = \mathbf{B}_s = \mathbf{0}$ (zero static fields)

Then (19) becomes [7, 8]:

$$\begin{aligned} &\hat{s} \equiv 1, \quad \hat{\mathbf{u}}^{\perp} = \frac{-q\boldsymbol{\alpha}^{\perp}}{mc^2}, \quad \hat{u}^z = \frac{\hat{\mathbf{u}}^{\perp 2}}{2}, \quad \hat{\gamma} = 1 + \hat{u}^z \\ &\hat{z}(\xi) = \int_{\xi_0}^{\xi} dy\, \frac{\hat{\mathbf{u}}^{\perp 2}(y)}{2}, \quad \hat{\mathbf{x}}^{\perp}(\xi) = \int_{\xi_0}^{\xi} dy\, \hat{\mathbf{u}}^{\perp}(y). \end{aligned} \tag{20}$$

The solutions (20) induced by two x-polarized pulses and the corresponding e^- trajectories in the zx plane are shown in fig. 2. Note that:

- The maxima of γ, $\alpha^{\perp}$ coincide (and approximately also of $\epsilon(\xi)$, if $\epsilon(\xi)$ is slowly varying).
- Since $u^z \geq 0$, the z-drift is positive-definite. Rescaling $\boldsymbol{\epsilon}^{\perp} \mapsto a\boldsymbol{\epsilon}^{\perp}$, $\hat{\mathbf{x}}^{\perp}, \hat{\mathbf{u}}^{\perp}$ scale like a, whereas $\hat{z}, \hat{u}^z$ scale like a^2 (hence the trajectory goes to a straight line in the limit $a \to \infty$). This is due to magnetic force $q\boldsymbol{\beta} \wedge \mathbf{B}$.
- **Corollary** The final u and energy gain read

$$\mathbf{u}_f^{\perp} = \hat{\mathbf{u}}^{\perp}(\infty), \quad u_f^z = \mathcal{E}_f = \frac{1}{2}\mathbf{u}_f^{\perp 2} = \gamma_f - 1; \tag{21}$$

Both are very small if the pulse modulation ϵ is slow [extremely small if $\epsilon \in \mathcal{S}(\mathbb{R})$ or $\epsilon \in C_c^{\infty}(\mathbb{R})$].

Recall the *Lawson-Woodward* Theorem [10–13] (an outgrowth of the original Woodward-Lawson Theorem [14, 15]): in spite of large energy variations during the interaction, the final energy gain $\mathcal{E}_f$ of a charged particle $\mathcal{P}$ interacting with an EM field is zero if:

i) the interaction occurs in $\mathbb{R}^3$ vacuum (no boundaries);
ii) $\mathbf{E}_s = \mathbf{B}_s = \mathbf{0}$ and $\boldsymbol{\epsilon}^{\perp}$ is slowly modulated;
iii) $v^z \simeq c$ along the whole acceleration path;
iv) nonlinear (in $\boldsymbol{\epsilon}^{\perp}$) effects $q\boldsymbol{\beta} \wedge \mathbf{B}$ are negligible;
v) the power radiated by $\mathcal{P}$ is negligible.

Our Corollary, as Ref. [9], states the same result if we relax iii), iv), but the EM field is a *plane* travelling wave.

To obtain a non-zero $\mathcal{E}_f$ one has to violate some other conditions of the theorem, as e.g. we see in next cases.

3.2 Case $\mathbf{E}_s = 0$, $\mathbf{B}_s = B_s^z\mathbf{k}$

Then (19) becomes $\hat{s} \equiv 1$ and

$$\begin{aligned} &(\hat{x} + i\hat{y})(\xi) = \int_0^{\xi} d\zeta\, e^{ib(\zeta-\xi)}(w^x + iw^y)(\zeta), \quad \hat{\mathbf{u}}^{\perp} = \hat{\mathbf{x}}^{\perp\prime}, \\ &\hat{u}^z = \hat{z}' = \frac{\hat{\mathbf{u}}^{\perp 2}}{2} = \mathcal{E} = \hat{\gamma} - 1, \quad \hat{z}(\xi) = \int_0^{\xi} d\zeta\, \frac{\hat{\mathbf{u}}^{\perp 2}(\zeta)}{2}. \end{aligned} \tag{22}$$

(22) reduces to the solution of [16, 17] for monochromatic $\boldsymbol{\epsilon}^{\perp}$. This leads to *cyclotron autoresonance* if $-b = k = \frac{2\pi}{\lambda} \gg \frac{1}{l}$: for circular polarization $w^x(\xi) + iw^y(\xi) \simeq e^{ik\xi}\mathrm{w}(\xi)$,

$$(\hat{x} + i\hat{y})(\xi) \simeq iW(\xi)e^{ik\xi}, \qquad W(\xi) \equiv \int_0^{\xi} d\zeta\, \mathrm{w}(\zeta) > 0$$

where $\mathrm{w}(\xi) \equiv q\epsilon(\xi)/kmc^2$; clearly $W(\xi)$ grows with ξ. In particular if $\boldsymbol{\epsilon}^{\perp}(\xi) = \mathbf{0}$ for $\xi \geq l \equiv$, then for such ξ

$$\hat{z}'(\xi) \simeq \frac{k^2}{2} W^2(l) \simeq 2\mathcal{E}_f, \qquad \frac{|\hat{\mathbf{x}}^{\perp\prime}(\xi)|}{\hat{z}'(\xi)} \simeq \frac{2}{kW(l)} \ll 1;$$

3.3 Case $\mathbf{E}_s = E_s^z\mathbf{k}$, $\mathbf{B}_s = \mathbf{0}$

Then the solution (19) reduces to $\hat{s}(\xi) = 1 - \kappa\xi$,

$$(\hat{x} + i\hat{y})(\xi) = \int_0^{\xi} dy\, \frac{(w^x + iw^y)(y)}{1 - \kappa y}, \quad \hat{z}(\xi) = \int_0^{\xi} \frac{dy}{2}\left\{\frac{1 + \hat{v}(y)}{[1 - \kappa y]^2} - 1\right\}; \tag{23}$$

If $\boldsymbol{\epsilon}^{\perp}$ is slowly modulated the energy gain (13) $\mathcal{E}_f$ is negative if $\kappa > 0$, positive if $\kappa \leq 0$ and has a unique maximum

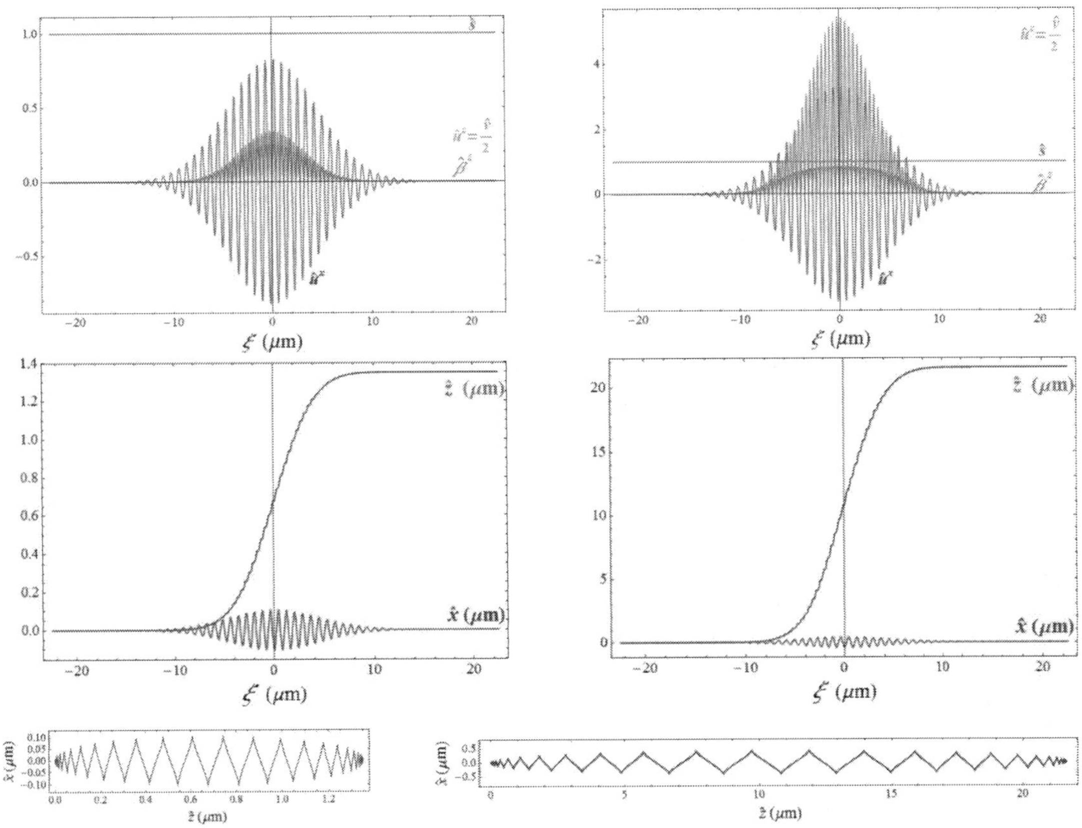

Figure 2. Solutions (20) and e^- trajectories in the zx plane induced by two x-polarized pulses with carrier wavelength $\lambda = .8\mu$m, gaussian modulation $\epsilon(\xi) = a\exp[-\xi^2/2\sigma]$, $\sigma = 20\mu\text{m}^2$, $ea\lambda/mc^2 = 4, 15$ (left, right).

point $\kappa_M < 0$ if $\epsilon(\xi)$ has a finite support with a unique maximum. Here is an acceleration device based on this solution: at $t = 0$ the particle initially lies at rest with $z_0 \lesssim 0$, just at the left of a metallic grating G contained in the $z=0$ plane and set at zero electric potential; another metallic plate P contained in a plane $z=z_p>0$ is set at electric potential $V = V_p$. A short laser pulse $\epsilon^\perp$ hitting the particle boosts it into the latter region through the ponderomotive force; choosing $qV_p > 0$ implies $\kappa = -qV_p/z_pmc^2 < 0$, and a backward longitudinal electric force qE_s^z. If qV_p is large enough, then $z(t)$ will reach a maximum smaller than z_p, then is accelerated backwards and exits the grating with energy $\mathcal{E}_f$ and negligible transverse momentum. A large $\mathcal{E}_f$ requires extremely large $|V_p|$, far beyond the material breakdown threshold, what prevents its realization as a static field (namely, sparks between G, P would arise and rapidly reduce $|V_p|$). A way out is to make the pulse itself generate such large $|E_s^z|$ within a plasma at the right time so as to induce the it slingshot effect, as sketchily explained at the end of next section.

4 Plane plasma problems

Assume that the plasma is initially in hydrodynamic conditions with all initial data [velocities, densities n_h, EM fields of the form (2)] not depending on $\mathbf{x}^\perp$. Then also the solutions for $\mathbf{B}, \mathbf{E}, \mathbf{u}_h, n_h, \Delta\mathbf{x}_h \equiv \mathbf{x}_h(t,\mathbf{X})-\mathbf{X}$ (displacements) do not depend on $\mathbf{x}^\perp$. Here $\mathbf{x}_h(t,\mathbf{X})$ is the position at t of the h-th fluid material element with initial position $\mathbf{X} \equiv (X,Y,Z)$; $\mathbf{X}_h(t,\mathbf{x})$ is the inverse (at fixed t). More specifically, we consider the impact of an EM plane wave with a pump of the type (3.a) on a cold plasma at equilibrium (figure below); the initial conditions are:

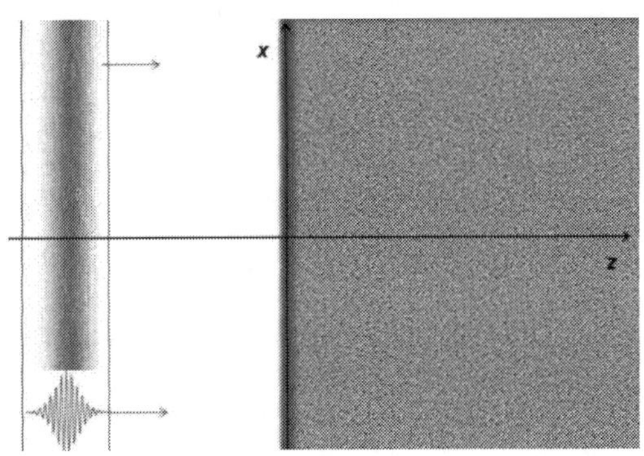

$$\mathbf{u}_h(0,\mathbf{x})=\mathbf{0}, \qquad n_h(0,\mathbf{x})=0 \qquad \text{if } z\leq 0,$$
$$j^0(0,\mathbf{x})=\sum_h q_h n_h(0,\mathbf{x})\equiv 0, \tag{24}$$
$$\mathbf{E}(0,\mathbf{x})=\boldsymbol{\epsilon}^\perp(-z), \quad \mathbf{B}(0,\mathbf{x})=\mathbf{k}\wedge\boldsymbol{\epsilon}^\perp(-z)+\mathbf{B}_s.$$

Then Maxwell eq.s $\nabla\cdot\mathbf{E}=4\pi j^0$, $\partial_t E^z/c+4\pi j^z=(\nabla\wedge\mathbf{B})^z=0$ (the current density is $\mathbf{j}=\sum_h q_h n_h\boldsymbol{\beta}_h=\sum_h q_h n_h\frac{\mathbf{u}_h}{\gamma_h}$) imply [8]

$$E^z(t,z)=4\pi\sum_h q_h\widetilde{N}_h[Z_h(t,z)], \tag{25}$$

where $\widetilde{N}_h(Z)\equiv\int_0^Z d\zeta\, n_h(0,\zeta)$: we thus reduce by one the number of unknowns, expressing E^z in terms of the (still unknown) longitudinal motion. $\mathbf{A}^\perp$ is coupled to the currents through $\Box\mathbf{A}^\perp=4\pi\mathbf{j}^\perp$ (in the Landau gauges). Including (24) this amounts to the integral equation

$$\mathbf{A}^\perp-\alpha^\perp-\frac{\mathbf{B}_s}{2}\wedge\mathbf{x}=2\pi\int ds\,d\zeta\,\theta(ct-s-|z-\zeta|)\theta(s)\,\mathbf{j}^\perp\!\left(\frac{s}{c},\zeta\right). \tag{26}$$

The right-hand side (rhs) is zero for $t\leq 0$, because $t=0$ is the beginning of the laser-plasma interaction. Within *short* time intervals $[0,t']$ (to be determined *a posteriori*) we can approximate $\mathbf{A}^\perp(t,z)\simeq\alpha^\perp(ct-z)+\frac{\mathbf{B}_s}{2}\wedge\mathbf{x}$; we also neglect the motion of ions with respect to that of electrons. Then the Hamilton equations for the electron fluid with 'time' ξ and the initial conditions amount to (9) and

$$mc^2\hat{s}_e'(\xi,Z)=4\pi e^2\left[\widetilde{N}(\hat{z}_e)-\widetilde{N}(Z)\right]+e(\hat{\mathbf{x}}_e^{\perp\prime}\wedge\hat{\mathbf{B}}_s^\perp)^z,$$
$$mc^2\hat{\mathbf{u}}_e^{\perp\prime}(\xi,Z)=e\alpha^{\perp\prime}-e(\hat{\mathbf{x}}_e'\wedge\hat{\mathbf{B}}_s)^\perp, \tag{27}$$

$$\hat{\mathbf{x}}_e(0,\mathbf{X})=\mathbf{X}, \quad \hat{\mathbf{u}}_e(0,\mathbf{X})=\mathbf{0} \quad\Rightarrow\quad \hat{s}_e(0,\mathbf{X})=1. \tag{28}$$

this is a family parametrized by Z of *decoupled ODEs* which can be solved numerically. The approximation on $\mathbf{A}^\perp(t,z)$ is acceptable as long as the so determined motion makes $|\text{rhs}(26)|\ll|\alpha^\perp+\frac{\mathbf{B}_s}{2}\wedge\mathbf{x}|$; otherwise rhs(26) determines the first correction to $\mathbf{A}^\perp$; and so on.

If $\mathbf{B}_s=\mathbf{0}$, again $(27)_2$ is solved by $\hat{\mathbf{u}}_e^\perp(\xi)=e\alpha^\perp(\xi)/mc^2$, while, setting $v=\hat{\mathbf{u}}^{\perp 2}$, $(9)_2$, $(27)_1$ take [20] the form of (17)

$$\Delta\hat{z}_e'=\frac{1+v}{2\hat{s}^2}-\frac{1}{2},\qquad \hat{s}_e'=\frac{4\pi e^2}{mc^2}\left\{\widetilde{N}[\hat{z}_e]-\widetilde{N}(Z)\right\}. \tag{29}$$

If $n_e(0,\mathbf{X})=n_0=$const for $Z\geq 0$, then as long as $\hat{z}_e(\xi,Z)>0$ (29), (28) reduce to the *same* Cauchy problem *for all* Z:

$$\Delta'=\frac{1+v}{2s^2}-\frac{1}{2},\qquad s'=M\Delta, \tag{30}$$
$$\Delta(0)=0,\qquad s(0)=1 \tag{31}$$

with $M\equiv\frac{4\pi e^2 n_0}{mc^2}$. In fig. 4 we show the solution if $\boldsymbol{\epsilon}^\perp$ is as in fig. 3 and $n_0=2\times10^{18}\text{cm}^{-3}$; $s(\xi)$ is indeed insensitive to the fast oscillations of $\boldsymbol{\epsilon}^\perp$ (see section 2.1). After the pulse is passed it becomes periodic: a plasma travelling-wave of spacial period $\xi_H\simeq 49\mu$m follows the pulse. The other unknowns are obtained through (16). Replacing in rhs(26) we find that $\mathbf{A}^\perp\simeq\alpha^\perp$ is verified at least for $t\leq 5\xi_H/c$

The above results are based on a laser spot size $R=\infty$ (plane wave). When including corrections due to the finite R (based on causality and heuristic estimates), they imply: the impact of a very short and intense laser pulse on the surface of a cold low-density plasma (or gas, ionized into a plasma by the pulse itself) may induce (for carefully tuned R), beside a wakefield propagating behind the pulse [23, 24], also a backward acceleration and expulsion of surface electrons [20, 21] (*slingshot effect*), as schematically depicted in fig. 5. For reviews see also [22].

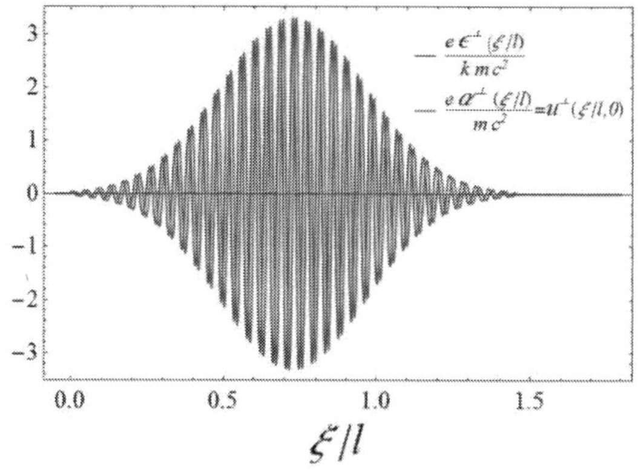

Figure 3. Normalized pump $\boldsymbol{\epsilon}^\perp$ with carrier wavelength $\lambda=0.8\mu$m, gaussian modulation $\epsilon(\xi)=a\exp[-\xi^2/2\sigma]$, $\sigma=20\mu\text{m}^2$, $ea\lambda/mc^2=15$, average pulse intensity 10^{19}W/cm^2, linear polarization. $l\simeq 27\mu$m is the length of the interval where the pump amplitude ϵ overcomes the ionization threshold for the cold gas (here helium) yielding the plasma; under such pulses the thresholds for 1st and 2nd ionization are overcome (i.e. Keldysh parameters become smaller than 1) almost simultaneously [18, 19].

References

[1] G. Fiore, arXiv:1607.03482; submitted to J. Phys. A.

[2] D. Strickland, G. Mourou, Opt. Commun. **56** (1985), 219.

[3] G. A. Mourou, T. S. Tajima, V. Bulanov, Rev. Mod. Phys. **78** (2006), 309.

[4] G. Mourou, S. Mironov, E. Khazanov, A. Sergeev, Eur. Phys. J. **ST 223** (2014), 1181.

[5] T. Tajima, J.Dawson, Phys.Rev.Lett. **43** (1979), 267.

[6] T. Tajima, K. Nakajima, G. Mourou, Riv. N. Cim. **40** (2017), 34.

[7] L.D. Landau, E.M. Lifshitz, *The Classical Theory of Fields*, 2nd edition, Pergamon Press, 1962.

[8] G. Fiore, J.Phys.A: M.Th. **47** (2014), 225501

[9] A. L. Troha et al., Phys. Rev. **E 60** (1999), 926.

[10] J. D. Lawson, Eur. J. Phys. **5** (1984), 104.

[11] R. B. Palmer, in *Frontiers of Particle Beams*, Lecture Notes in Physics **296**, p. 607, Springer-Verlag, 1988.

[12] R. B. Palmer, AIP Conf. Proc. No. 335 (1995), pp. 90-100; and references therein.

[13] E. Esarey, P. Sprangle, J. Krall Phys. Rev. **E 52** (1995), 5443.

[14] P.M. Woodward, J. Inst. Electr. Eng. **93** (1947), 1554.

[15] P. M. Woodward, J. D. Lawson, J. I.E.E. **95**, Part III (1948), 363; and references therein.

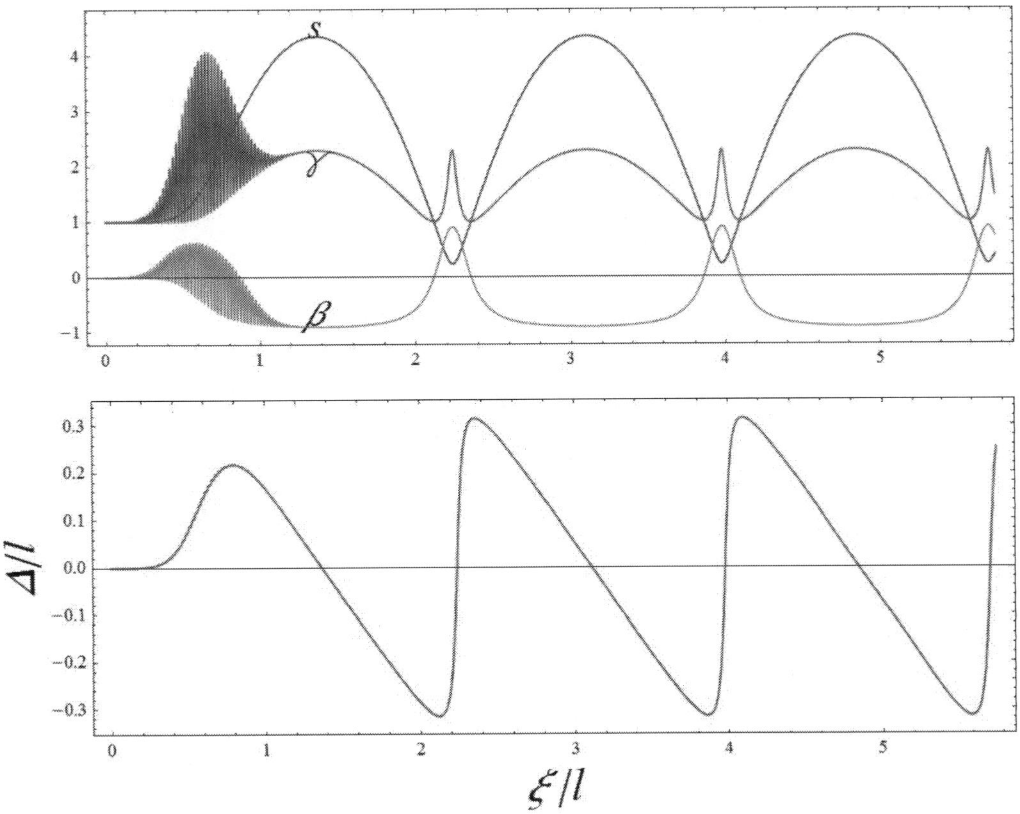

Figure 4. Solution of (29-8) corresponding to the pulse of fig. 3, initial density $\widetilde{n_{e0}}(Z)=n_0\theta(Z)$, $n_0=2\times10^{18}\mathrm{cm}^{-3}$.

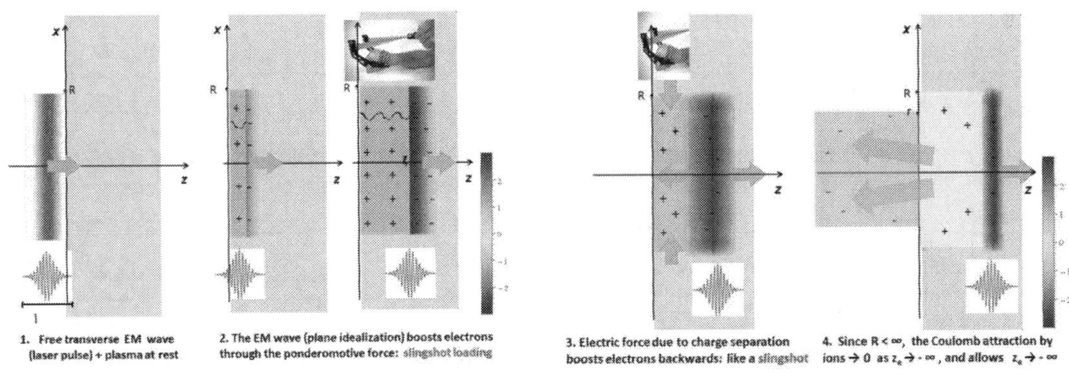

Figure 5. Schematic stages of the slingshot effect.

[16] A.A. Kolomenskii, A.N. Lebedev, Sov. Phys. Dokl. 7 745 (1963); Sov. Phys. JETP 17 179 (1963).

[17] V. Ya. Davydovskii, Sov. Phys. JETP 16 629 (1963).

[18] A. Pukhov, *Rep. Prog. Phys.* **65** (2002), R1-R55.

[19] D. Jovanović, R. Fedele, F. Tanjia, S. De Nicola, L. A. Gizzi, *Eur. Phys. J.* **D66** (2012), 328.

[20] G. Fiore, S. De Nicola, PRAB **19** (2016), 071302; NIMA **829** (2016), 104.

[21] G. Fiore, R. Fedele, U. de Angelis, Phys. Plasmas **21** (2014), 113105.

[22] G. Fiore, Acta Appl. Math. **132** (2014), 261; Ricerche Mat. **65** (2016), 491-503.

[23] G. Fiore, *On the impact of short laser pulses on cold diluted plasmas*, in preparation.

[24] G. Fiore, *On cold diluted plasmas hit by short laser pulses*, submitted to NIMA proceedings.

Diagnostics of Particles emitted from a Laser generated Plasma: Experimental Data and Simulations

Giuseppe Costa[1,*] and *Lorenzo* Torrisi[1]

[1]Dipartimento di Scienze Fisiche, MIFT, Università di Messina, V.le F.S. D'Alcontres31, 98166 S. Agata, Messina, Italy

Abstract. The charge particle emission form laser-generated plasma was studied experimentally and theoretically using the COMSOL simulation code. The particle acceleration was investigated using two lasers at two different regimes. A Nd:YAG laser, with 3 ns pulse duration and 10^{10} W/cm^2 intensity, when focused on solid target produces a non-equilibrium plasma with average temperature of about 30-50 eV. An Iodine laser with 300 ps pulse duration and 10^{16} W/cm^2 intensity produces plasmas with average temperatures of the order of tens keV. In both cases charge separation occurs and ions and electrons are accelerated at energies of the order of 200 eV and 1 MeV per charge state in the two cases, respectively. The simulation program permits to plot the charge particle trajectories from plasma source in vacuum indicating how they can be deflected by magnetic and electrical fields. The simulation code can be employed to realize suitable permanent magnets and solenoids to deflect ions toward a secondary target or detectors, to focalize ions and electrons, to realize electron traps able to provide significant ion acceleration and to realize efficient spectrometers. In particular it was applied to the study two Thomson parabola spectrometers able to detect ions at low and at high laser intensities. The comparisons between measurements and simulation is presented and discussed.

1 Introduction

Laser Generated Plasma (LGP) in vacuum is a relevant topic in many scientific fields, due to its large number of applications, such as preparation of ion sources, ion acceleration method, production of nuclear reactions, generation of neutron emission [1, 2].
In many applications ion beams with adequate characteristics are required, including energy and charge state distributions, emittance, high electron and ion currents.
Magnetic fields can be applied to LGPs in order to focus the charged particles emitted from the irradiated target or to select particles with definited energy using longitudinal or transversal magnetic fields, respectively. These two cases are performed at the Plasma Physics Laboratory in Messina University, using permanent magnetic fields with maximum values of about 0.3 T. A simulation software called "COMSOL Multiphysics" [3] was employed in this study, and to compare model results with experimental data.
When electric fields were associated to magnetic fields, they allow to get a better resolution for the mass-to-charge ratio measurements and for the energy measure of the particles emitted from the target.
Magnetic fields can be also employed to realize electron traps acting on the flux and on the energy of the plasma accelerated charged particles, according to the literature [4].
Electro-magnetic (EM) fields are used to realize the appropriate spectrometers known as the "Thomson Parabola Spectrometer" (TPS) [5].
In this paper, we focus on particles emitted from targets irradiated by low and medium laser intensity, i.e. from 10^{10} up to 10^{16} W/cm^2 intensities, respectively.

2 Materials and methods

In this work we have considered three systems that use magnetic and electrical fields to study a non-equilibrium plasma generated by laser pulses:

1. **Longitudinal Magnetic Field**: consisting of a magnetic field having a symmetry axis parallel to the main plasma plume emission, and hence orthogonal to the irradiated target.
2. **Transverse Magnetic Field**: consisting of a magnetic field having lines orthogonal to the main propagation direction of the non-equilibrium plasma, and hence parallel to the target surface.
3. **Transverse Magnetic and Electrical Field**: called Thomson Parabola Spectrometer (TPS), consisting of a magnetic and an electrical field transverse to the main propagation direction of the plasma, i.e. parallel to the target surface.

* Corresponding author: gcosta@unime.it

2.1 Longitudinal Magnetic Field

A Nd:YAg laser, with fundamental wavelength 1064 nm, maximum pulse duration 3 ns, and energy variable from 1 to 300 mJ, was employed to irradiate dielectric and metallic targets, in single shot mode. The laser beam is focused on the target, placed in a vacuum chamber at a pressure of about 10^{-6} mbar, through an optical lens having a focal distance of 50 cm. The spot area on the target surface is about 0.5 mm^2. The laser beam incidence angle on the target surface is 45°. The irradiated targets are constituted by polyethylene (-CH_2 monomer) and pure aluminium. A scheme and picture of the experimental setup are reported in Figure 1a and 1b, respectively [6].

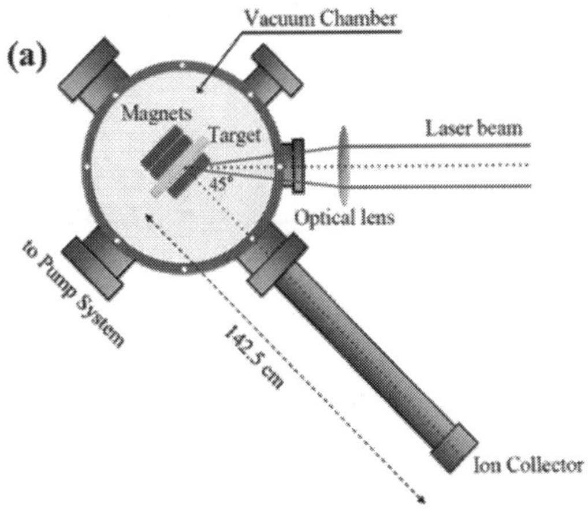

Fig. 1. Scheme (a), and picture (b) of the experimental setup for use of Longitudinal Magnetic field.

The magnet rings are permanent ferrite magnets. A single magnet has an outer diameter of 100 mm and an internal of 60 mm; its thickness is 20 mm. The produced magnetic field is axial (0.12 T on the surface of the magnet and 0.035 T at the centre of it). The axis of the magnets is placed along the normal to the target surface, in correspondence of the laser spot position.

The experiment was conducted without magnets, and with one circular magnet placed in front of the target and with one, two or four magnets placed behind the target.

The magnetic field module at the centre of magnets, accurately measured with a suitable Gaussmeter (Hirst GM08), ranges from 0.071 T to 0.12 T.

The detector used to collect the ions emitted from the plasma, is an ion collector (IC), placed at 142.5 cm distance from the target. The IC is polarized to -50 V with respect to the grid in order to reduce the secondary electron emission. This device is used in time of flight technique (TOF), which consists to evaluate the time that ions take to travel a known distance from the target to the detector, measured using a fast storage oscilloscope, as reported previously [6].

2.2 Transverse Magnetic Field

With the same laser as for polyethylene we irradiated a pure aluminium target. The particles emitted from the target cross a collimation system consisting of two pinholes with a smaller radius of 1.5 mm. The collimated particle beams pass through the centre of a system of two permanent magnets. The magnetic field module is about 0.38 T and it has a length of 10 mm, as shown in the plot of Fig. 2b. A group of 32 aligned Faraday Cups distributed at 90° angle, and placed on a movable chest, as shown in Fig. 2a, reveal the deflected particles crossing the permanent magnet.

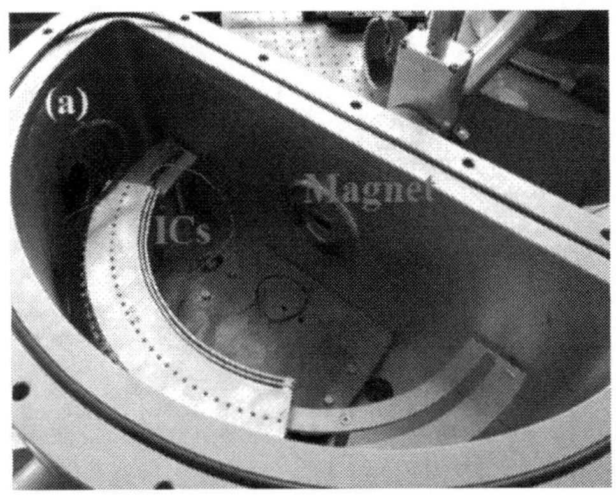

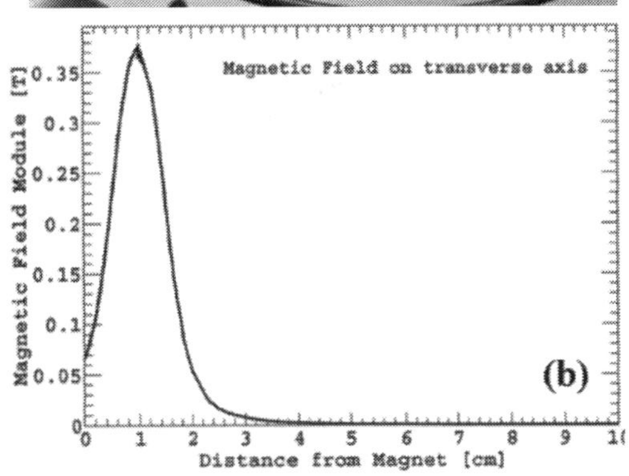

Fig. 2. Picture of the experimental setup (a). Magnetic Field Module, at the centre of the system (b).

2.3 Transverse Magnetic and Electrical Field

In this case, two types of lasers are considered to produce a non-equilibrium plasma in high vacuum.
In the first case, a Nd: YAG laser with 10^{12} W/cm^2 intensity, is used to irradiate a target of titanium in vacuum. The produced plasma is accelerated with a longitudinal post-acceleration system up to 30 keV per charge state, as reported previously [7]. The post-accelerated ion beams go through a magnetic and electrical fields orthogonal to their propagation direction, relative to the TPS system, which separates them according to their charge-to-mass ratio, projecting the ions on an orthogonal shield on which they produce parabolic curves. The used TPS scheme is shown in Fig. 3a. The TPS consists of two pinholes with diameters of 1 mm and 1 μm respectively, which have the role to make the beam tight and collimated. The ions then pass through a transverse magnetic field with module of 0.2 T and a transverse electric field with module 1 kV/cm, each 2.5 cm long, as shown in Fig. 3b. The ions travel 15 cm before being detected by a MCP (MicroChannel Plate) with phosphor screen. Finally the parabola are recorded through a glass windows with a fast CCD camera placed externally to the vacuum chamber.
This study was carried out at the National South Laboratories-INFN (LNS) in Catania [8].

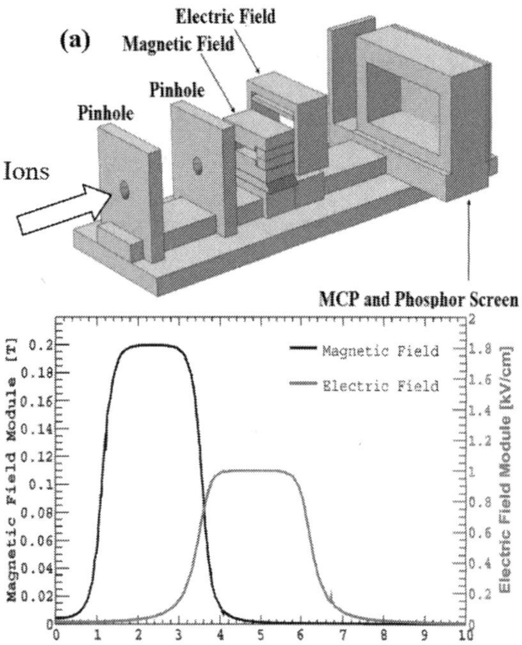

Fig. 3. Schematic (a), and magnetic and electric field module (b) applied at TPS of LNS (Catania).

In the second case, an Iodine laser with 10^{16} W/cm^2 intensity is used to irradiate a target of deuterated polyethylene. The laser beam forms a 0° angle with the target, and the interaction regime is TNSA (Target Normal Sheath Acceleration) [9]. The plasma generated by laser accelerates charge particles towards a system of two pinholes, used as collimators and accurate alignment of the incident ion direction. The collimated ions cross a magnetic field with module of 0.1 T and an electrical field with module of 2 kV/cm, each 8 cm long, as shown in Fig. 4b. The deflected particles travel 16.5 cm before impacting on a MCP with phosphor screen, as shown in Fig. 4a. This study was carried out at Prague Asterix Laser System (PALS), in Czech Republic [10].

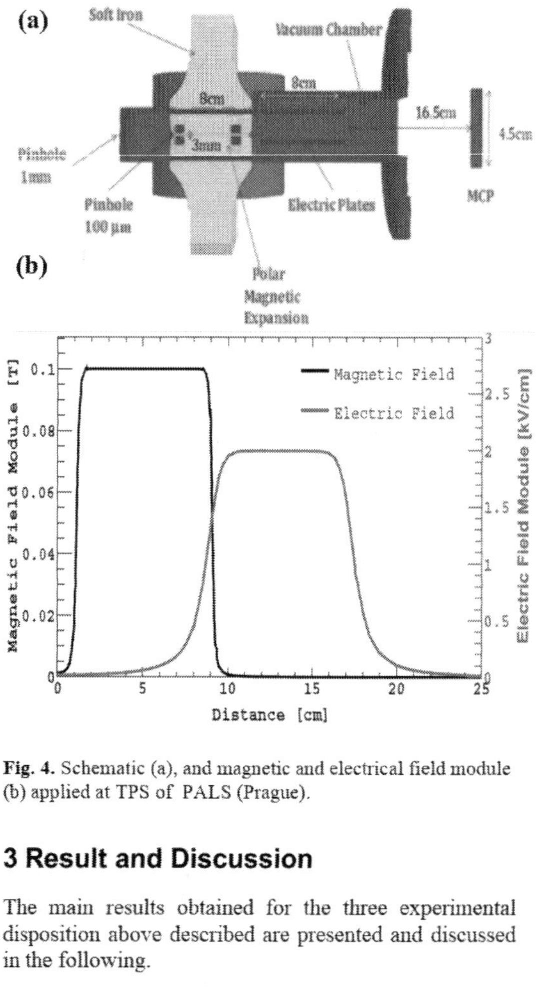

Fig. 4. Schematic (a), and magnetic and electrical field module (b) applied at TPS of PALS (Prague).

3 Result and Discussion

The main results obtained for the three experimental disposition above described are presented and discussed in the following.

3.1 Longitudinal Magnetic Field

In Figure 5 are shown the IC-TOF spectra obtained by irradiating targets of aluminium and glassy carbon, with and without the use of longitudinal magnetic fields.

Measurements demonstrate that the system of axial magnetic fields focuses the plasma ions enhancing the ion yield collected with the axial IC detector. The measurements of the ion current (yield) demonstrate that it increases when the axial magnetic field is applied by about a factor of 3.5 for aluminium target and by a factor of 2.5 for the glassy carbon target. Also experiments with the polyethylene target demonstrate an yield enhancement similar to that observed for carbon.

https://doi.org/10.1051/epjconf/201816704005

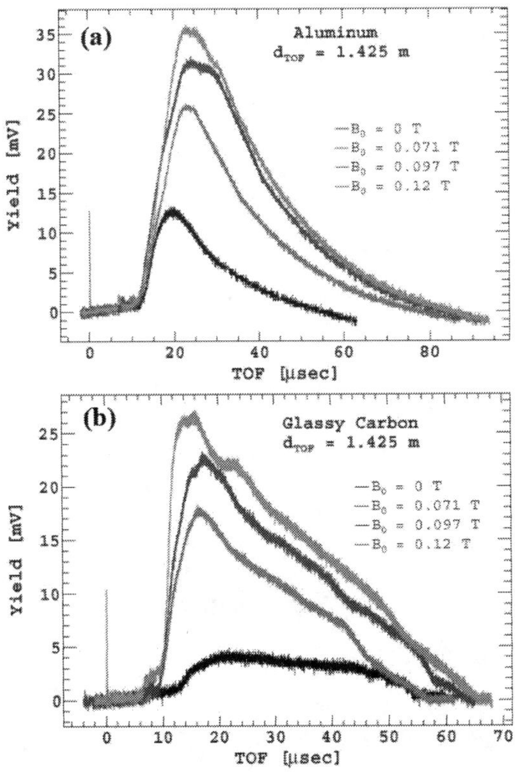

Fig. 5. Experimental data for pure Aluminium (a) and Glassy Carbon (b).

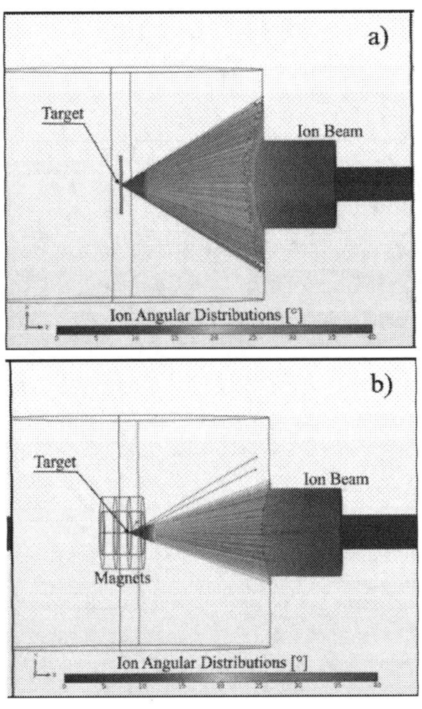

Fig. 6. Simulation with COMSOL for ion beam in the system without (a) and with (b) Magnets.

The increase of the ion yield was evaluated not only experimentally but also theoretically through the COMSOL-Multiphysics simulation program. An example is reported in Fig. 6 [6].

The simulation data have shown an excellent agreement with the experimental ones. In particular, the obtained experimental data and simulations show an increase in ion energy and ion yield when the axial magnetic field is applied. The energy increment for H^+ ions measured from the TOF spectra of Fig. 5 is of about 92% for carbon target and about 99% for aluminium target.

To explain the increase of measured ion energy it is necessary to know the dynamics of the plasma produced electrons in the system of interest. With the simulation software COMSOL it is possible to simulate the motion of the electrons in the magnetic fields with axial symmetry, as reported in Figure 7 [6].

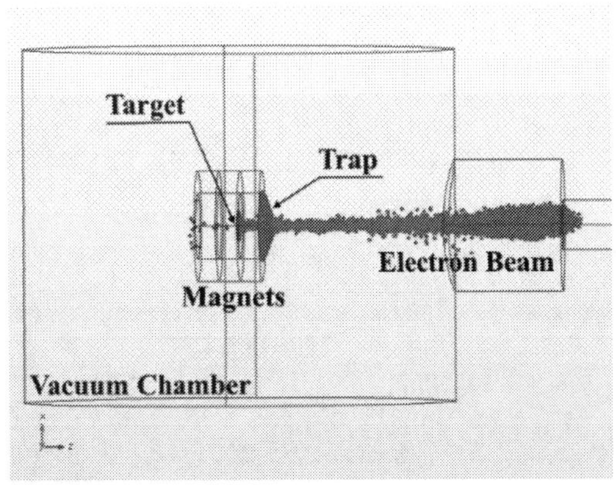

Fig. 7. Simulation with COMSOL for electron beam in the system with longitudinal Magnets.

In presence of the magnets, the kinetic electrons produce two traps, they form two negative clouds placed in front of the target surface that are responsible for an increase of the plasma potential and of the consequent ion acceleration. The perturbation of the emitted electron beams, due to the axial magnetic field, is responsible for the increase of plasma potential and consequently of the increase of the ion energy. The ion energy increases as a result of the electron density of the irradiated target and of the electron traps generated in front of the target surface [4].

3.2 Transverse Magnetic Field

In Fig. 8 are shown the highest ion yields displayed vs. the detection angle for a pure aluminium target.
Since the energy distribution for the particles emitted by the target is the Coulomb-Boltzmann shifted [11]:

$$f(E) = \frac{A}{\sqrt{2m}}\sqrt{\frac{1}{(\pi k_B T)^2}}E\exp\left[-\frac{1}{k_B T}\left(\sqrt{E}-\sqrt{\frac{\gamma k_B T}{2}}-\sqrt{zeV_0}\right)^2\right] \quad (1)$$

where A is a normalization constant, m is the mass of the ion considered, γ represents the adiabatic expansion coefficient, ze is the ion charge state, V_0 is the equivalent potential in plasma, k_B is the Boltzmann constant, and finally k_BT is the equivalent plasma temperature in eV, the TOF spectra obtained using the spectrometer shown in Fig. 2a were investigated. Again, using the COMSOL simulation program, and implementing energy particle distribution of eq. (1), a comparison between experimental data and simulation can be done, as shown in Fig. 8.

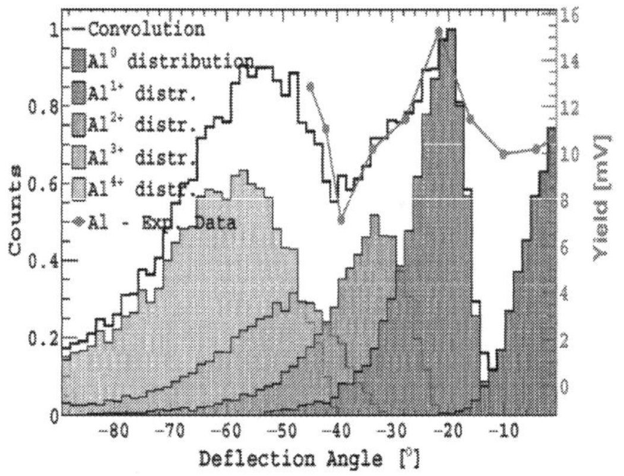

Fig. 8. Comparison of experimental spectrum and convolution of angular distributions of ionic species from 1+ to 4+ of aluminium.

From evaluating the number of deflected particles at a certain angle by the magnetic field, for each charge state from 1+ to 4+ of Al, the count-angle histogram is obtained. Each count represent the normalized number of particles deflected at a given angle. From the convolution of the simulations from 1+ to 4+ charge states of Al-ions, and considering the neutral components Al^0 distribution, the ion angular distribution measured and calculated by COMSOL were compared to each other, as shown in Fig. 8. From the excellent agreement between experimental data and simulation results, it was possible to estimate a mean plasma temperature of about 38 eV, and an average plasma potential of about 57 V.

3.3 Transverse Magnetic and Electrical Field

In this case, two spectrograms were analyzed, that were obtained, with the TPS at LNS in Catania, and with the TPS at PALS in Prague.

The spectrogram at LNS was obtained irradiating with 10^{10} W/cm^2 laser intensity a titanium target. It was achieved using a post ion acceleration from 1kV up to 30 kV and Gaf-Chromic films as ions detector [8].

The spectrogram at PALS was obtained irradiating with 10^{16} W/cm^2 laser intensity a deuterated polyethylene and using a Micro Channel Plate coupled to a phosphor screen as ions detector [12].

The overlapping between experimental spectrograms and simulation of the particle trajectories, calculated with COMSOL, allowed to determine the species analyzed for the measurements obtained at LNS shown in Fig. 9a, and to determine the species and energies for the spectrogram obtained at PALS, as shown in Fig. 9b. In the latter case, energy of 3 MeV for charge status was estimated, according to previous papers [13].

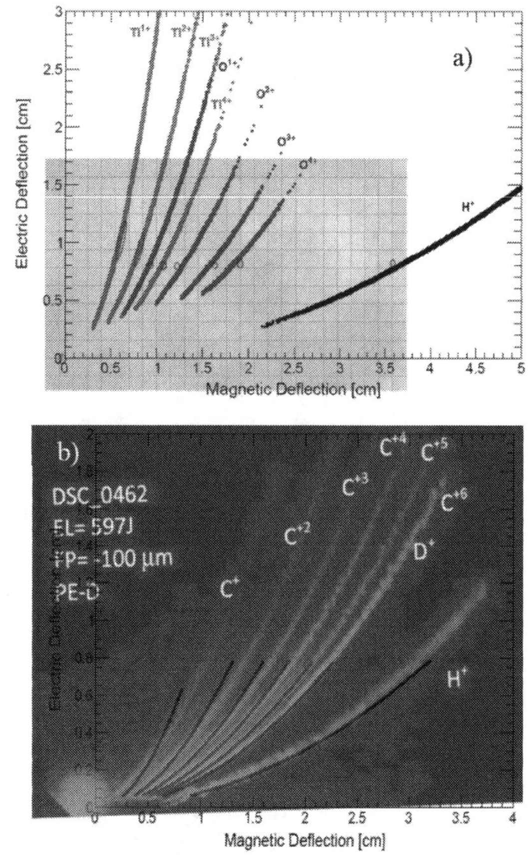

Fig. 9. Comparison between experimental spectrum and simulation, for spectrogram at LNS in Catania (a) and for spectrogram at PALS in Prague (b).

Fig. 9a shows the identifications of parabolas due to protons, titanium and oxygen ions, emitted from the target and post accelerated from 1 keV up to 30 keV per charge state. Due to the maximum charge state of 4+, it means that the maximum ion energy is 120 keV for oxygen and titanium and 30 keV for protons.

Fig.9b shows the identifications of parabolas due to protons, deuterons and carbon ions emitted form the hot plasma of PALS. In this case the acceleration was of about 3 MeV per charge state, thus the maximum kinetic energy was 18 MeV for Carbon ions and 3 MeV for protons. Deuterium ions are indistinguishable from C6+ ions due to the same mass-to-charge ratio.

4 Conclusions

In this work, we have emphasized the role of simulation programs, like COMSOL Multiphysics, which can be used in the analysis of the dynamics of a non-equilibrium plasma generated by high intensity laser pulses [3]. The simulation code can be used to characterize the plasmas and to study the kinetic energies of the accelerated charged particles.

Indeed, thanks to the present study, it was possible to explain the increase in detected yield and energy observed using magnetic fields with axial symmetry; in this case the simulations carried out with COMSOL Multiphysics have shown that the electrons remain trapped for times of the order of hundreds of nanoseconds; this is sufficient to accelerate ions at higher velocity.

Another result of this study is the analysis of the angular distribution of ions deflected by a magnetic field perpendicular to their motion, which made it possible to estimate the average plasma temperature and potential, both relevant for plasma diagnostics.

Finally, thanks to the study of ions trajectories deflected from magnetic and electrical fields, it was possible to characterize the species that constitute the plasmas, and to evaluate their energy using TPS spectrometers.

All of these studies can result in great benefits to innumerable applications where high intensity laser pulses are used to accelerate ions from solid matter.

This work was supported by the "Research and Mobility" project of Messina University No. 74893496, scientifically coordinated by Professor L. Torrisi.

References

1. K. Takahaschi, et al., Plasma and Fusion Research, **8**, 1206005 (2013)
2. N. Izumi, Y. Sentoku, H. Habara,1, K. Takahashi, et al., Physical Review E. **65**(3), 036413 (2002)
3. COMSOL Multiphysics (2017). Actual website: https://www. comsol.it/comsol-multiphysics
4. L. Torrisi, D. Margarone, S. Gammino, L. Andò, Laser and Part. Beam, **25**(3), 453-464 (2007)
5. D.C. Carroll, P. Brummitt, D. Neely, F. Lindau, et al., Nucl. Instr. and Meth. in Phys. Res. Sec. A : Acc., Spectr., Detec. and Assoc. Equip., **620**(1), 23-27 (2010)
6. L. Torrisi and G. Costa, Laser and Part. Beam, **35**(2), 202-209 (2017)
7. L. Torrisi, S. Cavallaro, L. Giuffrida, S. Gammino and L. Andò, Rad. Eff. and Def. in Sol. : Inc. Plasma Sci. & Plasma Tech., **165**(6-10), 509-520 (2010)
8. L. Torrisi et al., LNS Activity Report : *Ion Sources and Accelerators*, 176-179 (2010)
9. M. Passoni, L. Bertagna, and A. Zani, New J. Phys. **12**, 045012 (2010)
10. M. Cutroneo, L. Torrisi, et al., Acta Polytechnica, **53**(2), 138-141 (2013)
11. L. Torrisi, Rad. Eff. and Def. In Sol., **171**(1-2), 34-44 (2016)
12. L. Torrisi, M. Cutroneo, et al., EPJ Web of Conferences, **96**, 01032 (2015)
13. L. Torrisi et al., LNS Activity Report : *Ion Source and Accelerators*, 223-226 (2012)

On the origin of negative target currents during laser ablation of polyethylene

Domenico Delle Side[1,2,⋆], *Anna Paola* Caricato[1,2], *Josef* Krása[3], and *Vincenzo* Nassisi[1,2]

[1] *Dipartimento di Matematica e Fisica "E. De Giorgi", Universitá del Salento – Lecce, Italy*
[2] *Istituto Nazionale di Fisica Nucleare, Sezione di Lecce – Lecce, Italy*
[3] *Institute of Physics, ASCR – Praha 8, Czech Republic*

Abstract. The exposure of a target to a focused laser beam results in the occurrence of a time-varying current between the target itself and the grounded vacuum chamber. This current is composed by three distinct phases, namely the *ignition phase*, in which the laser pulse drives the electron emission, while electrons coming from the ground through the target holder balance the positive charge generated on the target. The *active phase* appears at post-pulse times and it is characterized by the presence of peaked structures in the time-resolved current, representing characteristics of the target composition. Lastly, the *afterglow phase* is determined by a current of electrons flowing from the target to the ground. During the active phase of the target current resulting from polymers ablation with an UV KrF laser, negative target current peaks are observed, whose origin is still unknown. We investigate the dependence of these current structures on the dimensions of the target, using ultra-high molecular weight polyethylene disks of different thickness.

1 Introduction

The interaction of an intense laser pulse with a solid target give rise to a series of complex phenomena that eventually ends with ejection of vaporized and ionized material from the target surface. Obviously, depending on the the laser pulse length, the physical processes underlying the ejection could change [1]. Neverthless, ejecta are readily created after the first moments of the laser pulse, both as neutral particles and ions, electrons being the first. Unless electrically insulated, the escape of charged particles from the target generates a return current from ground that balances such process. Target current (TC) is a relatively unexplored phenomenum in the field of laser-matter interaction. The initial studies started in the late seventies [2, 3] and continued in a patchy fashion across the following decades [4, 5] until recently.

In the last years, ultra-intense laser systems (from TW to PW powers) become available, leading to the discovery of the emission of hot electrons and energetic ions from the interaction of such laser with thin (from 1 to 10 μm) foils [6]. This renewed the interest in TC as tool to obtain insights both for the ion emission and the generated electromagnetic pulses [7–12].

Despite of the intriguing possibilities it offers, e.g. experimental study of electrons overcoming the plasma potential barrier and escaping the plasma, the physical description of TC in the context of such experiments is difficult, since currents reach values in the kA range, leading often to extremely complicated experimental setups. For such reasons, in 2014 we started an experimental campaign to analyze TC in the simpler framework of nanosecond laser ablation, in order to gain insight on the underlying process in a "relaxed" physical picture. We showed, in particular, that TC is composed by three distinct phases [13], namely the *ignition phase*, in which the laser pulse drives the electron emission, while electrons coming from the ground through the target holder balance the positive charge generated on the target. The *active phase* appears at post-pulse times and it is characterized by the presence of peaked structures in the time-resolved current, representing characteristics of the target composition. Lastly, the *afterglow phase* is determined by a current of electrons flowing from the target to the ground. Furthermore, changing the target material, we showed that the occurrence in time of current peak in the active phase is strongly related to the charge-to-mass ratio of the generated ion species [14], i.e.

$$t_{PEAK} \propto \left(\frac{A}{Z}\right)^{1/2}, \quad (1)$$

where Z represents charge state, while A the mass number. The duration of target neutralization with ground electrons reflects the duration of electron escape from the expanding plasma [14]. The TC occurs when the laser pulse starts to produce the plasma [11, 13].

Target current registered during ablation of plastic targets revealed an unxpected and unexplained behavior. Indeed, during ablation of conductive targets, TC readily passes from zero to mA peak values during the ignition and active phases, switching to negative values during the afterglow phase. In plastic targets, instead, we found that TC assumes negative values during the ignition phase [15, 16]. Furthermore, we showed that such negative

⋆e-mail: domenico.delleside@le.infn.it

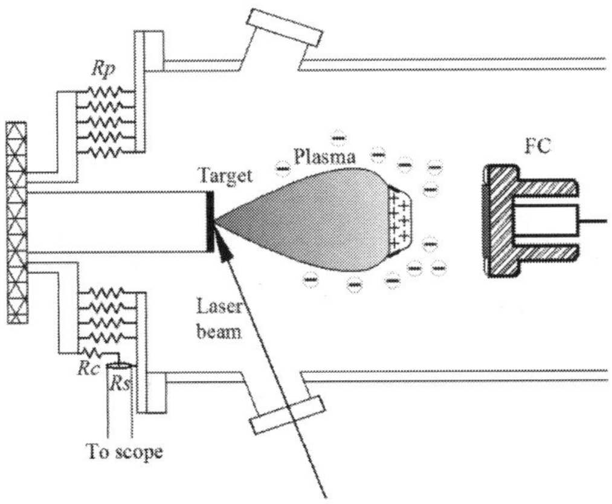

Figure 1. Sketch of the experimental setup: R_p parallel resistors, R_c single 60 Ω resistor, R_s scope impedance, FC is a Faraday cup.

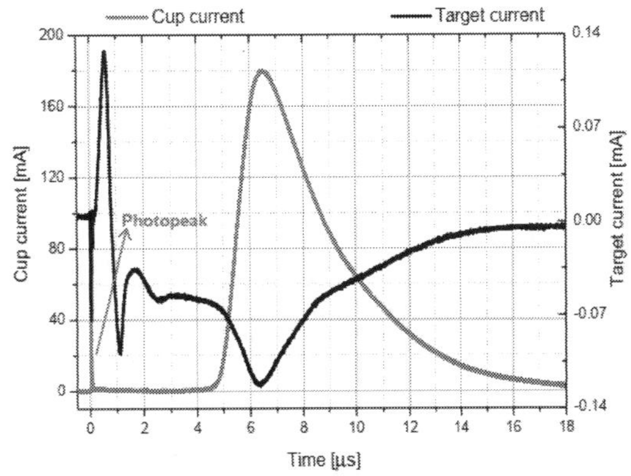

Figure 2. Representative results of both target current and the corresponding cup current measured on FC (target thickness is 0.2 mm)

current strongly depends on the target-holder-to-sample-surface ratio. This article is devoted to a deeper investigation of this obscure phenomenon.

2 Materials and methods

The experiments described here took place at LEAS Laboratory of Dipartimento di Matematica e Fisica "*E. De Giorgi*", Universitá del Salento, in Lecce (Italy). We used the PLATONE setup [17–19], a nanosecond Laser Ion Source operating in the UV range. Indeed, it uses a KrF excimer laser with pulse length of 23 – 25 ns and an intensity ranging from 10^8 to 10^{10} W/cm^2 (laser spot diameter $\approx 180\,\mu$m), operated in single shot mode. During the experiments, we used a fluence of 40 J/cm^2, corresponding to an intensity of $\sim 1.7 \times 10^9$ W/cm^2. The target holder, initially insulated, was connected to ground through an array R_p of 10 resistors (110 Ω). One of such resistors was the series between a further 60 Ω resistor (R_c) and a digital oscilloscope (see Figure 1). Consequently, the voltage V_S registered on the scope is related to the potential jump on the laser target (V_T) through the formula

$$V_T(t) = \frac{R_c + R_s}{R_s} V_S(t), \qquad (2)$$

with R_s being the scope internal impedance (i.e. 50 Ω). To diagnose the ion beams resulting from laser-matter interaction, we used a negatively biased Faraday cup (FC) with a grounded input mesh. FC was placed in front of the target (T) at a distance of 285 mm. Through a dedicated filter circuit, FC was connected to the same oscilloscope used to record the TC.

In order to gain insight on the origin of the negative peak observed in plastic materials, we used as targets commercial Ultra High Molecular Weight Polyethylene (UHMWPE) disks (20 mm diameter) of 4 different thickness: 1.2, 0.5, 0.2 and 0.075 mm.

3 Results and Discussion

The laser-matter interaction with the targets resulted in both TC and FC spectra, for all targets, that are consistent with the observation performed in the past [15, 16] (see Figure 2). As expected, the TC presented negative peaks for all the target thickness used. High resolution spectra of TC showed that in all samples, the negative part is composed of two distinct peaks (P1 occurring near 20 ns and P2 after 55 ns, see Figure 3). Despite of this, the temporal structure and the amplitude of the signals changed significantly from sample to sample. First of all, considering only signal amplitude, it appears that it exists an optimal thickness able to increase the height of the resulting TC peaks, as shown in Figure 3. Indeed, the maximum amplitude happens for a thickness of 0.5 mm, decreasing moving both to the left and right of this value. We'll explain such behavior below.

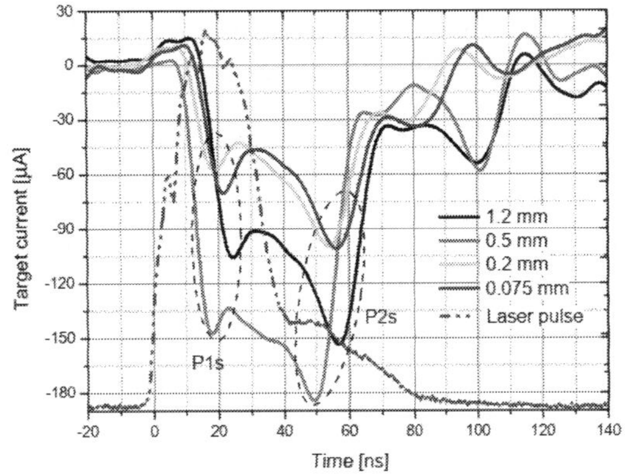

Figure 3. Details of the negative current peaks for all the targets thickness used, together with a waveform representing the laser pulse. The two ellipses identify P1 and P2 for each curve.

 https://doi.org/10.1051/epjconf/201816704006

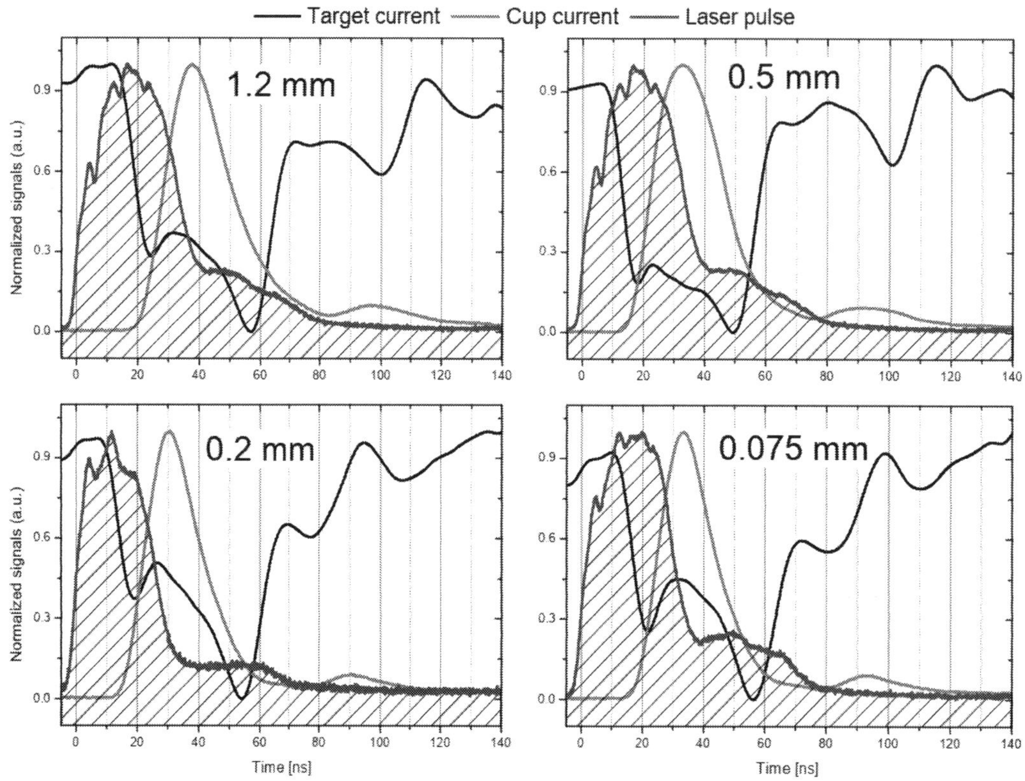

Figure 4. Normalized signals of TC, FC and laser pulse for the 4 targets under exam. It is worth noticing that for practical reasons, TC refers to the left scale, while FC current and the laser pulse to the right. Both scales run from 0 to 1, but the left, to maximize the TC peaks, stops slightly before 1.

The temporal occurrence of the negative peaks suggests a possible involvement of the laser-plasma XUV and soft X-rays that give rise to the photopeak signal on FC. For this reason, we plot on the same figure the results obtained for the 4 targets, normalizing TC, the photopeak and the laser pulse to the same [0, 1] scale (Figure 4).

Such comparison shows that the first peak occurs within the end of the laser pulse and the onset of the photopeak. This suggests that P1 is intrinsically related with the laser-matter interaction process. Such behavior could be explained considering the role of target electrons heated during the action of the laser pulse. Indeed, electrons absorbing the 248 nm laser radiation which penetrates into the plastic foil and also plasma radiation are liberated in the target via internal photoemission. These electrons could then diffuse to the target holder, flowing to ground and giving rise to negative currents. This explanation is also at the origin of the "optimal thickness" effect observed on the top of this section. Such thickness resulted about 0.5 mm in our experiments. Due to the thermal nature of P1, when target is too thin, it cools too rapidly, leading to smaller current amplitudes. Conversely, when the thickness is too large, the electron stream reaching the target holder is reduced, since they encounter more recombination centers. Also in the latter case, we obtain smaller current amplitudes. The nature of P2, instead, remains still unclear from the comparison. In order to obtain a clearer picture of the situation, we performed measurements of TC changing the negative bias on FC, leaving the grounded input mesh in its position. The increase (in absolute value) of polarization induced a sensible increase on P2's amplitude and a shift of its position towards P1, as shown in Figure 5.

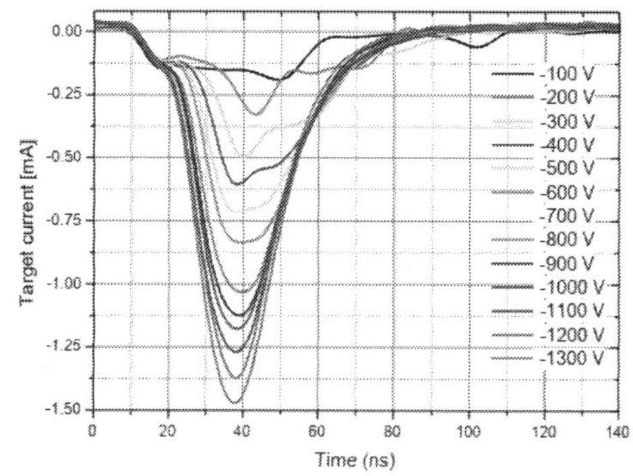

Figure 5. Dependence of ignition phase's negative TC on FC polarization. Bias voltages are reported in absolute value. Values obtained with a 0.5 mm target thickness.

As a further clarification, we connected the input mesh to a negative bias larger than that on the FC, in order to understand if P2 increase due to electrons hitting the targets from the photopeak. Indeed, secondary electrons extracted from FC are suppressed using a small negative potential between the mesh and the FC plate, such that electrons extracted from energetic photons are pushed back to the metal. Such measurements confirmed the increase in P2's amplitude increasing (in absolute value) the bias on the input mesh (data not shown).

This eventually means that P2 is due to free electrons attracted by target's positive charging during the formation of P1. Moreover, this process is further enhanced by the electric field created in the space between target and FC. The cup (and/or the input mesh), indeed, is negatively polarized to attract positive ions, consequently a static electric field is superimposed to the transient fields developed during the laser ablation process. This static field is able to push negative charges towards the target, originating the observed current spectra. Indeed, considering the area enclosed by P1 and P2, it easy to obtain by numerical integration the total charge reaching the target in the first 14 ns. Such values scale linearly with FC bias (Figure 6).

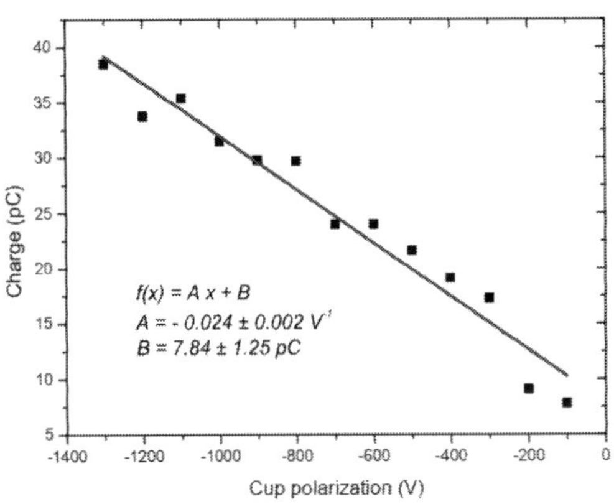

Figure 6. Charges reaching the target as a function of FC polarization. Ordinates are reported in absolute value.

The linear interpolation finds a 7.8 pC peak charge in absence of any polarization. Such value is roughly the double one would expect if only P1 was present. Indeed, a rough extimation of the P1 area from Figure 3 for 0.5 mm thickness suggests it contributes for a charge near 3.0 – 4.0 pC. This confirms that charge unbalance on the target surface is able to attract electrons also in the absence of any electric field between T and FC. Consequently P2 exist irrespectively of any bias voltage on FC, as reported in [14]. This explains also because the "optimal thickness" affects also P2. In effect, if P1 increase in amplitude, a proportional increase should appear also for P2 in order to balance the excess positive charge on the target. It has, indeed, already pointed out that a target-to-FC static electric could influences the properties of ion emission [20]. Consequently, it is reasonable to suppose that the same applies also for TC.

4 Conclusions

We reported evidence on the negative target current observed in non-conductive materials during ignition phase. High resolution spectra of TC show that it is composed by a double peaked structure that strongly depends on target thickness. The first peak depends on the laser-matter interaction process, leading electrons from target to reach the conductive holder. This produces a net charge unbalance that further attracts free electrons on its surface. Furthermore, the amplitude of this second peak is readily enhanced by the presence of a strong electric field between target and Faraday cup.

References

[1] B. Chichkov, C. Momma, S. Nolte, F. von Alvensleben, A. Tünnermann, Appl Phys Mater Sci Process **63**, 109 (1996)
[2] J.S. Pearlman, G.H. Dahlbacka, Applied Physics Letters **31**, 414 (1977)
[3] R.F. Benjamin, G.H. McCall, A.W. Ehler, Phys. Rev. Lett. **42**, 890 (1979)
[4] J.L. Borowitz, S. Eliezer, Y. Gazit, M. Givon, S. Jackel, A. Ludmirsky, D. Salzmann, E. Yarkoni, A. Zigler, B. Arad, Journal of Physics D: Applied Physics **20**, 210 (1987)
[5] Y.A. Bykovskii, I.Y. Konyukhov, V.D. Peklenkov, Quantum Electronics **31**, 45 (2001)
[6] A. Macchi, M. Borghesi, M. Passoni, Rev. Mod. Phys. **85**, 751 (2013)
[7] J.L. Dubois, F. Lubrano-Lavaderci, D. Raffestin, J. Ribolzi, J. Gazave, A.C.L. Fontaine, E. d'Humières, S. Hulin, P. Nicolaï, A. Poyé et al., Phys. Rev. E **89**, 013102 (2014)
[8] A. Poyé, J.L. Dubois, F. Lubrano-Lavaderci, E. D'Humières, M. Bardon, S. Hulin, M. Bailly-Grandvaux, J. Ribolzi, D. Raffestin, J.J. Santos et al., Phys. Rev. E **92**, 043107 (2015)
[9] J. Cikhardt, J. Krása, M. De Marco, M. Pfeifer, A. Velyhan, E. Krouský, B. Cikhardtová, D. Klír, K. Rezác, J. Ullschmied et al., Review of Scientific Instruments **85**, 103507 (2014)
[10] M. De Marco, J. Krása, J. Cikhardt, M. Pfeifer, E. Krouský, D. Margarone, H. Ahmed, M. Borghesi, S. Kar, L. Giuffrida et al., Journal of Instrumentation **11**, C06004 (2016)
[11] J. Krása, M. De Marco, J. Cikhardt, M. Pfeifer, A. Velyhan, D. Klír, K. Rezác, J. Limpouch, E. Krouský, J. Dostál et al., Plasma Physics and Controlled Fusion **59**, 065007 (2017)
[12] M. De Marco, J. Cikhardt, J. Krása, A. Velyhan, M. Pfeifer, E. Krouský, D. Klír, K. Řezáč, J. Limpouch, D. Margarone et al., Nukleonika **60**, 239 (2015)
[13] J. Krása, E. Giuffreda, D. Delle Side, V. Nassisi, D. Klír, J. Cikhardt, K. Rezác, Laser and Particle Beams **35**, 170–176 (2017)

[14] J. Krása, D. Delle Side, E. Giuffreda, V. Nassisi, Laser and Particle Beams **33**, 601–605 (2015)

[15] E. Giuffreda, D. Delle Side, J. Krása, V. Nassisi, Journal of Instrumentation **11**, C05004 (2016)

[16] E. Giuffreda, D. Delle Side, V. Nassisi, J. Krása, Nuclear Instruments and Methods in Physics Research Section B: Beam Interactions with Materials and Atoms **406**, 225 (2017)

[17] L. Velardi, M.V. Siciliano, D. Delle Side, V. Nassisi, Review of Scientific Instruments **83**, 02B717 (2012)

[18] V. Nassisi, L. Velardi, D. Delle Side, Applied Surface Science **272**, 114 (2013)

[19] L. Velardi, D. Delle Side, J. Krása, V. Nassisi, Nuclear Instruments and Methods in Physics Research Section A: Accelerators, Spectrometers, Detectors and Associated Equipment **735**, 564 (2014)

[20] M.M. Bialkowski, G.S. Hurst, J.E. Parks, D.H. Lowndes, G.E.J. Jr., Journal of Applied Physics **68**, 4795 (1990)

Nanoparticles generated by laser in liquids as contrast medium and radiotherapy intensifiers

Nancy Restuccia[1] and *Lorenzo* Torrisi[1]

[1]*Dottorato di Ricerca in Fisica, Dip.to di Scienze Fisiche-MIFT, V.le F. S. D'Alcontres 31, 98166 S. Agata (ME), Italy*

Abstract. The synthesis of Au and Ag nanoparticles (NP) though laser ablation in liquids as a function the laser parameters is presented. Spherical NPs with diameter distribution within 1 and 100 nm were prepared by laser ablation in water. The nanoparticles characterization was performed using optical spectroscopy and electronic microscopy (SEM and TEM) measurements. Studies of the possible use of metallic nanoparticles as intensifier of diagnostics imaging contrast medium and absorbing dose from ionizing radiations in traditional radiotherapy and protontherapy are presented. Examples of in vitro (in tissue equivalent materials) and in vivo (in mice), were conducted thank to simulation programs permitting to evaluate the enhancement of efficiency in imaging and therapy as a function of the NPs concentrations and irradiation conditions.

1 Introduction

In recent years, technological development has allowed us to prepare, manipulate and characterize, more nano-sized materials. Metallic nanoparticles (NP), based on Ti, Au, Ag, Bi and others, have numerous fields of applications: from electronics (detectors, solar cells, optoelectronics devices, sensors, nanoacoustic) to telecommunications (nonlinear optics, nanophotonics, nanothermic), biology and medicine (photothermal therapy, radiation and biological imaging diagnostics) and physical-chemistry (catalysts, preparing targets).
Metallic nanoparticles embedded in an insulating medium behave like ionizing radiation absorbing centres, being able to show certain absorption resonances bands at specific wavelengths, thank to the Surface Plasmon Resonance (SPR) absorption effect [1]. Their size distribution in a fluid changes the electronic and mass density and the equivalent atomic number of the material. The choice of gold as a starting material to prepare nanoparticles is not accidental. Gold, in fact, is stable, not oxidable and biocompatible for humans and, more in general, also for other living beings. In the light of these results, the medicine demonstrates a particular interest to the gold and silver nanoparticles for different aspects concerning prosthesis, interface adhesion, contrast medium for X-ray images and high Z-element for high absorbent radiations.
In the present study it was determined experimentally the behaviour of Au and Ag nanoparticles (Au NP and Ag NP) in therapy and in diagnostics, in particular by calculating the ratio of the dose delivered to a given depth and the dose delivered to the surface in various media (water, adipose tissue and compact bone) in presence or not of the nanoparticles. The anticancer potential of the gold and silver nanoparticles is generated by the different physical-chemical properties advantages. Numerous studies have demonstrated the safety and biocompatibility of Au and Ag both in vitro and in vivo, suggesting that the gold nanoparticles can be administered safely with minimal inflammatory activation and few local or systemic side effects [2-4] enhancing the equivalent Z-value of the treated tissues at significant nanoparticle concentrations.

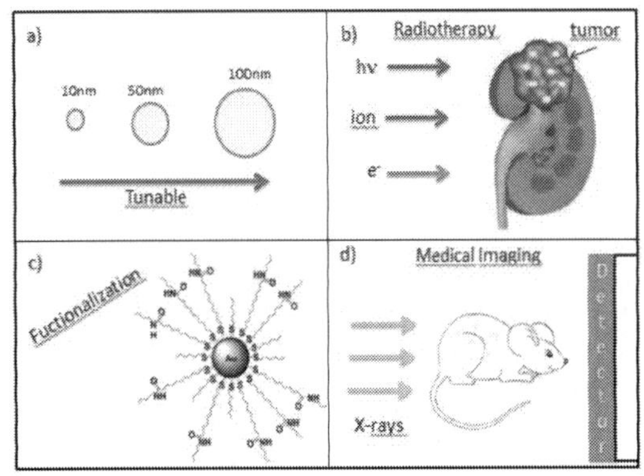

Fig. 1. Au NP with various shapes and size (a) and bio-functionalized with various biomolecules (c). The NP have the ability to enhance radiation therapy of tumours (b), as well as high-Z imaging contrast agents (d).

Fig. 1 shows Au NP of various shapes and sizes (a), their functionalization with particular molecules which allow the transfer to the tumour sites via biological processes (c) or use of drugs or by localized injection into the tumour zone. The NP so attached to the desease tissues

* Corresponding author: ltorrisi@unime.it

https://doi.org/10.1051/epjconf/201816704007

have the ability to improve radiotherapy efficiency (by X-rays, ions and electron beams) (b), and the imaging contrast of localized tissues (d).

The high atomic number of gold (Z=79) allows a high absorption and enhancement of ionizing radiation, as well as the X-ray attenuation and X-ray characteristic line fluorescence emission for higher imaging contrast applications [5]. Moreover, Au NP permit to enhance the electron and ion stopping power increasing the local release of ionizing energy in the cells enriched of Au_NP fluids.

2 Materials and Methods

Laser ablation of metal in water describes a process in which atoms are removed from a solid surface by the irradiation of a high-intensity laser beam and are dispersed in the liquid where aggregate as nanoparticles. Pulsed laser ablation in a liquid medium is a method used to produce a clean, stable, and relatively homogeneous dispersion of nanoparticles in a liquid. Our experimental set-up consists of a pulsed laser, Nd:Yag, operating at fundamental frequency (1064 nm) with a pulse of 200 mJ energy and 3 ns duration, a pulse repetition rate of 10 Hz, and a set of focusing optics. The Fig. 2 shows the scheme (a) and a photo (b) of the experimental set up used to prepare the liquid solution containing the Au NPs.

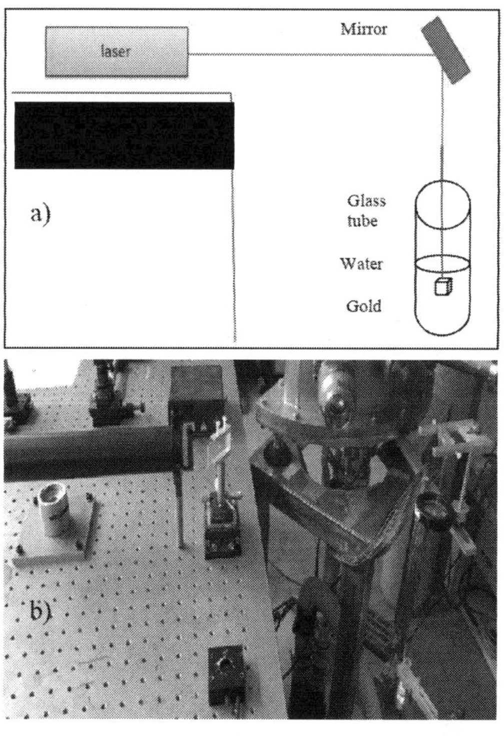

Fig. 2 . Laser ablation in liquid set up scheme (a) and photo of set up in our laboratory (b).

The Fig. 3 shows the prepared solution at 0 minutes, 5 minutes and 20 minutes of Au laser ablation in water. The solution changes its color with the irradiation time and assumes from a mild pink color up to a red color with the NP concentration increment. As the nanoparticles are released from the sample, the solution takes on a different color depending on the size of the nanoparticles, shape and nature of the liquid.

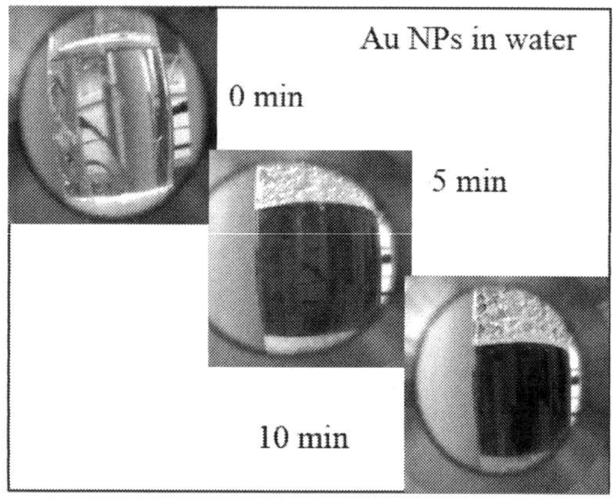

Fig. 3. Solution of Au Np in glass tube produced by laser ablation at different times.

A new compact X-ray imaging system emitting photons at a maximum energy of 45 keV with different exposition times, recording the images on a fast CCD digital recorder (Bruker-In vivo imaging system MS-FX PRO) has been employed in our study.

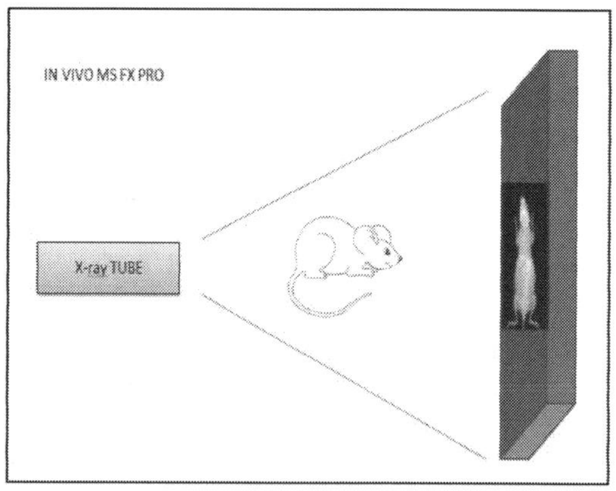

Fig. 4. Scheme of the X-rays imaging system.

The system uses X-rays at energy selectable between 20 and 45 keV, a maximum electron current of 500 mA, spot diameter of 60 mm, with a beryllium window filter. The exposition time can be in single capture of 1.2 s, or in multiple captures overlapping up to 32 image captures. Fig. 4 shows a scheme of the X-ray absorption image used in our laboratory [6-9].

The output image is typically grayscale and may be obtained thought of as a shadow of the object in the path of the X-ray beam. In a classic X-ray image with film, areas that appear white represents dense tissues (like

bone or inflammation and infected lung tissue that can obstruct the passage of X-ray energy). Meanwhile, tissue that is not dense (like the lungs, which are filled with air) will have contrast at the opposite end of the grayscale spectrum and appear black. The new used X-ray image system records images on a CCD detector though different filters, such as beryllium, aluminum and other thin films, in order to improve the signal-to-noise ratio.
This tool has allowed us to carry out animal measurements to verify the effectiveness of high-Z metallic nanoparticles as a potential contrast medium injectable by intravenous, peritoneal or ingestion way. It permits to record X-ray images not only using transmitted X-rays but also using secondary X-ray characteristic fluorescence emission lines, as obtained using Au and Ag NPs.

3 Results

The SEM images of the solution containing Au NP, deposited as a drop on a Si substrate and dried at 50 °C, indicate the presence of spherical shapes with a narrow size distribution, with about 100 nm in diameter. The optical properties of the spherical gold nanoparticles are highly dependent on the nanoparticle diameter, shape and nature of the used liquid. Smaller Au nanospheres primarily absorb light near 520 nm, while larger spheres exhibit increased scattering and have absorption bands that broaden significantly and shift towards longer wavelengths (phenomenon known as red-shifting).
Fig. 5 reports the SEM images of Au NPs at low magnification (a) and high magnification (b) deposited on a silicon substrate as a liquid drop.
Larger spheres scatter more light both because they have larger optical cross sections, and also because their albedo (a ratio of scattering to total extinction) increases with size. The optical properties of the gold and silver nanoparticles change when particles aggregate and the conduction electrons near each particle surface become delocalized and are shared amongst neighboring particles. The gold nanoparticles absorb and scatter visible light with extraordinary efficiency. The absorbance is however restricted to a specific band of wavelengths. It depends on the size of NP, on the shape and on the surrounding medium, as reported in literature [1]. Their strong interaction with light occurs because the conduction electrons on the metal surface undergo collective oscillation when excited by light at specific wavelengths. This oscillation is known as a surface plasmon resonance (SPR), and causes the absorption and scattering of visible light or near UV range or in the IR.
When this occurs, the surface plasmon resonance shifts to lower energies, causing the absorption and scattering peaks to red-shift to longer light wavelengths. UV-Visible spectroscopy can be used as a simple and reliable method for monitoring the stability of nanoparticle solutions.
The optical properties of spherical gold nanoparticles are highly dependent on the diameter of the nanoparticles.
Fig. 6 shows the effect of light illumination of metallic nanoparticles which becomes electric dipoles induced by the incident light with an electric field opposite to the incident one producing light attenuation [10].

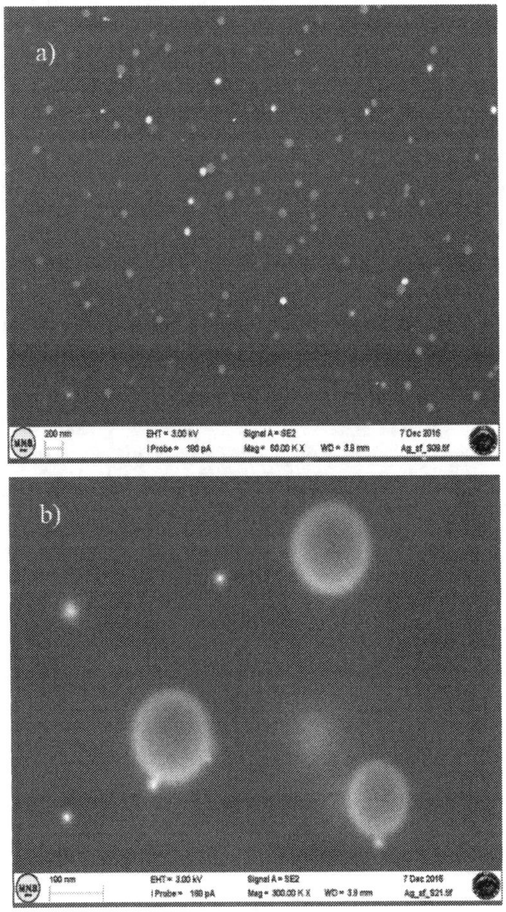

Fig. 5. SEM image of gold nanoparticles.

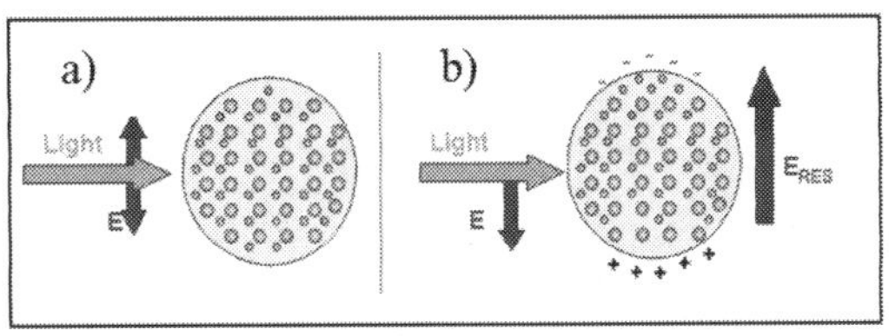

Fig. 6. Dipole induced by incident light on metallic nanoparticles and generation of opposite dipole electric field producing light attenuation.

Fig. 7 shows the result of the optical characterization of the Au NPs in water which absorbance can be measured as a function of the wavelength in order to evidence the SPR absorption bands typical of the prepared solution. Fig. 7a reports the absorbance as a function of the wavelength and of the diameter of the spherical Au NPs in water. By increasing the diameter, the absorption band shift towards higher wavelengths. Fig. 7b shows the different absorbance in presence of separated nanoparticles and of aggregated nanoparticles, up to 1 micron in size, in water. As the particles destabilize, the original extinction peak will decrease in intensity (due to the depletion of stable nanoparticles), and often the peak will broaden or a secondary peak will form at longer wavelengths (due to the formation of aggregates) and shifted towards high wavelengths.

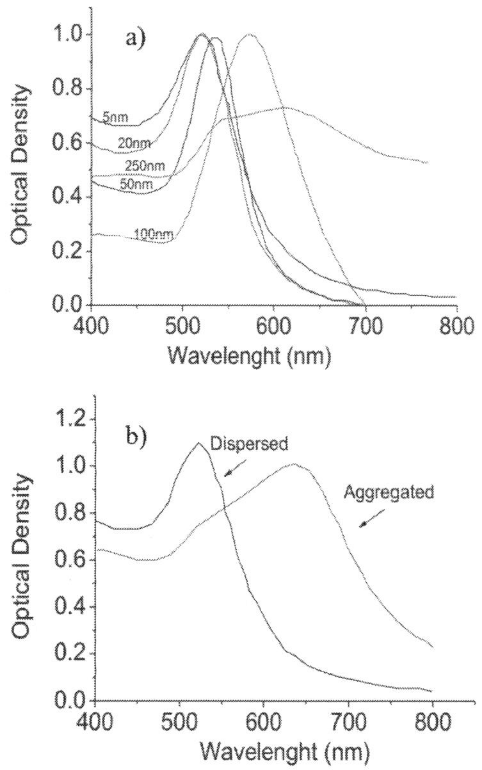

Fig. 7. UV-Vis spectrum of gold nanoparticles of different sizes (a) and extinction spectra of dispersed and agglomerated gold nanoparticles (b).

In order to avoid the nanoparticle aggregation phenomenon, in water was added a biocompatible solution of 1 μg/ml of sodium citrate that permits their uniform dispersion minimizing the collateral effect of coalescence [11-13].

Water solutions containing gold and silver nanoparticles, both generated by ablation in water and analyzed through SEM and TEM microscopy, giving an average diameter of about 50 nm, were injected in healthy mice. The administration of 0.01 ml solution containing nanoparticles per gram of mouse weight was given through intraperitoneal way. The Fig. 8 shows the differences between the X-rays of fluorescence images obtained using gold and silver nanoparticles injected in mice. Once injected gold nanoparticles are quickly absorbed and after about 120 minutes they are deposited in the colon (middle images), after 24 hours there is a remainder accumulated in the livers (bottom images). Instead, measurements demonstrated that silver nanoparticles are absorbed more slowly and the first visible results are found after 24 hours with a significant build up in the colon of the mouse.

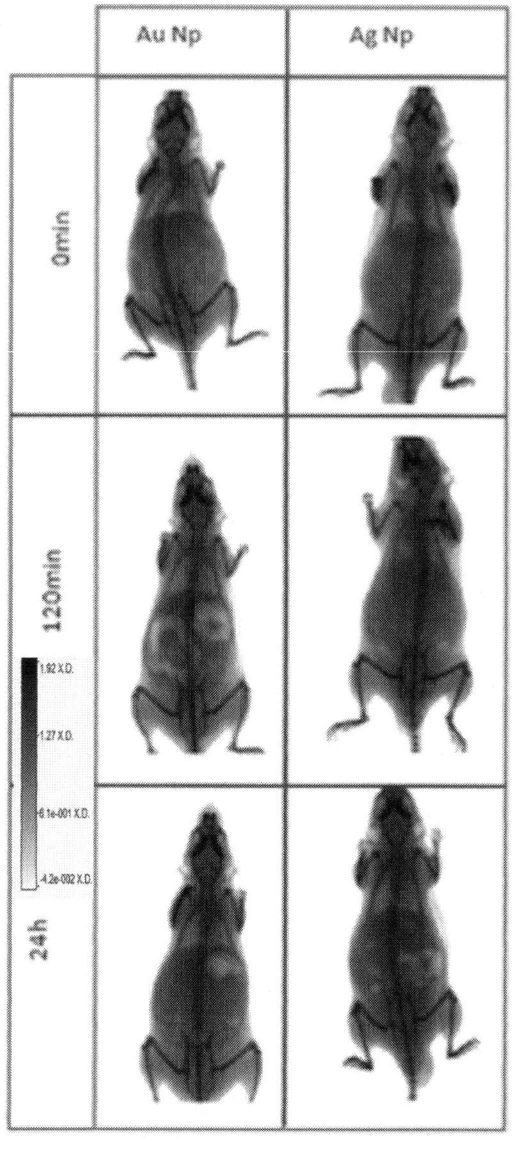

Fig. 8. X-rays fluorescence images of Au and Ag in mice as a function of the time from the peritoneal injection.

This result leads us to conclude that blood circulation of silver nanoparticles produces a slower up-take than that relative to gold nanoparticles. Thus the two kind of nanoparticles have a different metabolic up-take effect.

The high ionizing radiation absorption of high Z metal nanoparticles, such as gold, can be used to improve traditional radiotherapy with photons and electrons and the innovative one using protons and carbon ions. By inserting high-Z NPs into the site of the tumor it is possible to increase the dose released on that particular site. To this aim, theoretical studies have been carried out and are in progress.

For example, for 100 MeV protons in the soft tissue, the stopping powers at 3 cm deep (S_{prof}) and at surface (S_{sup}), can be calculated using NIST database and SRIM Code [14-15]. Their ratio (R_{tissue}) is increased by a factor 7% when Au NPs are added, at a concentration of 100 mg/ml, at 3 cm depth where the tumor is localized, as described by the comparison of the equations (1) and (2) reported in the following.

100 MeV protons:

$$R_{tissue} = \frac{S_{prof}}{S_{sup}} = \frac{605,3}{7,495} = 80,76 \quad (1)$$

$$R_{tissue+Au} = \frac{S_{prof\ tissue+Au}}{S_{sup\ tissue}} = \frac{647,24}{7,495} = 86,36 \quad (2)$$

Another example is evaluated for 100 keV photons in soft tissue with the tumor located at 3 cm deep and with 1 cm tumor thickness. The Lambert-Beer formula is used to calculate the transmitted intensity (I_T) of the incident one (I_0) in the tissue thickness Δx on the base of the energy massive attenuation coefficient (μ_e) and density (ρ). By adding Au nanoparticles to the tumor site, the transmitted intensity decreases indicating an increment of the absorbed radiation dose. The calculations indicate a dose increment of about 85% with the use of Au NPs at 100 mg/ml concentration, as reported in the calculations of the equations (3) and (4).

100 keV photons:

$$I_T = I_0 \cdot e^{-\frac{\mu_e}{\rho}\rho\cdot\Delta x} = 84,14 \cdot e^{-0,169\cdot 1,02\cdot 1} = 70,79keV \quad (3)$$

$$I_{T+Au} = I_0 \cdot e^{-\frac{u_n}{\rho}\cdot\rho\cdot\Delta x} = 70,79 \cdot e^{-0,68\cdot 2,85\cdot 1} = 10,56keV \quad (4)$$

Here the results are reported quantitatively; we immediately notice that thanks to the use of gold nanoparticles, radiotherapy becomes much more effective because the whole dose of radiation should be directly given to the tumor site and not to the surrounding healthy tissue. Reported results are in agreement with literature data [2, 3, 6, 7].

Conclusions

The use of high-Z metallic nanoparticles in the field of radiotherapy and diagnostics promises great results. It improves imaging quality as contrast medium and increases the efficiency of radiotherapy by localizing the released dose to the nanoparticles at tumor site. Actual experiments on mice are performed thank to the collaboration with biologists following the work scheme reported in Fig. 1.

This work was supported by the "Research and Mobility" project of Messina University No. 74893496, scientifically coordinated by Professor L. Torrisi.

References

1. M. A. Garcia "Surface Plasmon in metallic nanoparticles: Fundamental and applications." Journal of Physics: Applied Physics 44, 28 (2011).
2. F. Jay et al. "Gold nanoparticles in radiation research: potential applications for imaging and radiosensitization" Transl Cancer Res 2013;2(4):280-291.
3. Khlebtsov N, Dykman L. Biodistribution and toxicity of engineered gold nanoparticles: a review of in vitro and in vivo studies. Chem Soc Rev. 2011;40:1647–71.
4. Connor EE, Mwamuka J, Gole A, et al. Gold nanoparticles are taken up by human cells but do not cause acute cytotoxicity. Small. 2005;1:325–7.
5. L. Torrisi, et al."Effect of metallic nanoparticles in thin foils for laser ion acceleration" Phys. Scripta 9 (2015).
6. D. R. Cooper et al "Gold nanoparticles and their alternatives for radiation therapy enhancement" Frontiers in Chemistry (2014).
7. Vega and U.O. Häfeli "Utilization of nanoparticles as X-ray contrast agents for diagnostic imaging applications", Contrast Media Mol. Imaging 10 (2015).
8. J. Zou et al., "Micro CT visualization of silver nanoparticles in the middle and inner ear of rat and transportation pathway after transtympanic injection", Journal of Nanobiotechnology (2015).
9. K. Saha, S.S. Agasti, C. Kim, X. Li, and V.M. Rotello, Gold Nanoparticles in Chemical and Biological Sensing, Chem Rev. 112(5), 2739–2779 (2012).
10. S.K. Ghosh and T. Pal, "Interparticle Coupling Effect on the Surface Plasmon Resonance of Gold Nanoparticles": From Theory to applications, Chem. Rev. 107, 4797–4862 (2007).
11. L. Torrisi , "Radiotherapy Improvements by Using Au Nanoparticles", Recent Patents on Nanotechnology, 9(2), 114-125 (2015).
12. L. Torrisi N. Restuccia, I. Paterniti "Gold Nanoparticles produced by laser ablation in liquids for improvements of diagnostic imaging", Recent Patents on Nanotechnology, (2017).
13. L. Torrisi, N. Restuccia, S. Cuzzocrea, I. Paterniti, I. Ielo, S. Pergolizzi, M. Cutroneo and L. Kovacik, "Laser-produced Au nanoparticles as X-ray contrast agents for diagnostic imaging", Gold Bullettin 50(1) pp 51–60 (2017).
14. J. Ziegler, SRIM, The Stopping and Range of Ions in Matter, actual website 2017: http://www.srim.org/
15. NIST database 2017 https://www.nist.gov , J. H. Hubbell and S. M. Seltzer Radiation Physics Division, PML.

Zinc oxide nanocolloids prepared by picosecond pulsed laser ablation in water at different temperatures

Luisa D'Urso[1,*], *Salvatore* Spadaro[2], *Martina* Bonsignore[2], *Saveria* Santangelo[3], *Giuseppe* Compagnini[1], *Fortunato* Neri[2], and *Enza* Fazio[2*]

[1]Dip. di Scienze Chimiche, Università di Catania, Catania, Italy
[2]Dip. di Scienze Matematiche e Informatiche, Scienze Fisiche e Scienze della Terra (MIFT), Università di Messina, 98166 Messina, Italy
[3]Dip. di Ingegneria Civile, dell'Energia, dell'Ambiente e dei Materiali (DICEAM), Università Mediterranea, 89122 Reggio Calabria, Italy

Abstract. Zinc oxide with wide direct band gap and high exciton binding energy is one of the most promising materials for ultraviolet (UV) light-emitting devices. It further exhibits good performance in the degradation of non-biodegradable pollutants under UV irradiation. In this work, zinc oxide (ZnO) and zinc oxide/gold (ZnO/Au) nanocolloids are prepared by picosecond pulsed laser ablation (ps-PLA), using a Zn and Au metallic targets in water media at room temperature (RT) and 80°C. ZnO and Au nanoparticles (NPs) with size in the 10–50 nm range are obtained at RT, while ZnO nanorods (NRs) are formed when water is maintained at 80°C during the ps-PLA process. Au NPs, added to ZnO colloids after the ablation process, decorate ZnO NRs. The crystalline phase of all ZnO nanocolloids is wurtzite. Methylene blue dye is used to investigate the photo-catalytic activity of all the synthesised nanocolloids, under UV light irradiation.

1 Introduction

Photo-catalysis has been emerged as a viable route to decrease the damage caused by dye pollution to environment and humans [1]. The removal of persistent and toxic organic compounds (TOCs) from contaminated water occurs through oxidation and reduction processes. Several kinds of semiconductor-based photo-catalysts, such as titania (TiO_2), zirconia (ZrO_2), tungsten trioxide (WO_3), tantalum pentoxide (Ta_2O_5), hematite (Fe_2O_3) and zinc oxide (ZnO), have been applied in wastewater treatment to convert non-biodegradable TOCs into non-toxic chemicals under UV irradiation [2,3].

Among them, ZnO looks as a very promising candidate for environmental applications because it is cheap, and has a great photo-catalytic activity, a strong oxidation ability, a direct and wide band-gap energy in the near-UV spectral region (3.3 eV at 300K), higher quantum efficiency compared to other wide band-gap semiconductors, and a large free-exciton binding energy (60 meV) so that exciton emission processes can persist at or even above the room temperature. Moreover, being compatible with living organisms, ZnO is currently employed in a broad range of daily applications without risks to human health and environmental impact [4].

Many works in the literature deal with the utilisation of ZnO for UV lasers, high power light emitting diodes, solar cells, heterogeneous catalysis and antibacterial treatments and the effective degradation of harmful and toxic pollutants present in the wastewater [5].

Nonetheless, the photo-catalytic activity of ZnO nanomaterials is limited under the sun irradiation because the solar spectrum contains only 5−7% of the UV component. The relatively high recombination rate of generated electron-hole pairs also reduces the efficiency of this photo-catalyst. A large number of factors, such as crystallinity, crystallite size, crystal phase, and porosity, as well as specific surface area, can further affect the performance of ZnO [6].

Different strategies have been employed in order to improve the photo-catalytic performance of semiconductor materials. The nanostructuration, resulting in a high surface-area/volume ratio, enhances the amount of the photo-generated charge carriers, whereas the addition of metal NPs on the photo-catalyst surface reduces recombination rate of the excited photoelectron-hole pairs. Gold NPs have been successfully utilised by several groups. The addition of Au NPs shifts the optical absorption into the visible region, hence enabling a more efficient utilisation of the solar energy. Au NPs can act as a sink for photo-induced charge carriers, thus promoting interfacial charge-transfer processes. Moreover, receiving electrons from the Au NPs surface, oxygen can trap electrons and becomes the active O_2^- species with consequent enhancement of the photo-catalytic activity [7].

In this paper, we propose a simple and cheap physical approach to the synthesis of ZnO and Au-decorated ZnO nanorods with no environmental impact.

* Corresponding author: ldurso@unict.it; enfazio@unime.it

 https://doi.org/10.1051/epjconf/201816704008

Materials produced are tested as catalysts for environmental applications, showing relatively high photo-degradation efficiency of methylene blue (MB) in aqueous solution under UV light irradiation.

2 Experimental

2.1 Synthesis of the ZnO-based nanocolloids

Figure 1a displays the experimental setup utilised for the synthesis of nanocolloids. The second harmonic (532 nm) of a laser operating at 100 kHz repetition rate with a pulse width of 6–8 ps was used for the ablation experiments (Fig. 1b). The laser beam was focused to a spot of about 70 μm in diameter on the surface of the target with a galvanometric scanner having a telecentric objective with a focal length of 163 mm.

Fig. 1. (a) Experimental setup utilised for the synthesis of nanocolloids, (b) during laser irradiation.

Gold and metallic zinc targets with a thickness of 3 mm (99.99% purity) were positioned on a holder within a glass vessel and ablated in distilled water. In an experiment, the water was maintained at RT, in the other one it was heated at 80°C before the ps-PLA process. ZnO and Au nanocolloids were produced by separate ablation processes. The ZnO and Au targets were irradiated, at a typical laser power density of 2.0 W/cm^2, for 30 min and 5 min, respectively. Au decorated ZnO nanostructures were obtained by mixing ZnO and Au colloidal solutions, in 1:1 volume ratio, immediately after their production. After preparation, all the solutions were stirred at RT in an ultrasonic bath for about 15 min.

2.2 Nanocolloids characterisation and photo-catalytic activity evaluation

The morphology of the samples was investigated by scanning electron microscopy (SEM). SEM analyses were carried out with a Zeiss-Gemini 2 electron microscope, operating at 150 kV. The apparatus was coupled with a Quantax EDX spectrometer to carry out energy dispersive X-ray (EDX) analysis. The EDX detected pear-shaped dimension is about 0.7 μm. The crystal structures of the ZnO and ZnO/Au nanocolloids were investigated by X-ray diffraction (XRD) analysis. XRD patterns were recorded by using using the CuK$_{\alpha 1}$ wavelength (1.5405 Å) by means of a Bruker D8 Advance diffractometer. UV–vis absorption response of the Au and ZnO nanostructures was investigated by means of a Perkin-Elmer Lambda 750 UV–vis spectrometer in the 190–1100 nm range. Analyses were carried out in the colloidal phase immediately after the ablation process. Methylene blue (MB), from Sigma Aldrich, was used as a probe molecule for the preliminary evaluation of the photocatalytic activities of ZnO and ZnO/Au nanostructures. The reaction was conducted in an aqueous solution containing MB at a $1.5 \cdot 10^{-5}$ M concentration. The photocatalytic reaction was carried out, under stirring, using an UV lamp (100W), emitting mainly at 365 nm. The reaction beaker was located at a distance of about 15 cm from the irradiation source. The lamp energy density, measured by a power meter source, is of 30mW/cm^2; 0.5 mg of catalyst were dispersed in 3 ml of MB aqueous solution. The photo-degradation was monitored by recording the UV-Vis spectrum during the irradiation and following the optical density at 662 nm.

3 Results and discussion

Figure 2 displays the morphology of the samples as resulting from SEM analysis. Regardless of the water temperature, rounded Au NPs, with size narrowly distributed around 9 nm, are obtained (Fig.2a). Conversely, the morphology of ZnO is found to depend on the water temperature. NPs are formed at RT (not shown for briefness), whereas NRs are obtained at 80°C (Figs. 2b–c). Au NPs decorate the NRs in ZnO/Au colloids (Fig. 3).

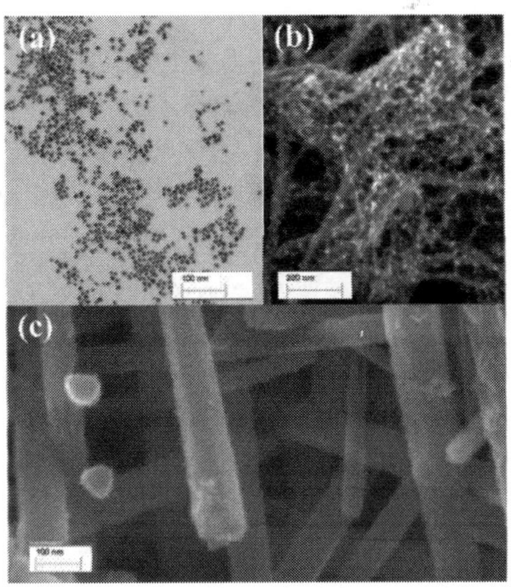

Fig. 2. SEM images of (a) Au colloids prepared at RT. Similar features characterize Au colloids synthesized at 80°C. (b) ZnO/Au colloids synthesized at 80 °C; (c) a morphological detail of the sample prepared at 80°C at higher magnification.

https://doi.org/10.1051/epjconf/201816704008

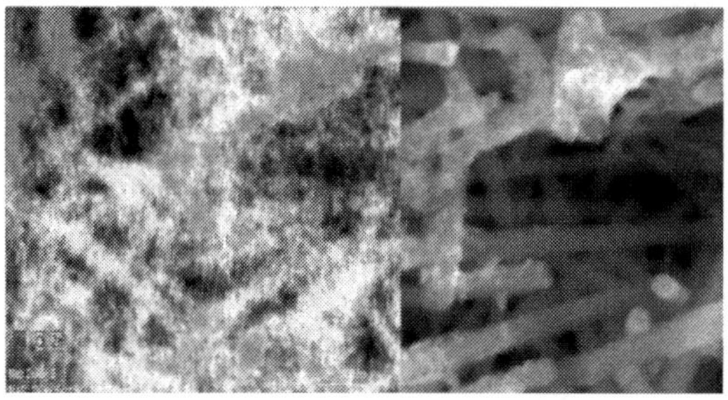

Fig. 3. Elemental dispersion in ZnO/Au colloids synthesised at 80 °C.

The samples high crystallinity, regardless of the water temperature, is evidenced by strong intensity peaks appearing in the XRD patterns (Fig. 4). The diffraction peaks can be indexed to the hexagonal wurtzite ZnO structure (JCPDS card No.36–1451). The mean crystallite size calculated from the most intense (101) peak by using the Scherrer's formula is 35.0 nm.

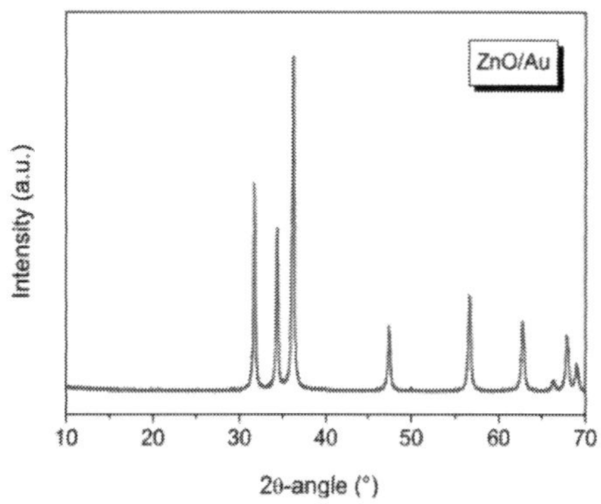

Fig. 4. XRD pattern of ZnO/Au colloids synthesised at RT.

The optical gap (E_g) of the synthesised colloids was estimated from their UV–Vis absorption spectra (Fig. 5), by applying to the Tauc relationship, namely $(\alpha h\nu)^{1/n} = C(h\nu - E_g)$, where α is the absorption coefficient, C is a constant, $h\nu$ is the energy of incident photons, and n an exponent whose value depends on the type of transition (the value of 2, typical of direct allowed transitions, can be used for n in the case of ZnO and ZnO/Au). Accordingly, by plotting $(\alpha h\nu)^2$ as a function of $h\nu$, a linear region is observed, which denotes the onset of absorption. The extrapolation of this region to the abscissa axis yields the energy value of the optical band gap of the material. By this procedure the values of 3.31 and 3.24 eV were found for ZnO and ZnO/Au colloids, respectively. The presence of surface and intrinsic defects might be responsible for the E_g lowering with respect to the bulk ZnO (3.37 eV). Moreover, no substantial difference emerges from the comparison between ZnO based colloids prepared in water at RT or 80°C.

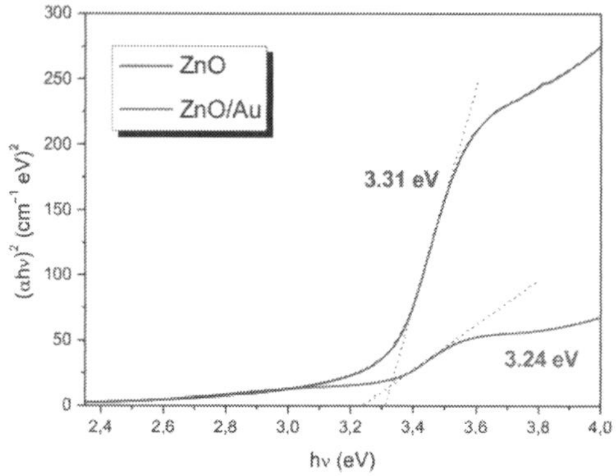

Fig. 5. Plot of $(\alpha h\nu)^2$ as a function of hν in colloids prepared at RT.

In order to explore the photocatalytic activity of as-synthesized ZnO and ZnO/Au nanostructures towards organic pollutants, an aqueous solution containing MB at a $1.5{\cdot}10^{-5}$ M concentration was employed as standard model water-contaminant. For this study, samples were irradiated by an UV light source emitting at 365 nm. Under UV irradiation, the MB aqueous solution undergoes degradation and its optical absorption spectrum changes. Figure 6 shows the spectral changes produced by the ZnO/Au photo-catalyst prepared by ps-PLA at 80°C water temperature.

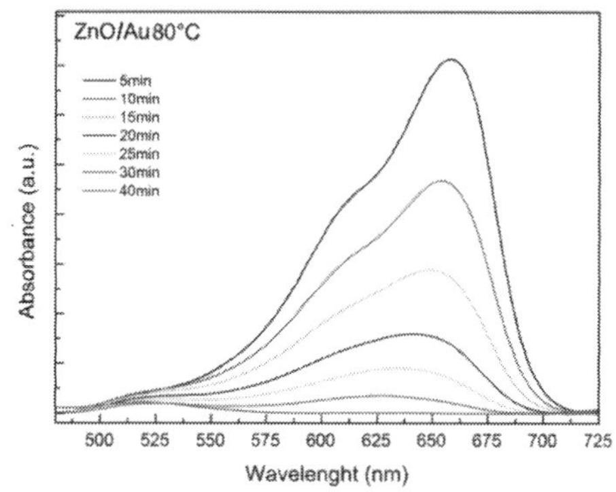

Fig. 6. Photo-degradation of $1.5{\cdot}10^{-5}$ M MB in presence of ZnO/Au (water temperature at 80°C) nanoparticles under UV lamp.

The characteristic absorption peak of MB appears at 662 nm, whereas the optical response of the mixed ZnO/Au colloids gives rise to a shoulder at 530 nm, which is associated to Au characteristic plasmon resonance peak. Under UV irradiation, a decrease and a blue shift of the maximum of the absorbance take place with increasing irradiation time. The absorbance decrease is indicative of the ability of catalyst to reduce MB, whereas the energy shift is probably due to the concomitant de-methylation of MB [8]. The photocatalytic process was monitored by measuring the absorbance spectra for all samples as a function of the irradiation time. Figure 7a gives the UV irradiation time dependence of the relative absorbance (A_t/A_0), where A_0 and A_t respectively denote the initial absorbance and the absorbance at an irradiation time t, both measured at a wavelength of 662 nm. In particular, from A_t/A_0 the MB degradation percentage, $D\%$, was calculated as:

$$D\% = (\frac{A_t}{A_0} - 1) \times 100.$$

In addition, the semilogarithmic plot of A_t/A_0 as a function of the irradiation time t (Fig. 7b) evidences the existence of a satisfactory linear correlation between $\ln(A_t/A_0)$ and t in our experiments. This indicates that the kinetics of photo-bleaching can be represented as a pseudo-first order law and can be expressed by the Langmuir-Hinshelwood equation [9] as follows:

$$\ln(\frac{A_t}{A_0}) = -k_a t.$$

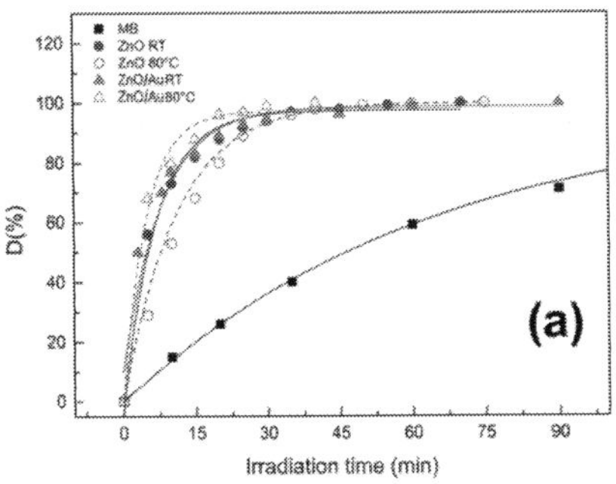

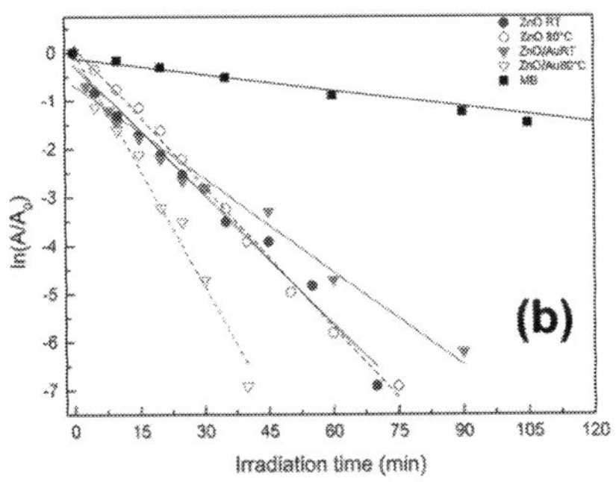

Fig. 7. (a) Photo-degradation percentage of 1.5· 10^{-5} M MB under UV irradiation; (b)The linear fitting of $\ln(A_t/A_0)$ versus irradiation time (t).

From the linear fitting to the $\ln(A_t/A_0)$ versus t data (Fig. 7b), the rate constant (k_a) can be directly obtained. The values of k_a inferred from the fitting procedure as well as the degradation percentage of MB after 15 min of reaction and the time necessary to reach the 50% of the MB degradation ($t_{1/2}$), are reported in Table 1. The data reported in Table 1 prove that, under UV irradiation, all the synthesised nanostructures show a photo-catalytic activity towards the selected dye molecules with the highest k_a value (0.162 min^{-1}), pertaining to Au-decorated ZnO nanorods (ZnO/Au 80°C). This value is 14 times higher than that obtained for the MB photo-degradation in absence of any catalysts, and 1.6 times higher than for ZnO nanorods (ZnO 80°C).

Table 1. Photo-degradation rate constants of MB with different catalysts using $\ln(A_t/A_0) = k_a t$ as fitting equation model, R-square value R^2, degradation percentage (D%) of MB and time ($t_{1/2}$) for D%=50%.

Sample	k_a (min^{-1})	R^2	D% (t=15min)	$t_{1/2}$ (min)
MB	0.011	0.99	19.96	46
ZnO RT	0.089	0.98	81.61	5
ZnO 80°C	0.097	0.99	68.23	7
ZnO/Au RT	0.064	0.97	82.49	3
ZnO/Au 80°C	0.162	0.98	89.21	3

The calculated k_a values confirm that Au NPs act as effective catalysts in the reduction of MB under UV irradiation. Moreover, the presence of Au NPs on the surface of ZnO nanorods (as proven by STEM images) is probably responsible for the improved photo-catalytic activity (i.e. increase of photo-degradation rate) of the ZnO/Au sample. This behaviour is explained in terms of a "doping action" of Au nanoparticles on the photo-catalytically active metal oxides. The accepted mechanism for the enhancement is the transfer of the excited electrons from the ZnO conduction band to the Au nanoparticles. Trapping of the electrons on the Au particles then slows down the recombination process [10].

4 Conclusion

ZnO and Au nanocolloids were successfully synthesised by picosecond pulsed laser ablation, using Zn and Au metallic targets, in water media at RT and 80°C. ZnO and Au nanoparticles were formed at RT with size in the 10–50 nm range, while ZnO nanorods were obtained when water was maintained at 80°C during the ablation process. By adding Au NPs to ZnO colloids after the ablation process, Au-decorated ZnO nanorods were attained. Under UV irradiation, all the produced nanostructures show a photo-catalytic activity towards the used dye molecule. An improvement of the catalytic efficiency was observed in the presence of Au NPS, specially when ZnO sample was characterized by well-defined nanorod structures.

References

1. V. Binas, D. Venieri, D. Kotzias, G. Kiriakidis, Journal of Materiomics, **3** (1), 3 (2017)
2. M. Anijum, R. Miandad, M. Waqas, F. Gehany, M.A. Barakat, Ar. J. of Chemistry, (to be published)
3. Y.S. Song, N.I. Cho, M.H. Lee, B.Y. Kim, D.Y. Lee, J Nanosci. Nanotechnol. **16** (2), 1831 (2016)
4. C. Wang, J. Lu, L. Zhou, J. Li, J. Xu, W. Li, L. Zhang, X. Zhong, T. Wang, PLoS One. **11**(10), e0164434 (2016)
5. M. T. Amin, A. A. Alazba, and U. Manzoor, AdV in Mat. Sci. and Eng., **24,** 6 (2014)

6. C. Chen, J. Liu, P. Liu, B. Yu, Advances in Chem.Eng. and Sci., **1**, 9 (2011)
7. J. Manna, T.P. Vinod, K.Flomin, R. Jelinek, J. Colloid and Interface Sci. **460**, 113 (2015)
8. T. Zhang , T. Oyama, A. Aoshima , H. Hidaka, J. Zhao, N. Serpone J. Photochem. Photobio. A Chem **140**, 163 (2001)
9. A. Nezamzadeh-Ejhieh, H. Z. Mobarakeh, J. of Ind. and Eng.Chem., **20** (4) 1421 (2014).
10. K.-J. Kim, . B. Kreider, C.-H. PChang, C.-M. Park, H.-G. Ahn, J. Nanopart. Res. 15, 1 (2013).

Molybdenum oxide nanocolloids prepared by an external field-assisted laser ablation in water

Salvatore Spadaro[1*], *Martina* Bonsignore[1*], *Enza* Fazio[1], *Francesco* Cimino[2], *Antonio* Speciale[2], *Domenico* Trombetta[2], *Francesco* Barreca[1], *Antonina* Saija[2], *Fortunato* Neri[1]

[1]Dip. di Scienze Matematiche e Informatiche, Scienze Fisiche e Scienze della Terra (MIFT),Università di Messina, 98166 Messina, Italy
[2]Dip. di Scienze Chimiche, Biologiche, Farmaceutiche ed Ambientali, Università di Messina, Italia

Abstract. The synthesis of extremely stable molybdenum oxide nanocolloids by pulsed laser ablation was studied. This green technique ensures the formation of contaminant-free nanostructures and the absence of by-products. A focused picosecond pulsed laser beam was used to ablate a solid molybdenum target immersed in deionized water. Molybdenum oxide nearly spherical nanoparticles with dimensions within few nanometers (20-100 nm) are synthesized when the ablation processes were carried out, in water, at room temperature and 80°C. The application of an external electric field during the ablation process induces a nanostructures reorganization, as indicated by Scanning-Transmission Electron Microscopy images analysis. The ablation products were also characterized by some spectroscopic techniques: conventional UV-vis optical absorption, atomic absorption, dynamic light scattering, micro-Raman and X-ray photoelectron spectroscopies. Finally, NIH/3T3 mouse fibroblasts were used to evaluate cell viability by the sulforhodamine B assay

1 Introduction

Over the last decade, laser ablation in liquid (PLAL) is gradually becoming an irreplaceable technique to synthesize metal oxide nanostructures which represent a new type of analytical tools for biotechnology and life science, including diagnosis and therapy of human diseases [1]. Metal oxide nanostructures can act as nanoprobes for biomedical imaging, drug delivery carriers, or as therapeutic agents by themselves. Moreover, they have shown to be excellent antioxidants *in vitro* and *in vivo* models, even if their toxicity emerges in some conditions. This is determined by the different physico-chemical properties of metal oxide nanostructures synthesized using mainly chemical approaches and by the testing conditions (i.e. the experimental environment or the kind of cells). For example, Mo exists in three oxidation states and thus can readily participate in redox reactions. Mo oxide nanostructures showed to be effective against cytotoxicity and oxidative stress induced by H_2O_2 in human breast MCF-7 and fibrosarcoma HT-1080 cells [2]. On the other hand, Mo oxide nanostructures exhibit excellent antimicrobial activity against *Candida albicans* and *Aspergillus niger* and potent cytotoxicity against lung and breast cancer cells (MCF-7 and HEP G2 cell lines) [3].

It is well known that nanosecond, picoseconds and femtosecond PLAL influences the active species differently in plasma, cavitation bubbles and droplets formation which, in turn, determines different nanocolloids characteristics, especially in terms of size and size distribution. PLAL is an excellent preparation method with respect to chemical approach such as sol-gel techniques because of the possibility to prepare high pure ligand-free nanoparticles in a liquid phase ready for manipulation or processing, like incorporation into different matrices or functionalization with biomolecules [4-6]. Unfortunately, the low content of nanoparticles obtained by laser ablation have limited the application of this method as a widely used or even industrial technique. Studies of PLAL have led to the development of various strategies in order to increase nanoparticle productivity (i.e. the amount of nanoparticles fabricated per unit of time). Recently, some efficient approaches for nanostructure assembly have been presented such as electrochemically assisted PLAL, laser writing of nanopatterns in liquids, temperature, magnetic and electric-field PLAL [7]. Laser ablation in liquids with an externally applied electric field extends the traditional laser ablation in liquid approach, allowing a better control of the morphology, size, chemical composition and structure of the oxide nanostructures as well as an increase of the nanoparticles concentration [8]. Firstly, the electric field is believed to transport the charged particles produced in the plasma plume at the electrodes, in a process that resembles electrophoresis [9]. On the other hand, the electric field allows an efficient removal of the bubbles formed both in the volume and on the target. The

* Corresponding author: salspadaro@unime.it (S. Spadaro); martinabonsi@virgilio.it (M. Bonsignore)

delivery of laser pulses onto the target is more effective and so the ablation rate in liquid results to be higher by a factor of 3 (generally, about 300 mg/30 min) compared to ablation without electric field (about 100 mg/30 min).

Here, we report about the synthesis of water nanocolloids of molybdenum oxide by ablating in water a molybdenum target, using a focused picosecond pulsed laser beam. The effect of an external electric field, applied during the ablation process, was also exploited in order to check for modifications of the nanoparticles surface morphology and chemical bonding configurations. Hence, in view of potential biological applications, the molybdenum oxide nanoparticles cytotoxicity towards the fibroblast cell line NIH/3T3 has been investigated.

2 Experimental

Pulsed laser ablation of high purity (99.9%) molybdenum solid target in deionized water (H_2O) was carried out. The second harmonic (532 nm) of a laser operating at 100 KHz repetition rate with a pulse width of 6-8 picoseconds was used. The target was irradiated with a typical laser power of 2.5 W and an irradiation time of 30 min. The laser beam was focused to a spot of about 75 μm in diameter on the surface of the target with a galvanometric scanner having a telecentric objective with a focal length of 163 mm. During the ablation process two Pt electrodes were immersed in the liquid in such a way to face their basal planes (see Fig. 1). The two electrodes were polarized by applying a DC potential of 100V (the estimated electric field is 20V/cm taking into account the electrode distance of 5 cm) and, between the electrodes, the plume was left to develop during the ablation. The experiments were performed at different water temperature (from room temperature (RT) up to 80 °C) and its values were monitored both before and after the ablation. During the overall process only a moderate increase was found (less than 10°C in 30 min), compatible with the thermal energy release due to the ablation phenomena.

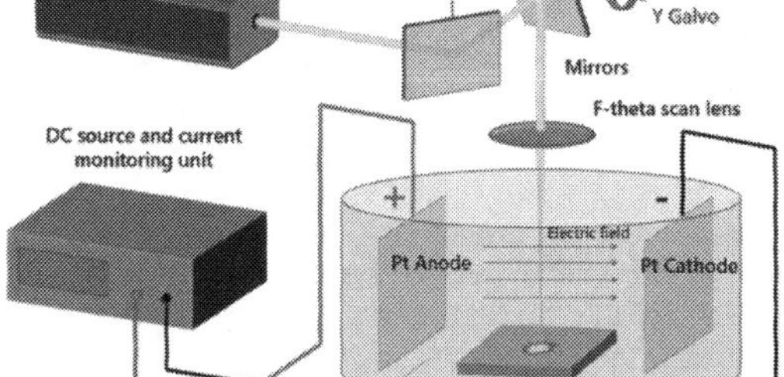

Fig. 1. Scheme of external field-assisted laser ablation setup.

In order to characterize the produced molybdenum oxide nanostructured materials by X-ray photoelectron and micro-Raman spectroscopies, the colloids were transferred on ultra-flat CaF_2 and silicon substrates by an ultrasonic spray-casting deposition method. The thickness of the films, measured by a conventional Tencor Alpha-Step 500 profilometer, is about 300 nm. The optical transmission of the molybdenum oxide colloids was analyzed by means of a Perkin–Elmer Lambda 750 UV–vis spectrometer in the 190-900 nm range, using quartz cells. Colloids zeta potential was measured with the laser Doppler method, based on the principle of electrophoretic mobility under an electric field, using the Horiba NanoParticle Analyzer SZ-100. A fraction of the colloids was deposited on nickel grids to carry out Scanning Transmission Electron Microscopy (STEM) characterization. STEM images were taken by a scanning electron microscope (ZEISS; model Merlin Gemini 2) operating at an accelerating voltage of 30 kV and at a working distance of 4 mm. The micro-Raman responses of the materials were investigated after the deposition of some drop of the water colloids on a CaF_2 substrates. Raman spectra were excited by the 532 nm radiation of a 30 mW diode laser, for an integration time of 20 sec. The backscattered radiation, collected by an Olympus BX 40 microscope optics using a 100X objective lens, was analyzed by an XploRA 600 cm^{-1} monochromator equipped with a Peltier CCD sensor. X-ray photoelectron spectroscopy (XPS) spectra were acquired using a K-Alpha system from Thermo Scientific, equipped with a monochromatic Al-Kα source (1486.6 eV) and operating in constant analyzer energy (CAE) mode with a pass energy of 20 eV for high resolution spectra and a spot size of 400 μm. Molybdenum content in colloids was determined by graphite furnace atomic absorption spectroscopy (GF-AAS) using a Varian 220/Zeeman atomic absorption spectrometer (Mulgrade, Victoria, Australia), equipped with a single-element hollow cathode lamp and a Varian PSD autosampler. For the analysis, samples were diluted 1:5000 (v:v) with 0.2% HNO_3 (AAS grade). The quantification of molybdenum was carried out using the external standard method; the reported data are the average of three determinations. The analytical method was validated according to the ICH guidelines (International Council for Harmonization of Technical Requirements for Pharmaceuticals for Human Use, 2005). The linearity was > 0.999, the precision (expressed as relative standard deviation) 0.552, and the detection limit 5.0 ng/ml. The biocompatibility of the colloids on NIH/3T3 cells was investigated using sulforhodamine B (SRB; a dye binding to cellular proteins) assay, as described by Vichai et al. [10] with some modifications. The fibroblast cell line NIH/3T3 was obtained from the American Type Culture Collection (Rockville, MD, USA), and cultured in Dulbecco's modified essential medium (DMEM) supplemented with 10% fetal bovine serum (FBS), 4 mM L-glutamine, streptomycin and penicillin. Cells were maintained in an incubator with humidified atmosphere containing 5% CO_2 at 37°C. For the SRB assay, cells were plated in 24-wells cell plates at an initial density of $6.5x10^4$ cells/well. After 24 h, semi-confluent monolayers were treated for 24 h with the colloid solutions added to the cell culture medium; control cells were exposed to the same volumes of the vehicle alone (water). Then, cells were fixed using 10% trichloroacetic acid for 1 h at 4°C. After fixation, cells were washed twice with water and incubated with SRB (0.4% w/v in 1% acetic acid) for 30 min at RT, followed by four washes with 1% acetic acid. The bound dye was solubilized in 1 mL of 10 mM Tris base solution and the absorbance was measured at 565 nm. Cell viability results are expressed as percentage of viable cells in treated samples with respect to control cells. All the experiments were performed in triplicate and repeated three times. Results are expressed as mean ± SD from three experiments and statistically analyzed by a one-way ANOVA test, followed by Tukey's HSD, using the statistical software ezANOVA (http://www.cabiatl.com/mricro/ezanova/). Differences in groups and treatments were considered significant for $P < 0.05$.

3 Results and discussion

The main processes involved in a PLAL process are: generation, transformation, and condensation of plasma plume. Nanoparticles formation is due to competition of these several processes which, in turn, affect nanoparticles properties. We expected that the high repetition rate of ps lasers is advantageous to enhance the production of nanoparticles while the morphology and composition of PLAL-generated nanomaterials can be readily controlled by changing the external environment. Specifically, we expected that the electric field assisted laser ablation leads to the generation of nanoparticles with a narrow size and a well defined distribution. For this purpose, samples were analyzed using different techniques such as STEM, XPS and optical conventional techniques. In Fig. 2 are shown electron microscopy images of the colloids prepared in water at RT and T=80°C, by applying an external DC electric field. Moreover, the STEM images of the samples prepared without applying the electric field are shown for comparison. Nanoparticles size and their distribution changes as a function of water temperature and electric field. Oblong nanostructures mixed with spherical shaped ones are obtained applying the DC electric field. Particularly, smaller nanoparticles characterize colloids obtained in hot water (c,d) respect to the ones obtained at room temperature (a,b). Moreover, for this sample, applying the external electric field, the amount of nanoparticles with size less than 15 nm increases. The optical characteristics and an exemplificative photo of the colloids are shown in Fig. 3. The optical absorption spectra in the 190-900 nm are generally not representative of the molybdenum oxide formation, since there is no well-defined absorbance band in this spectral region. However, in presence of the electric field, the broad band at higher wavelength could be indicative of the presence of positively charged structural defects, where electrons can be captured by the oxygen vacancies [11]. On the overall, the optical absorption spectra give indirectly semi-quantitative information about the nanoparticles concentrations.

The highest nanoparticles concentration is obtained at the water temperature of 80°C and applying the electric field. Mo content was determined by atomic absorption spectroscopy on the basis of three determinations. Moreover, the colloids are chemically and morphologically stable as indicated by the negative zeta potential values (see Table 1).

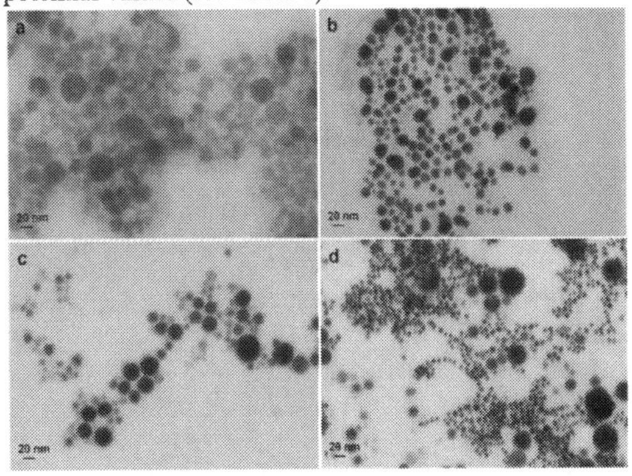

Fig. 2. STEM images of the colloids prepared by an external field-assisted laser ablation in water at (a) RT; (b) 80°C. For comparison, the images of the colloids prepared without the application of the electric field (c,d) are shown.

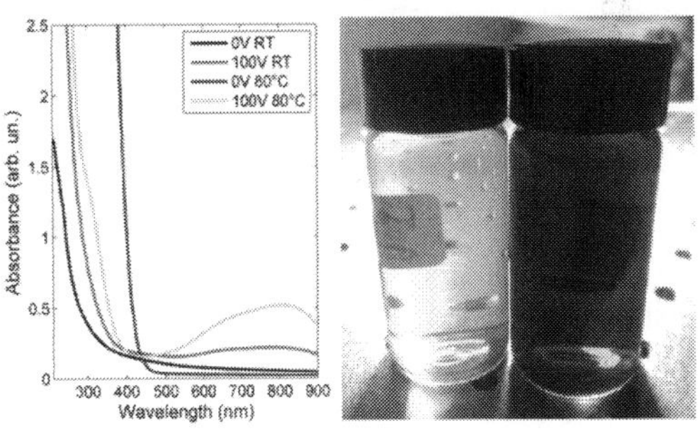

Fig. 3. UV-vis optical absorbance spectra of all samples (left). PLAL colloids exhibit different coloration: yellow without electric field and blue with electric field applied during ablation (right).

Table 1. Zeta potential values and Mo content determined by DLS and atomic spectroscopy, respectively. Data of Mo content are expressed as mean ± SD of three determinations.

Sample	Zeta potential (mV)	Mo content (µg/ml)
Mo_0V_RT	- 48.9	96.96 ± 4.8
Mo_0V_80°C	- 49.4	113.43 ± 1.72
Mo_100V_RT	- 57.5	155.75 ± 3.59
Mo_100V_80°C	- 42.5	171.39 ± 1.87

The core level spectra of the samples show the characteristic $Mo3d_{3/2}$ and $Mo3d_{5/2}$ doublet caused by spin-orbit coupling. The experimental XPS data were reproduced adopting a fitting model with three contributions ascribed to Mo different oxidation states: Mo^{6+}, Mo^{5+} and Mo^{4+}, respectively. The metallic phase is also take into account [12].

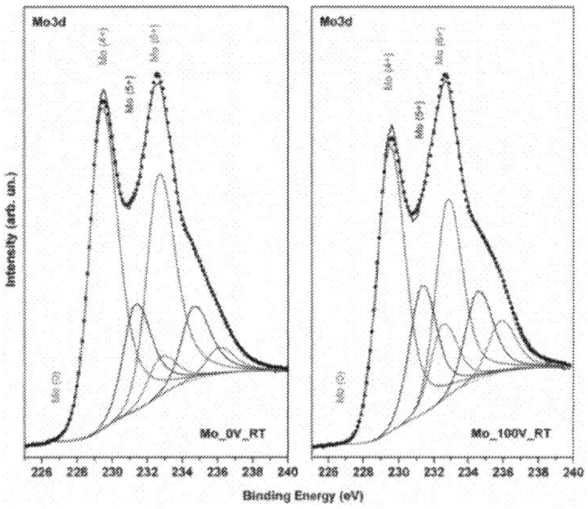

Fig. 4. XPS spectra of the samples prepared at RT but at different electric field (0V and 100V).

In Fig. 4 are shown the XPS high resolution spectra of the samples prepared at RT without and applying the electric voltage (namely, 0V and 100V). The application of the external electric field during the ablation process favors the growth of Mo^{5+} and Mo^{6+} oxidation phases with respect to the other ones. Similar trend was observed for the samples prepared at the water temperature of 80°C. The relative abundance of the identified Mo bonding configurations is: Mo^{0+} (0.4%), Mo^{4+} (68.7%), Mo^{5+} (22.7%) and Mo^{6+} (8.2%) for the sample prepared without electric field while Mo^{0+} (0 %), Mo^{4+} (58.4%), Mo^{5+} (26.2%) and Mo^{6+} (15.4%) in presence of electric field, respectively. This behaviour may be ascribed to non-equilibrium in nanoparticles charge distribution caused by excessive electrons emitted by cathode, during the plasma expansion. Nanoparticles produced by means of laser ablation in liquids are known to carry negative charge [13]. First, this fact accounts for repulsive forces between nanoparticles and provides their stability against agglomeration, even in absence of surface-active substances. On the other hand, the charge of nanoparticles and therefore potential of their interaction may change if the solution contains a source of charged particles, e.g., electrons. The increase of ion yield and ion charge state by application of an external electric field, as in our case, are most probably due to the trapping of electrons in front of the target surface, which boosts up the electron impact ionization process [14]. All these mechanisms may contribute to define Mo different oxidation states of the investigated samples observed by XPS characterization.

In Fig. 5 are shown the Raman spectra of these samples. The main Raman peaks, belonging to the α-MoO_3 orthorhombic structure, centered at about 284, 666, 820 and 995 cm^{-1} are evident as well as the peaks, associated to the Mo-O stretching mode of the monoclinic β phase, located at 776, 850 and 900 cm^{-1}. In detail, the contribution at 284 cm^{-1} represents the bending mode for double bond (Mo=O) vibration, the peak at 666 cm^{-1} is assigned to triply coordinated oxygen (Mo_3-O) stretching mode, which results from edge-shared oxygen atoms in common to three adjacent octahedral, the band at 820 cm^{-1} is ascribed to doubly coordinated oxygen (MO_2-O) stretching mode, which results from corner-sharing oxygen atoms common to two octahedral. Finally, the peak at 995 cm^{-1} is assigned to terminal oxygen (Mo^{6+}=O) stretching mode, which results from an unshared oxygen, singly coordinated oxygen [15]. The observed vibrational modes confirm the presence of several Mo-O coordinations in good agreement with XPS results.

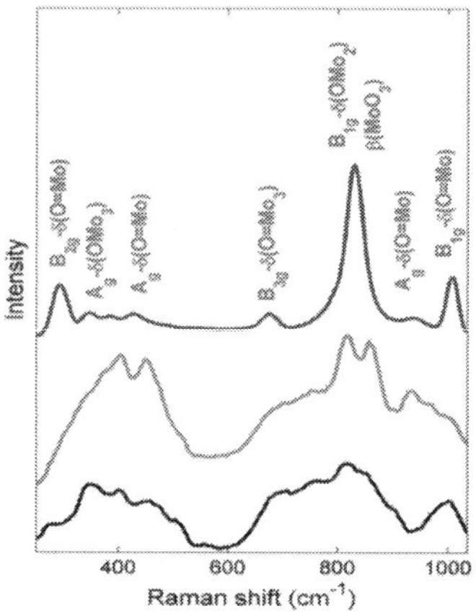

Fig. 5. Raman spectra in different points of the sample prepared at 80°C applying the electric field. Similar features were observed for all other samples.

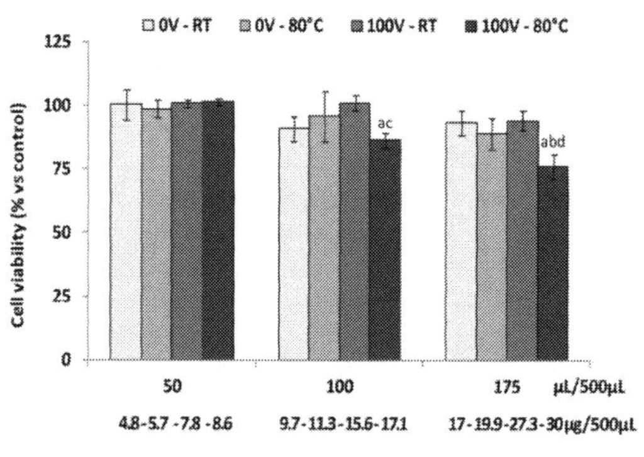

Fig. 6. SRB assay – NIH/3T3 cells were treated for 24 h with the colloid solutions added to the cell culture medium (500 μL total volume). Cultures treated with the same volumes of the vehicle alone (water) were used as controls (CTR). Cell viability data represent percentage of viable cells in treated samples vs CTR. Data are reported as function of the colloid concentration (μL/500 μL) or as Mo concentration (μg/500 μL). *a* $P < 0.05$ vs 50μL of the corresponding colloid; *b* $P < 0.05$ vs 100 μL of the corresponding colloid; *c* $P < 0.05$ vs 100μL of the colloid 100V-RT; *d* $P < 0.05$ vs 175μL of the other three colloids.

The surface morphology and the chemical bonding configurations of the synthesized nanoparticles make them an interesting material for biomedical applications. Nevertheless, to become a viable solution, it is mandatory to check their biocompatibility. Thus, the colloid solutions were tested by SRB assay on NIH/3T3 cells. Results showed that only colloids prepared at 80°C and applying an external electric field were able to affect, in a statistically significant way, cell viability, and this effect was dose-dependent. On the contrary, no cell toxicity was observed for the other colloids tested. Moreover, the size of the molybdenum oxide nanoparticles could play an important role to determine the cell response. Thus, further studies are needed for a deeper understanding of the interaction mechanisms between 3T3 cells and molybdenum oxide nanoparticles by a fine control of the compositional-structural properties of these materials.

4 Conclusion

A high efficient and "green" technique based on an external field-assisted laser ablation process is used to prepare water molybdenum oxide nanocolloids. Spherical nanoparticles with dimensions in the 20-100 nm range are obtained while a nanostructures reorganization is induced by the electric field action. The colloids prepared at 80°C and applying an external electric field show dose-dependent cytotoxic effect in NIH/3T3 fibroblasts cells. These results open new prospective for innovative biological applications, such as a potential anticancer complex that has an effect on cellular redox activity.

Authors gratefully acknowledge A.B.A.L. onlus Messina (Italy) (http://www.abalmessina.it) for the use of the XploRA Raman spectrometer.

References

1. H.S. Tuli, D. Kashyap, S.K. Bedi, P. Kumar, G. Kumar, S.S. Sandhu, Life Sci. **15**, 143 (2015).
2. M.J. Akhtar, M. Ahamed, H.A. Alhadlaq, A. Alshamsan, M.A. Majeed Khan, S. A. Alrokayan, J. Coll.and Int. Sci,. **457**, 370 (2015).
3. A. Fakhri, P.A. Nejad. J. Photochem. Photobiol B. **159**, 211 (2016)
4. E. Fazio, A.M. Mezzasalma, G. Mondio, F. Neri, R. Saija, Appl. Surf. Sci. **272**, 30 (2013).
5. E. Fazio, M. Santoro, G. Lentini, D. Franco, S. P. P. Guglielmino, F. Neri, Colloids and Surfaces A: Physicochem. Eng. Aspects **490**, 98 (2016).
6. S. Spadaro, M. Santoro, F. Barreca, A. Scala, S. Grimato, F. Neri, E. Fazio, Frontiers of Physics 13(1), 136201 (2018).
7. J. Xiao, P. Liu, C.X. Wang, G.W. Yang, Progress in Materials Science, **87**, 140–220 (2017).
8. Y. Li, O. R. Musaev, J-M- Wrobel, M.B. Kruger, J. of Laser Applications **28**, 022004 (2016).
9. G. Compagnini, M. Sinatra, P. Russo, G. C. Messina, O. Puglisi, S. Scalese, Carbon **50**, 2362 (2012).
10. V. Vichai, K. Kirtikara, Nat. Protoc. **3**, 1112 (2006).
11. H.H. Afify, S.A. Hassan, A. Abouelsayed, S.E. Demian, H.A. Zayed, Thin Solid Films, 623, 40 (2017).
12. J.-G. Choi, L.T. Thompson, Appl. Surf. Sci. **93**, 143 (1996).
13. J.-P. Sylvestre, S. Poulin, A.V. Kabashin, E. Sacher, M. Meunier, J.H.T. Luong, J. Phys. Chem. B **108**, 16864 (2004).
14. C. Pfeiffer, C. Rehbock, D. Huhn, C. Carrillo-Carrion, D. Jimenez de Aberasturi, V. Merk, S.Barcikowski, W. J. Parak J. R. Soc. Interface **11**:20130931, 1-13 (2014).
15. M. Dieterle, G. Mestl, Phys. Chem. Chem. Phys. **4**, 822 (2002).

Laser modification of graphene oxide layers

Petr Malinský[1,2*], *Anna* Macková[1,2], *Mariapompea* Cutroneo[1], *Jakub* Siegel[3], *Marie* Bohačová[4], *Kateřina* Klímová[4], *Václav* Švorčík[3] and *Zdeněk* Sofer[4]

[1]Institute of Nuclear Physics of CAS, v.v.i., Husinec - Řež 130, 250 68 Řež, Czech Republic
[2]Department of Physics, Faculty of Science, J. E. Purkinje University, Ceske mladeze 8, 400 96 Usti nad Labem, Czech Republic
[3]Department of Solid State Engineering, University of Chemistry and Technology, Technická 5, 166 28 Prague 6, Czech Republic
[4]Department of Inorganic Chemistry, University of Chemistry and Technology, Technická 5, 166 28 Prague 6, Czech Republic

Abstract. The effect of linearly polarized laser irradiation with various energy densities was successfully used for reduction of graphene oxide (GO). The ion beam analytical methods (RBS, ERDA) were used to follow the elemental composition which is expected as the consequence of GO reduction. The chemical composition analysis was accompanied by structural study showing changed functionalities in the irradiated GO foils using spectroscopy techniques including XPS, FTIR and Raman spectroscopy. The AFM was employed to identify the surface morphology and electric properties evolution were subsequently studied using standard two point method measurement. The used analytical methods report on reduction of irradiated graphene oxide on the surface and the decrease of surface resistivity as a growing function of the laser beam energy density.

1 Introduction

Graphene oxide (GO) is a single layer of oxidized graphite with structure similar to a graphene [1, 2]. The GO is described as a random distribution of oxidized areas with oxygen-containing functional groups, combined with non-oxidized regions where most of the carbon atoms preserve sp^2 hybridization typical of pure graphene sheets [2]. The basal planes of GO are decorated with epoxy and hydroxyl groups, while the GO edges are bonded with carboxylic and carbonyl groups [3, 4]. The main graphene and GO difference is that the GO is electrical insulator or low conductive because the most of carbon atoms are sp^3 hybridized and graphene is an excellent conductor because carbon is sp^2 hybridized [5]

GO reduction and deoxygenation provides an alternative path to graphene-like sheets (G) [6, 7]. GO can be reduced by low-temperature chemical reactions with reducing agents or thermal treatment, photo-irradiating processes, photo-thermal reduction or selective reduction by direct laser writing [2, 8]. The reduction way of GO is crucial because different reduction processes result in different properties of the materials or devices composed of reduced graphene oxide (RGO) [9].

The laser reduction process of the electrically and thermally insulating GO permits local formation of conductive domains with enhanced optical properties, which can be applicable in microelectronics and photo-electronics, as in organic photovoltaic devices, organic solar cells, LED, batteries, sensors etc. [10-12]. Laser direct writing is most attractive due the impurity free control of concentration carbon/oxygen ratio and simultaneously local patterning with micrometer resolution and high speed [8]. The ability of laser GO reduction allows the facile and non-toxic writing of RGO-GO-RGO patterns in various configurations to build electrical double layer capacitor or super-capacitors [13]. In [14] authors demonstrated the enhancement of antibacterial activity of laser irradiated graphene oxide in dependence on the experimental parameters (GO concentration, irradiation conditions).

In the present work the linearly polarized laser irradiation with low energy density is used on the modification of structure and composition of GO foils as the promising way to the GO deoxygenation and cleaning that does not require some chemicals or thermal annealing. The laser configuration used in our work was selected according to the previously realized surface treatment of biocompatible polymers [15, 16] and because the studies of the low energy density laser interaction with graphene based materials are not comprehensive. GO modified layers were investigated from the compositional and structural points of view, as well as the electrical properties were discussed in connection to the laser irradiation parameters used. The decline of oxygen functionalities concentration proportional to the used laser energy density is observed on the surface of GO foils after irradiation with pulsed laser that is connected with appropriate enhancement of surface conductivity.

2 Experimental

The used graphene oxide (GO) foils were prepared by graphite oxidation utilizing the permanganate oxidation method, as was reported elsewhere [17]. Briefly, the

* Corresponding author: malinsky@ujf.cas.cz

graphite (99.9995 %, Alfa Aesar) was subjected to oxidation under the action of potassium permanganate in sulfuric and phosphoric acid. Afterwards, the reaction mixture was quenched in ice and the formed GO was separated by centrifugation. The GO foils with final thickness ~10 µm were prepared by suction filtration using polycarbonate track etched membrane as a template (pore size 400 nm). The assembled membrane with GO layer was dried at 80 °C for 48 hours and subsequently the polycarbonate support was removed. The different degree of oxidation between the pristine GO foils used in the present experiment and pristine GO foils used in our previous work [17] originate from different method of graphene foil preparation and subsequent processing.

The prepared GO foils were irradiated in ambient atmosphere using KrF laser with following experimental conditions. The laser operated at wavelength 248 nm with pulse duration 20 ns, 6000 pulses and repetition rate 10 Hz was used. The incoming laser beam was perpendicular to the sample surface and laser energy density ranging from 6 mJ.cm^{-2} to 12 mJ.cm^{-2}.

The elemental composition of pristine and laser irradiated GO foils were analyzed using Rutherford Backscattering Spectrometry (RBS) and Elastic Recoil Detection Analysis (ERDA). The RBS and ERDA spectra were measured using a beam of 2.0 MeV He^+ ions. An Ultra-Ortec PIPS detector recorded He^+ ions backscattered at a laboratory scattering angle of 170°. The primary beam during ERDA comes at an angle of 75° with respect to the foil surface normal and hydrogen atoms recoiled at a scattering angle of 30° were registered with the detector covered by a 12 µm Mylar foil. The typical ion current used during the RBS and ERDA analysis was 5 nA. To reduce effects of the sample degradation during the RBS analysis, several particular spectra were measured on different beam spots and the final spectrum was obtained by summing the individual spectra. RBS and ERDA spectra were evaluated using SIMNRA code [18].

The surface morphology changes were followed using AFM morphological study before and after the laser irradiation. The AFM NTEGRA Spectra from NT-MDT was used in a tapping mode. The Average Roughness (R_a) of the profile height deviations from the center plane was calculated.

The current-voltage (I-V) characteristics of the pristine and irradiated GO foils were studied by the standard 2-point method utilizing the Keithley 6221 current source and Keithley 2128A nano-voltmeter. The Au contacts (50 nm thick) were sputtered on the surface of GO foils for the electrical resistance measurement.

Raman spectroscopy, ATR-FTIR and XPS measurements were used to get information about the GO foil structural and elemental modification caused by laser irradiation. An inVia Raman microscope (Renishaw, England) operated in backscattering geometry with a CCD detector. An Nd-YAG laser (532 nm, 50 mW) with 50× magnification objective was used for measurements. Instrument calibration was achieved with a silicon reference which yields a peak position at 520 cm^{-1}. No more than 5% of the total 50 mW laser power and beam spot diameter in order of µm was used in order not to damage the sample. Samples were drop-casted on silicon wafer from an isopropanol suspension (1 mg.mL^{-1}) in order to perform the measurements. Attenuated total reflectance Fourier transform infrared spectroscopy (ATR-FTIR) measurements were performed on a NICOLET iS50R FTIR spectrometer (Thermo Scientific, USA). A Diamond ATR crystal and a DTGS detector were used for all measurements, which were carried out in the range of 4000–400 cm^{-1} at a resolution of 4 cm^{-1}. High resolution X-ray photoelectron spectroscopy (XPS) was performed with an ESCAProbeP (Omicron Nanotechnology Ltd, Germany) spectrometer using a monochromatic aluminium X-ray radiation source (1486.7 eV). A wide-scan survey of all elements was performed, with subsequent high-resolution scans of the C 1s and O 1s core level spectra. For the evaluation of the carbon-to-oxygen (C/O) ratios from the survey spectra relative sensitivity factors were used. Prior to measurement, samples were applied onto conductive carbon tape. To eliminate sample charging during measurement (1–5 V) the electron gun was used and acquisition time of all XPS measurement were reduced to minimize the possibility of the surface damage by X-ray.

3 Results

RBS and ERDA analysis served for elemental depth profiling which results are presented in Figures 1 and 2 showing oxygen and hydrogen depth profiles in pristine and laser irradiated GO foils. Concentrations and relative abundance of C, O, H on the surface of analyzed graphene oxide foils are summarized in Table 1. In the as-prepared GO foils we detected additional S (~1.2 at. %) and Mn (~0.15 at. %) that originate from the synthesis procedure of GO. In the Table 1 one can see the significant decrease of O and H concentration on the irradiated GO surface that is proportional to the increasing laser energy density. The C/O ratio on the GO foil surface increases from 3.7 for pristine sample to 6.3 for GO treated with energy density 12 mJ.cm^{-2}. The surface C/H ratio is growing function of the laser energy density also and the values are varying from 8.5 to 12.6. The H/O ratio in the pristine GO foil is ~0.43 and after the laser modification using energy density 6 mJ.cm^{-2} slightly increases to value ~0.53, this value then stays similar for all other used laser energy densities. Hydrogen is in GO generally present in the form of hydroxyl functional groups (the H/O ratio is ~1), carboxylic groups (H/O ~0.5) or epoxides (H/O ~0) [19] and we can assumed the presence of carboxyl groups in the used GO structure. The O and H concentration increases with increasing depth and the concentration of these two elements achieves the pristine sample values in the depth about 180 nm (see Fig. 1 and 2). It should be said, that the information depth of RBS/ERDA methods using 2 MeV He^+ ions is about 1 µm in GO and that the absorption coefficient for 248 nm light in GO is about α~5x10^{-4} cm^{-1} [20]. The light penetration depth in materials is defined from Beer-Lambert law as the inverse of the absorption coefficient and for 248 nm laser the light penetration depth in GO is about ~200 nm that is in agreement with the depth of laser modified layer as was determined using RBS and ERDA. The increasing O and H concentration with increasing depth is connected to the decreasing light intensity according to the Beer-Lambert law. The C concentration enhancement on the surface of irradiated samples suggests the possible deoxygenation and dehydrogenation that is more pronounced for the increasing laser density and with respect to the following Raman, FTIR and XPS analysis we probably can talk about two important processes that occur during laser irradiation of GO; about reduction and cleaning, when highly oxidized fragments with low molecular weight are separated from the bulk GO sample [21].

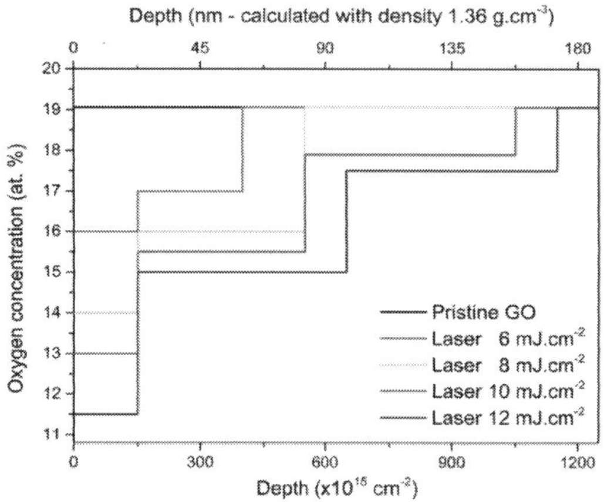

Fig. 1. The oxygen depth profiles in pristine and irradiated GO foils using various laser energy density.

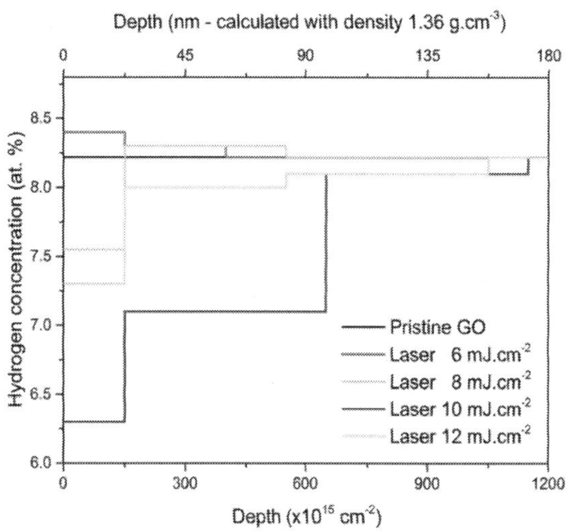

Fig. 2. The hydrogen depth profile in the GO foils before and after laser irradiation.

Table 1. The elemental composition and elemental ratios on the surface of used GO foils determined using RBS and ERDA.

Sample	The GO foils composition [at. %]					
	C	O	H	C/O	C/H	H/O
Pristine	70.0±1.6	19.1±0.5	8.2±0.5	3.7	8.5	0.43
6 mJ/cm^{-2}	72.8±1.6	16.0±0.5	8.4±0.5	4.7	8.7	0.53
8 mJ/cm^{-2}	75.7±1.6	14.1±0.5	7.5±0.5	5.4	10.1	0.53
10 mJ/cm^{-2}	76.9±1.6	13.0±0.5	7.3±0.6	5.9	10.5	0.56
12 mJ/cm^{-2}	79.4±1.6	11.5±0.5	6.3±0.6	6.9	12.6	0.55

Raman spectroscopy is a powerful and widely used method to study structure, disorder and defects in graphene-based materials [22, 23]. Fig. 3 represents the typical Raman spectra obtained for un-irradiated and irradiated GO. The spectra were normalized to the G peak. The two major peaks corresponding to the D (1350 cm^{-1}) and G (1590 cm^{-1}) bands and the multi-peak in the area between 2500—3500 cm^{-2} that consist from three peaks (2D ~ 2700 cm^{-2}, D+D' ~2900 cm^{-2} and 2D' ~3200 cm^{-2}) are observed in all Raman spectra (see Figure 3) and were described in literature [24, 25]. The D band is defect induced Raman feature that originates from structural defects, edge effects and appearance of dangling sp^2 bonds breaking symmetry [26]. The G band is associated with the in plane stretching motion of sp^2 bonded carbon atoms in graphene [19, 27, 28] and the broad multi-peak with wavenumber above 2400 cm^{-2} is typical for multilayer graphene structures [29]. No significant changes are observed in the peaks shape and position after laser irradiation in Figure 3. The I_D/I_G ratio exhibits value around 0.9 with only slight change for all used laser energy densities and all peaks keep the position. These findings can be interpreted that the laser irradiation with energy density in range from 6 to 12 mJ.cm^{-2} does not lead to the gradual changes in the crystallinity and to the used GO foil structure disorder and amorphization increase. Moreover, from the unimpressive decrease of the I_D/I_G ratio can be assumed that the carbon ring cluster sizes slightly increase [30]. This finding is also in agreement with the XPS results, where only slight decrease of C-O bond and slight increase of C-C bond can be observed. The similar results, but with higher degree of GO reduction, can be observed after light (He, H) ion irradiation with MeV energy and fluencies up to ion fluence 1.0x10 ^{14} cm^{-2} [17]. The ion irradiation with higher fluencies and heavy ions leads to the more pronounced sample amorphization and distortion of the sp^2 carbon bond angle, giving origin to non-six membered rings [30]. Higher ion irradiation fluencies caused the considerable number of collision cascades that especially in case of heavy ions with significant nuclear stopping lead to atoms knocking and creation of large defected zones and vacancies. Contrary the laser irradiation with low energy densities do not provide significant heating of substrate and leads only to the photo-chemical elimination of oxygen spices without sp^2 healing [8].

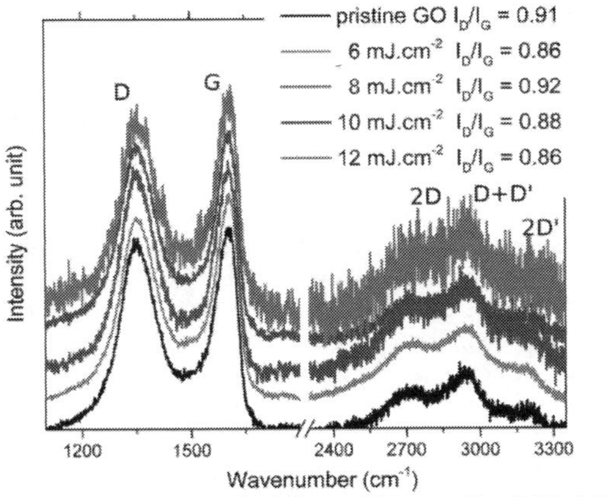

Fig. 3. Raman spectra of pristine and laser irradiated GO foils normalized to the G peak intensity.

The ATR-FTIR spectra (Fig. 4) of pristine GO foil show characteristic bands confirming presence of functional groups, such as O–H (3347 cm^{-1}), C=O (1726 cm^{-1}), C–OH (1176 cm^{-1}), and C–O (1223 and 1053 cm^{-1}) [30]. A broad characteristic peak appears at 3000–3600 cm^{-1} which represents the O—H stretching of hydroxyl and carboxyl groups. The oxygen in the pristine sample can be confirmed mainly by the absorption bands at ~1726 cm^{-1} (C=O stretching vibration) and by the presence of a broad band at 1000–1200 cm^{-1} attributed to C–O alkoxy and epoxy stretching vibration [27, 32, 33]. The presence of the absorption peak at ~1630 cm^{-1} can be attributed to the stretching vibration of C=C of graphene skeleton [34]. The presence of hydroxyl groups results in the formation of hydrogen bond which contributes to the hydrophilic nature of graphene oxide. The C=O/C=C intensity ratios are presented in the Fig. 4, where one can see the slight decrease of intensity of this peak compared to the C=C one. In a similar way slightly decrease the C-O intensity peak can be observed in the Fig. 4. The O-H intensity decrease is more pronounced compare to the carbon to oxygen bonds peak that indicate the C-OH and COOH groups removing after laser irradiation using even so low laser energy densities as in our experiment [35].

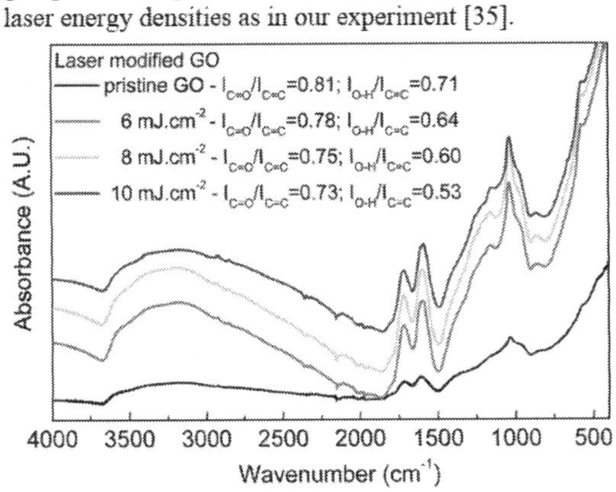

Fig. 4. The ATR-FTIR spectra and attached C-O/C=C and C=O/C=C intensity ratios measured on the pristine and laser irradiated GO foils.

The carbon chemical states in the surface of the GO foils before and after laser irradiation were investigated by means of C 1s peak deconvolution measured using XPS (Fig. 5.). One can indicate at the Fig. 5 presence of the five different carbon bonding states; C=C (284.4) and C-C (285.4) in the sp^2, C-O (286.3), C=O (288.0) in the carbonyl C and O-C=O (289.0) in the carboxylate C [9]. In the pristine GO foil dominates the C=C peak in sp^2 hybridized carbon, whereas the C-O peak that in GO commonly dominates is in the current sample reduced [17]. These indicate the significant degree of reduction of as-prepared GO foil that can be connected with warm drying during GO foil preparation. After laser irradiation there is evident slight increase of intensity of the C-C peak. This C-C mildly increase continues with the increasing laser energy density simultaneously with only C-O peak slight decrease. This phenomenon demonstrates the laconic reduction of the oxygen containing groups and creation of the carbon groups with increasing laser energy density, as has also been found using ATR-FTIR analysis. The RBS shows more pronounced deoxygenation compared to the XPS, but it should be emphasized that deconvolution of high resolution XPS C 1s peak provides information about oxygen bonded to carbon and that RBS detect oxygen in all chemical states in analysed matrix. It is therefore obvious that after laser irradiation mainly the groups obtaining the O-H bonds are removed as predicted by FTIR and can be concluded that the removal of the highly oxidized debris adsorbed on the surface of GO, so called cleaning, is more likely than GO reduction [21, 36].

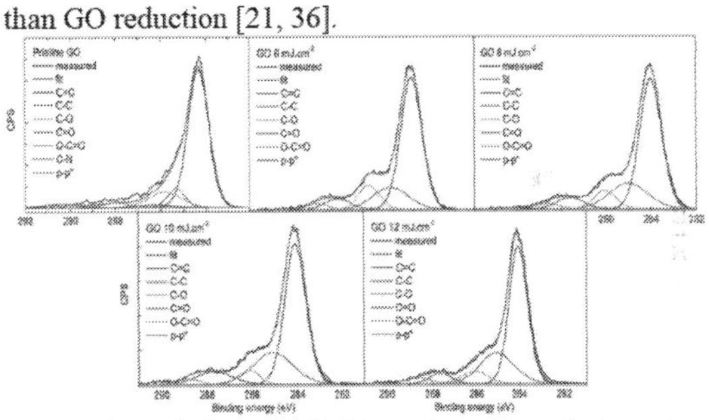

Fig. 5. The evolution of XPS C 1s peak correspondingly to the laser energy density enhancement.

The GO foil surface morphology investigated using AFM is presented in Fig. 6. All samples show a very rough and irregular surface with a maximal height in order of μm. It is evident, that roughness maximal height value decreases after ion irradiation from 2.1 μm for the pristine GO foil to 1.2 μm for irradiated GO foil with the laser energy density 12 mJ.cm^{-2}. As well as, the Average Roughness (R_A), shown in the figure, decreases with the increasing laser energy density. This roughness decline, caused by laser irradiation was referred previously using of 663 nm/80 mW power laser and can be connected to etching of graphene oxide porous surface by laser in the ambient atmosphere [4].

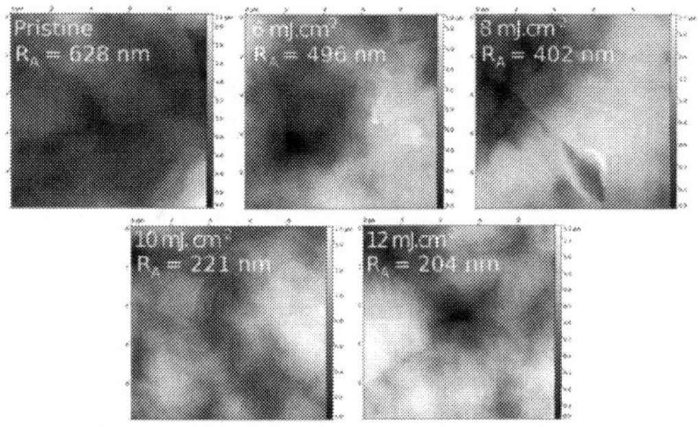

Fig. 6. The AFM images of rough graphene oxide foils surface before and after laser irradiation. The measured Average Roughness (R_A) is attached within.

To gain insight into the electric properties changes after the laser irradiation, the current-voltage (I-V) characteristic was measured using standard two points method. The current was applied in the range from -1500 nA to 1500 nA. The pristine GO foil exhibited a clear non-linear slope of I-V curve that remains after laser irradiation with energy density up to 10 mJ.cm^{-2}. With subsequent increasing of laser energy density the I-V characteristic curve becomes the ohmic one. The laser irradiation of pristine GO foil with energy density 6—12 mJ.cm^{-2} leads to a current response enhancement compared to the pristine one that is more pronounced for higher laser energy density. The changes of the electric properties are attributed to the decrease of the oxygenation and hydrogenation level of irradiated GO as proved using RBS simultaneously with slight chemical reduction confirmed using XPS and ATR-FTIR. There is known that the electric resistance of GO is related to the C/O ratio enhancement and overall layer quality and density of defects [37].

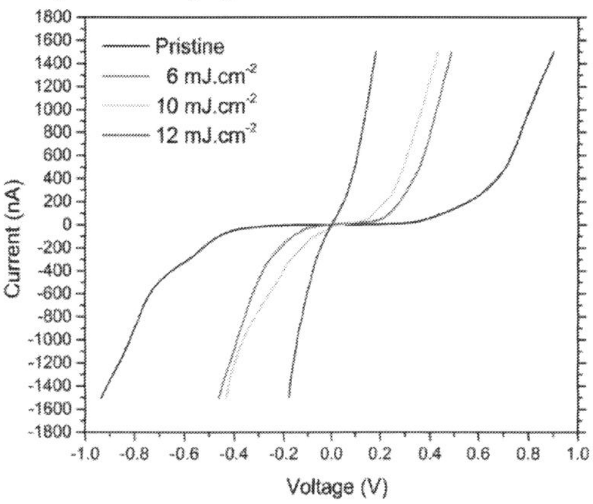

Fig. 7. The current-voltage characteristic of untreated and laser irradiated GO foils measured using standard two point methods.

4 Conclusions

We have demonstrated the graphene oxide foil properties modification induced using 248 nm pulsed linearly polarized KrF laser with energy density range from 6 mJ.cm^{-2} to 12 mJ.cm^{-2}. The untreated and laser irradiated GO foils were analysed using different analytical methods. From the ion beam analysis (RBS, ERDA) is apparent that the laser irradiation with increasing energy density leads to deoxygenation and dehydrogenation of the GO foils surface accompanied with C concentration enhancement. XPS and FTIR testify only about fine decrease of oxygen functionalities in irradiated GO and the Raman spectroscopy show also only very slight change in structure disordering, crystallinity and amorphization after laser irradiation. These on-going changes on the GO foils surface layers, along with decline of surface roughness and significant growth of surface electrical conductivity can testify about the moderate GO surface reduction accompanied by the significantly stripping of highly oxidized low molecular weight fragments and that intercalated water molecules are diminishing from the GO surface.

Finally, it can be concluded that removal of oxygen and hydrogen atoms from graphene oxide sheets using pulsed linearly polarized laser with the correctly selected energy density is a simple, rapid and efficient method for GO electric conductivity enhancement and that the influence of laser irradiation is comparable with light ion irradiation using energy in order of MeV. Difference is depth of the modified layer. The laser irradiation influenced layer is in order of tens of nm and light ions range with MeV energy is in order of µm.

The work was supported by Czech Science Foudation (GACR No. 16-05167S and GACR No. 17-05421S). Z.S., M.B. and K.K. were supported by specific university research (MSMT No. 20-SVV/2017). This work was created with the financial support of the Neuron Foundation for science support. This work was supported by the project Advanced Functional Nano-robots (reg. No. CZ.02.1.01/0.0/0.0/15_003/0000444 financed by the EFRR). The research has been carried out at the CANAM (Centre of Accelerators and Nuclear Analytical Methods) infrastructure LM 2015056. This publication was supported by OP RDE, MEYS, Czech Republic under the project CANAM OP, CZ.02.1.01/0.0/0.0/16_013/0001812.

References

1. D. A. Sokolov, Investigation of graphene formation from graphite and silicon graphite, Thesis, Georgia Institute of Technology 2013.
2. S. F. Spano, G. Isgro, P. Russo, M. E. Fragala, G. Compagnini, App. Phys. A **117**, 19 (2014).
3. S. Stankovich, D. A Dikin, R. D. Piner, K. M. Kohlhaas, A. Kleinhammes, Y. Jia, Y. Wu, S. T Nguyen, R. S Ruoff, Carbon **45**, 1558 (2007)
4. Y. Zhou, Q. Bao, B. Varghese, L. A. L. Tang, Ch. K. Tan, Ch. H. Sow, K. P. Log, Adv. Mater. **21**, 1 (2009).
5. Ch. Punckt, F. Muckel, S. Wolff, I. A. Aksay, C. A. Chavarin, G. Bacher, W. Mertin, Appl. Phys. Lett. **102**, 023114 (2013).
6. G. Eda, Y. Y. Lin, S. Miller, C. W. Chen, F. W. Su, M. Chhowala, Appl. Phys. Lett. **92**, 233305 (2008).

7. S. Stankovich, D. A Dikin, G. H. B. Dommett, K. M. Kohlhaas. E. J. Zimney, E. A. Stach, R. D. Piner, S. T. Nguyen, R. S Ruoff, Nature **442**, 282 (2006).
8. I. I. Bobrinetskiy, A. V. Emelianov, S. A. Smagulova, I. A. Komarov, N. Otero, P. M. Romero, Materials Letters 187, 20 (2017).
9. S. Pei, H. M. Cheng, Carbon **50**, 3210 (2012).
10. R. Trusovas, K. Ratautas, G. Raciukaitis, J. Barkauskas, I. Snakeviciene, G. Niaura, R. Mazeikiene, Carbon **52**, 574 (2013).
11. E. Kymakis, K. Savva, M. M. Stylianakis, C. Fotakis, E. Stratakis, Adv. Mater. **23**, 2742 (2013).
12. E. Kymakis, C. Petridis, T. D. Anthopoulos, E. Stratakis, IEEE Journal of Selected Topics in Quantum Electronics **20**, 106 (2014).
13. W. Gao, N. Singh, L. Song, Z. Liu, A. L. M. Reddy, L. Ci, R. Vajtai, Q. Zhang, B. Wei, P. M. Ajayan, Nature Technology **6**, 496 (2011).
14. M. A. Buccheri, D. D'Angelo, S. Scalese, S. F. Spano, S. Filice, E. Fazio, G. Compagnini, M. Zimbone, M. V. Brundo, R. Pecoaro, A. Alba, F. Sinatra, G. Rappazzo, V. Privitera , Nanotechnology **27**, 245704 (2016).
15. P. Slepicka, O. Nedela, J. Siegel, R. Krajcar, Z. Kolska, V. Svorcik, Express Polym. Lett., **8,** 459 (2014).
16. J. Siegel, J. Heitz, V. Svorcik, Surf. Coat. Technol., **206,** 517-521 (2011).
17. P. Malinsky, A. Mackova, R. Miksova, H. Kovacikova, M. Cutroneo, J. Luxa, B. Strochova, Z. Sofer, Phys. Chem. Chem. Phys. **19**, 10282 (2017).
18. M. Mayer, SIMNRA User's guide, Report IPP 9/113, Max-Planck-Institut fur Plasmaphysik, Garching, Germany, 1997.
19. O. Jankovsky, P. Simek, J. Luxa, D. Sedmidubsky, I. Tomandl, A. Mackova, R. Miksova, P. Malinsky, M. Pumera, Z. Sofer, ChemPlusChem **80**, 1399 (2015)
20. A. P. del Pino, E. Gyorgy, C. Logofatu, A. Duta, J. Phys. D: Appl. Phys. **46,** 505309 (2013).
21. H. R. Thomas, S. P. Day. W. E. Woodruff, C. Valles, R. J. Young, I. A. Kinloch, G. W. Morley, J. V. Hanna,N. R. Wilson, J. P. Rourke, Chem. Mater. **25**, 3580 (2013)
22. P. T. Araujo, M. Terrones, M. S. Dresselhaus, Materials Today **15**, 98 (2012)
23. G. Sobon, J. Sotor, J. Jagiello, R. Kozinski, M. Zdrojek, M. Holdynski, P. Paletko, J. Boguslawski, L. Lipinska, K. M. Abramski, Opt. Express **20**, 19463 (2012)
24. R. Beams, L. G. Cancado, L. Novotny, J. Phys. Condens. Matter. 27, 082002 (2015)
25. X. Diez-Betriu, S. Alvarez-Garcia, C. Botas, P. Alvarez, J. Sanchez-Marcos, C. Prieto, R. Menendez, A. de Andres, J. Mater.Chem. C **1**, 6905(2013)
26. M. A. Pimenta, G. Dresselhas, M. S. Dresselhaus, L. G. Cancado, A. Jorio, R. saito, Phys. Chem. Chem. Phys. **9**, 1276 (2007).
27. M. M. Viana, M. C. F. S. Lima, J. C. Forsythe, V. S. Gangoli, M. Cho, Y. Cheng, G. G. Silva, M. S. Wong, V. Caliman, J. Braz. Chem. Soc. **26**, 978 (2015).
28. E. D. Dikio, F. T. Thema, A. M. Farah, N. D. Shoto, Mater. Sci.-Poland **31**, 59 (2013).
29. A. Ambrosi, A. Bonanni, Z. Sofer, J. S. Cross, M. Pumera, Chem. Eur. J. **17**, 10763 (2011).
30. D. L. Babtista, F. C. Zawislak, Diam. Relat. Mater. **13**, 1971 (2004).
31. L. S. Bai, X. M. Gao, X. Zhang, F. F. Sun, N. Ma, Tetrahedron Lett. **55**, 4545 (2014).
32. K. Hareesh, R. P. Joshi, B. Shateesh, K. Asokan, D. Kanjilal, D. J. Late, S. S. Dahiwale, et al., J. Phys. D: Appl. Phys. **48**, 365105 (2015)
33. O. Jankovský, P. Šimek, K. Klimová, D. Sedmidubský, S. Matějková, M. Pumera, Z. Sofer, Nanoscale **6**, 6065 (2014)
34. N. Cao, Y. Zhang, J. Nanomater. **2015,** 16812 (2015).
35. V. L. Borgne, H. Bazi, T. Hayashui, Y. A. Kim, M. Endo, M. A.E. Khakani, Carbon **77**, 857 (2014).
36. X. Chen, B. Chen, Environ. Sci. Technol. **50**, 8568 (2016).
37. D. A. Sokolov, C. M. Rouleau, D. B. Geohegan, T. M. Orlando, Carbon **53**, 81 (2013).

Raman investigation of laser-induced structural defects of graphite oxide films

Valentino Romano[1*], *Lorenzo* Torrisi[1], *Mariapompea* Cutroneo[2], *Vladimir* Havranek[2] and *Giovanna* D'Angelo[1*]

[1]Dipartimento di Science Matematiche ed Informatiche, Scienze Fisiche e Scienze della Terra, Università di Messina, Viale F. Stagno d'Alcontres 31, S. Agata, 98166 Messina, Italy.
[2]Nuclear Physics Institute, CAS, 25068 Rez, Czech Republic.

Abstract. Since the beginning of intensive studies on graphene and graphitic materials, Raman spectroscopy has always been used as a characterisation technique. This is due to two main reasons: the non-destructive nature of this experimental technique and its ability to distinguish between the plethora of existing carbon materials. One of the most challenging research activities concerns the production of graphene microcircuits. To address this issue, a possible strategy is to directly reduce and pattern graphite oxide (GO) film by laser irradiation. The objective of this study is to evaluate the laser irradiation-induced structural changes on thin GO films by using Micro-Raman spectroscopy. We used as a source a Nd:YAG laser (1064 nm) and different laser fluences: 15 J/cm^2, 7.5 J/cm^2 and 5 J/cm^2. We have analyzed the modifications of the main Raman contributions of these graphitic materials: the *D* band (defect induced band), the *G* band (band due to sp^2 hybridized carbon atoms) and the *2D* band (D band overtone). In particular, we found out that our figure of merit (FOM) parameters, i.e. the intensity ratio I_D/I_G (for the *D* band and G band) and I_{2D}/I_G (for the *2D* band and *G* band), change with the laser fluences, revealing a different effect induced by the laser irradiation. The best results are found in the sample irradiated with 5 J/cm^2, suggesting that higher fluences do not lead to better results.

1 Introduction

Raman spectroscopy is a vibrational technique that is extremely sensitive to geometric structure and bonding within molecules. This sensitivity is useful for the study of the various allotropes of carbon (i.e. diamond, carbon nanotubes [1], buckminster fullerenes [2], graphene nanoribbons [3], etc.) where the different forms differ only in the relative position of their carbon atoms and the nature of the chemical bonding. It is well known that graphene is characterised by properties of great potential: breaking strength (≈ 40 N/m), room temperature thermal conductivity (≈ 5000 W m^{-1} K^{-1}), Young modulus (≈ 1 TPa) and so on [4]. Furthermore the electron mobility of graphene (1−2×10^{5}cm^2/Vs) is a hundred times higher than that of silicon [5], so graphene is a material of great potential for use in micro-electronics [6, 7]. However the manufacture of integrated circuits needs a transfer printing method on small size of graphene films [8]. Therefore, the quest for a suitable production method arised. One possibility is to reduce graphite oxide (GO). This material is obtained by oxidizing graphite by means of strong oxidizing compounds (like $KMnO_4$) [9]. This procedure induces profound changes in the structure of graphite because epoxy and hydroxyl groups occur in the basal plane, while carboxylic groups arise at the edges [10, 11]. These defects of the graphitic structure are responsible for the insulating or semiconducting properties of GO. The reduction of graphite oxide can be accomplished by means of several methods like dehydration, thermal reduction and UV-assisted photocatalysis [12]. Another option is to locally reduce and pattern graphite oxides (GO) by using laser ablation [13]. In fact high intensity laser beams (of the order of 10^{10} W/cm^2) induce change in the composition (due to hydrogen, oxygen and other impurities vaporisation), foil thickness, mechanical properties and morphology [14]. These are due to the high electron density and the high thermal conductivity of the carbon layers. In this context, it is also very important to understand and quantify the consequences of disorder induced by pulsed laser technologies on the electronic profile of GO. To evaluate the role of disorder, defects can be introduced in graphite by ion or laser irradiation [15]. In this work we present the results of a Micro – Raman spectroscopy investigation on GO samples irradiated by laser pulses with different fluences.

2 The Raman spectra of pristine and defected graphene

The Raman spectrum of pristine graphene is characterised by two main contributions: the *G* band (1582 cm^{-1}) and

* Corresponding authors: gdangelo@unime.it , vromano@unime.it

the *2D* band (around 2700 cm^{-1}) [16]. (Fig. 1, top, shows an example of this spectrum, taken from [18]).

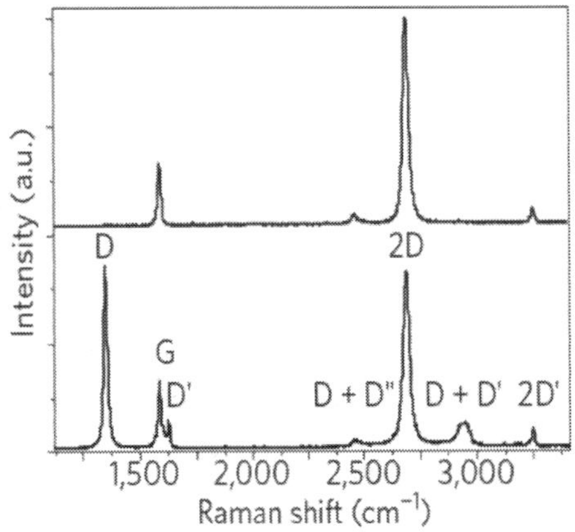

Fig. 1. Raman spectra of pristine (top) and defected (down) graphene [18].

The *G* peak corresponds to the E_{2g} phonon at the Brillouin zone center. The *D* peak is due to the breathing modes of sp_2 rings and requires a defect for its activation by double resonance. The *2D* peak is the second order of the *D* peak. This is a single peak in monolayer graphene, whereas it splits in multi-layer graphene, reflecting the evolution of the band structure. The *2D* peak is always seen, even when no *D* peak is present, since no defects are required for the activation of two phonons with the same momentum, one backscattered from the other [17]. In defected graphene and graphitic materials many other contributions occur at: 1350 cm^{-1} (*D* band), 1620 cm^{-1} (*D'* band), 2445 cm^{-1} (*D+D"* band), 2925 cm^{-1} (*D+D'* band) and 3160 cm^{-1} (*2D'* band) [17]. A thorough analysis of the Raman spectra provides many information about the structure in graphene samples. For example the Raman spectrum of pristine graphene shows a single *G* peak and a sharp *2D* peak, roughly four times more intense than the *G* peak. It was observed that in few layers graphene the *2D* region is characterised by a multi-band (centered at about 2700 cm^{-1}) and the insurgence of the medium frequency defect-induced contributions [18]. Thus, the intensity ratio for the *D*, *G* and *2D* Raman bands is often used as a criterion to assess the graphene quality. Minimum of I_D/I_G and maximum of I_{2D}/I_G are the indicators of the highest quality of graphene and, in the study case, provide information on the best conditions for the GO-to-graphene reduction. In fact, the first ratio characterizes the in-plane crystallite sizes [17], while I_{2D}/I_G is used to distinguish a single layer, from bilayer from few (less than 5) layers [18].

3 Experimental

3.1 Material synthesis

The graphite oxide (GO) foil was prepared by graphite oxidation adopting the permanganate oxidation method. Graphite (99.995% and 2-15 μm) was purchased by Alfa Aesar and mixed with H_2SO_4 (96 wt%) and H_3PO_4 (85 wt%). After that, $KMnO_4$ was added and the reaction mixture was heated to 50°C for 12 hours. Afterwards, the reaction mixture was quenched in ice (400 g) with hydrogen peroxide (30 wt%), and the formed graphite oxide was separated by centrifugation. Nucleopore 0.45 μm was used to obtain GO foil by suction filtration of graphite oxide suspension [19].

3.2 Laser treatment

A Nd:YAG laser operating at fundamental frequency (1064 nm), 3 ns pulse duration, 300 mJ maximum pulse energy, 1 mm^2 spot size focalization, about 10^{10} W/cm^2 intensity, operating in single pulse or up to 10 Hz repetition rate was employed in our investigations. Laser was employed to irradiate GO thin foils placed in the sample chamber using window and an external lens with 50 cm focal length for the light focalization. In order to reduce the ablation effect producing perforation at higher energies, the laser pulse energy was reduced to 150 mJ, 75 mJ and 50 mJ, employed with a fluence of 15 J/cm^2, 7.5 J/cm^2 and 5 J/cm^2, respectively.

The laser irradiation was monitored in-situ using a Joule-meter for the laser pulse energy control, an ion collector (IC) connected in time-of-flight configuration to a fast oscilloscope and a mass quadrupole spectrometer (MQS) for the gas degassing diagnostics [20]. The last two techniques have detected the presence of carbon and oxygen ions emitted from the laser ablation process. A scheme of the experimental set up is reported in Fig. 2.

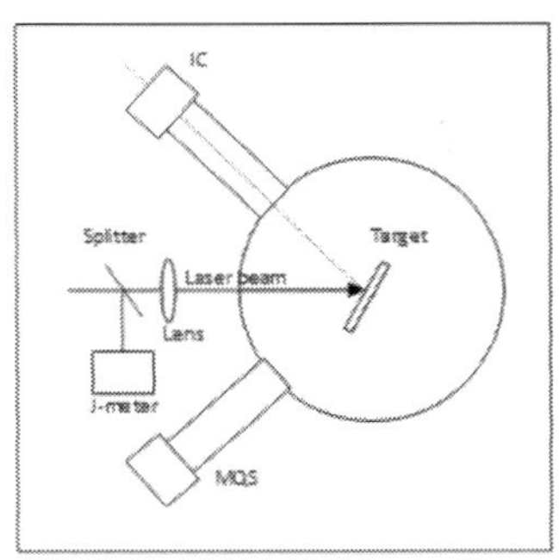

Fig. 2. Experimental set-up for graphite oxide irradiation.

3.3 Characterisation

Raman microscopy was used to investigate the structural transformations in GO samples. Raman measurements were performed with 532 nm excitation light by using the LabRam HR800 (Horiba Jobin Yvon) spectrometer with the 1800 gr/mm grating and a CCD camera cooled with liquid nitrogen. The laser power used was 1 mW and the spectra were taken with the 50x objective lens. The overall integration time was 100 s.

 https://doi.org/10.1051/epjconf/201816704011

4 Results and discussion

The Tandetron 4130 MC at the Tandetron laboratory of the Nuclear Physics Institute (Czech Republic) is compact, multipurpose linear electrostatic tandem accelerator system produced by High Voltage Engineering Europa (HVEE). The beam passes through an analysing magnet (providing mass and charge state of incoming ions) and ions are driven towards the beam line where the RBS-ERDA systems are located.

Rutherford Backscattering Spectrometry (RBS) and Elastic Recoil Detection Analysis (ERDA) were employed to evaluate the composition of GO foil. A probe of He^+ ions with energy of 1.9 MeV and incident angle of 0° with respect to the foil surface normal, was used for the measurements. During RBS analysis, an Ultra-Ortec PIPS detector detected He^+ ions backscattered at a laboratory scattering angle of 160°. In ERDA spectrum, hydrogen atoms that recoiled at a scattering angle of 35° were registered using a detector covered by a 4 μm Polypropylene foil to avoid the detection of the primary ions scattering. The ion current was 5.4 nA. RBS and ERDA spectra were evaluated using SIMNRA code [21]. During RBS and ERD analyses, several spectra are collected on different beam spots to avoid degradation effects of the sample and the final spectrum is obtained as the sum of the single collected spectra. The elemental compositions of GO obtained by RBS and ERD analyses are shown in Fig. 3 and presented in Tab. I.

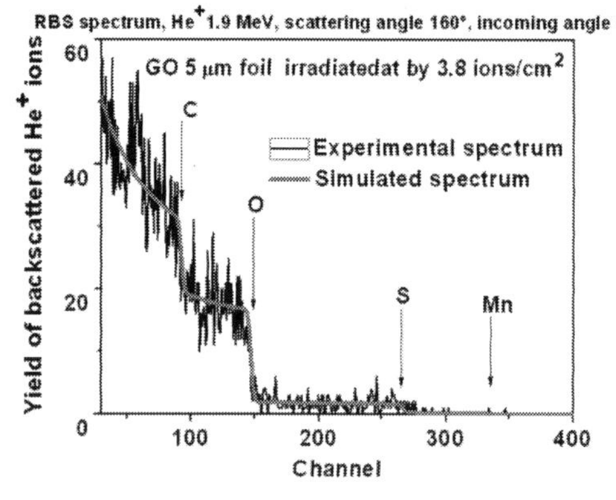

Fig. 3. Spectrum related GO foil obtained by Rutherford backscattering spectrometry.

Tab. I Elemental compositions of GO foils.

Target/ Thickness	Foils composition (at %)				
	H	C	O	S	Mn
GO 5 μm	18.4	47.6	33.1	0.75	0.046

In this study we irradiated the GO samples using different laser fluences (5 J/cm^2, 7.5 J/cm^2 and 15 J/cm^2). The irradiation procedure was performed in air atmosphere. During laser irradiation IC and MQS have monitored a high desorption of oxygen gas from the GO target, together emission of carbon and hydrogen, such as ions and neutral atoms. The ablated samples were investigated by means of Raman microscopy, the related spectra are shown and compared with the reference spectrum in Fig. 4.

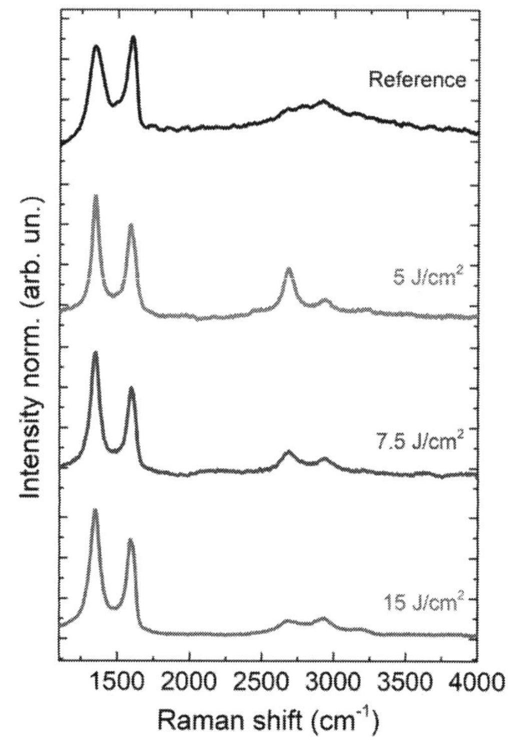

Fig. 4. Raman spectra of GO reference sample (black line), 5 J/cm^2 irradiated area (red line), 7.5 J/cm^2 irradiated area (blue line), 15 J/cm^2 irradiated area (green line). The spectra are normalized with respect to the *G* peak intensity.

These spectra can be separated in two parts: the medium-Raman-shift region (below 2000 cm^{-1}) and the high-Raman-shift region (above 2000 cm^{-1}). The first one is characterised by the *D* and *G* peaks, while the high Raman shift region is characterised by a broad bump, centred around the *2D* peak position, which is due to the appearance of the defect-induced contributions discussed in one of the previous sections. As concerns the reference spectrum, the *D* and *G* peaks have similar intensities while the bump around 2700 cm^{-1} has a low intensity. This can mean that the sample is made up by several graphene layers and that defects (e. g. structural, chemical, etc) are present. The spectra in the irradiated samples have a *D* intensity which is higher than the *G* one, which means that a more defects are present, although the *2D* peak is more intense than in the reference case.

To make these observations more quantitative, we used the intensity ratios I_D/I_G and I_{2D}/I_G as figures of merit (FOM) of our study. Minimum of I_D/I_G and maximum of I_{2D}/I_G underline that the ablation procedure leads to a better quality material. The values of the intensity ratios computed for the spectra of Fig. 4 are reported in Tab. II.

Tab. II Computed values of the intensity ratios I_D/I_G and I_{2D}/I_G for the Raman spectra shown in Fig. 4.

Sample	I_D/I_G	I_{2D}/I_G
Reference	0.96	0.68
5 J/cm^2	1.14	0.78
7.5 J/cm^2	1.17	0.69
15 J/cm^2	1.17	0.69

These values confirm that the irradiated GO samples are characterized by a higher number of defects, with respect to the reference sample, but they still are made up by a lower number of layers. We believe that there are two possible reasons for this observation. On the one hand, since the ablation procedure was carried out in air atmosphere, further oxide species can be created (CO, CO_2) inducing more structural and chemical defects [16]. On the other hand the irradiation causes the removal of several layers (so the I_{2D}/I_G ratio increases) but more edge defects are produced. These effects cause the formation of smaller crystalline regions, because edge defects and chemically different species limit the dimensions of the crystallites. The Raman results can be used to get some insight in this problem. In fact Tuinstra and Koenig demonstrated that the intensity ratio I_D/I_G is inversely proportional to the in plane crystallites size and proposed a formula which is valid for the 514.5 nm excitation wavelength only [22, 23]. A general equation, that can be used for every wavelength, is given in [17]:

$$L_a\,(nm) = (2.4 \times 10^{-10})\,\lambda^4_{laser}\,(I_D/I_G)^{-1} \qquad (1)$$

where L_a denotes the crystallite size and λ_{laser} the laser excitation wavelength. In Fig. 5 we report the values that we computed on the basis of our Raman spectra.

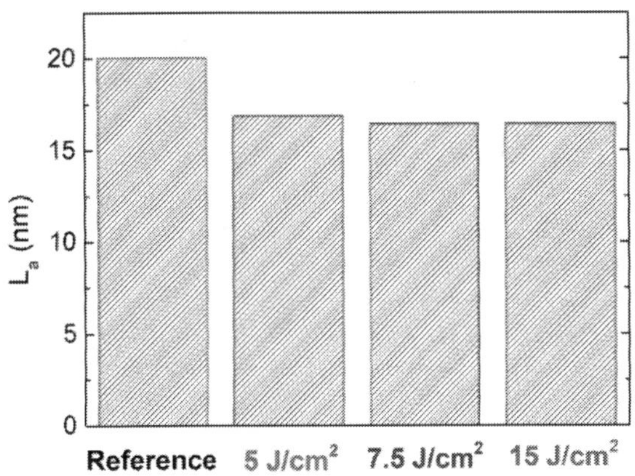

Fig. 5. Computed values of the crystallite size L_a in nanometers.

These results confirm our hypothesis: the higher the laser fluence (i.e. the damage) to the GO sample is, the smaller the crystallites size become.

5 Conclusions and perspectives

To the best of our knowledge, this is the first time that Raman analysis of nano-second laser ablation of graphite oxide is investigated. Up to know only pico and femto-second laser pulse sources have been used. Here we show that some improvements are still observed in this time domain. The best results concern the region irradiated with a 5 J/cm^2 laser fluence. However the ablation procedure leads to both a greater I_{2D}/I_G intensity ratio and an increase in the I_D/I_G intensity ratio. We believe that this is due to the formation of carbon-oxide species, since the laser irradiation has been performed in air atmosphere. These results lead us to the idea of performing the same kind of study in inert atmosphere (trying to optimise this procedure). In this way we could be able to demonstrate if nano-second laser ablation can compete with pico and femto-second sources.

This work was supported by the "Research and Mobility" project of Messina University No. 74893496, scientifically coordinated by Professor L. Torrisi. The research has been also carried out at CANAM infrastructure LM2015056 and Czech Science Foundation (GACR No. 16-05167S).

References

1. M. S. Dresselhaus, G. Dresselhaus, R. Saito, A. Jorio, *Phys. Rep.* **409**, 47-49 (2005).
2. D. S. Bethune, G. Meijer, W. C. Tang, H. J. Rosen, W. G. Golden, H. Seki, C. A. brown, M. S. de Vries, *Chem. Phys. Lett.* **179**, 181-186 (1991).
3. L. Xie, H. Wang, C. Jin, X. Wang, L. Jiao, K. Suenaga, H. Dai, *J. Am. Chem. Soc.* **133**, 10394-10397 (2011).
4. A. K. Geim, *Science* **324**, 1530-1534 (2009).
5. C. Jacoboni, C. Canali, G. Ottaviani, A. A. Quaranta, Sol.-St. Elect., **20**, 77-89 (1977).
6. H. B. Heersche, P. Jarillo-Herrero, J. B. Oostinga, L. M. K. Vandersypen, A. F. Morpurgo, *Nat.* **446**, 56-59 (2007).
7. G. Eda, M. Chhowalla, *Nano Lett.* **9**, 814-818 (2009).
8. M. J. Allen, V. C. Tung, L. Gomez, Z. Xu, L. M. Chen, K. S. Nelson, C. W. Zhou, R. B. Kaner, Y. Yang, *Adv. Mater.* **21**, 2098-2102 (2009).
9. J. Chen, B. Yao, C. Li, G. Shi, *Carbon* **64**, 225-229 (2013).
10. G. Eda , M. Chhowalla, *Adv. Mater.* **22**, 2392-2415 (2010).
11. D. R. Dreyer , S. Park , C. W. Bielawski , R. S. Ruoff, *Chem. Soc. Rev.* **39**, 228-240 (2010).
12. F. Bonaccorso, A. Lombardo, T. Hasan, Z. Sun, L. Colombo, A. C. Ferrari, *Materials Today* **15**, 564-589 (2012).
13. Y. Zhang, L. Guo, S. Wei, Y. He, H. Xia, Q. Chen, H. Sun, F. Xiao, *Nano Today* **5**, 15-20 (2010).
14. P. Kumar, K. S. Subrahmanyam, and C. N. R. Rao, *Int. J. Nanosci.* **10**, 559-566 (2011).

 https://doi.org/10.1051/epjconf/201816704011

15. L. Tapaszto, G. Dobrik, P. Nemes-Incze, G. Vertesy, P. Lambin, L. P. Biro, *Phys. Rev. B* **78**, 23 (2008).
16. R. Trusovas, K. Ratautas, G. Račiukaitis, J. Barkauskas, I. Stankevičienė, G. Niaura, & R. Mažeikienė, *Carbon* **7**, 574-582 (2012).
17. M. A. Pimenta, G. Dresselhaus, M. S. Dresselhaus, L. G. Cancado, A. Jorio, R. Saito, *Phys. Chem. Chem. Phys.* **9**, 1276-1290 (2007).
18. A. C. Ferrari, D. M. Basko, *Nat. Nano.* **8**, 235-246 (2013).
19. Ch. K. Chua and M. Pumera, *Small* **11**, 1266–1272 (2015).
20. L. Torrisi, S. Gammino, A. M. Mezzasalma, A. M. Visco, J. Badziak, P. Parys, J. Wolowski, E. Woryna, J. Krasa, L. Laska, M. pfeifer, K. Rohlena and F.P. Boody, *Appl. Surf. Sci.* **227**, 164-174 (2004).
21. M. Mayer, SIMNRA version 6.06, Max-Planck-Institut fur\Plasmaphysik Garching, Germany, 2006, available at: http://home.mpcdf.mpg.de/~mam/].
22. F. Tuinstra, J. L. Koenig, J. *Chem. Phys.* **53**, 3 (1970).
23. F. Tuinstra, J. L. Koenig, J. *Compos. Mater.* **4**, 4 (1970).

EPJ Web of Conferences **167**, 05001 (2018) https://doi.org/10.1051/epjconf/201816705001
PPLA 2017

Aneutronic pB-Reaction in Magnetized Inertial Confined Plasma of Laser-Accelerated Particles

Sergey Gus'kov[1,2*] *Philipp* Korneev[1,2]

[1]LPI RAS, 119991, Moscow Leninskii pr. 53, Russian Federation
[2]NRNU MEPhI, 115409 Moscow, Kashirskoe sh. 31, Russian Federation

Abstract. A brief review on studies aimed at initiating a neutronless nuclear reaction between protons and boron nuclei is presented. The least energy-cost of a currently known approach to ignition of *pB* reaction is proposed using a petawatt laser for producing a magnetized plasma of laser-accelerated ions and electrons as well as using a terawatt laser for inertial confinement of such a plasma.

1 Introduction

Promising aneutronic nuclear power based on the reaction of a proton with boron, which gives three α particles with a total energy of about $\varepsilon \approx 8.6$ MeV, is currently under active discussion. As compared to other aneutronic reactions, *pB* reaction has the largest cross section in the region of the main resonance, $\sigma_{max} \approx 1$ b (at the energy of the relative motion of the nuclei $E_{rel} \approx 675$ keV), which is nevertheless much smaller than the maximum cross section for the reaction of deuterium with tritium (6 b at $E_{rel} \approx 60$ keV). The initiation of the *pB* reaction in the plasma is experimentally studied with the use of laser methods of creation of a high-temperature plasma and a proton beam [1–4]. The maximum detected yield of α particles in the *pB* reaction was about 10^9 [3]. Due to a rather small cross section, the initiation of an energetically profitable *pB* reaction requires a significant initial energy expense much higher as compared to the fusion reaction between hydrogen isotopes. Nevertheless, several approaches were proposed to solve this problem. They are (i) the use of an external magnetic field for the quasi stationary interaction between the colliding proton beam and boron nuclei [5], (ii) initiation of the reaction in a degenerate plasma with a superhigh density of about 10^5 g/cm^3 [6], (iii) action of laser-accelerated proton beams on a preliminarily compressed boron-containing target [7], and (iv) the collision between ponderomotively accelerated thin layers of matter [8]. All these approaches are devoted to reduce the effect of Coulomb breaking of protons and boron nuclei initially accelerated to MeV energies.

Here we discuss the least energy-cost of currently known approach to ignition of *pB* reaction, which has been proposed in [9]. It is based on the inertial confinement of a magnetized plasma of laser-accelerated charged particles. The features of this approach are, first, the creation of the magnetized *pB* plasma in the inner cavity of the target in the form of a spherical or cylindrical shell owing to the irradiation of its inner surface by a picosecond petawatt laser pulse and, second, the compression and confinement of such a plasma owing to the irradiation of the outer surface of the target by a nanosecond terawatt laser pulse. In order to match the processes of formation and compression of the plasma, a plasma producing pulse should be applied at the final stage of acceleration of the target under the action of a compressive pulse. It is proposed to use targets similar to the spherical [10, 11] or cylindrical [12] target designed for fast ignition [10, 13] of the preliminarily compressed plasma at the inertial confinement (see Fig. 1). In both cases, compression occurs in the radial direction. The spherical target is compressed along the conical guiding channel and the cylindrical target with the shape of a coil is compressed along the end walls. Holes in the vertex of the conical channel and end walls serve for the introduction of the plasma-producing laser pulse.

*Corresponding author: guskov@sci.lebedev.ru

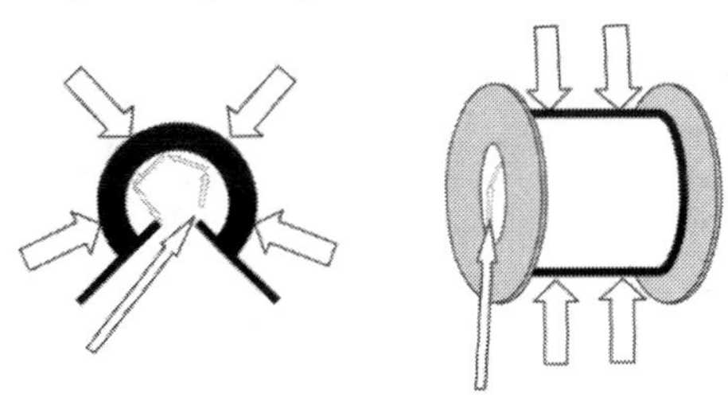

Fig. 1. Spherical (left) and cylindrical (right) targets for the compression and inertial confinement of the magnetized plasma formed at the grazing incidence of a relativistic laser pulse on the surface of the inner cavity of the target. The thin arrows correspond to the beams of petawatt laser, the thick ones corresponds to the beams of terawatt laser.

The creation of the magnetized plasma of laser accelerated ions and electrons was predicted in [14] on the basis of the PIC simulation of the interaction of a relativistic laser pulse with an intensity of 10^{19} W/cm^2

with the inner surface of a quasicylindrical target. Radiation tangentially incident on the curved surface of the target generates surface currents of fast electrons. They generate a flow of fast ions and discharge currents of thermal electrons. In the region between the two currents, a high magnetic field is generated and is carried together with the plasma of laser-accelerated and thermal particles into the cavity of the target; i.e., the ablation of the magnetized plasma inside the target occurs. The numerical calculations in [14] indicate the formation of a plasma with MeV particles, where the density exceeds the critical plasma density (10^{21} cm^{-3}). The time of existence of the magnetic field with a strength of several hundred megagauss was much longer than the duration of the pulse. The kinetic energy of the shell at the time of the end of the creation of the plasma should be enough to compress the plasma to a density necessary for breakeven, i.e., for the release of the energy in the pB reactions equal to the energy of the plasma. This requirement determines the relations between the parameters of the plasma-producing and compressive pulses.

2 Magnetized plasma of laser-accelerated Mev ions

To numerically simulate the formation of the pB plasma a model substance as a mixture of three parts of hydrogen and one part of boron was used in PIC simulations. The shape of the target was specified by the expression $r(\theta) = r_0(1 + \theta\delta r/2\pi r_0)$, $\theta \subset (0, 2\pi)$. The angle $\theta = 0$ corresponds to the upward direction along the y axis in Fig. 2 and the parameters were $\delta r = 50$ μm and $r_0 = 98$ μm. The two-dimensional geometry, limited computation region, and limited duration of calculation (9.5 ps) were taken according to the available computational capabilities. A helical shape of the target was used for the two-dimensional simulation of the penetration of a laser pulse into the cavity of the target after its tangential incidence and multiple reflections from the walls of the target, which is the most important phenomena here.

The laser pulse had an intensity of 10^{19} W/cm^2 and a duration of 1.7 ps, the wavelength of radiation was 1.06 μm, and the radius of the beam was 20 μm. The electron density $n_e = 4.5\cdot 10^{22}$ cm^{-3} corresponded to 45 critical densities. The spatial and time steps were 20 nm and 0.07 fs, respectively. Figure 2 shows the spatial distributions of electron fluxes and the strength of the magnetic field at the time of the end of the laser pulse t = 1.7 ps. The initial shape of the target is also presented and the trajectory of the laser pulse, as well as the region filled with the plasma, are qualitatively shown.

The characteristic plasma density in the region of interaction of the laser beam with the plasma near the boundary of the cavity was $4\cdot 10^{21}$ cm^{-3}. A more detailed description of the spatial distributions of the parameters of the plasma and magnetic field, as well as of their time evolution, was given in [9]. Figure 3 shows the energy distributions of electrons at t = 1.7 and 9.5 ps, as well as of protons and boron nuclei at t = 9.5 ps in the inner region of the target. The energy of protons and boron nuclei in the central region of the target is several MeV, and the strength of the field, which is "frozen" in the plasma at such a high energy of particles, is about 100 MG. In such a field, the Larmor radii of boron nuclei, protons, and 3-MeV α particles from the pB reaction are about 10, 3, and 50 μm, respectively.

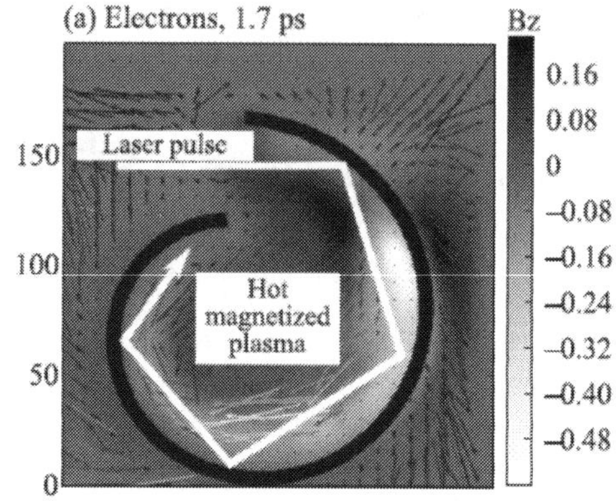

Fig. 2. Spatial distribution of the electron flows and magnetic field at a time of 1.7 ps. The momenta of the electrons are demonstrated by arrows in units of m_ec, the strengths of the magnetic field Bz in the direction perpendicular to the figure plane are shown as the background of the figure according to the color scale at the same times.

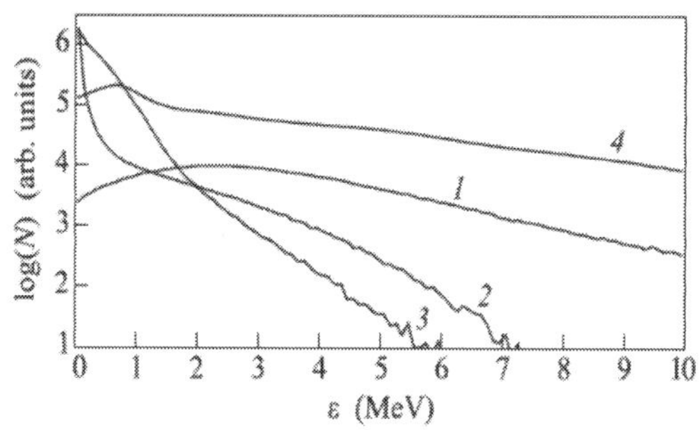

Fig. 3. Energy spectra of electrons at a time of 1.7 ps (line 1) and 9.5 ps (line 2) as well as of protons (line 3) and boron nuclei (line 4) at a time 9.5 ps in the inner region of the target inside a distance of 50 μm from the center of the target.

3 Lawson's criterion for pB-plasma

In view of the possibility of the producing of the magnetized plasma with MeV light nuclei in the inner cavity of the target, we estimate the conditions for the initiation of the pB reaction in the following formulation of the problem. A plasma-producing laser pulse with the intensity I_1, energy E_1, and duration τ_1 is guided through a transport channel to the inner surface of the spherical pB shell in the final stage of its acceleration under the action of a laser pulse with the respective parameters I_2, E_2, and τ_2 ($\tau_2 >> \tau_1$). The velocity of flight of the shell to the center u and the radius of its inner surface R_0 at the beginning of the action of the plasma-forming pulse correspond to the condition under which the time of the convergence of the shell to the center $t_h = R_0/u$ is much larger than the duration of the plasma-producing pulse: $t_h >> \tau_1$. Although the frequencies of collisions involving laser-accelerated particles are very low, as will be shown below, the electron–ion and, the more so, ion–ion energy relaxation processes occur in the time of the compression of the plasma. Then, under the assumptions that (i) the average energy of particles of the plasma formed at the interaction between flows of laser-accelerated and thermal particles is $\varepsilon \approx 500$ keV (corresponds to the kinetic energy of 675 keV of nuclei in the center-of-mass system), (ii) the number of protons is equal to the number of boron nuclei, and (iii) the plasma is quasineutral, the ratio of the energy released in nuclear reactions to the energy of the plasma is $G \approx \varepsilon_r n_c \sigma_{max} v t_r / 8\varepsilon \approx 2.1 \cdot 10^{-15} n_c t_r$, where $v = (2\varepsilon/m_p)^{1/2}$ is the relative velocity of nuclei, which is approximately equal to the velocity of protons; m_p is the mass of the proton; n_c is the density of nuclei in the plasma in cm^{-3}; and t_r is the time of the confinement of the plasma in seconds. Consequently, the condition of breakeven, when the requirement $G>1$ (Lawson' criterion) is satisfied for the pB reaction has the form

$$n_c t_r > 4.5 \cdot 10^{14}\ \text{s/cm}^3 \qquad (1)$$

4 Compression and inertial confinement of magnetized pB-plasma of laser accelerated particles

Now the requirements to the degree of compression and the duration of confinement of magnetized 500-MeV pB-plasma under the action of terawatt laser pulse will be determined which are necessary to achieve Lawson's criterion (1). The degree of compression of the plasma can be estimated in the adiabatic approximation from the condition of the transfer of the kinetic energy of the shell to the internal energy of the plasma decelerating it

$$\frac{n_c}{n_0} \equiv \left(\frac{R_0}{R_c}\right)^3 \approx \left(\frac{\eta_2 E_2}{\eta_1 E_1}\right)^{1/(\gamma-1)} \qquad (2)$$

where

$$n_0 = \frac{3\eta_1 E_1}{32\pi R_0^3 \varepsilon} \approx 3.7 \cdot 10^{11} \frac{\eta_1 E_1}{R_0^3} \qquad (3)$$

is the density of nuclei in the plasma produced by the petawatt pulse; n_c and R_c are the density of the compressed plasma and corresponding radius of the shell, respectively; η_1 is the fraction of the energy of the plasma-producing laser pulse spent on the formation of the plasma; η_2 is the hydrodynamic efficiency of energy transfer [15], which is the ratio of the kinetic energy of the shell to the energy of the compressive laser pulse; and E_1 and R_0 are measured in joules and seconds, respectively. In view of Lawson's criterion (1), the time of confinement of the plasma in the shell with the mass M_c is given by the expression

$$t_r \approx R_c \left[\frac{(\gamma-1)\eta_1 E_1}{M_c}\right]^{-1/2} \approx 1.73 \frac{R_0}{u} \qquad (4)$$

The substitution of Eqs. (2)–(4) into Eq. (1) gives

$$\frac{\eta_1 E_1}{R_0^2 u}\left(\frac{\eta_2 E_2}{\eta_1 E_1}\right)^{3/2} > 7 \cdot 10^2 \qquad (5)$$

where u is measured in centimeters per second.

For further analysis, we introduce a number of additional constraints. To simplify the focusing of the plasma-producing beam with the radius R_1 on the inner surface of the shell at the final stage of its acceleration, we set $R_0/R_1 = 5$. Below, for the stability of compression, we take the relatively small ratio $R_0/R_c = 5$, which corresponds to the degree of compression $n_c/n_0 = 125$. In this case, according to Eq. (2) at $\gamma=5/3$, $\eta_2 E_2/\eta_1 E_1 = 25$. Taking into account the accepted relations, we estimate the initial radius of the cavity as $R_0 \approx 1.8 \cdot 10^3 u^2/\eta_1 I_1$ according to Eq. (5). Taking $\eta_1 = 1$, the velocity $u = 300$ km/s as a typical value for inertial confinement fusion, and the intensity $I_1 = 10^{20}$ W/cm^2, we estimated the radius of the cavity as $R_0 \approx 180$ μm, the time of compression as $t_h \approx 600$ ps, and the other parameters of the plasma-forming pulse as $R_1 = 36$ μm, $\tau_1 \approx 15$ ps, and $E_1 = 50$ kJ. According to Eqs. (3) and (2), the initial and final densities of nuclei in the pB plasma are, respectively, $n_0 \approx 3.4 \cdot 10^{21}$ cm^{-3} and $n_c \approx 4.2 \cdot 10^{23}$ cm^{-3} (larger than the density of solid boron by a factor 3) and the mass of the plasma is $m \approx 1.6$ μg. The calculation shows that the ion–ion energy relaxation time (about 2 ps) is much shorter than the plasma compression time, whereas the electron–ion energy relaxation time (about 1 ns) is close to this time.

Energy losses to cyclotron radiation in a time of about 1 ns in a field of 100 MG are no more than a few percent. The above analysis was performed in the approximation of full magnetization of the electron heat conduction to the shell compressing the plasma. For this reason, the field does not appear in the expression for the ignition criterion. A more accurate calculation should certainly include the dependence on the field caused by the field dependence of energy losses owing to the electron heat conduction. The parameters of the laser pulse required for the compression of a spherical inertial confinement fusion target were chosen such that approximately 80% of the mass of the shell is evaporated in the process of acceleration. In this case, the hydrodynamic efficiency is close to its maximum possible value of about 30%. Then, with the use of the parameters $\eta_2 = 0.3$, $\eta_2 E_2/E_1 = 25$, and $E_1 = 50$ kJ, the energy of the compressive pulse should be $E_2 = 4.1$ MJ. The final mass and velocity of the spherical shell at laser ablation are (see, e.g., [15])

$$u_c \approx 2u_{av} = 2\left(\frac{2}{3}\right)^{1/3} \alpha^{1/2} V_s \quad (6)$$

$$M_c = M_0\left(1 - \frac{2}{3}\alpha^{1/2}\right) \quad (7)$$

respectively. Here, u_{av} is the average velocity of the shell, $V_s = (I_2/3\rho_{cr})^{1/3}$ is the speed of sound in the fully ionized plasma of the evaporated part of the target, and $\alpha = \rho_{cr}R_s/\rho_s\Delta_s$ is the dimensionless acceleration parameter, where R_s, Δ_s, and ρ_s are the initial radius, thickness, and density of the material of the shell, in microns, respectively; and $\rho_{cr} \approx 3.6\cdot10^{-3}\ \lambda_{\mu m}^{-2}$ is the critical plasma density, where $\lambda_{\mu m}$ is the wavelength of radiation in microns.

For the pulse of the third harmonic of the Nd laser for which the critical density is 0.032 g/cm^3 and for the target with $\rho_0 \approx 2.34$ g/cm^3 at $M_c/M_0 = 0.2$ and $u_c = 3\cdot10^7$ cm/s, estimates by Eqs. (6) and (7) give $\alpha \approx 1$, $R_s/\Delta_s \approx 70$, and $I_2 \approx 5\cdot10^{13}$ W/cm^2. This value of the intensity I_2 at the duration of laser pulse equal to the time of shell implosion $\tau_2 = 2R_s/u_c$ corresponds to the radius of the target $R_s = 0.46$ cm. At the aspect ratio $R_s/\Delta_s \approx 70$, this radius corresponds to the thickness of the shell of 65 μm, the final mass of the shell $M_c \approx 8$ mg, and the duration of the pulse $\tau_2 = 30$ ns. Summarizing the parameters of the compressive pulse of the third harmonic of the Nd laser are $I_2 \approx 5\cdot10^{13}$ W/cm^2, $E_2 = 4.1$MJ, and $\tau_2 = 30$ ns. The total energy $E_2 + E_1 = 4.15$ MJ is more than five times higher than the energy of the laser pulse necessary for the compression and ignition of the traditional inertial confinement fusion target with the DT fuel (see, e.g., [16]). Such a high energy is due primarily to a significantly higher energy of the shell required for the compression of the *pB* plasma preliminarily heated by a relativistic pulse.

The ratio of the final mass of the shell to the mass of the compressed plasma M_c/m is about $5\cdot10^3$. The total gain G_t equal to the ratio of the energy released in *pB* reactions to the total spent energy $E_1 + E_2 \approx E_2$ at the propagation of combustion to the compressed material of the shell can be larger than $G_t \approx M_cE_1/mE_2 \approx 60$, which is sufficient for the positive energy cycle including the efficiency of the laser. However, the initiation of a combustion wave with allowance for the features of the problem under consideration requires separate analysis.

4 Conclusion

The presented justification of the fundamental possibility of the least energy-cost of currently known approach to ignition of *pB* ignition of the magnetized *pB* plasma produced in the inner cavity of the compressed inertial confinement fusion target implies a further study of the dependences of the parameters of the *pB* plasma, the strength of the spontaneous magnetic field, and the time of conservation of the magnetized plasma state on the parameters of the plasma-producing pulse.

This work was supported by the Russian Foundation for Basic Research (project no. 16-52-50019 YaF).

References

1. V. S. Belyaev, A. P. Matafonov, V. I. Vinogradov, V. Krainov, P. Lisitsa, V. S. Roussetski, A. S. Ignatyev, and G. N. Andrianov, Phys. Rev. E **72**, 026406 (2005)
2. C. Labaune, S. Depierreux, C. Goyon, G. Loisel, V. Yahia, and J. Rafelski, Nat. Commun. **4**, 2506 (2013)
3. A. Picciotto, D. Margarone, A. Velyhan, et al. (Collab.), Phys. Rev. **4**, 031030 (2014).
4. C. Baccou, S. Depierreux, V. Yahia, C. Neuville, C. Goyon, R. De Angelis, F. Consoli, J.E. Ducret, G. Boutoux, G. Loisel, J. Rafelski and C. Labaune, Laser Part. Beams **33**, 117 (2015)
5. N. Rostoker, M. Binderbauer, and H. Monkhorst, Science **278**, 1419 (1997).
6. S. Son and N. J. Fisch, Phys. Lett. A **329**, 76 (2004).
7. S. Eliezer and J. M. Martinez-Val, Laser Part. Beams **16**, 581 (1998).
8. H. Hora, G. H. Miley, M. Ghoranneviss, B. Malekynia, and N. Azizi, Opt. Commun. **283**, 4124 (2009).
9. S.Yu. Gus'kov, Ph. A. Korneev, JETP Letters **104**, 1 (2016)
10. N. G. Basov, S. Yu. Gus'kov, and L. P. Feoktistov, J. Sov. Laser Res. **13**, 396 (1992).
11. P. A. Norreys, R. Allott, R. J. Clarke, J. Collier, D. Neely, and S. J. Rose, Phys. Plasmas **7**, 3721 (2000).
12. S. Yu. Gus'kov, Laser Part. Beams **23**, 255 (2005).
13. M. Tabak, J. M. Hammer, M. E. Glinsky, W. L. Kruer, S. C. Wilks, J. Woodworth, M. E. Campbell, M. D. Perry, and R. J. Mason, Phys. Plasmas **1**, 1626 (1994).
14. Ph. Korneev, E. d'Humieres, and V. Tikhonchuk, Phys. Rev. E **91**, 043107 (2015).
15. Yu. V. Afanasiev and S. Yu. Gus'kov, *Nuclear Fusion by Inertial Confinement* (CRC Press, Boca Ration, FL, 1993)
16. V. B. Rozanov, C. P. Verdon, M. Decroisette, S. Yu. Gus'kov, J. D. Lindl, K. Nishihara, and H. Takabe, *Energy from Inertial Fusion* (IAEA, Vienna, 1995)

Synthesis by picosecond laser ablation of ligand-free Ag and Au nanoparticles for SERS applications

Enza Fazio[1], *Salvatore* Spadaro[1], *Marco* Santoro[1], *Sebastiano* Trusso[2], *Andrea.* Lucotti[3], *Matteo.* Tommasini[3], *Fortunato* Neri[1], *Paolo Maria* Ossi[4*]

[1] Dipartimento di Scienze Matematiche e Informatiche, Scienze Fisiche e Scienze della Terra (MIFT), Università di Messina, 98166 Messina, Italy
[2] CNR-Istituto per i Processi Chimico-Fisici, 98158 Faro Superiore, Messina, Italy
[3] Dipartimento di Chimica, Materiali ed Ingegneria Chimica "G. Natta", Politecnico di Milano, Milano, Italy
[4] Dipartimento Energia & Centre for NanoEngineered Materials and Surfaces -NEMAS, Politecnico di Milano, 20133 Milano, Italy

Abstract. The morphological and optical properties of noble metal nanoparticles prepared by picosecond laser generated plasmas in water were investigated. First, the ablation efficiency was maximized searching the optimal focusing conditions. The nanoparticle size, measured by Scanning Transmission Electron Microscopy, strongly depends on the laser fluence, keeping fixed the other deposition parameters such as the target to scanner objective distance and laser repetition frequency. STEM images indicate narrow gradients of NP sizes. Hence the optimization of ablation parameters favours a fine tuning of nanoparticles. UV-Visible spectroscopy helped to determine the appropriate laser wavelength to resonantly excite the localized surface plasmon to carry out Surface Enhanced Raman Scattering (SERS) measurements. The SERS activity of Ag and Au substrates, obtained spraying the colloids synthesized in water, was tested using crystal violet as a probe molecule. The good SERS performance, observed at excitation wavelength 785 nm, is attributed to aggregation phenomena of nanoparticles sprayed on the support.

1 Introduction

Over the last decade, noble metal nanoparticles (NPs) attracted strong interest due to their distinctive features that include their optically active plasmonic modes which originate intense absorption and scattering bands in the visible–near infrared interval; the extinction cross-section of NPs can be as high as 10^6 times that of the best organic chromophores and so is the local field amplification due to plasmon resonances at the basis of Surface Enhanced Raman Scattering (SERS); the easy surface functionalization with a broad range of organic molecules [1]. In particular, the use of noble metal nanostructures with plasmonic properties is rapidly spreading through biology and medicine [2]. In this respect, SERS technique is regarded as a promising analytical tool to analyse biological samples or drugs because it provides detailed spectroscopic information, which can be translated into imaging signal and adapted to an *in vivo* imaging system. A good basis for the intensity enhancement of SERS signals is a fine control of NP size, shape, spatial density and their distribution [3].

In the last few years, different preparation techniques have been used to prepare gold (Au) and silver (Ag) nanocolloids. However, wet chemical synthesis methods (i.e. precipitation and sol-gel techniques) leave residual reagents which, unfortunately, cause the deactivation of NP surfaces. In contrast, pulsed laser ablation in liquid (PLAL) is a physico-chemical synthesis method that produces ligand-free, high purity NPs with superior conjugation efficiency, grafting density, electroaffinity toward charged biomolecules [4-6]. Nevertheless, with this synthesis method, it is difficult to achieve the large amounts of NPs required for industrial applications, in a short time, keeping unaltered the morphological properties. Being the ablation phenomenon in liquid a combination of different, probably competing, mechanisms in the laser–target-liquid system, it is of paramount importance to use an ultrafast laser source to minimize thermal effects and, then, to determine the appropriate ablation parameters to obtain NPs with tailored size distributions and concentration.

In this work, we investigate the morphological and optical properties of ligand-free Ag and Au NPs synthesized by picosecond (ps) laser ablation. First, we assessed the effects of the focusing conditions on the ablated spot diameter which, in turn, determines the operative fluence. Once established the optimal focusing conditions, we examined the effects of the ps pulsed-laser fluence to obtain an higher NPs productivity with a

* Corresponding author: paolo.ossi@polimi.it

fine tuning of NPs size, that is essential to prepare appropriate SERS active substrates.

2 Experimental

Pulsed laser ablation of high purity (99.9%) Ag and Au solid targets in deionized water (H_2O) was carried out using the second harmonic (532 nm) of a laser operating at 100 KHz repetition rate with pulse width of 6-8 ps. The target was irradiated with laser powers of 1.5 W and 4.5 W for irradiation times of 10 min and 5 min for the Ag and the Au target, respectively. The laser beam was focused to a spot of about 80 μm in diameter on the surface of the target with a galvanometric scanner. The target surface was scanned on a 10 ×10 mm^2 area with a scan speed of 800 mm/s. In Fig. 1 we show the scheme of the ps-pulsed laser ablation setup and a detail of the raster-scan pattern made during the ablation process. We carried out profilometry measurements with a stylus scanning method (10 μm lateral resolution; 10 Å vertical resolution).

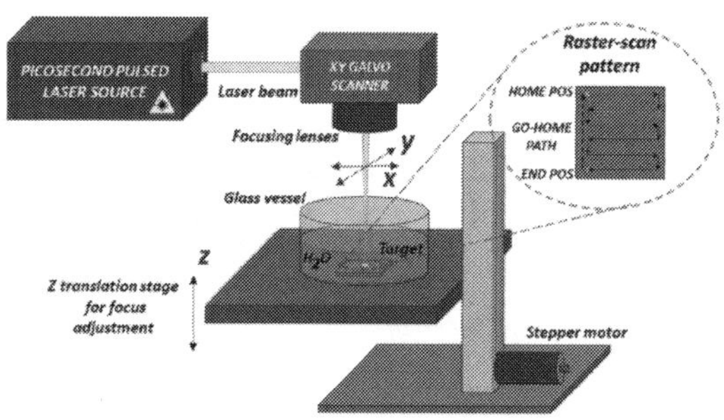

Fig.1: Schematic of picosecond pulsed laser ablation setup.

The optical absorbance spectra of produced colloids were acquired by means of a Perkin–Elmer Lambda 750 UV–vis spectrometer in the 190-900 nm range, using quartz cuvettes. By ultrasonic spraying we deposited a fraction of the colloids on nickel grids to carry out Scanning Transmission Electron Microscopy (STEM) characterization. Another fraction of the colloids was sprayed on glass or on silicon substrates. On the resulting films we carried out optical absorbance and SERS measurements, respectively. STEM images were taken by a scanning electron microscope (ZEISS; model Merlin Gemini 2) operating at an accelerating voltage of 30 kV and at a working distance of 4 mm. SEM images of the craters were acquired at an accelerating voltage of 5 kV and a working distance of 5.6 mm.

All SERS spectra were collected by a Jobin Yvon LabRAM HR800 Raman spectrometer equipped with a 785 nm solid-state laser (Laser XTRA, Toptica Photonics), a 600 grooves mm^{-1} grating and a Peltier-cooled CCD detector. Notch filters for 785 nm were used to suppress the Rayleigh scattering. The laser was focused on the sample through a 50× Olympus objective (NA=0.75). The nominal laser power at the sample was 0.5 mW, over a spot size of about 1 μm, and an exposure time of 30 s (average of 2 accumulations). The spectra were recorded in the region between 300 and 2000 cm^{-1}, which encompasses the Raman features of the reference analyte we used, namely crystal violet (CV). The SERS activity of the Ag and Au films was checked using a CV solution at the concentration of 8.9×10^{-6} M. The CV solution was prepared in deionized water. The substrates were dipped in these solutions for 20 min, then taken out for free drying in air, then the SERS spectra were collected.

3 Results and discussion

Fig. 1 shows representative crater shapes produced by laser ablation of the Ag target in water, with a laser power of 4.5 W for a fixed number of pulses of 500000. SEM images show that ps laser pulses produce on a target, immersed in the liquid medium, craters whose width slightly changes as a function of the scanner objective-target distance *d*.

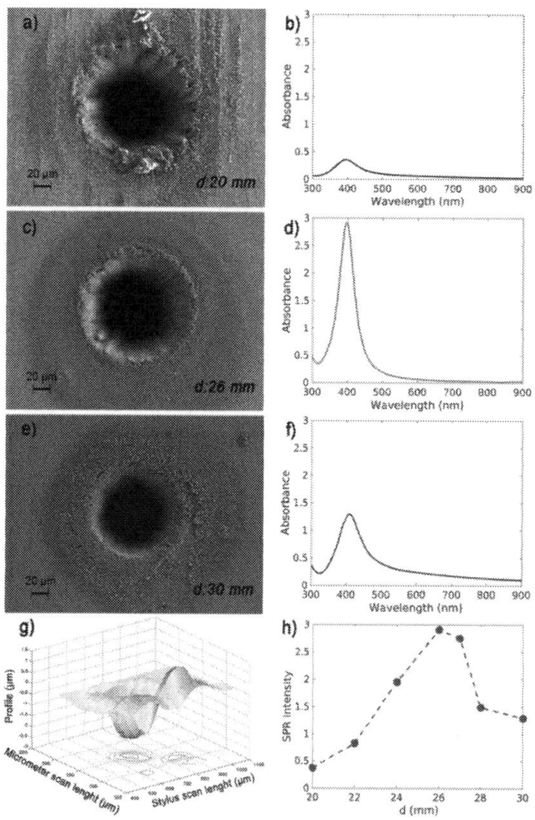

Fig. 2: SEM images of craters produced by ps laser ablation of the Ag target at a laser power of 4.5 W for 500000 pulses (a,c,e), varying the objective-target distance *d*. The optical absorbance spectra of Ag colloids (b,d,f); Crater profile for *d*= 26 mm (g) and effect of objective-target distance *d* on SPR intensity (i.e. on NP productivity) (h). The dashed line is a guide for the eyes.

Moreover, surrounding the edge of the crater, an accumulation of removed material can be envisaged.

This is more evident when the objective-target distance *d* is fixed at 20 mm and 30 mm, respectively (Fig. 2a,e,g). Otherwise, no ripples and a better defined shape of the crater is found at *d* = 26 mm (see Fig. 2c). For us, this distance represents the optimal focusing condition which, as shown below by optical properties, corresponds to higher ablation efficiency. These observations can be explained taking into account that the removal of mass from the irradiated zone can occur by both thermal and nonthermal mechanisms. The incident laser radiation on the metal creates a large population of highly excited nonequilibrium electrons in a region near the target surface. This electronic excitation can lead to bond breaking of the sample material or ''desorption,'' and subsequently can cause atomic-size particulate to be ejected from the surface. For a thermal mechanism, the electron energy is redistributed through lattice vibrations and consequently, heat is conducted into the sample. Such heat besides melting the sample can bring its local temperature to the vaporization temperature. Mass removal in the form of micron-sized droplets can also result from hydrodynamic instability of the molten liquid layer. The evaporating vapor can exert a recoil pressure on the melted surface, and molten mass can be pushed radially outward from the irradiated zone [7]. In focusing conditions, the crater has sharp borders and matches the laser spot. This occurs mainly with ultrafast localized laser energy transfer (the machining process occurs via non-linear absorption, enabled by the high intensity in the focus of radiation, resulting in the direct evaporation of the material, in principle, without activating any melting mechanism) when ablation is dominated by direct photoionization processes, thus by locally induced space-charge separation fields and by electron–ion collisions. The crater shape on the target surface is likely to be also affected by the amount of ablated micro/nano particles. This occurs since the laser beam arrives at the target surface after passing through the liquid where a high concentration of micro/nano sized ablated particles is present. In such a condition, a fraction of the laser beam energy gets absorbed by these particles before reaching the target surface, in turn affecting the mass removal efficiency from the irradiated zone [8].

Thus, a first control over the optical properties of Ag and Au NPs must be carried out modulating the beam intensity profile at the target (see Fig. 2g). The colloids, produced with laser power of 4.5 W and irradiation time of 10 min in the three selected focusing conditions show that: (1) the Surface Plasmon Resonance (SPR) peak position remains almost unchanged; (2) its intensity changes as a function of *d* (see Fig. 2b,d,f). On the overall, it emerges that the higher Ag NP amount was obtained for *d* value of 26 mm. Similar results were obtained using the Au target (not shown).

The NP productivity of ultra-short pulse laser ablation mainly depends on the pulse energy or fluence and on the pulse repetition rate used in processing. Both these factors are limited by physical processes. The pulse energy is limited by thermal diffusivity in the material, in turn affected by the quality of the processed specimen. For a given pulse energy the ablation rate can be increased by using a higher average power and a higher frequency. This allows to inject laser energy into the material without using unreasonably high pulse energies. The frequency is restricted by shielding caused by the ejected material, whether in the form of plasma, vapour, liquid or particles. In our case, the laser repetition frequency is not a relevant parameter limiting the NP productivity.

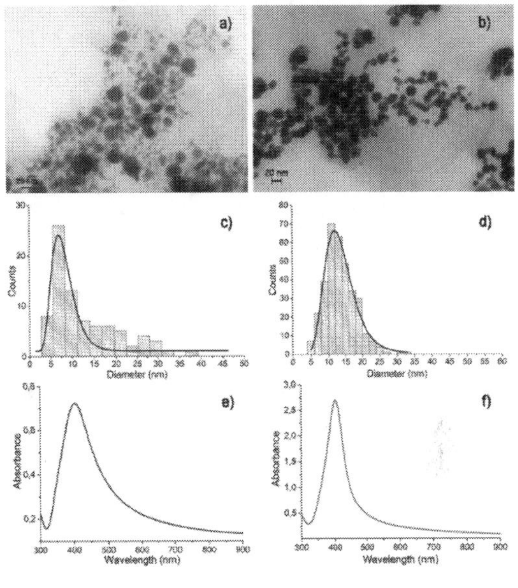

Fig. 3: STEM images, averaged size distribution and optical absorbance spectra of Ag colloids prepared at the laser fluence of 0.5 J/cm^2 ((a,c,e) and 1.5 J/cm^2 (b,d,f), respectively.

Keeping fixed the objective-target distance *d* at 26 mm, for which we found the highest NP productivity, we investigated the effects of the laser fluence on the optical/morphological properties of the colloids. In Fig. 3 are shown STEM images and optical absorbance spectra of Ag colloids prepared at the laser fluence of 0.5 and 1.5 J/cm^2, respectively. In the sample prepared at 0.5 J/cm^2, nearly spherical NPs with a diameter of about 8 nm are observed (see Fig. 3c). Few NPs larger than 40 nm are also evident. Increasing the laser fluence (1.5 J/cm^2) the spherical NPs, whose size is of about 15 nm, often agglomerate and overlap (see Fig. 3d). We show in Fig. 3e-f the trend of the UV-Vis absorption spectra as a function of the diff erent morphologies. We observe that the SPR peak intensity increases and the Full Width at Half Maximum (FWHM) decreases, when we increase the laser fluence, keeping fixed all other deposition parameters.

In Fig. 4 are shown a STEM representative image, the size distribution and the optical absorbance spectrum of Au colloids prepared at the laser fluence of 1.5 J/cm^2. A similar trend was observed for the colloids prepared at 0.5 J/cm^2 (not shown). A high NP productivity as well as a uniform and narrow size distribution (centred around 9 nm) emerges from STEM image analysis. These evidences are confirmed by the relatively narrow SPR lineshape centred at about 505 nm.

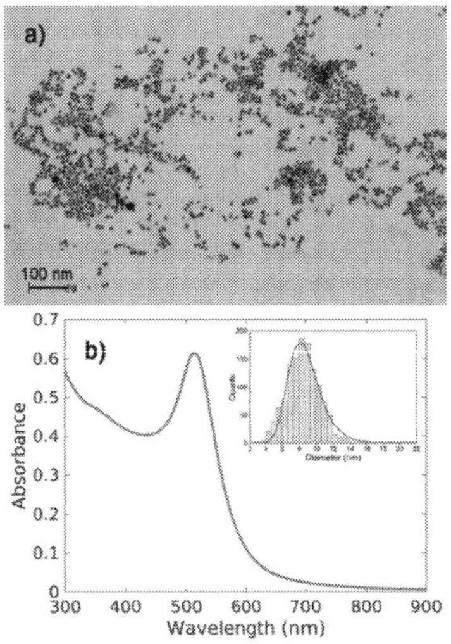

Fig. 4: Representative STEM image /(a) and optical absorbance spectrum (b) of Au colloids prepared at the laser fluence of 1.5 J/cm^2. In the inset of Fig.4b the NP colloid size distribution.

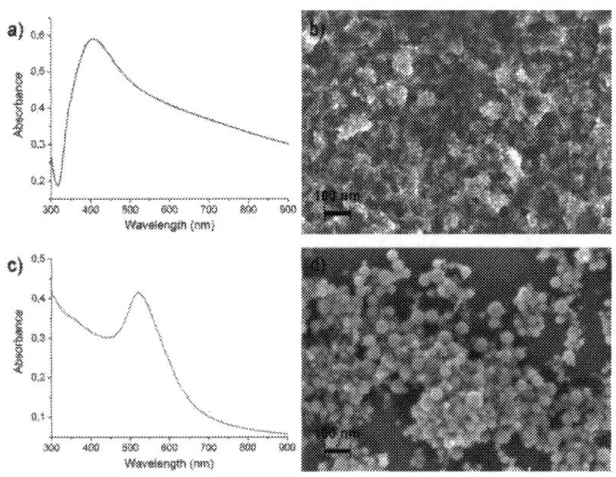

Fig. 5: UV-vis optical absorbance spectra (a,c) and SEM images (b,d) of the films obtained spraying Ag and Au colloids on a Si support (a,b).

Fig. 5a,c show the optical absorbance spectra of the films obtained spraying Ag and Au colloids on glass substrates. A red-shift and a broadening of the width of the SPR peak is observed, mainly for Ag sample. This behavior is ascribed to a clustering process occurring during the transfer of colloids onto the substrates. In fact, different surface morphologies, from isolated nearly spherical NPs, detected by STEM images, to larger islands (agglomerates) with smooth edges were observed by means of SEM the inhomogeneous films sprayed on the support (see Fig. 5b,d).

We tested the sprayed films for their SERS activity. We report in Fig. 6 preliminary SERS spectra measured from the Ag and Au films on silicon after exposition to an aqueous solution of CV (concentration 8.9×10^{-6} M). For both films we observe well resolved SERS spectra, showing all expected features of CV, *i.e.*, the main peaks located at about 420, 441, 725, 780, 915, 1173, 1392, 1587, 1619 cm^{-1} [9].

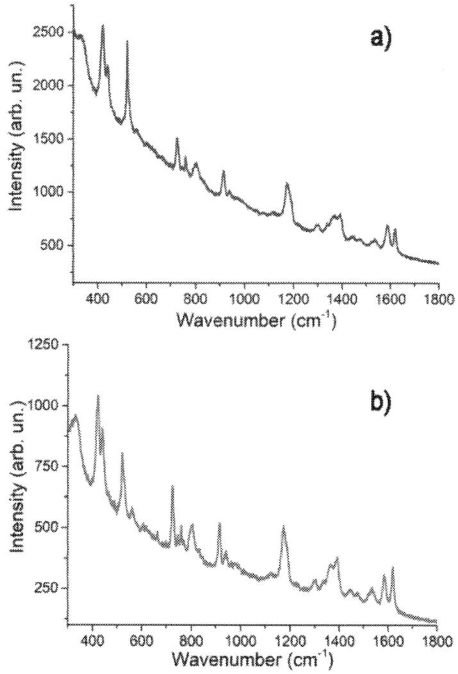

Fig. 6: Raman spectra of crystal violet in H_2O at the concentration 8.9×10^{-6} M adsorbed onto the surface of the Ag (a) and Au (b) samples, respectively. Exciting line wavelength: 785 nm.

4 Conclusion

In conclusion we used an highly efficient "green" technique based on a laser ablation process to prepare in water Ag and Au nanocolloids. Ag and Au NPs as active SERS substrates were successfully synthesized by a simple spraying procedure of the obtained colloids onto inert supports. The promising SERS action of the Au and Ag nanostructured films, that was checked against CV as the probe molecule using an excitation at 785 nm, is due to suitable plasmon characteristics induced by NP clustering.

References

1 V. Amendola, M. Meneghetti, Phys. Chem. Chem. Phys., **11**, 3805 (2009).

2 G.Lentini, E. Fazio, F. Calabrese, L. M. De Plano, M. Puliafico, D. Franco, M. S. Nicolò, S. Carnazza,

S. Trusso, A. Allegra, F. Neri, C. Musolino,, S. P.P. Guglielmino, Biosensors and Bioelectronics **74**, 398 (2015).

3 N. R. Agarwal, E. Fazio, F. Neri, S. Trusso, C. Castiglioni, A. Lucotti, N. Santo, P.M. Ossi, Cryst. Res. Technol. **46,** 836 (2011).

4 S. Petersen, S. Barcikowski, J. Phys. Chem. C **113**, 19830 (2009).

5 D. Zhang, B. Gokce, Appl. Surf. Sci. **392**, 991 (2017).

6 L. Gamrad, C. Rehbock, J. Krawinkel, B. Tumursukh, A. Heisterkamp, S.Barcikowski, J. Phys. Chem. C **118**, 10302 (2014).

7 A. Semerok, C. Chaleard, V. Detalle, J.-L. Lacour, P. Mauchien, P. Meynadier, C. Nouvellon, B. Salle, P. Palianov, M. Perdrix, G. Petite, Appl. Surf. Sci. **138**, 311 (1999).

8 V. Amendola, M. Meneghetti, Phys. Chem. Chem. Phys. **15**, 3027 (2013).

9 F. A. Harraz, A.A. Ismail, H. Bouzid, S. A. Al-Sayari, A. Al-Hajry, M.S.Al-Assiri, Appl. Surf. Sci. **331**,241 (2015).

Laser Irradiated Foam Targets: Absorption and Radiative Properties

Martina Salvadori[1,2,⋆], *Pier Luigi* Andreoli[2], *Mattia* Cipriani[2], *Fabrizio* Consoli[2], *Giuseppe* Cristofari[2], *Riccardo* De Angelis[2], *Giorgio* di Giorgio[2], *Danilo* Giulietti[1], *Francesco* Ingenito[2], *Sergey Yu.* Gus'kov[4], and *Alexander A.* Rupasov[3]

[1] *Pisa University, Physics Department "E.Fermi" and INFN*
[2] *ENEA Fusion and Nuclear Safety Department, C.R. Frascati, Via E. Fermi 45, 00044 Frascati, Italy*
[3] *Lebedev Physical Institute, Russian Accademy of Sciences, Moscow*
[4] *National Research Nuclear University, Moscow Engineering Physics Institute MEPhI*

Abstract. An experimental campaign to characterize the laser radiation absorption of foam targets and the subsequent emission of radiation from the produced plasma was carried out in the ABC facility of the ENEA Research Center in Frascati (Rome). Different targets have been used: plastic in solid or foam state and aluminum targets. The activated different diagnostics allowed to evaluate the plasma temperature, the density distribution, the fast particle spectrum and the yield of the X-Ray radiation emitted by the plasma for the different targets. These results confirm the foam homogenization action on laser-plasma interaction, mainly attributable to the volume absorption of the laser radiation propagating in such structured materials. These results were compared with simulation absorption models of the laser propagating into a foam target.

1 Introduction

Porous materials of light elements consist of randomly distributed solid fibers or membranes, with transversal size b_0 in the range of few microns up to hundreds of microns, separated by voids of similar dimensions. According to their pore size δ_0, these materials can be divided in two groups, large pore foams and small pore foams. The first one comprehends porous plastics ($[CH]_n$, $[CH_2]_n$), porous Beryllium and agar-agar ($C_{12}H_{18}O_9$), with pore size and thickness of solid elements ranging from $30\mu m$ to $100\mu m$ and from 0.1 to $1\mu m$ respectively; cellulose triacetate ($C_{12}H_{16}O_8$) and trimethylolpropane triacrylate ($C_{15}H_{20}O_6$) are in the small pore group, with pore size ranging from 1 to $10\mu m$ and the thickness of solid elements from 0.01 to $0.1\mu m$.

The average density of foams can vary from few mg/cm^3 to several hundreds. This wide variety in characteristic dimensions leads to different behavior under laser irradiation. In particular, according to their initial average density ρ_a, corresponding to an electron density $n_e = \frac{\rho_a Z_{eff}}{M}$ where Z_{eff} and M are respectively the effective state of charge and the mass of the atoms or molecules, and to the impinging laser wavelength λ, we can distinguish *overcritical* foams, characterized by n_e greater than the critical density n_{cr}, from *subcritical* foams, with $n_e < n_{cr}$, where

$$n_{cr} = \frac{1.1 \times 10^{21}}{\lambda^2_{\mu m}} \text{cm}^{-3} \tag{1}$$

⋆e-mail: martina.salvadori90@gmail.com

Regardless of this distinction, due to the internal structure of the material, at the initial stage of interaction, the laser radiation has the chance to penetrate up to a distance in both cases defined by the initial *transparency length*:

$$L_0 \simeq C(\frac{\rho_s}{\rho_a})^{\frac{1}{3}} \delta_0 \tag{2}$$

where ρ_s is the density of the solid elements and C is a constant which depends on their shape [1]. For our foam parameters, $\delta = 40\ \mu m$ $\rho_a = 10$ mg/cm^3, $\rho_s = 1050$ mg/cm^3, the estimated initial transparency length is equal to 500 μm.

Laser irradiation gives rise to an homogenization process leading to a fully homogeneous plasma. As the process goes on, the transparency length changes accordingly to density variations. In this phase laser radiation is absorbed, via inverse Bremsstrahlung mechanism, in a plasma of stochastically varying density created by the expansion of the solid parts under laser irradiation. The volume of overcritical regions increases, producing a progressive decrement in the depth of the absorption region. In both cases laser light absorption continues according to usual mechanisms for a homogeneous material. In overcritical foams the radiation is absorbed in the subcritical plasma developed in front of the target. In subcritical foams laser radiation is absorbed behind the ionization front propagating inside the target.[2].

These considerations show that the initial homogenization stage has a peculiar role in laser light absorption, allowing a deeper penetration in the target even with an overcritical average density. This explains the experimental observation of a deep absorbing region and an anomalously high absorption efficiency for overcritical foams up to 80%-90% of incoming radiation [3]. Moreover, these peculiarities lead to some interesting properties such as the reduction of the irradiation inhomogeneities [4] and the amplification of ablation pressure [5].

For all these reasons, the foams are of great interest for their applications in Inertial Confinement Fusion experiment, as they are excellent absorbers of laser radiation and inhibitors of detrimental hydrodynamic instabilities. The aim of this work is to collect further information about three aspects of the laser-foam interaction: The plasma density profile, the absorption mechanism and the emitted radiation. In this paper we present the main results obtained irradiating overcritical porous Polystyrene comparing them with the one obtained by the irradiation of solid Polystyrene and Aluminum.

2 Experimental setup

The experiments have been performed at ABC Laser, a Nd laser with a wavelength λ = 1054nm, a pulse energy $\simeq 40J$, a time duration $\tau_L \simeq$ 3ns with a rising time of about 1.5ns. An F = 1 optics allows to focus the laser beam down to $\sim$ 100μm spot diameter, corresponding to an intensity on target of 10^{13}W/cm^2. The targets were made of solid Polystyrene with density ρ = 1050mg/cm^3 and thickness δ = 1100μm; solid Aluminum with ρ = 2700 mg/cm^3 and δ = 510μm; different porous polystyrene targets with thickness ranging from 290 to 915μm and ρ_a = 10 mg/cm^3. Because of their density the foam samples are overcritical for ABC laser wavelength.

The main diagnostics in use were devoted to the investigation of density profile and radiation emission. For this purpose we use a Nomarski interferometer, an optical streak camera, silicon surface barrier detectors and diamonds detectors. The diagnostic layout is shown in figure 1. A frequency doubled diagnostic beam is employed in

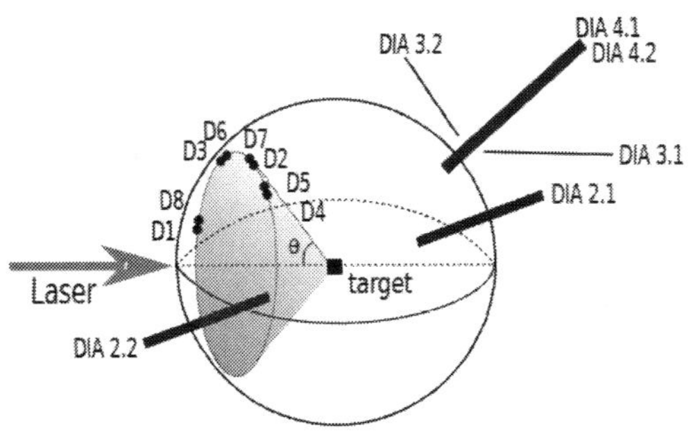

Figure 1: Experimental chamber layout, Vacuum diodes are placed on the base circumference of a cone whose vertex is on target. Thanks to the symmetry in the plasma radiation emission, these positions ensure to have comparable signals on each diode. Diamonds are placed at different distances and angles with respect to the target.

the Nomarski Interferometer which provides information about the electron density profile of the generated plasma. The optical streak camera allows to detect the plasma expansion normal to the target surface, with a temporal and spatial resolution of 20ps and 20μm respectively.

The emitted X-Ray radiation is observed through Diamond detectors, which are sensible to wavelength shorter than 250nm, and through silicon surface barrier detectors sensible also to visible light. Since we are interested in higher energy emission, we used Beryllium and Nickel filters of various thickness to select the proper spectral region. Diamond detectors were also employed for Time Of Flight (TOF) measurements. In table 1 the distances of the diamonds from the target and the diode's filters layout are reported.

Diode	Filter(μm)
D1	Be 25.4
D2	Ni 10.91
D3	Ni 4.59
D4	Ni 7.61
D5	Ni 9
D6	Be 50.8
D7	Ni 5.46
D8	Be 12.7

(a)

Diamond	Distance(mm)
DIA 2.1	1130
DIA 2.2	1120
DIA 3.1	899
DIA 3.2	899
DIA 4.1	1739
DIA 4.2	1739

(b)

Table 1: (a)Filtering layout of vacuum diodes. (b) Diamond detectors distances from target

3 Experimental Results and Discussion

Density Profile

Images obtained with Nomarski interferometer are shown in figure 2. The interference pattern gives no evidence of a sizable difference in the electron density distribution between the solid and the foam targets. Most probably this is due to the fringe blurring produced in the region of interest by the relative long pulse duration of the probe. Nevertheless this diagnostic channel provide an indirect information about the density gradients. In fact comparing the three images it is possible to see that in the case of irradiation of solid polystyrene and solid Aluminum the presence of second harmonic emission is apparent, meanwhile it is completely absent in the case of irradiation of porous polystyrene.

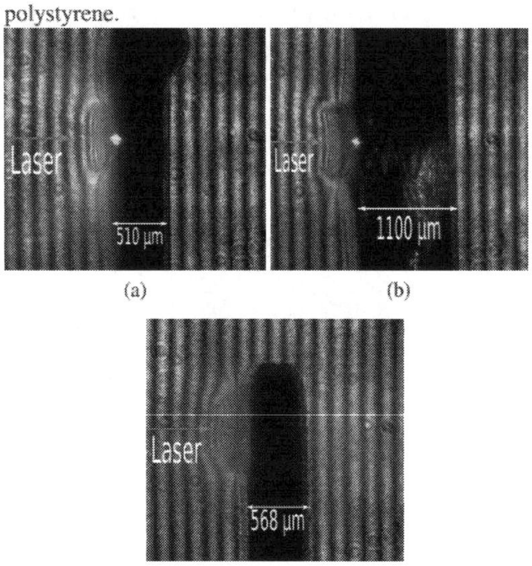

Figure 2: Images obtained by the Nomarski interferometer when irradiating target of different materials: (a) Aluminum 510μm thick (b)Solid Polystyrene 1100μm thick and (c)Porous Polystyrene 568μm thick

Since the second harmonic emission is related to the presence of plasma density gradients [6],

$$\mathbf{J}(2\omega) = \frac{in_e e^2}{4m^2\omega^3}\nabla(\mathbf{E}\cdot\mathbf{E}) + \frac{ie^2}{m^2\omega^3}\left[\frac{\nabla n_e\cdot\mathbf{E}}{1-\omega_p^2/\omega^2}\right]\mathbf{E} \quad (3)$$

where m is the electron mass, n_e is the plasma electron density, e is the electron charge, ω is the frequency of the laser, ω_p is the plasma frequency and $\mathbf{E}$ is the electric field, its absence can be related to small value of ∇n_e, hence to the production of a plasma characterized by a smoother density profile. For this reason we can assume that the irradiation of porous polystyrene produces a plasma whose density variations are smoother than in the case of irradiation of solid polystyrene and aluminum.

Another indirect evidence of plasma homogeneity comes from time of flight measurement. As we can see in figure 3 the diamond signals are characterized by a first higher peak, which is generated by radiation, and, at later times, by a series of subsequent lower peaks. These are ascribable to the incoming accelerated particles, in particular to protons and ions. The force providing ions acceleration arises from the electric field generated by the displacement of electrons which are accelerated trough the electron plasma waves. The subsequent velocity acquired by a charged particle is given by the equation 4.

$$\mathbf{v}(t) = \left(\frac{2\mathbf{E}eL_{acc}}{m_p}\frac{Z}{A}\right)^{1/2} = \mathbf{v}_p\left(\frac{Z}{A}\right)^{1/2} \quad (4)$$

where L_{acc} is the length in which the accelerating force acts, m_p the proton mass, Z and A are the state of charge and the number of mass of the ions respectively, v_p is the velocity acquired by a proton.

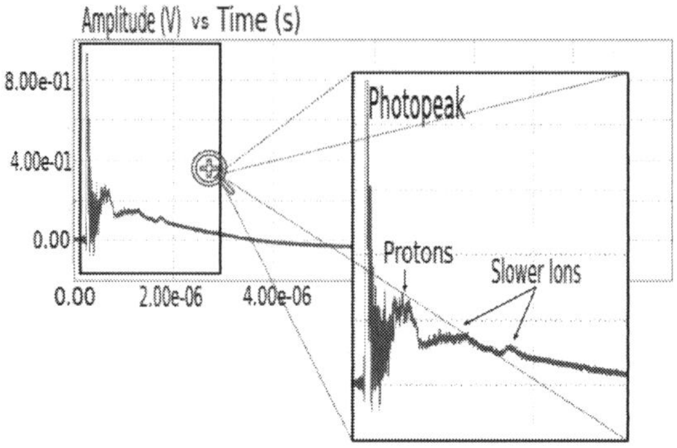

Figure 3: Typical signal obtained from diamond detectors and a magnification of its most interesting part. In order to avoid the dependence of time measurement from the delay of the oscilloscope, we took as reference the photopeak time and we have measured the time interval $t_i - t_\gamma$ where t_i and t_γ are the time at which ions and photons arrive on the diamond detectors.

From the equation 4 it is possible to see that the fastest particles will be those with a larger $\frac{Z}{A}$ ratio, hence we will detect first protons and than ions. In our experimental set up fast electrons signal is superposed with the one given by the photopeak. From these considerations we are able to identify in figure 3: the photopeak, the protons and the different ions of the target specimen. Comparing the results obtained for the velocity of protons generated by different materials, it turns out that irradiation of porous polystyrene produces slower ions than the irradiation of solid aluminum or polystyrene, as it is possible to see in figure 4. Since in our simple scheme, ac-

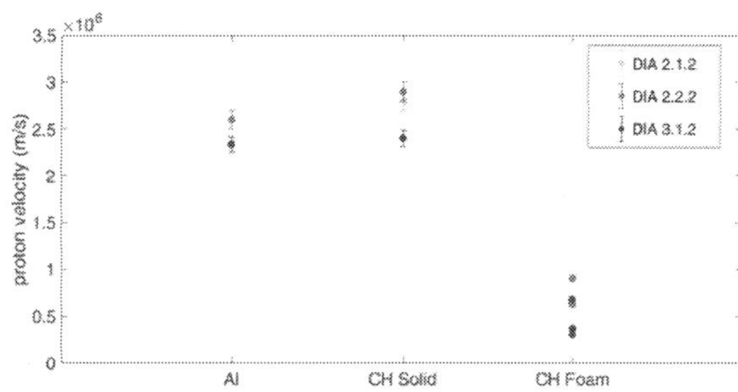

Figure 4: Velocities of protons generated by irradiation of different types of material. The velocities are obtained trough Time of Flight measurement performed with Diamond detectors.

celeration of ions is related to the amount of accelerated electrons, less energetic ions means a lower yield of accelerated electrons. In our experimental condition the main mechanism of electron acceleration is related to the presence of electron plasma waves produced by parametric instabilities which, to be generated, require a threshold laser intensity that could be reached by the activation of the filamentation instability. Since our irradiation parameters are under this threshold value, we think that in the case of solid material the phenomenon of filamentation is activated by the presence of inhomogeneities in laser irradiation (i.e. due to the presence of the hot spots) and contributes to reach the intensity threshold, while in the case of porous materials filamentation doesn't occur because of the homogenizing action of foams on laser irradiation inhomogeneities. In addition to intensity gradient, filamentation is facilitated also by the presence of high density gradients hence its absence can be related to the production of a plasma characterized by smoother variation in density [7].

Absorption and Emission Processes

Through the images obtained by the streak camera it has been possible to investigate on plasma emission. In figure 5 shots on polystyrene foams of different thickness are reported. The luminescence at the front side of the target arises from the plasma self emission and allow to follow its expansion. The luminescence at the rear side is due to the plasma generated by the hydrothermal wave propagating inside the target.

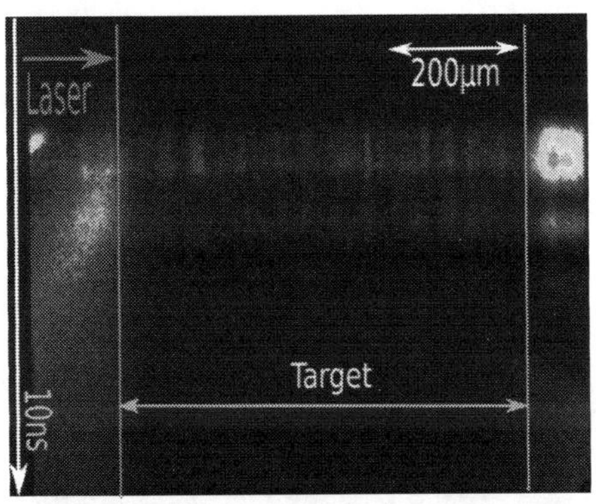

(a)

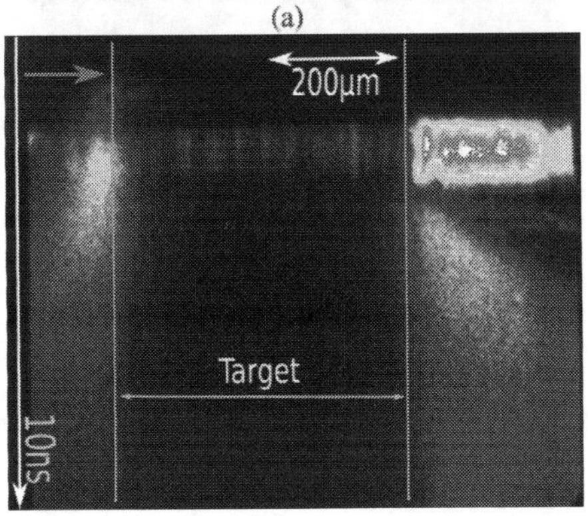

(b)

Figure 5: Images obtained by the streak camera in the case of irradiation of Porous polystyrene with thickness (a) about the initial transparency length(430μm) and (b)lower than the initial transparency length (330μm)

Streak camera images provides another useful information, in fact focusing the laser near to the external surface toward the streak camera direction,as shown in figure 6(a), it is possible to observe the emission coming from the internal region of the porous target as the laser propagates inside it, figure 6(b). This allows us to perform measurement of its propagation velocity. We obtain values consistent

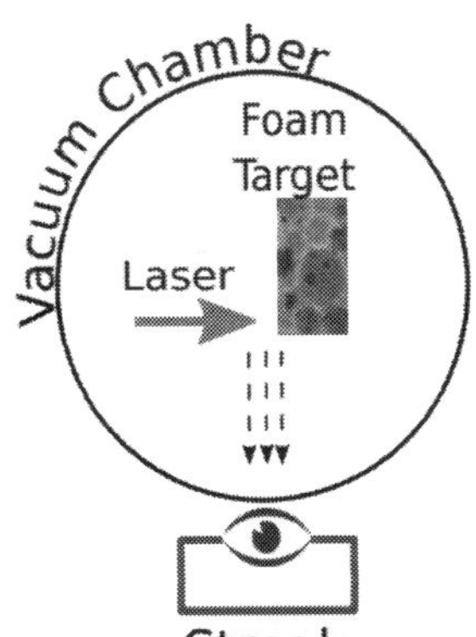

(a)

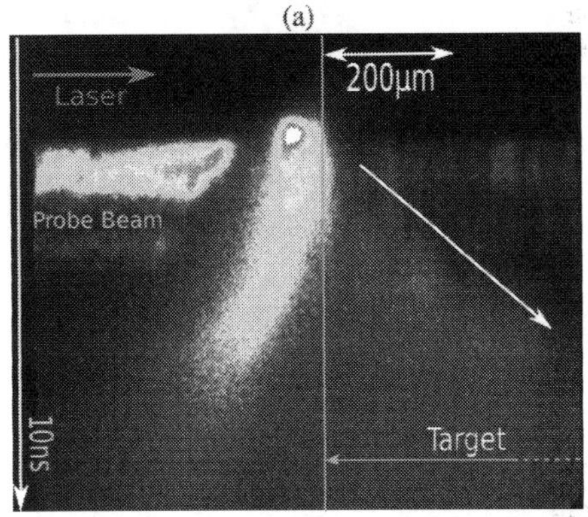

(b)

Figure 6: (a) Top view of the focusing scheme during the shot. (b) Image obtained from the optical streak camera, The white arrow highlights the propagation of ionization wave inside porous polystyrene. From its slope it is possible to evaluate the propagation velocity.

with theoretical simulation, $v = (1.2 \pm 0.1) \times 10^7$cm/s. A more accurate survey of the images in figures 5 and 6 provides information about the expansion velocity of plasma, that is of the order of the sound velocity in the medium given by the relation:

$$c_s = 3 \times 10^7 \sqrt{\frac{ZT_e(\text{KeV})}{A}}\frac{cm}{s}. \quad (5)$$

From this value it is possible to estimate the electron plasma temperature. As shown in figure 7, comparing the results for different materials, Porous Polystyrene shows higher temperature than solid one. This result suggest a greater emission of radiation at higher energy, to investigate this aspect we examine the data obtained from diamonds and silicon surface barrier detectors. The first, sig-

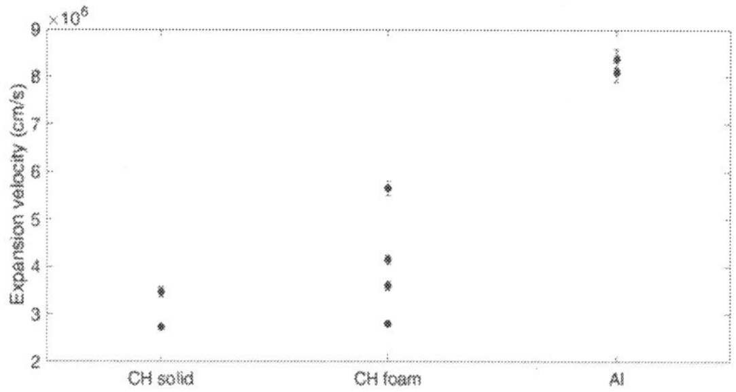

Figure 7: Plasma expansion velocity measurement obtained by streak camera images

nal is reported in figure 8, show an higher photopeak amplitude in the case of porous polystyrene than in the one of solid plastic, and similar photopeak amplitude for Aluminum and porous Polystyrene targets. Solid and Porous

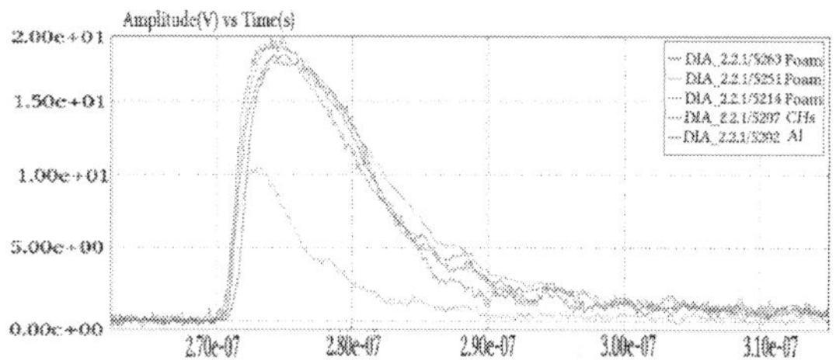

Figure 8: Photopeak amplitude obtained irradiating different types of material

polystyrene have the same chemical composition, hence this high difference in the emission behavior has to be attribute to their different internal structure. Since an higher emission is related to an higher absorption this result is consistent with the described peculiar absorption process. Due to the composition of plastic material we expect that the main contribution to the radiation emission comes from Bremsstrahlung process, while in the case of Aluminum both contribution of continuum and lines emission play a central role. Moreover we know that the contribution of Bremsstrahlung is proportional to the atomic number of the emitting element, from these consideration, in the case of solid material, it is evident the result shown in figure 8. On the other hand we suggest that the similar photopeak amplitude for Aluminum and porous polystyrene can be another evidence of an higher absorption in the case of porous materials.

Vacuum diodes confirm that irradiation of porous plastics provides an higher radiation emission than in the case of interaction with solid one. As it is possible to see in figure 9, this result is apparent when we consider the signals obtained from the diodes filtered with the thickener Nickel foil, whilst if we look at data obtained from diodes with a lower filtering, the differences in the amount of emitted radiation gradually disappears. Since an higher value of filtering allows only to higher energy radiation to be detected, this result suggest that the differences in emission of radiation arises from the high frequency region, which means that in the case of porous polystyrene we are dealing with an hotter plasma than in the case of solid polystyrene, result which is consistent with streak camera measurement.

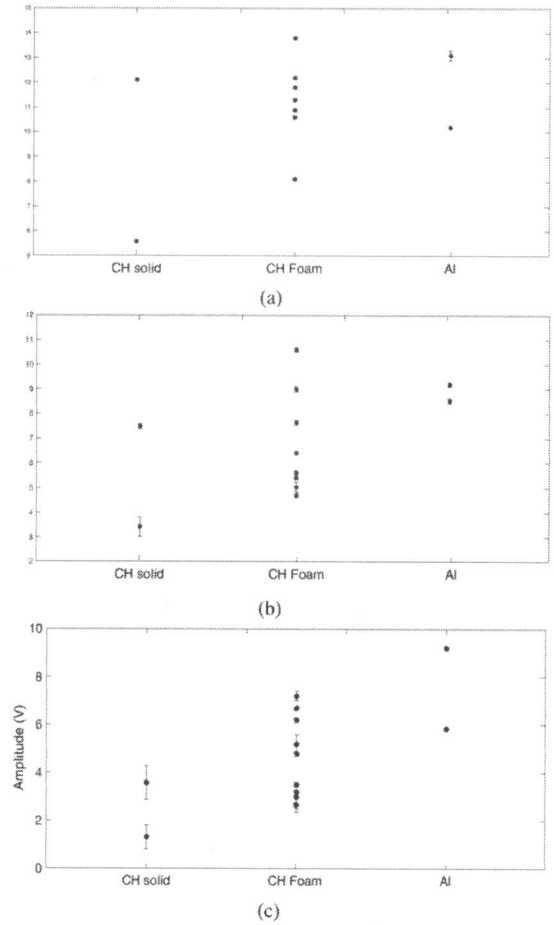

Figure 9: Comparison of the amplitudes of vacuum diodes signal when irradiating different types of material. Each graph is obtained with a different thickness of Nickel filter: (a) 4.59μm, (b) 9μm and (c) 10.91μm.

4 Conclusion

During our experiments we investigated the absorption and emission properties of targets of different chemical composition and internal structure. In particular, in foam targets, we highlighted an improvement of the laser radiation absorption and the subsequent larger X-Ray emission. Furthermore indirect measurements suggest that the irradiation of porous materials produces a more homogeneous plasma than the irradiation of solids.

This work was supported by the Program of Increase in the Competitiveness of the National Research of Nuclear

University MEPhI. The work of S. Gus'kov and A. Rupasov was also supported by Russian Foundation for Basic Research project No. 17 – 02 – 00059 – a. The experimental campaign was carried out at ENEA research centre in Frascati, in The ABC laboratory.

References

[1] S. Gus'kov, Journal of Russian Laser Research **31**, 574 (2010)

[2] S. Gus'kov, M. Cipriani et al., Plasma Physics and Controlled Fusion **57**, 125004 (2015)

[3] A Bugrov, S. Gus'kov and V Rozanov, Journal of experimental and theoretical physics **88**, 441 (1999)

[4] S. Gus'kov et al., Laser and Particle Beams **18**, 1 (2000)

[5] R. De Angelis et al., Physics of Plasma **22**, 072701 (2015)

[6] Y. R. Shen, *The Principles of Nonlinear Optics* (John Wiley and Sons, New York, 1984) 10

[7] D. Giulietti and L. Gizzi, Nuovo Cimento, **21**, 1-93, (1998)

Use of the Femtosecond Lasers in Ophthalmology

Anna M Roszkowska[1,2,*], Mario Urso[1], Alberto Signorino[1], Pasquale Aragona[1]

[1]Ophthalmology Section, Department of Surgery, University Hospital of Messina, 98124 Messina, Italy.
[2] Department of Ophthalmology, Andrzej Frycz Modrzewski Kraków University, Cracow, Poland.

Abstract. Femtosecond laser (FSL) is an infrared laser with a wavelength of 1053 nm. FS laser works producing photodisruption or photoionization of the optically transparent tissue such as cornea. Currently FS lasers have a wide range of applications in ophthalmic surgery. They are used above all in corneal surgery in refractive procedures and keratoplasty, and recently in cataract surgery. The use of the FSL in corneal refractive surgery includes LASIK flap creation, astigmatic keratotomy, Femtosecond Lenticule Extraction (FLEx), Small Incision Lenticule Extraction (SMILE) and channels creation for implantation of the intrastromal corneal rings. As to the corneal grafting, the FS lasers are used in laser-assisted anterior and posterior lamellar keratoplasty and customized trephination in the penetrating keratoplasty. FS Laser Assisted Cataract Surgery (FLACS) includes capsulorrhexis and nuclear fragmentation that enhance safety and efficacy of the procedure.

Introduction

Since the ocular damage subsequent to exposure to the intense sunlight during eclipse was noticed, the idea for the development of therapeutic lasers opened the new era in ophthalmology. Actually, different lasers are used in diagnostic and therapeutic ophthalmic procedures. It is due to their particular characteristics that distinguish them from conventional light sources such as their directivity, pulsed mode and monochromaticity, with different wavelength absorbed by different ocular structures that permits to obtain their final effects on tissues.
First lasers were applied to treat diabetic retinopathy. They worked in blue/green part of light spectrum (488-515 nm) being absorbed by posterior ocular layers such as pigmented retinal epithelium, retinal vessels and blood pigment. Successively other devices were introduced and nowadays numerous lasers work with different wavelengths reaching almost all ocular structures and treating various ocular disorders. The use of lasers in ophthalmology is based on their three fundamental tissue effects: thermal, photoablative and mechanical. Thermal effects, obtained with lasers working with 400-700 nm wavelength, are exploited to perform tissue photocoagulation in several retinal diseases, whereas photoablation, obtained with excimer lasers working with 193 nm wavelength, permits to modify the anterior corneal shape by evaporation of the corneal stroma and so is used successfully in corneal refractive surgery.
Mechanical effects are obtained by two type of lasers working with different modalities such as Nd:YAG and femtosecond laser. Nd:YAG emits light with a wavelength in the infrared of 1064 nm, and produces mechanical effect by plasma creation. It is mostly used to treat the secondary cataract practicing a posterior capsulotomy. However it is also used to perform peripheral iridotomy in acute angle closure glaucoma.
The femtosecond lasers work in deep infrared absorbed by superficial ocular layers that permits to use them in corneal and anterior segment surgery.

Femtosecond lasers.

The most recent lasers approved for ophthalmic procedures are the Femtosecond lasers (FSL), working in the deep infrared and characterized by ultra-short pulses (1028-1053 nm) in the domain of femtoseconds ($1\ fs = 10^{-15}\ s$). The final effect is a perfect cut of tissues obtained by cavitation. FS belongs to the category of ultrafast lasers or ultra short pulse lasers with beam diameter less than 8 micron. Characteristic of the absorption of FS wavelength permits to use them in corneal and anterior segment surgery. Their capacity to perform multiple geometric cuts, with horizontal, vertical or oblique pattern, permitted to use them in refractive, corneal and cataract surgery. Thousands of laser spots are applied uninterruptedly to form a resection plane, creating multiple cavitational bubbles and thus generating the surgical effects (Fig. 1). Through a photodisruption process that rapidly executes thousands of impulses close to each other, it creates a perfect cut. Ultra-short pulse duration delivers laser energy with preserving the surrounding tissues. Lasers energy is absorbed by targeted tissues with rapidly expanding plasma formation with creation of cavitation

* Corresponding author: aroszkowska@unime.it

bubbles separating the tissue plane. The cavity bubble results in a single spot with the size of about 5 microns and is constituted of carbon dioxide and water vapor. This bubble expands several times and the final dimensions depend on the applied energy (Fig. 2). The application of this technology in the field of ocular microsurgery has found various uses over time. Since the year 2000 the FSL have been used in lamellar and perforating corneal transplantation and in 2001 the FDA (USA Food and Drug Administration) approved it for the creation of the corneal flap in refractive surgery. Today it is also used to correct the corneal astigmatism by curve keratotomy, and in the management of keratoconus with intrastromal rings. The first use of the Femtolaser in cataract surgery had place in 2008 in Budapest and in 2009 the FDA approved its employment for this procedure. The platforms actually available for Femtolaser Assisted Cataract Surgery are: LenSx (Alcon, Fort Worth, Texas, USA); LensAR (Orlando, Florida, USA); Catalys (Abbott/ Optimedica, North Chicago, Illinois, USA); Victus (Bausch & Lomb Rochester, New York, USA); LDV Z8 (Ziemer, Port, Switzerland). The main differences between all laser platforms are the frequency and duration of the energy impulse emission. Victus works with frequency of 80 kHz, an average use power equal to about 7 μJ and duration of the spot of about 400-550 fs. Catalys works at 120 kHz with an energy average of 4 μJ and duration of 600 fs spot. The LDV Z8 uses an average frequency of 2 MHz, one energy impulse in the range of nano Joules (50 nJ-2500 nJ) and the duration of the unspecified spot; the LenSx uses frequencies equal to 50 kHz, a pulse of energy of 15 μJ and duration of about 600-800 fs. The LensAR laser emission frequency is equal to 80 kHz. Catalys allows to run grids in addition to radial cuts, Victus and LenSx have concentric cuts ring and radial, and recently LenSx also to grid pattern. LensAR performs grid and radial cuts with or without concentric ring cuts, while LDV Z8 can only perform radial cuts by default. The FSL can work in various steps of the operations, performing cutting modulated in extension and depth, thus simplifying and making the surgery safer. Generally the FSL are dedicated to only one type of surgery. FSL Visumax (Carl Zeiss, Jena, Germany) and FSL Intralase (Alcon, USA) are used in corneal procedures being the first one, the sole capable to carry out the SMILE technique. The lasers Victus (Bausch and Lomb,) and Femto LDV Z8 Ziemer, Switzerland) are used both in cataract and corneal surgery. FDA approved FSL for cataract surgery in 2010 with indication to make clear corneal incisions, capsulorhexis and lens fragmentation/softening. In the near future a progressive transitioning to the all laser cataract surgery could be expected [1].

FSL clinical applications

Refractive surgery

Femtosecond lasers are used in various refractive surgery procedures such as Femtolasik, Femtosecond lenticule extraction (FLEx) and Small-incision lenticule extraction (SMILE)[2]. In the Femtolasik two different lasers are used to perform two surgical stages. During the first one, a corneal flap is created using the FSL, successively the flap is lifted and the next stage consisting in excimer laser ablation is done. In this technique the FSL replaced the microkeratome cut used in the traditional LASIK method to execute the corneal flap (Fig. 3). To create the flap, the laser spot moves inside the cornea at programmed thickness, creating a perfect cut and once the lamella is created, the Femtolaser bundle continues along the flap circumference. This makes the procedure completely blade-free and permits the surgeon to adapt the procedure to the characteristics of each patient, with a better flap thickness predictability. ReLEx (Refractive Lenticule Extraction) is performed entirely with a FSL system. The femtosecond laser cuts a small lenticule of

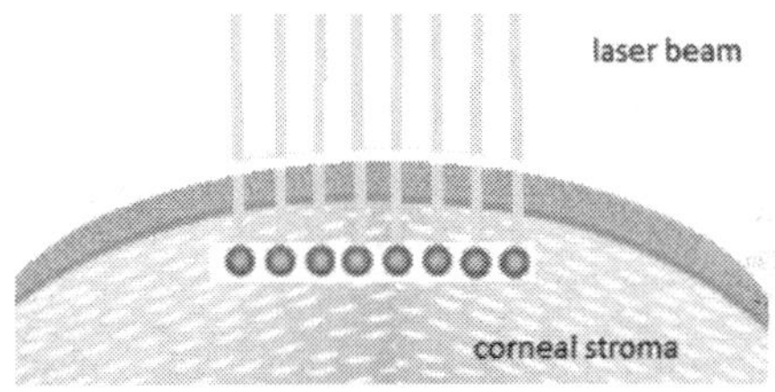

Fig. 1. Each pulse forms a microscopic bubble. The cut is achieved by placing thousand of pulses next to each other.

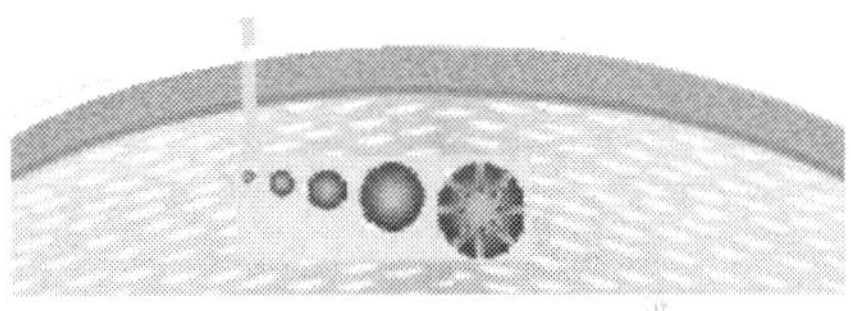

Fig. 2. Expansion of plasma creates cavitation bubbles separating the tissue plane.

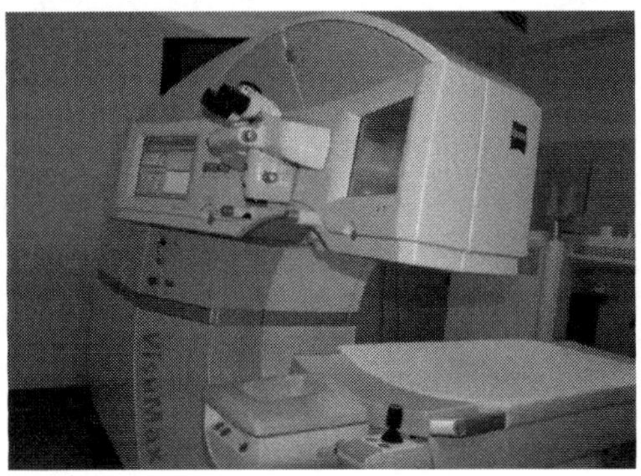

Fig.3. Femtolaser platform.

tissue within the center of the cornea and creates a flap, similar to the flap created in LASIK. The flap is lifted and the lenticule is removed, than the flap is repositioned. The removal of the lenticule reduces the corneal curvature, thereby reducing refractive error.

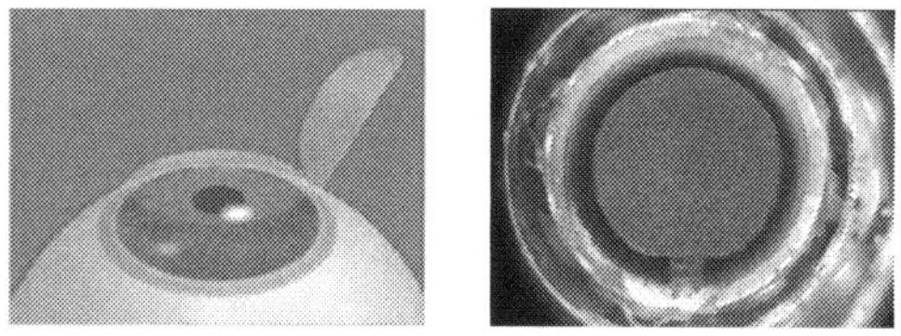

Fig. 3 A, B. Corneal lamellar flap performed by femtosecond laser.

Recently ReLEx was refined to the SMILE procedure that consists in the sculpting of intrastromal lenticule with characteristics adapted to correct the myopic or astigmatic refractive error. FSL performs the anterior and posterior dissection of both faces of the refractive lenticule followed by a small incision in the mid-periphery of the cornea, varying from about 2 to 4 mm, allowing the surgeon to extract safely the intrastromal lenticule.

Incision for ICRs

FSL are used to create intrastromal channels for the intra-corneal ring (ICR) segments implantation used to flatten the central cornea in high myopia or keratoconus. The incision for ICR offers the possibility to create even inclined cutting geometries and tunnel segments between 90° and 270°.

Corneal surgery

Corneal surgery assisted with FSL is related to different techniques of transplantation. In fact the laser cuts available the surgeon to create the perfectly matching donor-recipient buttons due to the high precision in corneal incisions. The femtosecond laser-assisted keratoplasty procedure has been developed to produce bladeless, precise, and predefined lamellar cuts with minimal tissue injury [3, 4]. Precise corneal incisions, with customized graft edges and lamellar planes for both donor and recipient corneas can be created by customized FSL trephinations in zig-zag or mushroom or hat conformation [5] (Fig 4).

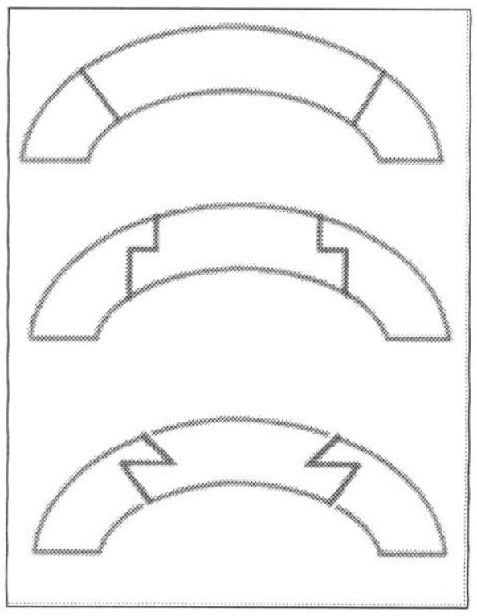

Fig. 4. Possibilities of corneal cut in keratoplasty. The mean central corneal thickness is about 540 μm and it ranges from 500 to 600 μm.

Cataract surgery

Femtosecond lasers represent an innovative technology and a new frontier in cataract surgery [6, 7]. Many potential benefits are offered by the introduction of the femtosecond laser into the field of cataract surgery. FSL is able to perform three important steps in cataract surgery such as corneal incisions, capsulotomy, and lens fragmentation (Fig. 5). There are currently several laser platforms used for Femto Laser Assisted Cataract Surgery (FLACS). All lasers enter in contact with ocular surface by so called, docking phase, after that, the eye structures are visualized [8]. Once the captured images are confirmed, all platforms are programmed to detect margins of ocular structures before the laser action, for a totally safe cutting. After laser circular capsulotomy, a nucleus fragmentation can be performed through different cutting patterns that are chosen by surgeon accordingly to cataract's hardness. The procedure than is completed manually by the surgeon. LenSx, LensAR and LDV Z8 let to run the whole operation on the same operating bed, moving simply the laser platform with micrometrical resolution.

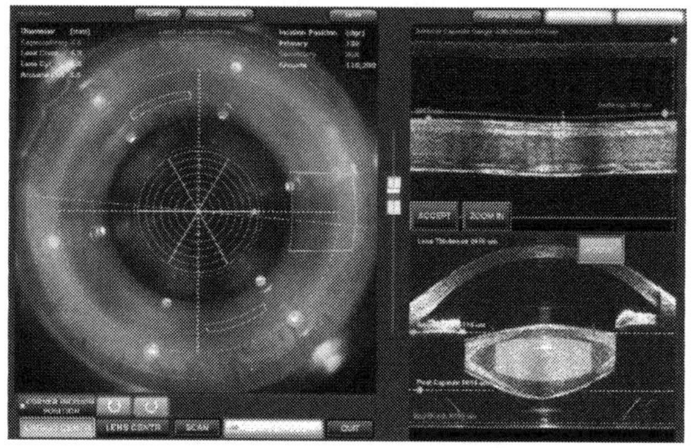

Fig 5. Working plan in FLACS. Right: eye view after docking. Left: Optical Coherence Tomography images of the lens capsule and anterior segment section with cornea and lens (yellow). (Courtesy of dr. F.D'Arrigo)

Discussion

Ophthalmology was the first medical specialty to utilize laser energy in patient treatment, and still more laser operations than any other specialty are done. The use of femtosecond laser in ocular surgery represents the most advanced and recent technology. This technology, in comparison to the traditional one, does not require the use of blades, therefore resulting less traumatic and ensuring a faster healing. Since their introduction and first human treatment in 2008, by Franz Fankhauser and Daniele Aron-Rosa, significant improvement was achieved. The circular opening of the capsule in the cataract surgery, is perfectly centered and round, thus resulting very important in the post-operative phase when enables a better intraocular lens centration and stability. The liquefaction of softer lenses and the fragmentation of harder lenses allow chopping the nucleus without the use of phaco-energy, in fact there were observed a reduction of 39% to 43% in the energy and a 51% decrease in time of phacoemulsification [9]. The most important advantage of femtolaser cataract technology consists in the planning and customization of some important surgical steps. Reduced trauma of the cornea lead to a decreased lost of endothelial cells as compared to conventional phacoemulsification [10-12]. Other advantages in FLACS are standardized corneal incisions, centered and round capsulorhexis, and lens nucleus fragmentation even in eyes with hard cataracts [13]. The laser precision is due to the real time OCT software programs, which cover the whole anterior segment until the posterior lens capsule. Quality of vision can be improved with less posterior chamber lens tilt, more centralized position of the lens, less endothelial damage, less macular edema, and less posterior capsule opacification formation [14-16]. The different cut modalities in corneal transplantation offer many advantages with respect to the traditional keratoplasty. The principle advantages of FSL in this field are faster wound healing, an earlier suture removal, a better donor-host matching and reduced astigmatism. Moreover, femtosecond laser assisted Deep Anterior Lamellar Keratoplasty (DALK) results to be easier to perform as compared to the manual procedure [17]. Corneal grafting with FSL cuts proved to be more predictable and safe technique with better clinical outcomes, due to the laser's precision and repeatable, well controlled, sizing of the donor and recipient corneal buttons [18]. This technique can offer a better centration, a thinner graft/host complex and an earlier corneal healing. As to refractive surgery different advantages of the femtosecond lasers used were reported. In Femtolasik a reduced incidence of flap complications like buttonholes, free caps, irregular cuts are observed [19]. A perfect control of the flap diameter and thickness, side cut angle, hinge position and length, better thickness predictability and increased precision of the flap creation, improved significantly the flap safety and decreased incidence of epithelial ingrowth [20-21]. In fact, several flap-related complications occurring when the conventional microkeratome were used actually are overcome by the femtosecond laser employment. Femtosecond laser creates corneal flaps that have better safety and thickness predictability, their adherence is stronger with respect to the classical manual cut and less influenced by trauma, minimizing the risk of flap displacement and less rate of epithelial ingrowth. SMILE offers additional advantages as no corneal flap is created, and this procedure presents less risk for post-surgical dry eye related to corneal nerves cut and lower incidence of ectasia with respect to LASIK [22-25].

Conclusion

Wide diffusion of femtosecond lasers is to be expected in ophthalmology in the next future. In fact this technology, working like a blade managed by sophisticated computer software allows the technique to be adapted to each individual case with extreme safety, precision, predictability and repeatability. Moreover FSL work with reduced trauma to the tissues, which results in less inflammation and an earlier recovery. Thanks to the significant impetus of technology, the surgery is moving toward a standardization of surgical procedures and customized approach to each patient.

References

1. GK. Marino, MR. Santhiago, SE. Wilson. Asia Pac J Ophthalmol. (Phila). **6**,5 (2017) G. Sutton, SJ. Bali, C. Hodge. Curr Opin Ophthalmol. 24, 1 (2013)
2. GK. Marino, MR. Santhiago, SE. Wilson. Asia Pac J Ophthalmol. (Phila). 6,5 (2017)
3. M. Taneja, VM. Rathi, B. Bagga. Optom Vis Sci. **92**, 2 (2015); 1.
4. A.Aristeidou, E.Taniguchi, M. Tsatsos, R. Muller, C. McAlinden, R. Pineda 4, E. Paschalis. Eye Vis (Lond). **14**, 2 (2015)
5. C. Chan, R. Ritenour, N. Kumar., W Sansanayudh., D. Rootman. Cornea. 29, 3 (2010).
6. G. Sutton, SJ. Bali, C. Hodge. Curr Opin Ophthalmol. 24, 1 (2013)
7. ZZ. Nagy. J Refract Surg. 28, 12 (2012).
8. MM. Gavris, R. Belicioiu, I. Olteanu, I. Horge, J. Rom. Ophthalmol. 59,1 (2015)
9. ZZ. Nagy, JP. Szaflik. Klin Oczna.;**114**, 4 (2012)
10. S. Ganesh, S. Brar. Int J Ophthalmol. **18**, 8 (2015)
11. J. Kacerovská, M. Kacerovský, R. Kadlec. Cesk Slov Oftalmol. **69** 5 (2013)
12. AJ. Kanellopoulos, G. Asimellis. Eye Vis (Lond). **29**, 3 (2016)
13. M. Pahlitzsch, N. Torun, ML. Pahlitzsch, MKJ. Klamann, J. Gonnermann, E. Bertelmann, T. Pahlitzsch. Semin Ophthalmol. **32**, 4. (2017)
14. A. Agarwal, S. Jacob. Curr Opin Ophthalmol. **28** 1 (2017).
15. H. Abouzeid. J Cataract Refract Surg. **42** 8 (2016)
16. Z. Zoltan Nagy. Clin Ophthalmol. **8**: 1157–1167. (2014)
17. S. Mariacher, M. Ebner, AM. Seuthe, K. Januschowski, C. Ivanescu, N. Opitz, P. Szurman, K. Boden. J Cataract Refract Surg. **42** 8 (2016)
18. Blériot A, Martin E, Lebranchu P, Zimmerman K, Libeau L, Weber M, Vabres B, Orignac I. J Fr Ophtalmol. **40**, 7 (2017)
19. J. Alio, A. Vega-Estrada, F. Soria, A. Abdou. Jaypee Brothers Medical Publishers. 75, 76 (2013)
20. T. Kohnen, OK. Klaproth, V. Derhartunian, D. Kook. Ophthalmologe. **107**, 5 (2010)
21. ZS. Bashir, MH. Ali, A. Anwar, MH. Ayub, NH. Butt. J Pak Med Assoc. 67 4 (2017)
22. D. Touboul, F. Salin, B. Mortemousque, P. Chabassier, E. Mottay, F. Léger, J. Colin. J Fr Ophtalmol. 28 5 (2005)

23. S. Taneri, S. Kießler, A. Rost, B. Dick. Klin Monbl Augenheilkd. 234 1 (2017)
24. I. Seven, A. Vahdati, IB. Pedersen, A. Vestergaard, J. Hjortdal, CJ. Roberts, WJ Jr Dupps. J Refract Surg. 33,7 (2017).
25. Burazovitch J, Naguzeswski D, Beuste T, Guillard M. J Fr Ophtalmol. 40, 6 (2017)
26. B. Spiru, S. Kling, F. Hafezi, W. Sekundo. Invest Ophthalmol Vis Sci. 58, 5 (2017)

α particle space distribution from fusion reactions in Boron irradiated by mono-energetic protons

Riccardo De Angelis[1], *Danilo* Giulietti[1,2,*], *Lucio* Calcagnile[3], *Gianluca* Quarta[3], *Pierluigi* Andreoli[1], *Mattia* Cipriani[1], *Fabrizio* Consoli[1], *G.* Cristofari[1], *Domenico* Delle Side[3], *Giorgio* Di Giorgio[1], *Francesco* Ingenito[1], and *Lucio* Maruccio[3]

[1]*ENEA, Fusion and Nuclear Safety Department, C.R. Frascati, Via E. Fermi 45, 00044 Frascati, Italy*
[2]*Physics Department of the University and INFN, Pisa, Italy*
[3]*CEDAD, Dipartimento di Matematica e Fisica "E. De Giorgi", Universitá del Salento – Lecce, Italy*

Abstract. This work presents results regarding the p + ^{11}B $\rightarrow$ 3α + 8.7 MeV aneutronic fusion reaction. We obtained such results by directing a 0.675 MeV proton beam from a Tandetron accelerator to a solid Boron sample, varying the incidence angle. Three different detectors were used to reveal the alpha particles emitted during the experiments. The evidence obtained leads us to propose an innovative scheme to investigate the yield of the aneutronic reaction when the proton beam is directed against a ^{11}B laser induced plasma.

1 Introduction

The fusion reaction p+^{11}B $\rightarrow$ 3α+8.7 MeV has been considered in the past as a possible fuel for future fusion reactors. The advantages over more conventional approaches (such as (D, T) and $(D,^3 He)$) lie in materials availability, absence of high energy neutrons and easiness of conversion to electrical energy. The possibility to produce this reaction in a plasma has been considered both for magnetically confined [1] and laser produced plasmas [2, 3], the latter being favored by the presence of strong non thermal energetic ions in laser-plasma interactions. Although the optimum cross section for this reaction occurs at very high colliding energies between the nuclei, the free electrons changing the screening of nuclei may result in the increase of the reaction rate.

At ENEA we have been studying the possibility to trigger aneutronic reactions such as p + ^{11}B $\rightarrow$ 3α + 8.7 MeV in laser produced plasmas. Experiments were performed in various laser facilities [4–6]. With the ABC, laser focusing 100J on plastic targets doped with Boron, the number of α particles detected is very low and the response of detectors is often uncertain. In order to test the behavior of our detectors we conducted conventional experiments by accelerating mono-energetic protons against a Boron target. The experiments, performed using three different detectors (CR39, imaging plates and Passivated Implanted Planar Silicon (PIPS)), allowed to detect reaction alfa particles and to measure the relative energy spectrum. Moreover the measurements represent an interesting and fruitful calibration of CR39 plates for proton and α particle detection. A pilot experiment was also planned and its feasibility at CEDAD evaluated. It consists of the study of the

Figure 1. The 3 MV Tandetron accelerator at CEDAD – Brindisi, University of Salento.

(p + ^{11}B $\rightarrow$ 3α) reaction in a B plasma. The experimental set-up allowed to direct to a B plasma a 0.675 MeV proton beam. The plasma was produced by focusing on the B target the third harmonic (355nm) of a ns pulsed Nd:YAG laser, while the proton beam was focused on that plasma up to a 50 μm diameter.

2 Materials and methods

In Figure 1 a schematic view of the Tandetron accelerator used in the experiment is reported.

The p + ^{11}B reaction was first produced At CEDAD by a proton beam $\Phi \approx$ 1 mm impinging normally on a solid Boron target (thickness δ = 1.14 mm). The particles emitted were measured by a PIPS detector located at the periphery of the vacuum chamber at 10° to the target normal as it is shown in 2. The PIPS detector has a 25 mm^2 active surface (3 mSr detection solid angle) with 15 keV energy resolution. In particular, we irradiated solid natural Boron with 0.675 MeV mono-energetic protons, detecting the α particles produced and the backscattered protons. For calibration purposes, the particles were also detected using

*e-mail: danilo.giulietti@unipi.it

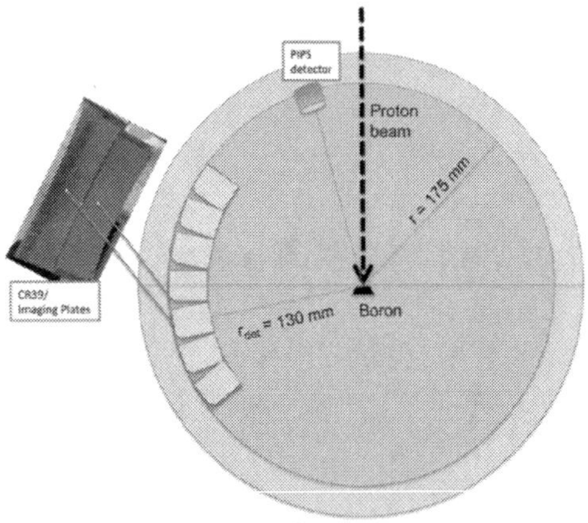

Figure 2. The vacuum chamber and the detector layout.

CR39 plates and imaging plates. Measurements were performed by irradiating the Boron sample at fixed impinging angle while varying the detection angle and afterward by keeping constant the angle between the proton beam and the detector view line and varying the proton beam incidence angle (i.e. by rotating the Boron sample).

The signal from the PIPS detector showed a clear evidence of the production of α particles with the expected energy spectrum, accompanied by backscattered protons. The energy of the incident proton was optimized to maximize the signal and the best value of E_p = 675 keV was found as expected.

The absence of artifacts was checked by substituting the natural B (80% ^{11}B + 20% ^{10}B) with pure ^{10}B or W targets. In Figure 3 we can see that no α particle are produced when the proton beam is directed to solid W target. It is worth noticing that energy measurements are reported in "channel". This corresponds to the analog-to-digital conversion operated by the detection electronics. Nevertheless, each channel corresponds to an energy of 3.4 keV.

In order to eliminate the contribution of backscattered protons, all the detectors were covered with 7 μm Al foils, which completely absorbed scattered protons. The spectra registered by the PIPS in the two cases (with and without the Al shield) is shown in Figure 4.

3 Results and Discussion

3.1 α yield Angular dependence

The angular dependence of the detected α particles was monitored and resulted consistent with the available literature. It is known indeed that the α particles yield has a weak dependence on the detection angle (θ). This dependence could be expressed in terms of a linear combination of the first 4 Legendre polynomials $P_n(\cos\theta)$ [7]. The coefficients of such expansion depend on the energy of the incident protons (E_p) and on the particular α particle emitted in the reaction, i.e.

$$\text{yield} \propto 1 + \sum_{n=1}^{4} a_n(E_p)P(\cos\theta)\,. \qquad (1)$$

Literature results show in particular that the two main contributions (namely α_0 and α_1) have a slightly different angular distribution [7]: α_1 is practically angle-independent, while α_0 weakly depends on θ.

In Figure 5 the dependence of the α energy spectrun on detection angle is reported. It is easy to see that the spectrum maximum displaces towards higher energies as the detection angle increases, whereas integrated α particle yield decreases. This is in good agreement with literature. In particular, we observe that the spectrum at the maximum is mainly the sum of the α_0 and α_1 contributions. Figure 5 can be interpreted as due to a contribution of the alpha particles at higher energies that remains practically constant, while the one of the lower energy alpha particles slightly decreases, leading to the observed displacement. We can see (Figure 6 left), in particular, that the maximum displacement is quite narrow, near 70 channels (i.e. less than 0.25 MeV). Moreover, integrating the plots of the Figure 5 we can ascertain that the yield remains always of the same order of magnitude (Figure 6 right).

3.2 CR39 and IP data

A set of CR39 track detectors and Imaging Plates (Fuji SR) was also deployed along the inner chamber wall. This allowed the calibration of the detectors' responses and to estimate the dimensions of the tracks produced by the α particles. Following standard etching procedures [8] (7 h at 70°C in NaOH 6.25 molar), α particles were easily recognized as tracks with a 4 – 8 μm diameter. By irradiating the Boron target in normal incidence condition, we found $\approx$ 1000 particles/mm^2 after a 900 s exposition of the B sample to continuous 2 nA proton beams. The CR39 analysis provided an angular pattern of alpha emission in accordance with PIPS results (Figure 7).

3.3 Protons on a Boron plasma

A preliminary attempt to produce the fusion reaction in a plasma was performed by sending the proton beam on a plasma pre-formed by focusing a 0.4J, 10ns, 10 Hz rep. rate Nd:Yag laser (3rd harmonic λ = 355 nm) on a Boron target. These measurements were conducted on a second chamber were the proton beam dimension was about Φ = 50 μm, see Figure 8. The α particle yield on the same solid target was preliminary tested and was within a factor 2 similar to that of the other chamber. The layout was such that the continuous proton beam was close to the target surface without touching it, so that no reaction occurs in such conditions. When the laser pulsed hit the target the plasma created interacts with the proton beam, allowing the occurrence of p-B reactions. Figure 9 shows one of the CR39 used in this setup. In these conditions, the average count was $\approx$ 50α/mm^2, with a 1 nA continuous proton beam for about 3000 laser pulses.

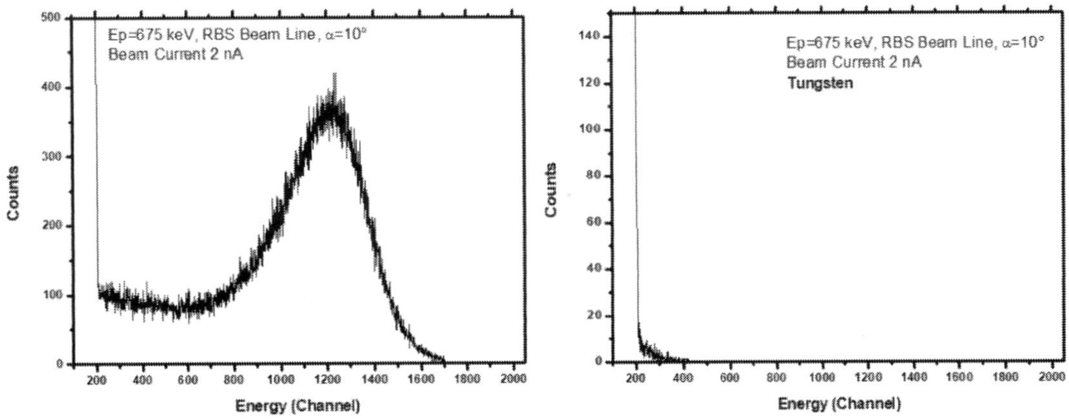

Figure 3. α particle yield when 0.675 MeV protons are directed on ^{11}B (left) and on W (right).

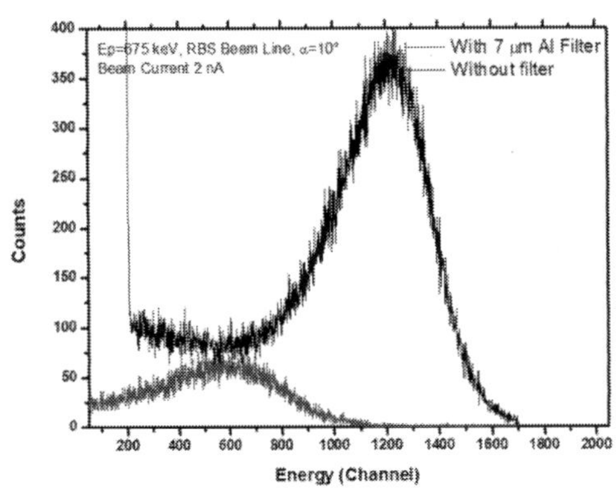

Figure 4. The spectrum of the α particle produced in the p + ^{11}B fusion reaction as detected with and without a 7 μm Al filter in front of the PIPS detector.

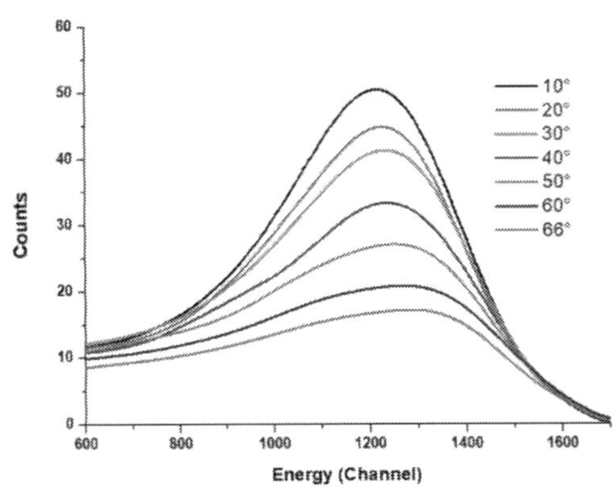

Figure 5. Dependence of the α energy spectrum on detection angle.

4 Discussion of the experimental results

We performed a first "back of the envelope" estimate of the number of alpha particles that should be produced in the interaction of the proton beam with both the ^{11}B sample in solid state and in the plasma state. In the first case, we direct a proton beam (2 nA current) for 900s onto a ^{11}B target, which corresponds to a total number of protons $N_p \approx 10^{13}$ sent to the Boron. Taking into account the penetration range of 700 keV protons in Boron ($L_{pB} \approx 10^{-4}$ cm), it is possible to estimate the number of fusion events (N_f)

$$N_f \approx n_B N_p \sigma_{pB} L_{pB} = 10^9 \text{ fusions} \qquad (2)$$

Where $n_B \approx 10^{24} cm^{-3}$ is the number of Boron atoms per volume unit; $\sigma_{pB} \approx 1$ barn is the cross section of the fusion process. Once estimated the number of fusion processes, the alpha produced is 3 times that number, i.e. $N_\alpha \approx 3N_f$. It is easy to get the number of alpha hitting on 1 mm^2 at a distance of 130 mm from the source, where the CR39 detector is located. We found: $N_\alpha/mm^2 \approx 1.5 \times 10^4/mm^2$, while the alpha detected by the CR39 were $N_\alpha/mm^2 \approx 10^3/mm^2$. Better agreement was obtained with the PIPS detector. In this case, the number of the alpha particle we estimate to reach the detector is near $2 \times 10^5/mm^2$, while the ones effectively counted are 1.6×10^5 . The disagreement of the CR39 measurements could be partly ascribed to the small differences observed in the angular distribution of the alphas. The CR39 plates were positioned at a larger detection angle than the PIPS, so they receive nearly half the number we computed. Consequently, we obtain $N_\alpha/mm^2 \approx 7 \times 10^3/mm^2$. Such estimates, indeed, neglect the fact that the alpha particles produced must in turn exit the Boron sample and the dependence of the cross section on energy (considering that protons lose energy as they penetrate the target).

In the second case, the estimate of the alpha particles produced when the proton beam interacts with the Boron plasma is much more complicated and the approximations even more rough. We can sketch the process as a well-defined proton beam that crosses a Boron plasma with

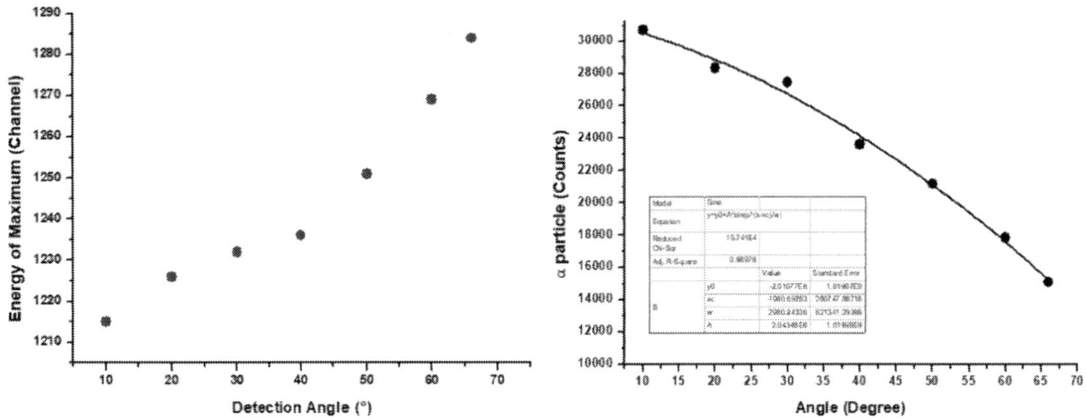

Figure 6. The maximum of the α particle spectrum (left) and the α reaction yield versus the detection angle (right).

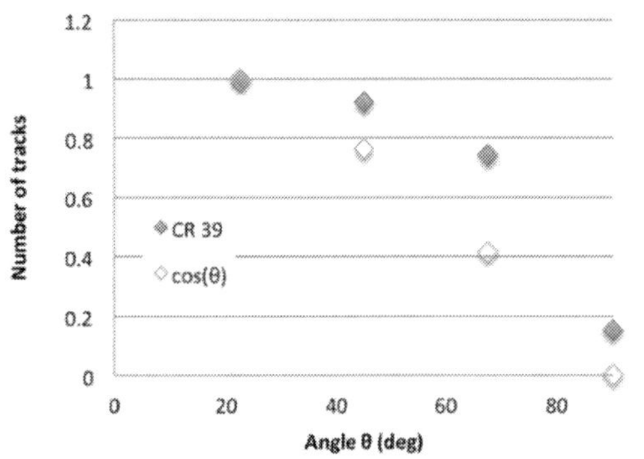

Figure 7. α particle yield versus detection angle, obtained by CR39 exposition.

Figure 9. The CR39 used in this setup. In these conditions the average count was $\approx 50\alpha/\text{mm}^2$, with a 1 nA continuous proton beam for about 3000 laser pulses.

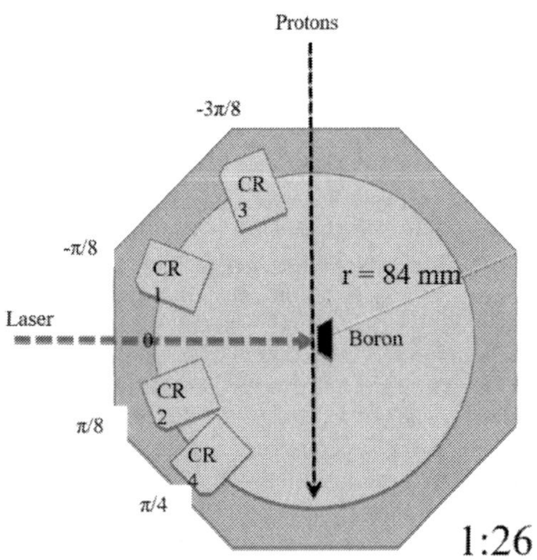

Figure 8. Experimental chamber during experiments with Boron laser induced plasma.

a density about the critical density for a $0.35\,\mu$m radiation ($n_{Bplasma} \approx 10^{22} cm^{-3}$). The plasma extends for about $L_{Bplasma} \approx 1mm$, which corresponds to the sound velocity at plasma temperature at about $T \approx 100eV$ for the duration of the laser pulse near 10 ns, while the proton flux, corresponding to a 1 nA current, is $F_p \approx 6\times10^9$ protons/s

$$\frac{N_f}{\text{shot}} \approx n_{Bplasma}\sigma_{pB}L_{Bplasma}F_p\tau = 0.1\ \frac{\text{fusions}}{\text{shot}} \quad (3)$$

Consequently, the number of the alpha particles impinging on 1 mm^2 at 84 mm from the alpha source after 3000 shots should be $N_\alpha/mm^2 \approx 10^{-2}\alpha/mm^2$. The number of alpha particle detected by CR39 is sensibly higher ($N_\alpha/mm^2 \approx 50\alpha/mm^2$). This extremely high yield probably cannot be attributed to an increased cross section for the fusion process in a plasma, but to some effects related to the particularly hasty set-up. There is therefore a need to repeat the experiment with a more accurate set-up. In fact in a plasma with an electron density $n_{Bplasma} \approx 10^{22}cm^{-3}$, at a temperature $T \approx 100\,eV$, the Debye length is $\lambda_D \approx 7.5\,$Å, about 10 times the Boron radius $r_B \approx 0.8\,\text{Å}$. So the electron screening is less effective of the electron cloud of the atom and we expect the cross section for the fusion process is smaller in such a plasma. The situation could be opposed in a denser ($n_{Bplasma} \approx 10^{24}\,cm^{-3}$) and cold plasma ($T \approx 10\,eV$) for which we have $\lambda_D \approx 0.23\,\text{Å}$, as one might think to realize by focusing an X-ray backlight radiation on a solid ^{11}B target.

5 Conclusions

The results obtained for solid Boron give a number of α particles per incident proton N_α/N_p lower than expected, although estimates of the reabsorption of particles inside the solid were not yet taken into account. Conversely for the Boron plasma test, N_α/N_p was higher than expected showing a possible effect of stray protons inducing unwanted reactions on the solid Boron. Further experiments have been already planned to confirm the obtained results.

References

[1] *TAE technologies, inc.*, https://trialphaenergy.com, accessed: 2017-09-30 and N.Rostoker, M.W. Binderbauer, H.J. Monkhorst, Science **278**, pp. 1419–1422 (1997)

[2] D. Moreau, Nuclear Fusion **17**, 13 (1977)

[3] V.S. Belyaev, A.P. Matafonov, V.I. Vinogradov, V.P. Krainov, V.S. Lisitsa, A.S. Roussetski, G.N. Ignatyev, V.P. Andrianov, Phys. Rev. E **72**, 026406 (2005)

[4] A. Bonasera et al., in Proc. Int. Conf. Fission and Prop. Neut. Rich Nuclei, Sanibel Island, USA (2007)

[5] C. Baccou, S. Depierreux, V. Yahia, C. Neuville, C. Goyon, R. De Angelis, F. Consoli, J. Ducret, G. Boutoux, J. Rafelski et al., Laser and Particle Beams **33**, 117–122 (2015)

[6] D. Giulietti, P. Andreoli, D. Batani, A. Bonasera, G. Boutoux, F. Burgy, M. Cipriani, F. Consoli, G. Cristofari, R.D. Angelis et al., Nuclear Instruments and Methods in Physics Research Section B: Beam Interactions with Materials and Atoms **402**, 373 (2017)

[7] R.E. Segel, S.S. Hanna, R.G. Allas, Phys. Rev. **139**, B818 (1965)

[8] F. Ingenito, submitted to the European Physical Journal Web Conferences

Directional Track Selection Technique in CR39 SSNTD for low-yield reaction experiments

Francesco Ingenito[1], *Pierluigi* Andreoli[1], *Dimitri* Batani[2], *Aldo* Bonasera[3], *Guillaume* Boutoux[2], *Frederic* Burgy[2], *Mattia Cipriani*[1], *Fabrizio* Consoli[1], *Giuseppe* Cristofari[1], *Riccardo* De Angelis[1*], *Giorgio* Di Giorgio[1], *Jean Eric* Ducret[2], *Danilo* Giulietti[4], and *Katarzyna* Jakubowska[2,5]

[1]ENEA, Fusion and Technologies for Nuclear Safety Department, C. R. Frascati, Via E. Fermi 45, 00044 Frascati (Roma), Italy
[2]Centre d'Etudes Lasers Intenses et Applications (CELIA), Université de Bordeaux, CNRS, CEA, 351 cours de la liberation 33405 Talence, France
[3]Cyclotron Institute, Texas AM University (TAMU), 120 Spence St, College Station, TX 77840, United States
[4]Physics Department of the University and INFN, Largo B. Pontecorvo 3, 57127, Pisa, Italy
[5]Institute of Plasma Physics and Laser Microfusion (IPPLM), Hery Street 23, 01-497 Warsaw, Poland, Poland

Abstract. There is a great interest in the study of p-^{11}B aneutronic nuclear fusion reactions, both for energy production and for determination of fusion cross-sections at low energies. In this context we performed experiments at CELIA in which energetic protons, accelerated by the laser ECLIPSE, were directed toward a solid Boron target. Because of the small cross-sections at these energies the number of expected reactions is low. CR39 Solid-State Nuclear Track Detectors (SSNTD) were used to detect the alpha particles produced. Because of the low expected yield, it is difficult to discriminate the tracks due to true fusion products from those due to natural background in the CR39.To this purpose we developed a methodology of particle recognition according to their direction with respect to the detector normal, able to determine the position of their source. We applied this to the specific experiment geometry, so to select from all the tracks those due to particles coming from the region of interaction between accelerated protons and solid boron target. This technique can be of great help on the analysis of SSNTD in experiments with low yield reactions, but can be also generally applied to any experiment where particles reach the track detector with known directions, and for example to improve the detection limit of particle spectrometers using CR39.

1 Introduction

Solid-State Nuclear Track Detectors (SSNTD) are samples of solid materials where exposition to ionizing radiation generates tracks due to local damaging of the detector [1]. In polymeric materials, such as CR39, tracks are caused by the breaking of the long polymer chains. Along these tracks the material is more susceptible to chemical attack. Therefore, these regions have much faster velocity of etching with respect to undamaged material. Etching of several hours allows for enlarging of the damage to micrometer dimensions, which can be observed with a microscope [1]. Information on particle energy can be determined by track dimension and etch-rate of track compared to that of the bulk. Moreover, circular pits are generated if the particles enter the surface at normal incidence, otherwise elliptical pits are generated, where ellipticity and orientation of the pit mouth are related to the direction of incidence.

These detectors are widely used in different fields and are fundamental in environments heavily affected by electromagnetic radiation of high intensity, where active electronic detectors will be not applicable [2-4]. In particular, they are the key diagnostics in investigation of low-yield nuclear reactions, such as the neutron-less p-^{11}B reactions generated by laser-matter interaction [5-12]. There is a strong interest in their application, both for energy production and for determining the reaction cross-sections at low energies for astrophysical studies. One of the main problems in this context is the discrimination of a very small number of fusion products with respect to the background.

2 Experiment description

In this work we show results of experiments performed at CELIA laboratories in Bordeaux. An accelerated particle beam was produced by Target Normal Sheath Acceleration (TNSA) [13], focusing the ECLIPSE laser beam on 10 μm spots on an aluminium thin foil of 6 μm thickness [12]. The laser parameters were: $\lambda = 800$ nm, up to 170 mJ energy in 35 fs of FWHM (Full Width Half Maximum), temporal contrast of $4{\cdot}10^{-9}$ at 12 ns and 10^{-5} at ASE (Amplified Spontaneous Emission) level, energy stability 1.5% RMS (Root Mean Square) shot to shot. The Intensity was $10^{18\text{-}19}$ Wcm^{-2}. As shown in Figure 1 and described in more details in reference [12], a proton beam

* Corresponding author: riccardo.deangelis@enea.it

with energies up to ~ 300 keV energy was generated in this configuration and directed to a CR39 plate shielded with a thick Al foil. A hole on this plate, placed in correspondence of that on the thick Al, allowed protons to reach the solid B sample. The expected reaction is $p+^{11}B \rightarrow 3\alpha$ [5-12]. The side of the CR39 detector facing the B target was unfiltered and so directly exposed to the alpha particles, but also to the recoil protons from target, which represented a source of noise to the detector.

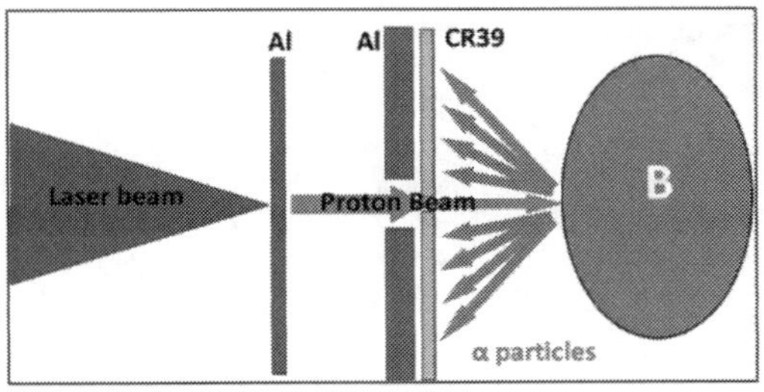

Fig. 1. Experimental set-up.

2.1 CR39 SSNTD analysis

The CR39 detectors exposed to the radiation were etched with a NaOH solution 6.25 M at 70°C for 7 hours. Under these conditions, tracks with diameter of about 20 μm are expected for alphas in the 2-4 MeV range generated by possible pB fusions [5-12].

After the etching, the samples were scanned with an automated procedure by a reflection microscope, and this led to reconstruct a square image of 20 mm side. This image was analyzed with the ImageJ program [14] and its built-in particle-counting routine.

Both the images and the particle counter results showed 'noise', in the form of a high number of equivalent little tracks with a diameter of a few microns. In some cases these noise tracks also formed clusters, which were recognized as a single big track by the counter program. A careful recognition process is thus necessary, which would require a human operator performing it. Of course, this is a time-consuming and tedious work, which limits appreciably the extension of the detector surface that can be analyzed. This is a drawback in low-rate reactions as pB. Moreover, a giga-pixel image was produced as results of microscope scanning. For all of these reasons a fully automatic and reliable selection method was needed.

3 The method

Track shape on SSNTDs is influenced by a number of parameters, some due to the etching conditions and others to the geometry of the particle collision [15]. In general, the pit mouth of the etched track can be described as an ellipse with the major axis along the particle propagation direction. We used this specific feature to discriminate tracks actually due to the reactions of our interest from those of background. In the scheme of Figure 1, possible alpha particles imping on the CR39 have a specific propagation direction on any given point of the detector. This Directional Track Selection (DTS) method was applied only to tracks having transversal dimensions compatible with those expected for alpha particles up to 4 MeV, and this greatly improved the procedure.

3.1 Determination of key parameters

Here we explain the steps followed in the procedure.

3.1.1 Determination of the particle direction

As already mentioned, the shape of the track is approximately an ellipse. Taking advantage of the specific ImageJ built-in function, each track is fitted with the proper ellipse and its major axis direction is determined by the φ angle with the horizontal axis of the image. From this information it is possible to infer the direction of the possible incident particle.

3.1.2 Determination of the expected particle direction

The expected particle direction is the actual direction of particles emitted by a given source. To find it, as a first step we projected the particle source on the detector plane, as indicated in Figure 2. The grey circle in that picture is the projection on the detector plane of the region of boron interacting with the proton beam, as shown in Figure 1. We set our reference system in the center of this circle. In this reference system the expected particle direction is represented by the segment of length d connecting the origin of the reference system to the center of the track image, and it is individuated by the θ angle it forms with the horizontal axis.

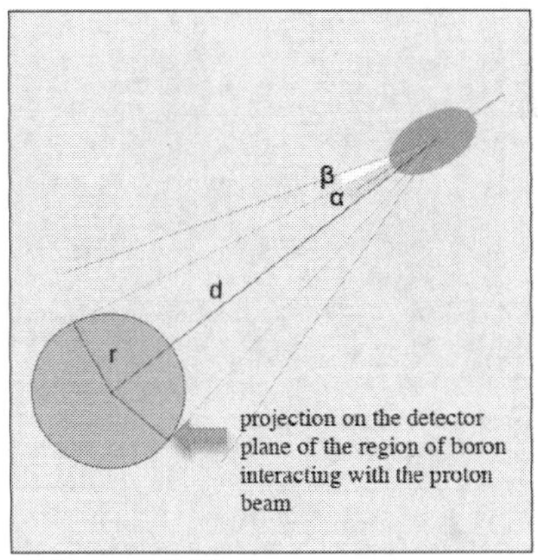

Fig. 2. Basic scheme for parameters of track selection.

3.1.3 Determination of the acceptance angle

Since the source has a specific extension, a particular track can be due to particles emitted by any point within source projection. Moreover, the uncertainties on the position, shape and extension of the source, on position of the track center and on the major-axis direction must be also taken into account. To this end, we define an overall acceptance angle, calculated as the sum of two contributions. As shown in Figure 2, the first term accounts for the possible provenience of particles from a generic point on the source:

$$\alpha = \arcsin (r / d) \qquad (1)$$

where r is the radius of the circle including the projection of the source zone on the detector plate. The second term β is numerically evaluated for each image from the tolerances on the track ellipsoid drawn by the imaging program and the tolerances in assessing the source dimension and center position. The resulting $\gamma = \alpha + \beta$ acceptance angle is schematically indicated in Figure 2.

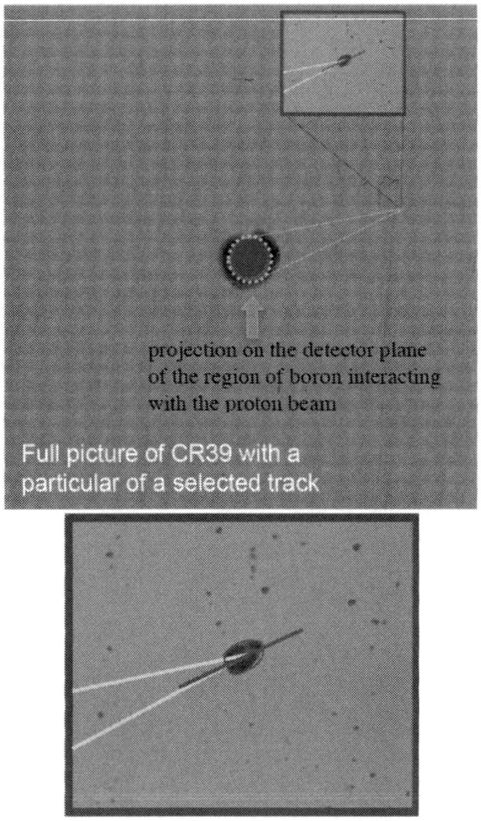

Fig. 3. In the top part of the picture there is the complete CR39 image obtained as an array of multiple microscope scans. A detail of one of these scans is shown, and it is also supplied in the lower part of the figure, enlarged with improved resolution (the size of the enlarged image is 180*225 μm). Particle direction, fitting ellipse and acceptance range are indicated.

3.2 Track selection

After the determination of the particle direction, the expected particle direction and the acceptance angle, the numerical process of selection can be performed. As previously mentioned, this is applied only to the tracks having dimensions compatible to alpha particles generated by p-B fusion events. Among these, the tracks having $|\varphi - \theta| \leq \gamma$ are selected. In this way the operator has a reduced number of tracks to control for possible further validation.

In Figure 3 we show a picture obtained by the microscope scan of a CR39 exposed to a cumulative experiment of 300 nearly identical shots on fresh portions of the target, with average energy on Al target of ~110 mJ, and good shot stability. This is the appearance of a recollection of many scanned images in a big one. At the picture center we see the hole of the CR39, with an annular blur due to heavy radiation damage, caused by boundary effects due to particle passage through the hole, backscattering of particles impinging on the B surface and X-ray flux coming from bremsstrahlung of particles on B surface. One of the pictures composing this array is enlarged, and also shown with improved resolution in the lower part of the same figure. It is possible to observe a specific track selected by the numerical process, with the blue ellipse resulting from the fitting procedure, the associated particle direction indicated by a blue segment, and the γ acceptance range represented by the two green lines.

As preliminary results of this technique we applied the selection procedure to all the detected traces. The first selection step, based only on the transversal dimension of the track, returned 112 cases compatible with emitted alphas from pB fusions. The application of the Directional Track Selection was able to reduce that number to only 12 traces, resulting in a more than 9 times decrease.

4 Conclusions

The great reduction of background indicated by the preliminary results of application of the presented selection technique allows for detailed individual analysis of the selected tracks by human operator with much lower effort. The technique is still in an early development stage, but allowed a far more efficient detector analysis. For this reason it can be of great importance when reactions with low yield (as in this p-^{11}B case) are studied. Moreover, the application of this methodology can be very important to improve the background reduction in many different experiment types where it is possible to have geometrical constraints for direction of particles reaching the track detector. For this reason this methodology can be very powerful for improving the detection limit when CR39 are used in magnetostatic, electrostatic or Thomson particle spectrometers.

This work has been carried out within the framework of the EUROfusion Consortium and has received funding from the Euratom research and training programme 2014-2018 under grant agreement No 633053. The views and opinions expressed herein do not necessarily reflect those of the European Commission. The authors acknowledge the support of the COST Action MP1208 "Developing the Physics and the Scientific Community for Inertial Fusion".

References

1. S. A. Durrani, R. K. Bull, *Solid State Nuclear Track Detection Principles, Methods and Applications*, (Pergamon Press, 1987).

2. J. Raimbourg, Rev. Sci. Instrum. **75,** 4234 (2004).
3. C.G. Brown Jr., *et al*, J. Phys.: Conf. Ser. **244**, 032001 (2010).
4. F. Consoli, *et al.*, Phys. Procedia **62,** 11 (2015).
5. V.S. Belyaev, *et al*, Phys. Rev. E **72**, 026406 (2005).
6. A. Bonasera, A. Caruso et al, *Proceedings of the 4th International Conference on Fission and Properties of Neutron Rich Nuclei*, 11-17 November 2007, Sanibel Island, USA.
7. S. Kimura, A. Anzalone, A. Bonasera, Phys. Rev. E **79**, 038401 (2009).
8. C. Labaune, *et al,* Nat. Comm. **4**, 2506 (2013).
9. A. Picciotto, *et al*, Phys. Rev. X **4,** 031030 (2014).
10. C. Baccou, *et al*, Laser Part. Beams **33**, 117 (2015).
11. C. Baccou, *et al*, Rev. Sci. Instrum. **86**, 083307 (2015).
12. D. Giulietti, *et al,* Nucl. Instrum. Meth. B **402**, 317 (2017)
13. A. Macchi, *et al*, Rev. Mod. Phys. **85**, 751 (2013).
14. Available at https://imagej.nih.gov/ij/
15. D. Nikezic, K.N. Yu, Mater. Sci. Eng. R-Rep. **46**, 51 (2004).

https://doi.org/10.1051/epjconf/201816705007

Research on Formation Mechanism of Dynamic Response and Residual Stress of Sheet Metal Induced by Laser Shock Wave

Aixin Feng[1,3], *Yupeng Cao*[2], *Heng Wang*[2] *Zhengang Zhang*[3]

[1]School of Mechanical and Electrical Engineering, Wenzhou University, Wenzhou, Zhejiang 325035, People's Republic of China
[2]School of Mechanical Engineering, Nantong University, Nantong, Jiangsu 226019, People's Republic of China
[3]School of Mechanical Engineering, Jiangsu University, Zhenjiang, Jiangsu212013,People's Republic of China

Abstract. In order to reveal the quantitative control of the residual stress on the surface of metal materials, the relevant theoretical and experimental studies were carried out to investigate the dynamic response of metal thin plates and the formation mechanism of residual stress induced by laser shock wave. In this paper, the latest research trends on the surface residual stress of laser shock processing technology were elaborated. The main progress of laser shock wave propagation mechanism and dynamic response, laser shock, and surface residual stress were discussed. It is pointed out that the multi-scale characterization of laser and material, surface residual stress and microstructure change is a new hotspot in laser shock strengthening technology.

1 Introduction

Laser shock processing (LSP) is widely used for surface modification of metallic materials owing to its characteristics like high strain rate and high pressure. The parameters of laser shock processing are closely associated with the size and distribution of surface compressive stress[1]. Presently, the multi - scale study concerning mechanism of laser shock wave propagation in the material and the deformation mechanism of the material under the high strain rate, the microstructure change, the residual stress distribution is still to be strengthened. Compared with the traditional mechanical strengthening technology such as shot blasting and sand blasting, the laser shock processing technology has the advantages of deep impact layer, controllable impact area and pressure, easy to maintain the surface roughness and dimensional accuracy and so on, which has a tremendous technical advantage in quantitative control surface residual stress. It is a research front and crucial for the current laser shock processing technology to understand the quantitative control of the residual stress on the surface of metal materials and breaking the conventional technology research of the impact of laser shock processing parameters on the material residual stress.

In this paper, the latest research trends in laser shock processing technology in regulating surface residual stress were investigated on the basis of three aspects: laser shock wave propagation mechanism and dynamic response, laser shock and surface residual stress, microstructure change[2,3]. It is pointed out that the multi-scale characterization of laser and material, surface residual stress and microstructure change is the current research hotspot.

2 Materials and Methods

The apparatus used for the of dynamic strain on surface (shown in Fig.1 (a)), consists of the black tape with 4mm×4mm×150μm closely adhered to the surface of the sample as the absorption layer, and the K9 glass with thickness of 4mm covered as the restraining layer on the black tape. PVDF piezoelectric sensor is located on the front of the sample, along with the laser spot radial direction and 2mm away from the center of the spot. Apparatus for the detection of dynamic strain consists of YAG solid state laser (Beamtech radium photonics Co.,Ltd., Canada),DL9140 digital oscilloscope (Yokogawa Group, Japan), the optical system, industrial computer, and workstations. Apparatus for the detection of dynamic strain on the back is shown in Fig.1 (b). The PVDF piezoelectric sensor is attached to the center of the laser spot at the back of the sample, and the size of the PVDF piezoelectric sensor is 30μm ×5mm×10mm.In the setting of laser shock parameters, the pulse width was 10ns, the wavelength was 1064nm, the pulse energy was 0.5J-1J, the spot diameter was 1mm, 2mm, 3mm. Under the restraining layer of K9 glass, the surface of the sample is shocked separately, and the clamping method of the constrained layer is shown in Fig.1 (c).

* Corresponding author: author@e-mail.org

 https://doi.org/10.1051/epjconf/201816705007

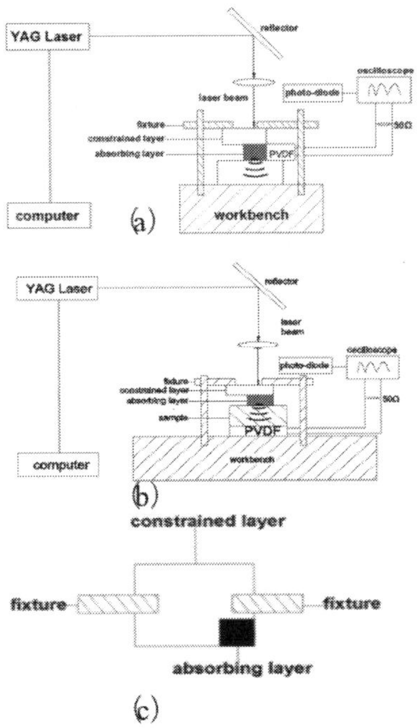

Fig. 1. Experimental apparatus of the detection of dynamic strain by laser impact the material sample. (a) surface (b)backside (c) the constraining layer was clamped by a fixture.

Based on the characteristics of dynamic strain of the thin metal plates, the correlation between laser shock processing parameters, dynamic response and distribution of surface residual stress was studied[4]. PVDF piezoelectric sensors were attached to back laser spot center and the surface of the sample along the radial, where the surface piezoelectric sensors were 3 mm away from spot center. The Gaia-R high energy pulse lamp pumped YAG laser system (France Thales Co., Ltd) was used. During the impact test, an aluminum foil of 6mm×6mm×150μm was used as an absorbent layer adhering to the surface of the 7050 aluminum alloy sheet and K9 glass as a restraining layer. The selected pulses energy were 2J、3J、3.89J、5.43J、8J, spot diameter 5mm, pulse width 10ns, wavelength 1064nm.

3 Results and Discussions

1.2 Shock wave propagation mechanism and dynamic response

The piezoelectric signal V (ε, t) was measured on the surface of the sample by means of the PVDF using fig.1 (a) device, as shown in Fig.2. V (ε, t) curve appeared the compressional wave whose waveform was the concave wave after the time point R1, and its amplitude was obviously increased to the surface Rayleigh wave. It can be inferred that when the pulse laser power density does not exceed 12.7GW / cm2, the surface Rayleigh wave and the shear wave did not produce coupling.

Using the device of Fig. 1 (b), the PVDF piezoelectric sensor measures the piezoelectric signal V (ε, t) on the back of the sample, as shown in Fig.3. The laser-induced elastic-plastic stress wave reached the back of the specimen to cause the first strain piezoelectric peak (point A1), and then reflected the elastic-plastic tensile wave propagating to the laser shock surface on the back of the sample, and caused a second strain piezoelectric peak (point B) on the back. Multiple shock waves propagating to the back of the specimen and the reverse tensile wave occurs unloading (C point), causing the back dynamic strain to fluctuate (point D, point E, point F) after the maximum peak. The stress wave decayed continuously in the reflection of the front and back surfaces of the specimen. Combined with the theory of impact dynamics, the experimental results of Zhang Yongkang, Wang Xuede and other, and the dynamic strain of the surface and the back of the sample, the "Laser Shock Wave Propagation Model 690 High-strength Steel Sheet" was established, as shown in Fig. 4. The "Propagation Model of Laser Shock Wave" describes the propagation processes of longitudinal compressional wave, reflected wave, transmitted wave, reverse stretching wave, shear wave, surface sparse wave and surface Rayleigh wave, comprehensively elaborating the propagation mechanism of laser shock wave in 690 high strength steel sheets[5]..

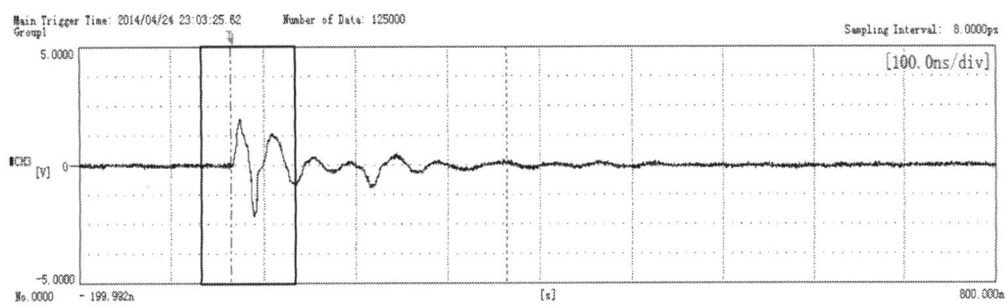

Fig.2. Dynamic strain of material at the area away from the edge of 0.5 mm when laser shock is loaded on the material surface with laser power densities of 12.7GW/cm^2

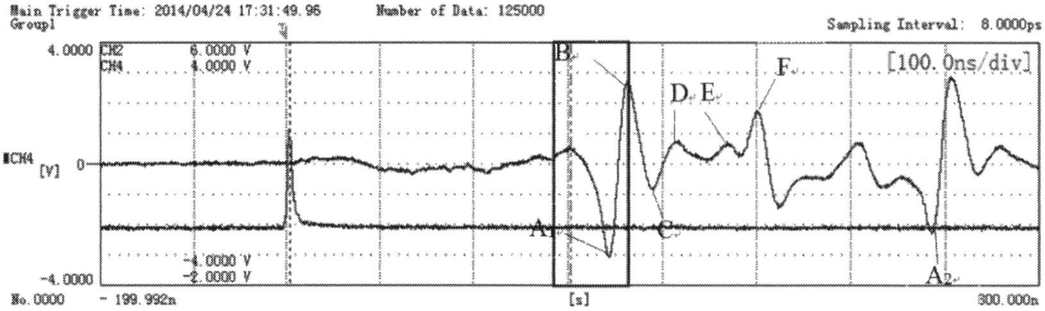

Fig.2. Piezoelectric waveform representation of the dynamic strain of sample backside when the laser shock is loaded on the thickness of 1mm material with a laser power density of 12.7GW/cm^2

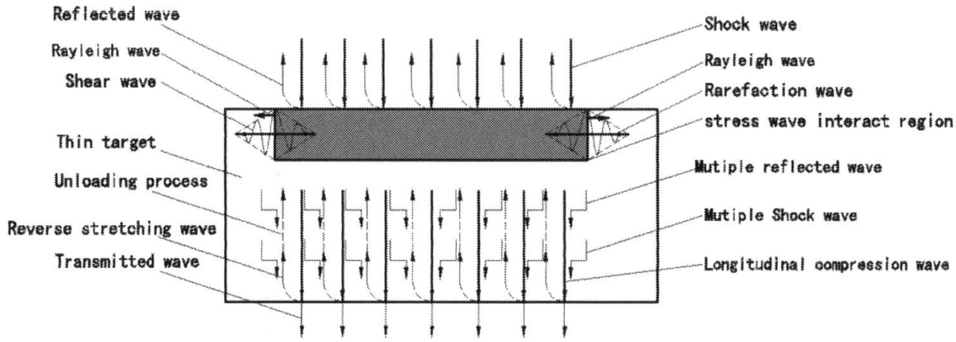

Fig.4. The laser shock wave propagation model of 690 high-strength steel sheet

3.2 Study on dynamic strain characteristics of metal sheet induced by laser shock

Based on the "Propagation Model of Laser Shock Wave", the results of numerical simulation and the boundary conditions of the surface dynamic strain test, a "Dynamic Strain Model of the Metal Sheet Surface Induced by Laser Shock" was built, and a testing method was also provided. The surface dynamic strains of 7050 aluminum alloy were measured by a PVDF piezoelectric sensor, as shown in Fig. 5 – Fig. 6. According to the attenuation law of shear wave in the 7050 aluminum alloy sheet, no coupling of shear wave and surface Rayleigh wave will be produced when the laser power density is less than or equal to 12.7GW/cm2, which is the boundary condition of surface dynamic strain model; the experimental data of 7050 aluminum alloy sheet loaded by laser shock showed that the "Dynamic Strain Model of the Surface of Metal Sheet Loaded by Laser Shock" is accurate and reliable. In the meantime, it has been also verified that the dynamic stress-strain curve of the 7050 aluminum alloy sheet under high strain rate has elasticity, yield and strengthening stages, similar to the static stress-strain curve under static tensile conditions[6]. The accuracy and reliability of the model were verified by experiments. In addition, the law of dynamic mechanical responses of metal sheets under high strain rate conditions was inventively discovered.

3.3 Residual stress formation mechanism of metal sheet surface induced by laser shock processing

Loading laser with different power densities shocks on 7050 aluminum alloy sheet, the maximum residual principal stress of the 7050 aluminum alloy sheet surface were biaxially distributed when power density was 1.53 GW/cm^2; when the power density was 1.98 GW/cm^2 and 2.77GW/cm^2, the maximum residual principal stress of the 7050 aluminum alloy sheet surface produced the "Residual Stress Hole" phenomenon; when the power density was 4.07GW/cm^2, the maximum residual principal stress of the 7050 aluminum alloy sheet was evenly distributed[3]. It can be seen from Fig. 7 (a) that when the power density was 1.53 GW / cm^2, the amplitudes of the obtained shear wave and the longitudinal wave were not large. The interaction between the surface rarefaction wave and the lateral deformation causes the maximum residual principal stress of the sample to be biaxial. As shown in Fig. 7 (b), no dynamic strain was measured by the PVDF piezoelectric sensor on the surface of the impact area after the shear wave and the Rayleigh wave was attenuated. The shock wave reflected back and forth in the thin plate sample and the rarefaction wave was propagated to the center of the spot, resulting in a large shear plastic strain in the central area of the sample, so the residual compressive stress in the central area of the laser spot were reduced, resulting in the phenomenon of "Residual Stress Hole ".With further increase of laser energy, the PVDF piezoelectric sensor pasted on the vicinity of the edge of the spot to detect the dynamic strain again, as shown in Fig.7 (c).

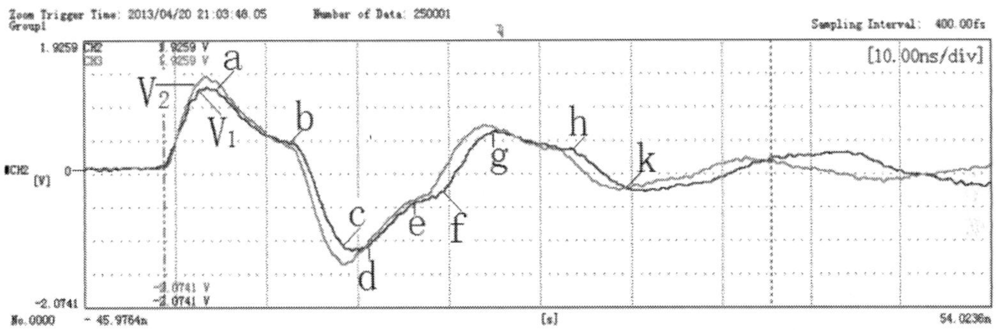

Fig.5.Piezoelectric waveforms (collected by the PVDF sensor) of the dynamic strain of the material when the laser shock is loaded with a laser power density of 12.7 GW/cm^2

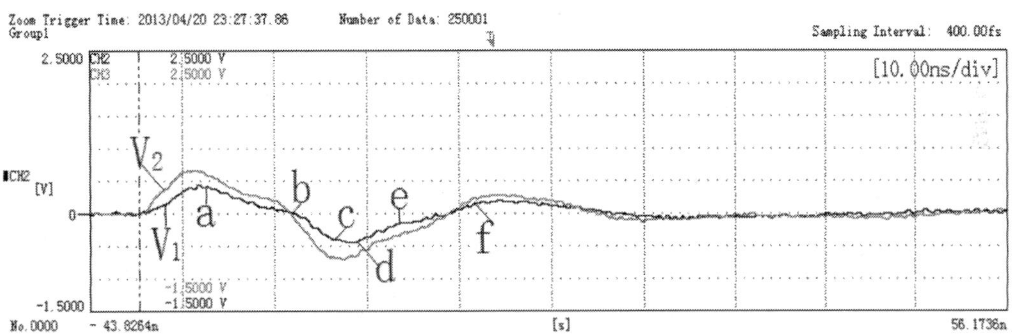

Fig.6.Piezoelectric waveform representation (collected by PVDF sensor) of dynamic strain of material when laser shock is loaded on the material surface with laser power density of 3.2 GW/cm^2.

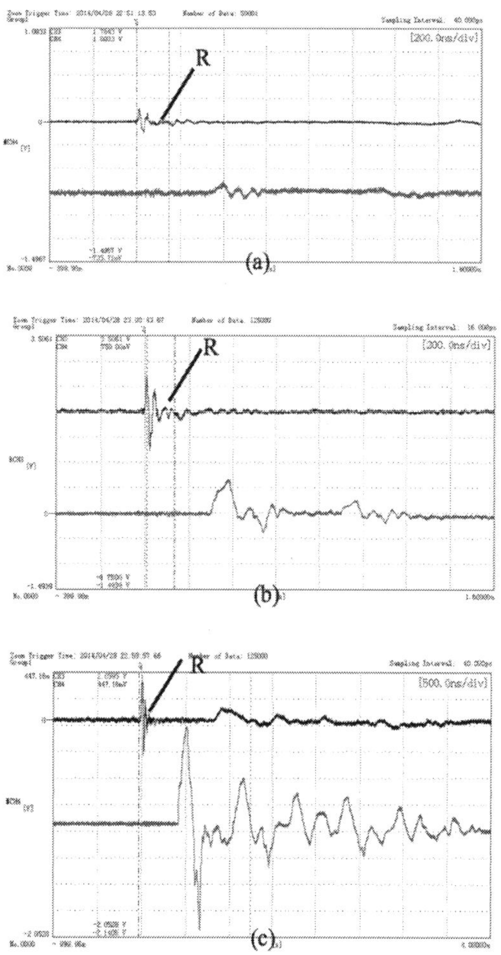

Fig. 7. Dynamic strain of material when laser shock is loaded on the material surface with laser power densities of (a)1.53 GW/cm^2; (b)2.77GW/cm^2; (c)4.07 GW/cm^2.

4 Conclusions Figures and tables

(1)The principle and method to measure laser shock wave characteristics were improved and optimized. Taking 690 high-strength steel sheet as a case, this research has, for the first time, established a "Propagation Model of Laser Shock Wave" by integrating dynamic responses on both sides of the plate. The "Propagation Model of Laser Shock Wave" described the propagation processes of longitudinal compressional wave, reflected wave, transmitted wave, reverse stretching wave, shear wave, surface sparse wave and surface Rayleigh wave, comprehensively elaborated the propagation mechanism of laser shock wave in 690 high strength steel sheets.

(2) It verified that the dynamic stress-strain curve of the 7050 aluminum alloy sheet under high strain rate has elasticity, yield and strengthening stages, similar to the static stress-strain curve under static tensile conditions.

(3) Based on the characteristics of dynamic strain of the thin metal plates, the correlation between laser shock processing parameters, dynamic response and distribution of surface residual stress was studied, and the formation mechanism of residual stress on the metal sheet surface was explored. When the power density was 1.53 GW/cm2, the interaction between surface rarefaction wave and transverse deformation resulted in equal-biaxial distribution of the maximum residual principal stress on the specimen surface of 7050 aluminum alloy thin plate; when the power density was 2.77GW / cm2, the shock wave on the back and surface of 7050 aluminum alloy sheet was reflected back and forth, and the influence of residual stress on the surface cannot be neglected, the phenomenon of "Residual Stress Hole" was caused by combined action of the shock wave reflected back and forth in the thin plate and the sparse wave converging to the center.

(4) Basic and applied research were carried out based on the multi-scale characterization of the propagation mechanism of laser shock wave in metal materials, the dynamic mechanical properties of metal materials under high strain rate, surface residual stress and microstructure change. This study provides a theoretical basis and technical support for the quantitative control of surface residual stress thin-walled induced by parts laser shock processing.

The authors are grateful for the support provided by the National Natural Science Foundation of China (Grant Nos. 51175237、51505236) and the natural science foundation of Jiangsu Province (BK20151271).

References

1. Feng Aixin,Nie Guifeng, Xue Wei ,et al.. Experimental research on laser shock wave loading mechanism of 2024 aluminum alloy sheet[J]. Acta Metallurgica Sinica, 2012, 48(2):205-210.
2. Cao Yupeng, Feng Aixin, Xue Wei ,et al..Experimental research and theoretical study of laser shock wave induced dynamic strain on 2024 aluminum alloy surface [J]. Chinese J Lasers, 2014,41(9):0903004.
3. Cao Yupeng, Xu Ying, Feng Aixin,e t al..Experimental study of residual stress formation mechanism of 7050 aluminum alloy sheet by laser shock processing[J]. Chinese J Lasers, 2016(7):133-140.
4. Cao Yupeng, Zhou Dongcheng, Feng Aixin,et al..Formation mechanism of residual stress hole on 7050 aluminum alloy sheet sample under laser shock[J]. Chinese J Lasers, 2016(11):78-87.
5. Cao Yupeng, Zhou Dongcheng, Feng Aixin,et al.. Simulation and experiment of transmission

* Corresponding author: author@e-mail.org

mechanism on laser shock wave loading 690 high-strength steel sheet[J]. Chinese J Lasers, 2016(11):129-138.

6. Cao Y, Feng A, Hua G. Influence of interaction parameters on laser shock wave induced dynamic strain on 7050 aluminum alloy surface[J]. Journal of Applied Physics, 2014, 116(15):775-1918.

Modification induced by laser irradiation on physical features of plastics materials filled with nanoparticles

Cristina Scolaro[1,*], *Annamaria* Visco[1,4], *Lorenzo* Torrisi[2], *Nancy* Restuccia[2] and *Eugenio* Pedullà[3]

[1]Department of Engineering, University of Messina, C.da Di Dio, 98166 Messina, Italy
[2]Dept. of MIFT, University of Messina, V.le F. S. D'Alcontres 31, 98166 Messina, Italy
[3]Dept. of General Surgery and Medical-Surgical Specialties, University of Catania, Via Plebiscito 628, 95124, Catania, Italy
[4]Istituto per i Processi Chimico-Fisici CNR –IPCF, Viale Ferdinando Stagno d'Alcontres, 37 - 98158 MESSINA

Abstract. The Thermal Laser Welding (TLW) process involves localized heating at the interface of two pieces of plastic that will be joined. Polymeric materials of Ultra High Molecular Weight Polyethylene (UHMWPE), both pure and containing nanostructures at different concentrations (titanium and silver nano-particles), were prepared as thin foils in order to produce an interface between a substrate transparent to the infrared laser wavelength and an highly absorbent substrate, in order to be welded by the laser irradiation. The used diode laser operates at 970 nm wavelength, in continuum, with a maximum energy of 100 mJ, for times of the order of 1-60 s, with a spot of 300 μm of diameter . The properties of the polymers and of nanocomposite sheets, before and after the laser welding process, were measured in terms of optical characteristics, wetting ability, surface roughness and surface morphology.

1 Introduction

Plastic material junctions are of great interest for polymeric industrial applications; for this reason welding technological processes needs to be studied and optimized in order to produce high aesthetic and resistant quality products. Moreover, the polymers can be welded with various experimental techniques. For some years, Trasmission Laser Welding (TLW) has a considerable role among the various junction methods, allowing to work with low mechanical and thermal loads, ensuring welds of excellent mechanical and aesthetic quality as no residues are present in the area of welding and the thermally altered area is very narrow.

All polymers commonly used for laser welding, under non-doping conditions, are more or less transparent in the visible and near infrared ranges. The most used lasers for the soldering process [1], are represented by continuous Diode laser, operating in the range 800-1000nm wavelength [2], based on semiconductor pn junction, and the nanosecond Nd:YAG laser operating at 1064 nm fundamental frequency [3-5]. The addition of filling materials, such as pigments (fillers), to the polymer matrix provides the absorption of laser energy by the doped polymer in specific wavelength ranges.

One of the most popular polymers in the world, widely daily used, is the polyethylene, which is the raw material for food film, plastic bottles, pipes, packaging, sports goods, fabrics, furnishing accessories, toys, electronic applications, medical products, etc..

The generic formula for polyethylene is $(\text{-}C_2H_4\text{-})_n$, where n is the degree of polymerization. Although it has a simple basic chemical composition, there are many varieties (*Low-Density Polyethylene* - LDPE, *High-Density Polyethylene* - HDPE, *Linear Low-Density Polyethylene* - LLDPE, *Ultra-High-Molecular-Weight Polyethylene* - UHMWPE, *Ethylene Acrylic Acid* - EAA, *Ethylene-Vinyl Acetate* - EVA), whose thermal and mechanical properties vary according to the structure of the molecular chains, that make up and determine their different use. In any form it is used, its excellent features of good insulating, eco-compatible, corrosion-resistant, and others, make it always reliable in any circumstance.

For these reasons and beyond the most immediate application [6], polyethylene is a precious material especially in the world's most advanced sectors, such as industry, health [7] and energy, where, alongside advances and innovations in techniques already known and experienced, interesting perspectives emerge around new research topics, some of which are being tested and concerning this work. Great industrial interest is in fact directed toward the welding of plastics with different concentrations of nano-materials and ones that must be transparent for aesthetic reasons [8].

This investigation focused on the realization of welded joints of high molecular weight UHMWPE polyethylene and UHMWPE polyethylene doped with various concentrations of nanostructures (titanium ad silver) to make the absorbent material. The joints have been welded using a 970 nm wavelength laser diode. The great advantage of the diode laser is that it can be used as a pen and the welding process is very easy to obtain. Therefore, compact design and high-efficiency diode

* Corresponding author: cscolaro@unime.it

laser make it very interesting for practical and industrial applications for the adhesion of thermoplastic polymers. The properties of the polymers before and after the laser welding process were measured in terms of optical characteristics, wet ability of biocompatible solutions, surface roughness, surface morphology and surface composition (UH pure/ nanocomposites).

2 Materials and Methods

2.1. Materials

UHMWPE (Ticona-GUR 1020, ρ = 930 kg / m3, Mw ≈ 3x106 g / mol) was mixed with silver nanostructures and titanium dioxide with spherical shape and diameter between 50-100 nm, used as fillers in a quantity of 1% in weight (for titanium dioxide nanoparticles, UHNTiO) and 4% in weigth (for silver nanoparticles, UHNAg).
Considering the mechanical response of joints, here not shown in detail for brevity, just the 1% in weigth of TiO_2 (UHNTiO1) and the 4% of Ag nanoparticles (UHNAg4) have been chosen as nanocomposites fillers.
These nanofiller amounts showed the best shear strength among all the samples, with a measured value of 107 N for titanium dioxide nanoparticles and of 78 N for silver nanoparticles [9].
The different nanofiller types (TiO_2/Ag) were dispersed in pure ethanol. The polymer-filler-solvent system was sonicated at room temperature for 2h. The solvent residue was then removed by stirring on a heat plate. The obtained nanocomposites were molded and pressed at T=200°C/20 minutes and P=200 atm, obtaining 60mm x 60mm and 0.5mm thick sheets using thin film in Teflon (123µm thick) as detachable agents.

2.2 Laser Irradiation

The laser used for laser welding of polymeric joints, is a Diode Laser (Lamba Scientific - D5-Doctor Smile) operating at a wavelength of 970 nm, CW (0.5-7 W), maximum power 100 mJ, for order times of 1-60 s, 300µm fiber optic diameter (Fig.1, left). No lens was used to focus the beam of the laser led to air at 25 °C and at 1 atm.

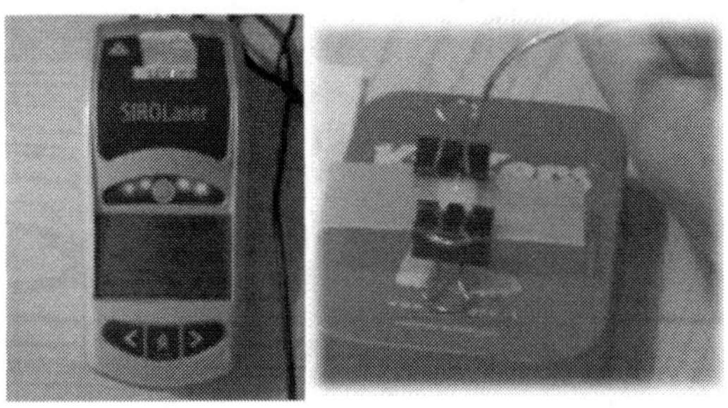

Fig. 1. Diode Laser (Lamba Scientific - D5-Doctor Smile)

The welded joints were obtained by partially overlapping polymeric sheets, white and black as in the sequence schematized in Fig.2.a (Single Lap, SL) and Fig.2.b (Double Lap, DL).
The external sheet is laser light transparent (pure UHMWPE) while the inner sheet laser is light absorbent (nanocomposite). Samples of laser irradiation have a rectangular geometry of 30mm x 20mm x 0.5mm thick (Fig.3). Transparent and absorbent polymer sheets (at different filler percentages) are partially superimposed (for 0.8 mm lengths) and blocked by the use of two clip-on tweezers (Fig.1, right), with the task of exerting the required force (30N) [10] for a good welding process according to the scheme shown in Fig.3.

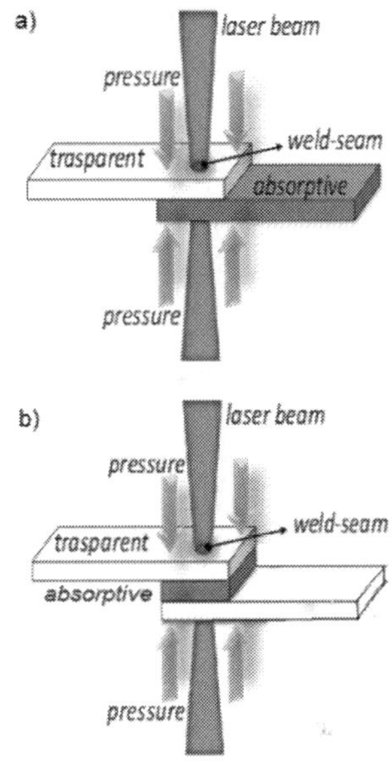

Fig. 2. a) Schematic of the system of laser welding in single lap (SL) or b) in double lap (DL).

A point geometry was created by the diode laser with an irradiation time of about 2 minutes per point for a total of 6 points (for both silver and titanium dioxide nanocomposites. Samples were investigated in the pure zone (zone 1, Fig.3) and just before the Heat Altered Zone (just before HAZ, zone 2 and zone 3, Fig.3).

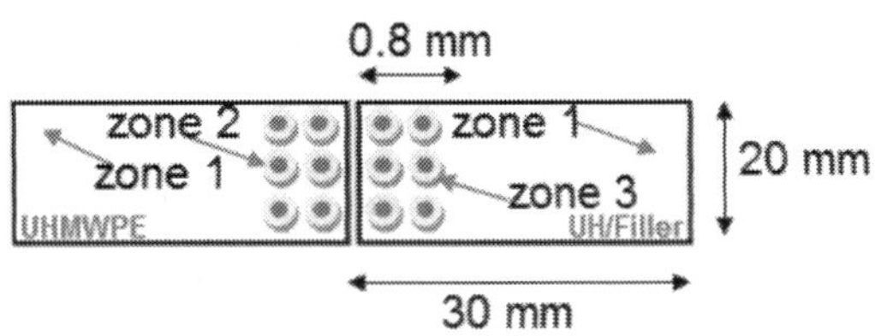

Fig. 3. Schematization of a point geometry created by diode laser and zones investigate.

2.3 Contact angle measurements

The contact angle was evaluated by the wet ability of the nano-composite surface. Wet ability measurements were performed using the experimental setup shown in Fig.4.

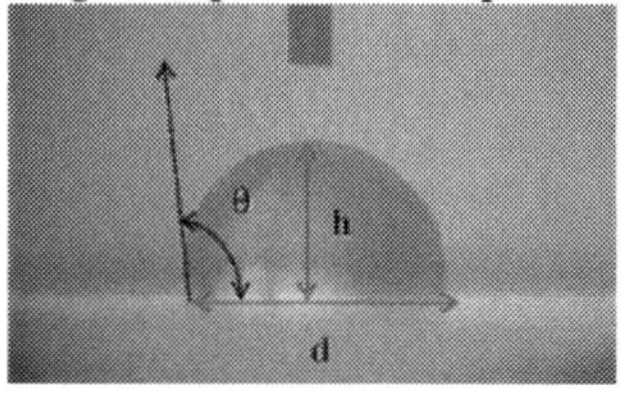

Fig. 4. Drop profile and contact angle determination.

By placing a drop of 1 μl of distilled water at room temperature on the horizontal surface of the sample using a microlitric syringe (Hamilton, 10 μl), a video camera connected to the computer captures the image of the drop just deposited on the material and define, by means of software [11], the measurement of the base line of the drop from which the diameter *d* (mm) and the height *h* (mm) of the drop are obtained which are used to calculate the contact angle *θ* by the following formula [12]:

$$\theta = 2 arctg(2h/d) \qquad (1)$$

At a temperature of T = 22°C, UR = 35% and P = 1atm, six measurements of the contact angle of the distilled water at different points for each single sample were performed and then the average value was determined arithmetically.

2.4 Roughness measurements

The roughness measurements on the polyethylene sample surface were carried by using the Tencor P-10 surface profilometer with a micrometric movable tip loaded with a force of 1 mg that scans the surface of the sample [13,14]. The system has a depth resolution of 10 nm, a lateral resolution of 1 μm and a scan surface speed of 100 ! m/s. The analysis is done using a diamond tip that touches the surface of the sample to follow the profile. A position transducer converts the tip movement into its height value. All information is transmitted to a computer connected to the instrument that shows the surface morphology of the sample. The tip of the instrument sweeps horizontally, depending on the requested sansion, and vertically due to a constant pressure within 1-5 mg. The vertically excursion recorded by an analogical-digital converter and plotted on the pc screen.

The surface roughness was measured using the average peak-to-peak distance of the roughness profile. The variations observed by measuring the contact angle of distilled water on nano-composites can be attributed to the heterogeneity of the surface determined by the roughness. The theory that explains the effect of surface roughness on the wet ability is described by Wenzel's equation [15]:

$$cos\theta_w = r cos\theta_Y \qquad (2)$$

Where θ_W is the angle of effective contact of a rough surface real, or angle of Wenzel, *r* is the ratio between the actual area of the surface of the solid and the nominal one, and θ_Y is the contact angle of equilibrium of Young observed on a ideally smooth surface.

3 Results and discussion

Prior to perform the diode laser irradiation, surface wet ability and roughness tests were carried out on the pure and doped polyethylene UHMWPE matrix.

Wet ability tests performed on pure UH polyethylene and on UH with 1wt% TiO_2 show a tendency for hydrophilic behavior at nano-composite concentrations.

The contact angle measurement showed a change in the contact angle of pure UHMWPE (Fig.5.a): its value of 96° decreases in both nano-composites. In fact it is of 94° in the nano-composite filled with titanium dioxide (UHNTIO1, Fig.5.b) and of the 77° in the nano-composite filled with silver nanoparticles (UHNAg4, Fig.5.c). These values depend on the surface roughness of the material that changes from 2.06 μm for pure UHMWPE (Fig.5.b), to 2.00 μm for UHNTiO1 (Fig.5.d) and to 3.31μm for UHNAg4 (Fig.5.f). These measurements are affected of the error of the order ten percent.

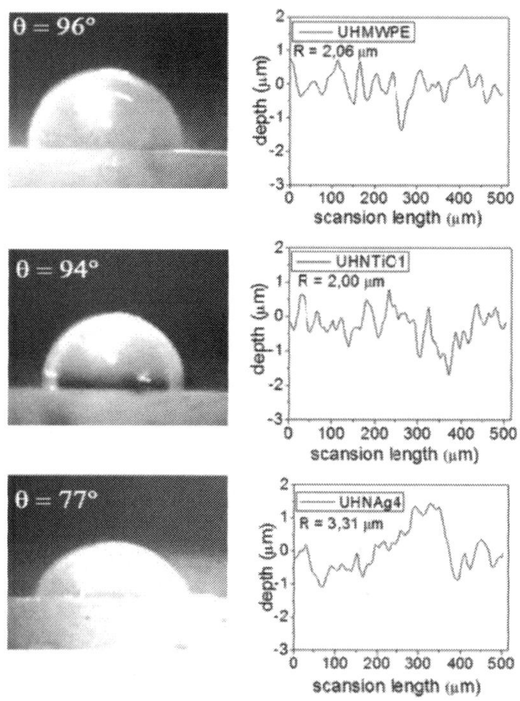

Fig. 5. Images of distilled water droplets and roughness profiles respectively of a-b) UHMWPE pure, c-d) UHNTiO1 and d-e) UHNAg4.

Generally, profilometric measures have shown that the greater the amount of filling, the greater the roughness of the polymeric nano-composite. As previously mentioned, the roughness has a fundamental influence

https://doi.org/10.1051/epjconf/201816705008

on the contact angle. In fact, the value of the contact angle θ decreases by increasing the surface roughness of the nano-composites, as shown in Fig.6.

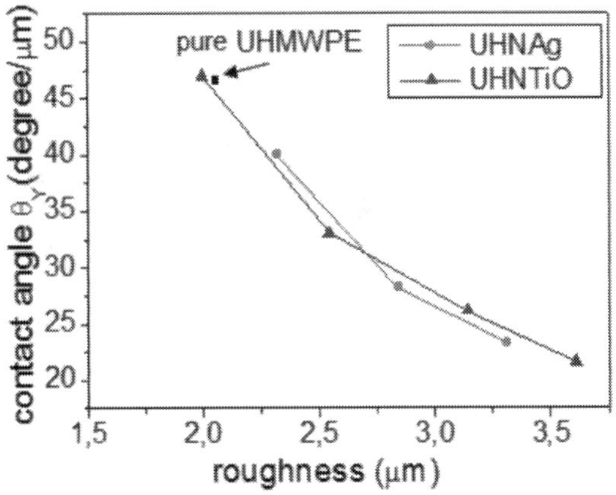

Fig. 6. Change the contact angle *vs* the roughness of the nanomaterial.

This result confirms that the growth of filler in the polymer nano-composite makes its surface more irregularly increasing its surface roughness. The latter lowers the interfacing tension of the liquid drop, favouring its wet ability.

Morphological observation by optical microscope of the spot welded area, using the TiO_2 nanoparticles, shows how laser irradiating the polymer melts the materials and changes its surface morphology (Fig.7).

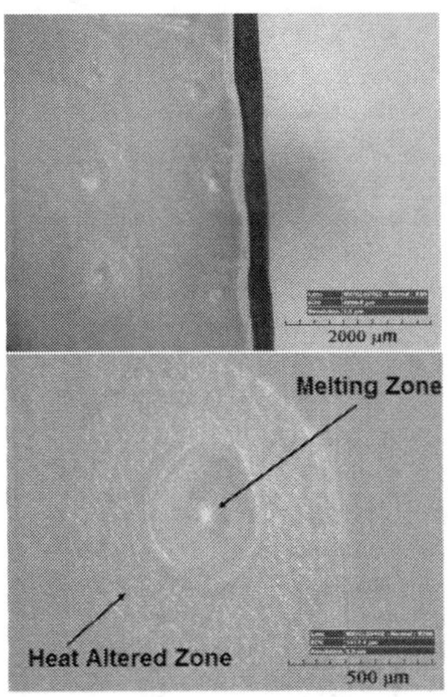

Fig. 7. Optical Microscope images, UHNTiO1 welding junctions (left) and a magnification particular of the melted and of the HAZ area (right).

Comparison of the contact angle pattern determined in the junction interface (UHMWPE pure and transparent and UH doped with fillers and absorbent) of the indicated materials ((NAg4 – NTiO1) are shown in Fig.8. Figures 8.a and 8.e show the pictures of distilled water droplets, UHNTiO1 and UHNAg4 junctions respectively created with the Diode laser and determined just before the HAZ (UH doped with fillers and absorbent). Experimentally determined contact angles are 110° for UHNTiO1 and 75° for UHNAg4. Surface roughness values found are 1.53 µm (UHNTiO1, Fig.8.b) and 3.04 µm (UHNAg4, Fig.8.f).

Figures 8.c and 8.g show the pictures of distilled water droplets, UHNTiO1 and UHNAg4 junctions respectively created with the Diode laser and determined on the heat altered zone (UHMWPE pure and transparent). Experimentally determined contact angles are 138° for UHNTiO1 and 94° for UHNAg4. Surface roughness values found are 1.2 µm (UHNTiO1, Fig.8.d) and 2.01 µm (UHNAg4, Fig.8.h).

The morphological study of the welded area shows that their physical properties depend on many factors.

The mechanical strength, uniformity, colour and roughness of the melted reinforcement, for example, depends on the time of irradiation, the thickness of the transparent polymer, the laser intensity, and the amount of fillers used in the absorbent polymer.

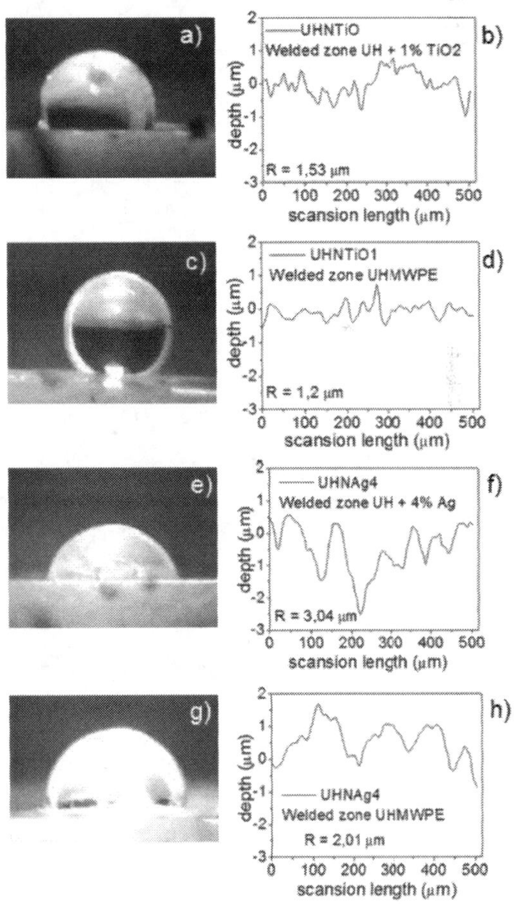

Fig. 8. Pictures of distilled water droplets and roughness profiles of the UHNTiO1 and UHNAg4 welding zones (just before the HAZ).

Tables 1 summarize the wet ability and roughness values of the fusing zone interface of the joints UHNAg4 and UHNTiO1 created with the diode laser.

Table 1. Contact angle and roughness of the UHNTiO1 and UHNAg4 joints created with the Diode laser just before the HAZ.

Contact angle and Roughness results for UHNTiO1 and UHNAg4						
Welded Joint	Contact angle Zone 1 (°)	Contact angle Zone 2 (°)	Contact angle Zone 3 (°)	Roughness Zone 1 (μm)	Roughness Zone 2 (μm)	Roughness Zone 3 (μm)
UHNTiO1	94	138	110	2.00	1.2	1.53
UHNAg4	77	94	75	3.31	2.01	3.04

The dependence of the contact angle (normalized to the surface roughness) by the roughness is shown in Fig.9. The angle values have been calculated according to the Wenzel equation (2) and measured just before the HAZ at the junction interface, both in the pure polyethylene and in the nano-composite. The contact angle decreases with increasing the surface roughness just before the HAZ for both materials. Furthermore, the lowest wet ability (hydrophobic feature) can be reached in the nano-composite with low amounts of filler particles and, hence, of surface roughness, and *viceversa*. Considering the filler amounts of the tested nano-composites, the sample UHNTiO exhibited a lower roughness and so a higher hydrophobic character than the UHNAg one (see table I).

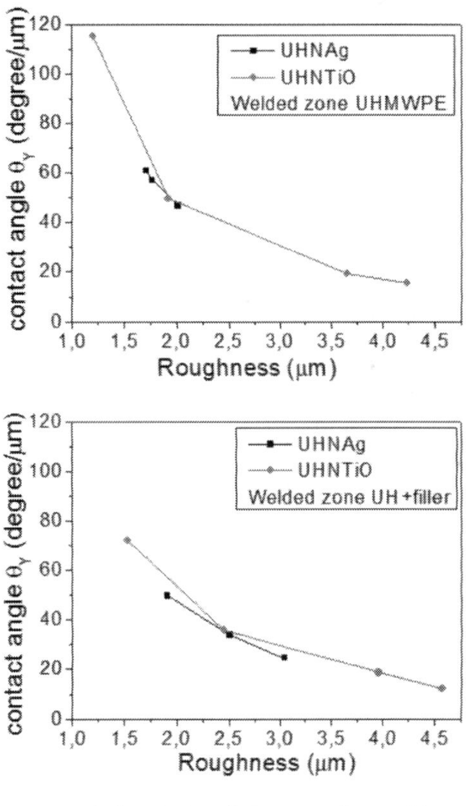

Fig. 9. Trend of contact angle θ_Y *vs* roughness of the UHNTiO1 and UHNAg4 just before the HAZ.

4 Conclusions

Polymeric single and double lap joints were successfully welded with diode lasers source. The laser absorbent sheets were doped with different amount of nano-materials (titanium and silver oxides). We have studied the surface features of the polymers, the optical absorbing properties, wetting ability and surface roughness. The tested nano-composite sheets, filled with titanium dioxide and silver nanoparticles can be usefull in order to have white color joint with also a low surface roughness and low damage and a quite hydrophobic character, also after the laser action. In our work we demonstrated that proposed technique is promising for application in the fields of the prosthesis, restorative plastic materials, food and water container and other industries.

This work was supported by the "Research and Mobility" project of Messina University No. 74893496, scientifically coordinated by Professor L. Torrisi.

References

1. Friedrich G. Bachmann, Ulrich A. Russek, *Laser Processing of Advanced Materials and Laser Microtechnologies*, 385, (2003); DOI:10.1117/12.515630.
2. M Visco, G. Galtieri , L. Torrisi , C. Scolaro, A. Cannavò, IV°PBSA 2014, 87-91, ISBN: 978-88-8305-107-4.
3. A.M. Visco, V. Brancato, M. Cutroneo, C. Scolaro, F. Caridi and L. Torrisi, III°PBSA 2012, 133-137, e-ISBN: 978-88-8305-088-6.
4. G. Galtieri , M Visco, D. Nocita, L. Torrisi, G. Ceccio, C. Scolaro, PPLA 2015, Journal of Scientific Instruments – IOPscience, **11**_ C04013, (2016) doi:10.1088/1748-0221/11/04/C04013.
5. V.V. Semak, R.J. Steele, P.W. Fuerschbach, J. of Physics D: Applied Physics, **33**, 1179-1185 (2000).
6. F. Caridi, A. Picciotto, L. Vanzetti, E. Iacob, C. Scolaro, Radiation Physics and Chemistry **115**, 49–54, (2015), DOI:10.1016/j.radphyschem.2015.06.009.
7. S.C. Veerla, S. Spriano, A. Cochis, F. Uberti, L. Rimondini, E. Bertone, A. Vitale, C. Scolaro, M. Ferrari, F. Cirisano, G. Gautier di Confiengo, S. Ferraris Materials Science and Engineering C **74**, 542–555(2017).
8. A.M. Visco, L. Torrisi, G. Galtieri, C. Scolaro, Journal of Thermoplastic Composite Materials, 1-18 (2016), DOI: 10.1177/0892705716662515.
9. A.M. Visco, C. Scolaro, T. Terracciano, R. Montanini, A. Quattrocchi, L. Torrisi, N. Restuccia, PPLA2017, submitted at EPJ Web Conference.

10. L. Torrisi, F. Caridi, A.M. Visco, N. Campo, Applied Surface Science, **257**, 7, 2567-2575, (2011) DOI:10.1016/j.apsusc.2010.10.023.
11. L.Torrisi, C. Scolaro, N. Restuccia, J Mater Sci: Mater Med (2017) 28:63, DOI 10.1007/s10856-017-5871-1.
12. Abraham Marmur, *A Guide to the Equilibrium Contact Angles Maze in Contact angle, Wettability and Adhesion*, K.L. Mittal (Ed), **6**:3-18, Vrill, Leiden (2009).
13. C. Scolaro, L. Torrisi, M. Cutroneo, L. Velardi, PPLA 2013, J. Phys.: Conf. Ser. **508** 012030, (2014).
14. L. Torrisi, C. Scolaro, Journal of Thermodynamics & Catalysis, **8**:2, 184, (2017). DOI: 10.4179/2160-7544.1000184.
15. Wenzel, R. N., *"Resistance of solid surfaces to wetting by water"*. Ind. Eng. Chem. **28**, 988–994, 1936. DOI:10.1021/ie50320a024.

Static and dynamic characterization of biomedical polyethylene laser welding using biocompatible nano-particles

Annamaria Visco[1,2,*], *Cristina* Scolaro[1], *Teresa* Terracciano[1], *Roberto* Montanini[1], *Antonino* Quattrocchi[1], *Lorenzo* Torrisi[3] and *Nancy* Restuccia[3]

[1]Department of Engineering, University of Messina, C.da Di Dio, 98166 Messina, Italy
[2]Istituto per i Processi Chimico-Fisici CNR –IPCF, Viale Ferdinando Stagno d'Alcontres, 37 - 98158 MESSINA
[3]Dept. of MIFT, University of Messina, V.le F. S. D'Alcontres 31, 98166 Messina, Italy

Abstract. In this research, polymeric joints of Ultra High Molecular Weight Polyethylene (UHMWPE) sheets were realized and welded by a diode laser operating at 970 nm wavelength. One of the polymer sheet was doped, at different concentrations, with nano filler (carbon, titanium and silver nano-particles) in order to enhance the absorption coefficient at the laser wavelength. Laser operated in repetition rate with a maximum pulse energy 100 mJ, for times of the order of 1-60 s transporting the light trough a fiber with 300 μm diameter. The laser light has been transmitted by the transparent first polymer and absorbed on the face of the second doped polymer. At the interface of the two polymer foils, 0.5 mm thickness each, the released energy induces melting, assisted by pressure, producing a fast and resistant welding. Single lap and double lap geometries have been performed and studied by means of mechanical static (shear stress) and dynamical analysis. Effect of the different particles nature on the mechanical features of the joints has been evaluated. Morphological observation of the jointed areas are presented and discussed. Joints could be useful in biomedical field for their special features.

1 Introduction

The growing development in many plastics industries and the increasing attention paid to the quality and aesthetics of products has defined new technological demands for components junction processes. Welds must be excellent in mechanical and aesthetic quality without residues in the welded area [1].
Among the plastic materials, polyethylene is a thermoplastic bio compatible material (for its chemical inertia), highly applied in the biomedical field [2,3]. For its high mechanical ductility and other important features, polyethylene is considered an unique material, used in several industrial sectors such as health and energy, and for advanced applications. Polyethylene is also a weld able polymer, and good quality joints can be obtained by employing diode or Nd:Yag lasers [4,5].
Mechanical characterization is a very important aspect of materials knowledge since polymers are often used in applications that involve stresses, such as that of polymeric joints. So we must study the effect that stress has on them. In the field of mechanical characterization, static testing (load/displacement curves) and fatigue test (lifetime or durability analysis) are commonly employed. In our case, the mechanical properties of our joints have been studied to verify the resistance features of the laser welding effect. In particular, mechanical static and dynamic (fatigue) tests have been performed. Life of polymeric and composite joints is an important and crucial aspect that needs attention. It has been researched in the last 40 years. Renton et al. [6] and Ferreira et al. [7] studied the fatigue behavior of polymeric and composite single lap joints, putting into evidence the effects of layer orientation, lap joint length and water immersion. In both papers, the failure appeared in the adhesive layer. The fatigue life is strongly influenced by the profile of the edges of the joint. Harris et al. [8] verified that the initiation of small cracks at the edges points represents the major part of the fatigue life. Moreover, temperature and humidity of the adhesive cure are extremely relevant for the damage accumulation. They produce significant changes in the stress state of lap joints [9].
In this paper we have considered a particular type of polyethylene, a biomedical grade Ultra High Molecular Weight-UHMWPE. It has a particular high viscosity and low friction coefficient, and it's largely used for bio-medical applications [2].
Polyethylene itself is transparent to laser light. So, we have prepared single lap or double lap joints (welded by means of a diode laser) by employing fillers (nano particles) to make laser absorbing one polymeric sheet of the joint. The nanofillers have been suitably selected for type and quantity to keep color as close to white as possible for aesthetic requirement. The joints have been

* Corresponding author: avisco@unime.it

mechanically analyzed by means of static and fatigue tests in order to verify their strength and durability.

2 Materials and Methods

2.1. Polymer/nanocomposite sheets preparation

Polymeric nanocomposite sheets, 30mm x 20 mm x 0,5 mm thick, were obtained by dispersing the carbon nanofiller, titanium dioxide and silver nano particles inside the matrix (biomedical grade UHMWPE GUR 1020) with a solvent.
The amount of fillers added to the UHMWPE polymer matrix was of 0.003-0.016-0.025-0.25-0.5-1.0% (for carbon nanoparticles, UH-NC code), 1-2-4-6 % (for titanium nanoparticles, UH-NTiO code) and 1-2-4% (for silver nanoparticles, UH-NAg code).
Polymeric nanocomposites of carbon are filled with very low filler amount of nanoparticles, (0.003%, 0.016%, 0.025%,) so that it anyway appears of white color.
The obtained nanocomposites were then molded into an hot press obtaining 60mm x 60mm and 0.5mm thick sheets. The preparation steps are shown in Fig.1.

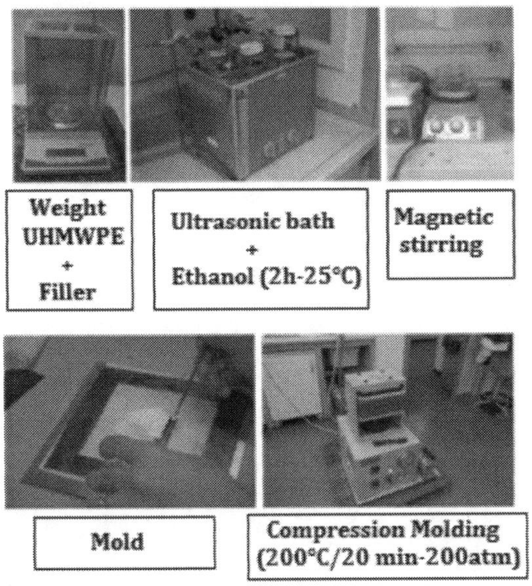

Fig. 1. Preparation of nanocomposite sheets

Single (SLJ) and double lap joints (DLJ) (Fig.2.a) were obtained by a diode laser operating at 970 nm, CW continuous wave (0.5-7W), maximum energy of 100 mJ, for times of the order of 1-60 s. Optical fiber thickness: 300 μm.
Two types of welding geometry were created: points and grid (Fig.2.b). The irradiation time to create welding by a diode laser varies from a dozen seconds (NC-doped material) to over 2 minutes (doped with NAg and NTiO).

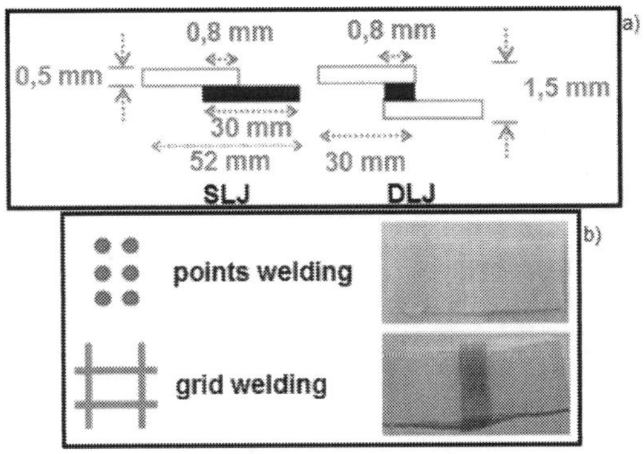

Fig. 2. a) Single-lap (SL) and double lap (DL) joint geometries; b) welding geometries and typical joint.

2.2. Physical Characterization

Optical absorption measurements: light absorption coefficients in pure and doped polymers with fillers were measured by using optical spectroscopy in the range of 150nm to 800nm. Absorption studies were performed by evaluating the absorption of characteristic peaks emitted by a Hg (Ar) lamp and detected by the high sensitivity optical spectrometer.
The sample is placed on a suitable support, the incident light is perpendicular to the sample and the emerging light is measured in intensity with respect to the incident. The light beam emerging from the sample is captured by a fiber interfaced with the optical spectrometer (Horiba Jobin) that detects wavelengths between 220 nm and 1100 nm.
The Lynear spectroscope software acquires 100 spectra per second and the average spectrum is then stored on a PC.
Absorption coefficients, μ, have been experimentally calculated using the Lambert law:

$$\mu=-(\Delta x)\cdot ln(I_t/I_o) \quad (1)$$

where Δx = thickness, I_0 = incident laser intensity and I_T = transmitted laser intensity.

Morphological Optical (MO) characterization: the Hirox Digital Microscopekh KH8700 optical microscope has been used for morphological observation of the nanocomposite materials surface with a low magnification, of 50x.

2.3 Mechanical Characterization

Static tensile tests have been conducted to evaluate how the presence of NC, NTiO and NAg at different concentrations in UHMWPE affects the mechanical properties of the final nanocomposite. The tests were performed according to the ASTM D638 standard with a Lloyd Universal Testing Machine (LR10K), imposing a traverse feed rate of 5 mm/min and a load cell of 500N.
Load /displacement curves of the joints shear stressed during the tensile motion. The data are from the average of six specimens.

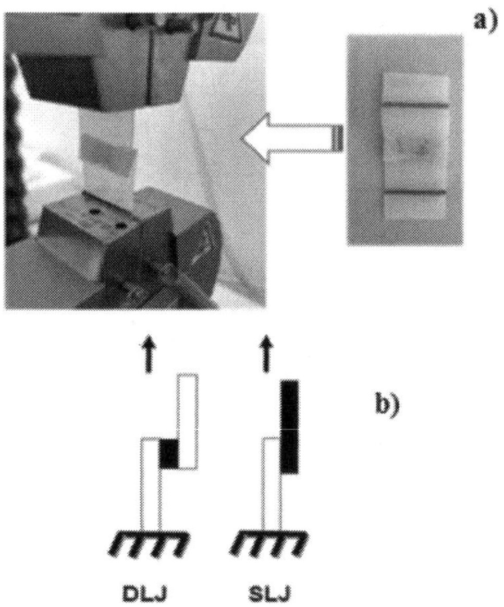

Fig. 3. a) Photograph of a joint and its disposition between the clamps during a mechanical tests and b) scheme of double and single lap joint's arrangement during the test .

Dynamic fatigue tests: the experimental setup allows to test the sample with fatigue loads, monitoring the mechanical and thermal behaviors. It consists of an electro mechanic testing system (Electropuls E300, Instron), equipped with two pneumatic grips (±3 kN max) able to guarantee a grip force of 4.5 kN, a calibrated load cell (±5 kN max), and an infrared camera (SC 7200, Flir). The fatigue tests were performed according to ASTM D7791 standard. They were carried out in load control at different maximum force values (fractions of the break load) with stress ratio (R) of 0.7 and a frequency of 1 Hz at room temperature (25±2 °C and 50±10 RH). The mechanical parameters were acquired at 100 Hz with a defined reduction of the analyzed cycles: from 0 to 1000th cycle all ones were captured, between 1000th and 100000th cycle 1 every 4, finally after 100000th 1 every 40. IR images were obtained at an acquisition frequency of 100 Hz, monitoring the thermal effect at the maximum applied force of the characteristic cycles and assuming as negligible the small variation of the temperature room (±2 °C) during the test.

3 Results and discussion

The presence of filler within the UHMWPE makes the nanocomposites less transparent to the laser radiation. Increasing the percentage of filler concentration the absorbance increases at the laser wavelength in 0.5mm thick target. Pure polyethylene (UHMWPE) instead is a transparent material to laser light.

Fig.4.a shows the variation of light absorption coefficient in pure and doped polymers at different percentages by weight (0.003%, 0.016%, 0.025%, 0.25%, 0.5%, 1%) of nano carbon particles. From Fig.4.b we observe how the light absorption coefficient in the polymer, for example, at 545 nm, grows with the growth of the amount of carbon nanostructures. A similar result has been obtained testing the other fillers.

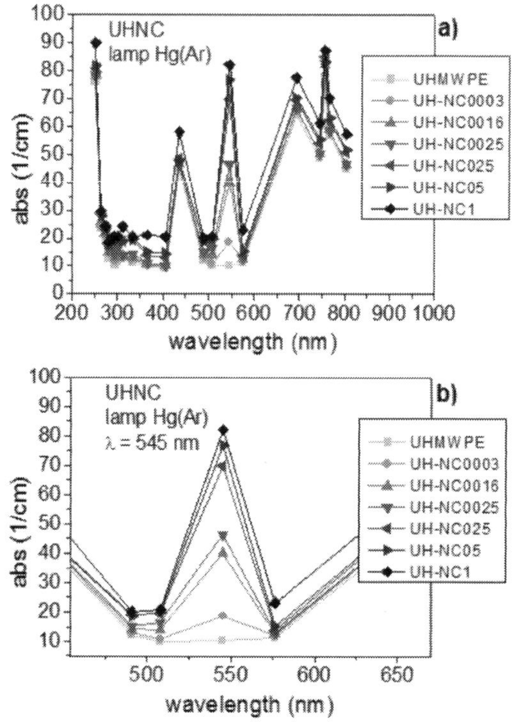

Fig. 4. a) Measure the absorption coefficient of pure polyethylene with doped with NC, b) measure the absorption coefficient at $\lambda = 545$ nm according to the NC filler percentage.

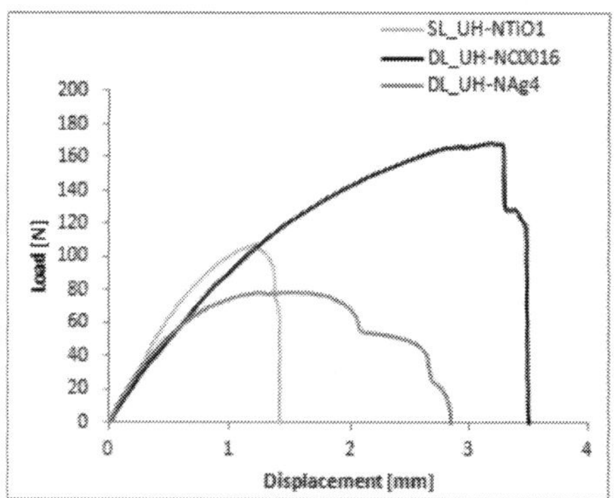

Fig. 5. Load /displacement curves of the joints with different filler types.

Regarding the static mechanical tests of the prepared joints, a previous experimentation (not shown for brevity) indicated to us that the optimal filler amount was of 0.016 wt% of CN, 1 wt% of TiO_2 and of 4 wt%

of Ag. The static load/displacement curves shown as the highest load of 169 N is reached in the NC filled joint and the lowest of 78 N in the Ag filled one (Fig.5). The values of static strength of these joints are anyway appreciable, compared to the mechanical behavior of other similar polymeric joints [2]. In Table 1 are listed the contact angle values measured in water and the surface roughness of above mentioned nanocomposites. Values indicated a more hydrophobic character of the materials filled with Ag and less in these filled with TiO_2 and NC, while surface roughness was within the range 2.0-3.3 micron.

Table 1. Mechanical and physical features of the joints.

Mechanical statical test results and surface features					
Joint code	Filler amount (%)	Load (N)	Displacement (mm)	Contact angle (°)	Roughness (μm)
UH-TiO_2	1	107	2.76	94	2.00
UH-NC	0.016	169	3.49	93	2.09
UH-Ag	4	78	2.16	77	3.31

The morphological observation of the welded area of the UH-NC0016 joint (that exhibited the best static mechanical strength) is shown in Fig.6. The image highlights the melted track with the cross line shape. The laser contact with the polymer melts the materials, changing the starting smooth morphology of polyethylene surface.

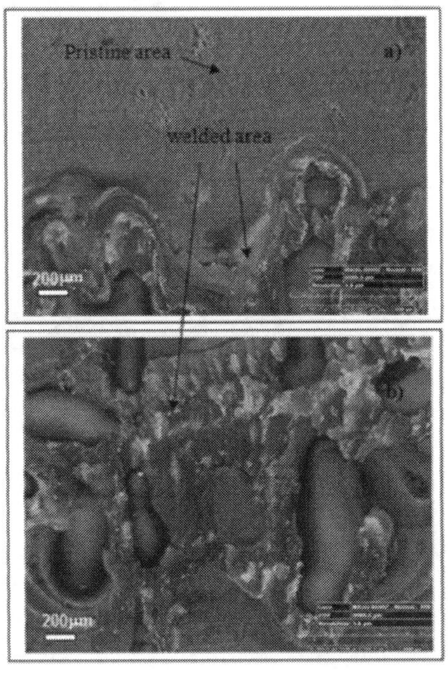

Fig. 6. Microscope images of UH-NC0016 welding junction a) of the pristine polymeric surface and b) of the laser irradiated surface.

The great morphological modification indicates a deep interpenetration between the polymeric sheets confirming a good welding action, in agreement with mechanical results. Fig.7 highlights the displacement of the fatigue machine actuator, and so the elongation of two welded joints of UH-NC0016 sample, subjected to 90% of break load (FR). Three different regions can be identified by observing the displacement-number of cycles trend:

(1) a first part with elastic deformation due to polymeric structure orientation for the normal load;
(2) a second step with stiffening of the basic material of the joints and progressive plastic deformation;
(3) finally, a rapid damage and a break of the welded zone.

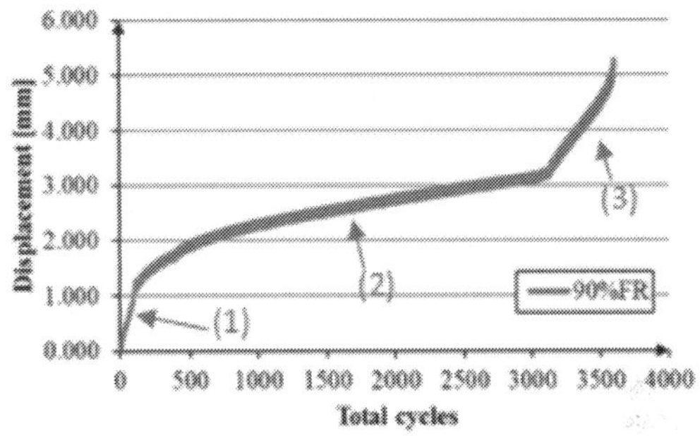

Fig. 7. Elongation of two welded joints

The fatigue limit of the UH-NC0016 joint is about 3500 cycles. Fig.8 shows the typical images of the thermal behaviors of the cyclic load at 90% FR, when the sample is subjected to the maximum load (F_{max} = 130 N) on the fatigue test.

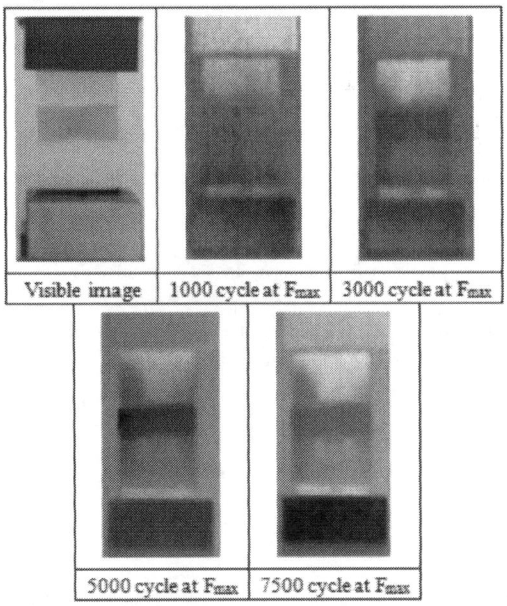

Fig. 8. Comparison between visible image and IR images at maximum load and specific cycles

In this condition, the thermal increase starts to the upper joint, locked at the machine fatigue actuator. Progressively, it evolves to the inferior joint and to the welded zone. Particularly, in the welded zone, the temperature growth is only placed closer to the melted area.

Fig.9 displays the effect of the applied load on the measured elongation of the two welded joints. Different trends for compression and tensile are observed, defining a hysteric cycle. The hysteresis area grows when the number of fatigue cycles increases from 50 cycles (Fig.9.a) to 3500 cycles (Fig.9.c), putting into evidence the plastic deformation of the joints whose displacement improves from ~ 1.45 mm (Fig.9.a), to ~ 2.65 mm (Fig.9.b) and to ~ 4.7 mm (Fig.9.c).

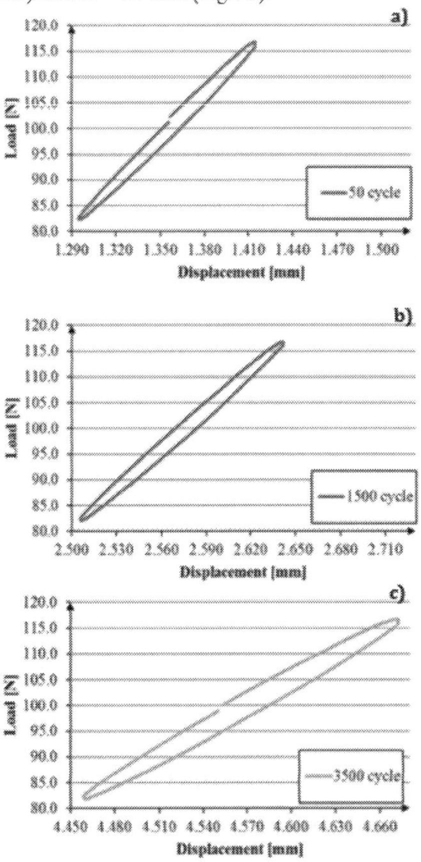

Fig. 9. Areas of hysteresis at 90%FR and different cycles a) 50, b) 1500, c) 3500.

4 Conclusions

Polymeric sheets were welded in two different geometries by means of a diode laser obtaining single lap and double lap SL joints. One of the sheets must be doped with an absorbent filler in order to absorb the laser energy, melting the overlapped sheets at their interface. Polymeric nanocomposites are filled with very low filler amount, (carbon, titanium dioxide and silver nanoparticles) so that it anyway appears of white color. This can be useful for aesthetic reasons. Melted area create a strong joint that exhibits a high mechanical resistance, during both static and fatigue tests.

The highest shear load has been shown by the 0.016 weight % of NC filled nanocomposite (169 N). Its fatigue cycles resistance (at 90% of maximum load) were of about 17000 while at 70% of maximum load were about 350. The other joints made by the nanocomposites filled with 1 weight % of titanium dioxide and 4 weight % of silver nanoparticles, reached a lower strength of 107 N and 78 N, respectively, due to a lower filler absorption power and, consequently, to a lower adhesion at the polymeric sheets interface. Experimental results performed with the dynamic fatigue test coupled with thermography camera, resulted successfully applicable in order to check the fatigue behavior and life of this type of polymeric joints, giving a good response during the tests. In fact fatigue tests highlighted an appreciable mechanical behavior of the polymeric joints: in particular, the fatigue limit of the UH-NC0016 joint (subjected to 90% of break load) was about 3500 cycles. Three different steps for damage of the single lap joints have been identified: elastic deformation for the basic material of the joint, plastic deformation of the sample and rapid damage and break for the welded zone. Hysteresis and IR images have been confirmed the previous observations: single lap joints failure is mainly due to the damage of the welded zone.

This joint exhibits appreciable mechanical and aesthetic features; next step will be the biomedical features evaluation of the materials in order to consider the possible biomedical applications for medical device technology, human care or domestic appliances (such food and beverage packaging).

This work was supported by the "Research and Mobility" project of Messina University No. 74893496, scientifically coordinated by Professor L. Torrisi.

References

1. S.T. Amancio-Filho, J.F. dos Santos, *Polymer Engineering and Science*, **49**, 1461-1476, (2009).
2. G. Galtieri, A. Visco, D. Nocita, L. Torrisi, G. Ceccio, C. Scolaro, PPLA 2015, Journal of Scientific Instruments, **11**_ C04013, (2016), DOI 10.1088/1748-0221/11/04/C04013.
3. A.Visco, V. Brancato, L. Torrisi, M. Cutroneo, Int. J.Polym. Anal. Charact. **19**:489-499, (2014).
4. A.Visco, G. Galtieri, L. Torrisi, C. Scolaro, Int. J.Polym. Anal. Charact. **20**:442-456, (2015).
5. L.Torrisi, C. Scolaro, N. Restuccia, J Mater Sci: Mater Med, **28**(4):28-63, (2017), DOI 10.1007/s10856-017-5871-1.
6. W.J. Renton et al., J Aircraft, **12**(5:):442–7, (1975).
7. J.A.M. Ferreira et al. Composites Science and Technology **62**, 1373–1379 (2002)
8.] Harris JA, Fay PA., Int J Adhes Adhes, **12**(1:):9–18, (1992).
9. Adams RD, Coppendale, J, Mallick, V, Al-Hamdan H, Int J Adhes Adhes, **12**(3):185–90 (1992).

Laser Techniques on Acoustically Levitated Droplets

Antonio Cannuli[1,*], *Maria Teresa* Caccamo[1], *Giuseppe* Castorina[1], *Franco* Colombo[1, 2] and *Salvatore* Magazù[1]

[1]Department of Mathematical and Informatics Sciences, Physical Sciences and Earth Sciences, Messina University, Viale F. Stagno D'Alcontres 31, 98166 Messina, Italy
[2]Italian Air Force Meteorological Service, Comando Aeroporto, Sigonella, Catania, Italy

Abstract. This work reports the results of an experimental study where laser techniques are applied to acoustically levitated droplets of trehalose aqueous solutions in order to perform spectroscopic analyses as a function of concentration and to test the theoretical diameter law. The study of such systems is important in order to better understand the behaviour of trehalose-synthesizing extremophiles that live in extreme environments. In particular, it will be shown how acoustic levitation, combined with optical spectroscopic instruments allows to explore a wide concentration range and to test the validity of the diameter law as a function of levitation lag time, i.e. the D^2 vs t law. On this purpose a direct diameter monitoring by a video camera and a laser pointer was first performed; then the diameter was also evaluated by an indirect measure through an OH/CH band area ratio analysis of collected Raman and Infrared spectra. It clearly emerges that D^2 vs t follows a linear trend for about 20 minutes, reaching then a plateau at longer time. This result shows how trehalose is able to avoid total water evaporation, this property being essential for the surviving of organisms under extreme environmental conditions.

1 Introduction

Laser techniques are largely employed on modern spectroscopic instruments. Due to their enormous potentials in terms of optical power, wavelength tenability, time and spatial coherence, narrow line-width, peak power and ultra-short pulse generation, the field of spectroscopy has been deeply widened from the introduction of lasers. Several laser sources can be employed in Raman and Infrared (IR) spectroscopies, such as fixed-wavelength solid-state lasers with relatively high power, ensuring an access to very wide wavelength regions. These spectroscopic techniques are largely employed for probing samples in order to get information on the structural and dynamical properties of substances. They have several advantages, such as ease-of-use, minimal preparation of samples and quick realization of spectra (in real time and in situ) [1-5]. In particular, Raman scattering occurs when there is an energy shift in frequency among the incident photons and the vibrational energy levels of the molecule, and in many cases, the magnitude of this shift is the same as that of IR adsorption; though some vibrational modes occur only in the Raman spectrum and vice versa. The differences between Raman and IR are also connected to the molecule symmetry, in fact, more symmetric is the molecule, greater will be the differences between the two spectra. An example is the study of chainlike disaccharide molecules, in which the vibrations of the carbon chain are easily analysed by using Raman spectroscopy, while the vibrations of the side groups can be characterized by using IR spectroscopy. These techniques can be applied in conjunction with levitation methods for the analysis of high concentration samples or when it is necessary to avoid heterogeneous nucleation [6-13]. Droplet free flotation has in fact the advantage of avoiding disturbs due to solid walls around the sample, increasing detection sensitivity. Different techniques for sample levitation have been previously developed and are employed in a variety of applications [14-25]. These include optical electro-magnetic, electrostatic gas-film aerodynamic and acoustic levitation [26-30]. Acoustic levitation combines biological compatibility with stable sample position and easy access to the sample, without requiring any specific physical properties (e.g. electric charge or a certain refractive index) for the sample. In order to benefit from the employ of acoustically levitated droplets, sensitive, remote and non-invasive detection analyses methods are necessary [31-42]. Acoustic levitation also allows to get high concentration samples starting from diluted solutions. In particular, the technique can furnish precious information on hydrogen-bonded systems of biophysical interest, such as, for example, aqueous solutions disaccharides proteins polymers and polyols [43-58]. For these systems, it is extremely important to address the experimental studies to the whole concentration range, starting from highly diluted to highly concentrated solutions [59-69].

In the present work the findings of an experimental study, where laser techniques are applied to acoustically levitated droplets in order to perform spectroscopic analyses are presented [70-75]. The investigated samples are trehalose aqueous solution, synthesized by

* Corresponding author: acannuli@unime.it

extremophiles that live in extreme environments. Furthermore the D^2 versus time law, for the droplet diameter, is tested by means of a direct monitoring performed by a video camera and by means of an indirect measure based on the analysis of the OH/CH band area ratio.

It will be shown that acoustic levitation, combined with optical spectroscopic techniques, allows to explore a wide concentration range and to test the validity of the diameter law as a function of levitation lag time, i.e. the D^2 vs t law. It clearly emerges that D^2 vs t follows a linear trend for about 20 minutes, reaching then a plateau at longer times. This result shows that by increasing disaccharide concentration the water evaporation is more and more reduced, this property being essential for the surviving of organisms under extreme environmental conditions.

2 Experimental section

Aqueous solutions of trehalose, purchased from Aldrich-Chemie, were investigated at different concentration values. The solutions were prepared using doubly distilled water. At first, drops of aqueous solutions of trehalose (of 1.50 mm of mean radius) were introduced into the acoustic levitator by means of a micropipette injection and the drop diameter decreased as a function of the time. In order to investigate the drying process of aqueous solutions of trehalose, Raman and Infrared data were collected. In particular a HORIBA Jobin-Yvon T64000 spectrometer was employed to collect Raman spectra at T=25°C; for the IR spectra, a FTIR Vertex 70V by Bruker Optics spectrometer was employed, to collect absorption spectra under vacuum, at T=25°C.

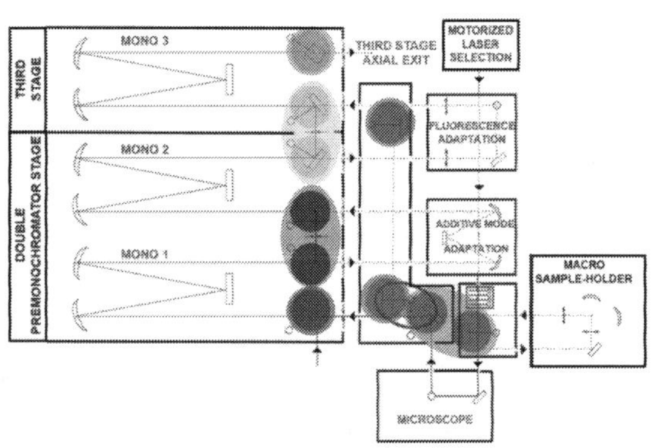

Fig. 1. Scheme of laser beam path in a Horiba Jobin-Yvon T64000 Raman spectrometer from source to detectors. Green section = micro/macro setup; violet section = triple/single configuration; brown section = lateral/axial entrance; blue section = additive/subtractive mode; light blue section = double monochromator-third stage entrance; orange section = axial/vertical exit.

Both Raman and IR spectra are affected by random noise and to reduce the spectral noise a wavelet analysis was performed using thresholding estimators by means of Matlab (R2016a, The MathWorks, Inc. Natick, MA, USA) [77-82]. A Single-axis Acoustic Levitator SAL^{TM}, by Materials Development, Inc (MDI), constituted by a levitation device, with two transducers placed on a vertical axis, generating two acoustic waves of 22 KHz has been combined with two spectrometers for collecting spectra in a wide wave-vector range.

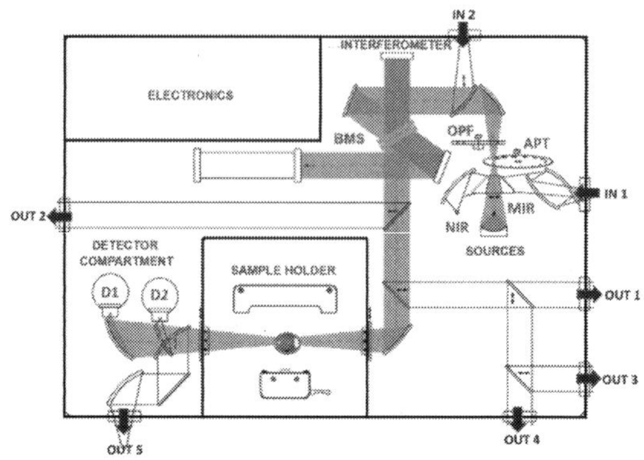

Fig. 2. Scheme of optical beam path in the Bruker Vertex 70V IR spectrometer from source to detectors. NIR, MIR: Beamsplitters; APT: Aperture wheel; OPF: Optical filter wheel; BMS: Interferometer beamsplitter; D1, D2: detectors; IN: IR beam inlet ports and OUT: IR beam outlet ports.

More precisely, Raman spectra were collected by the Horiba Jobin Yvon T64000 spectrometer, which works with a He-Ne laser with a mean source power of 60mW at a wavelength of 632.8 nm. As far as the IR spectra are concerned, a Vertex 70v by Bruker spectrometer, with a resolution of 4 cm^{-1} using an average of 128 scans and a spectral range of 400 ÷ 4000 cm^{-1} was employed. Fig. 1 and Fig. 2 show the laser beam and the path in the Raman and IR spectrometers respectively.

Fig. 3 reports the acoustic levitation system. It has dimensions of 30 x 45 x 60 cm, a total weight of 30 Kg, works at sound pressure levels lower than 160 dB and has a transducers distance of 15 cm.

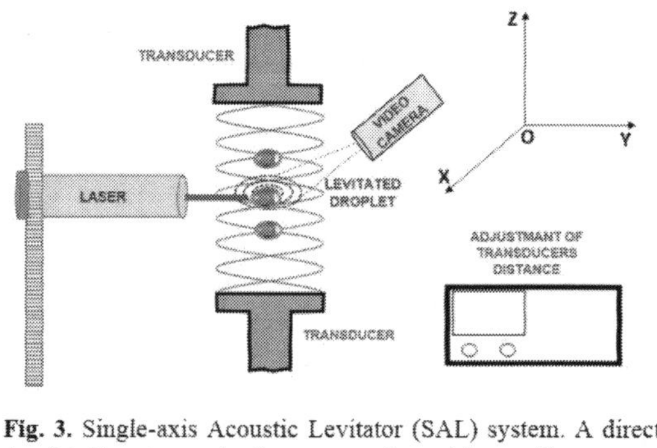

Fig. 3. Single-axis Acoustic Levitator (SAL) system. A direct droplet diameter evaluation was performed by a video camera and a laser pointer; the diameter was also evaluated by an indirect measure through the OH/CH band area ratio analysis of collected Raman and Infrared spectra.

3 Results and discussion

For testing the D^2 versus time law, a direct measure was performed by means of a video camera and a laser pointer provided with a micrometric screw for an exposition time of 1 μs. In addition, an indirect diameter evaluation was performed by the analysis of the OH/CH band area ratio of the solution spectra collected as a function of time. The D^2-law predicts that the square of the droplet diameter decreases linearly with time, in the initial stage. Few important assumptions must be considered for the droplet in suspension: i) heat and mass transfer between liquid and gas phase are diffusion controlled; ii) droplet spherical symmetry; iii) single phase component; iv) no thermal radiation effects take place; v) constant and uniform temperature. For the mass flux of vapour leaving the droplet surface, $\dot{m}_v$:

$$\dot{m}_v = \pi D \rho_v D_g Sh B_M \quad (1)$$

where D is the droplet diameter, ρ_v is the vapor density, D_g is the diffusion coefficient of the gas, Sh is the Sherwood number for mass transfer, and B_M represents the mass transfer Spalding number [83-84]. For the rate change of liquid droplet mass it is:

$$\dot{m}_d = -\rho_d \frac{dV}{dt} \quad (2)$$

ρ_d is the density of the droplet, and $\frac{dV}{dt}$ is the rate of the droplet volume change. Considering that $\dot{m}_v = -\dot{m}_d$, it is possible to obtain the equation for the diameter:

$$\frac{dD^2}{dt} = -4\frac{\rho_g}{\rho_d} D_g Sh \ln(1 + B_M) \quad (3)$$

where ρ_g is the density of the gas; for diffusion controlled evaporation the Sherwood number Sh takes the value of 2:

$$\frac{dD^2}{dt} = -8\frac{\rho_g}{\rho_d} D_g \ln(1 + B_M) \quad (4)$$

the integration of this equation at $t = 0$ gives the D^2-law that describes the time evolution of droplet surface of pure liquid:

$$D^2 = D_0^2 - \beta t \quad (5)$$

where β is the evaporation rate coefficient:

$$\beta = 8\frac{\rho_g}{\rho_d} D_g \ln(1 + B_M) \quad (6)$$

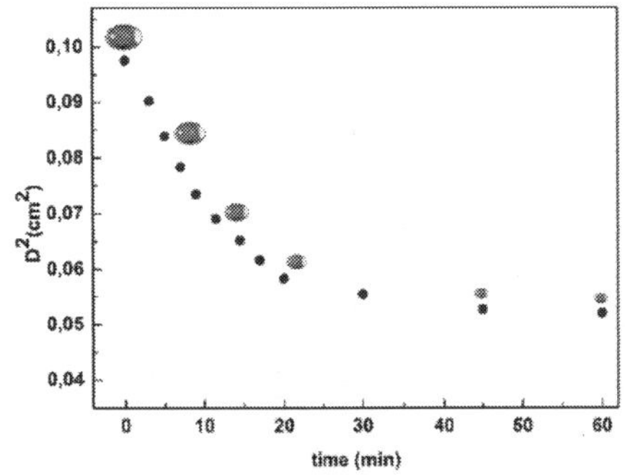

Fig. 4. Measured D^2-Law as a function of time. Droplets: diameter values as evaluated by means of the video camera. Blue dots: diameter values as evaluated by the analysis of the OH/CH band area ratio of the solution spectra. As it can be seen D^2 versus t decreases linearly in in the initial stage.

Fig. 4 reports the D^2 data, droplets as a function of time, obtained by a direct measure with a video camera. As it can be seen, D^2 vs t follows a linear trend for about 20 minutes, reaching a plateau at longer time.

This result shows how trehalose is able to avoid total water evaporation since a percentage of water is strongly bonded with the disaccharide. Such a property plays a key role in bioprotection.

Another analysis that allows to extract information on the drying process of aqueous solution of trehalose is to consider the intramolecular region of spectra. In particular, due to the presence of an isosbestic point in the IR spectra and of an isoskedastic point in the Raman spectra, it is possible to decompose the spectra into two contributes: an "open" contribution, centered around at 3210 cm^{-1} and a "closed" contribution, centered around at 3420 cm^{-1}.

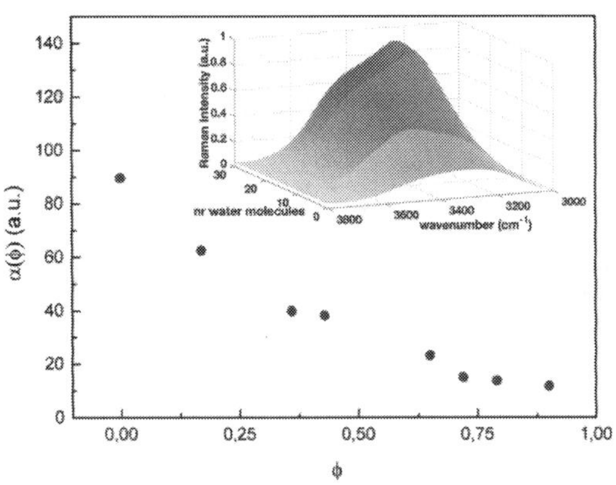

Fig. 5. Behavior of the Raman open area contribution as a function of trehalose weight fraction, i.e. ϕ = (grams of trehalose)/(grams of trehalose + grams of H2O), is reported. In the insert, 3D levitated Raman spectra of trehalose aqueous solutions as a function of number of water molecules and wavenumber are reported.

Fig. 5 shows the behavior of the total area, sum of the two contributions, as a function of concentration of trehalose, where ϕ is the trehalose weight fraction, i.e. ϕ = (grams of trehalose)/(grams of trehalose + grams of H_2O). In the insert, the 3D levitated Raman spectra of trehalose aqueous solutions as a function of number of water molecules and wavenumber at T=25°C are reported.

Fig. 6 reports the behavior of the open area contribution as a function of time is reported. In the insert, the 3D levitated IR spectra of trehalose aqueous solutions as a function time and wavenumber at T=25°C are reported.

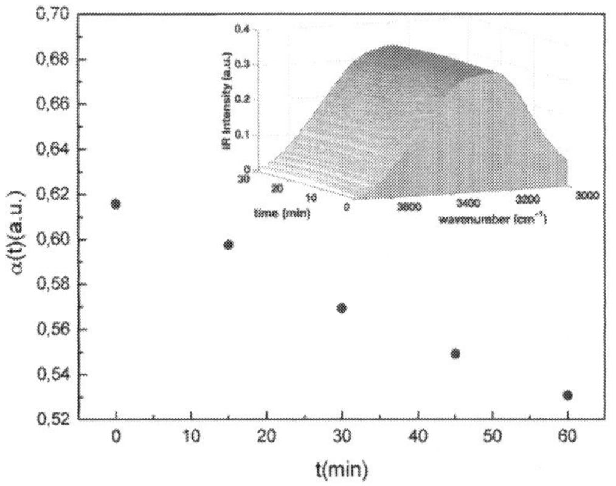

Fig. 6. Behaviour of the open area contribution of levitated IR spectra as a function of time. In the insert, the 3D levitated IR spectra of trehalose aqueous solutions as a function time and wavenumber at T=20°C are reported.

The mathematical procedure to evaluate concentration is a mapping of the OH/CH band area ratio as a function of different solution concentration values. In particular a OH/CH band area ratio for the acoustically levitated droplets for different levitation time was evaluated starting from the measured IR spectra; then, by using density versus concentration data, it was possible to determine, in an independent way, the droplet diameter evolution as a function of time. In such a way it is possible to test the D^2 law of the drying process theoretical model. Fig. 4 reports the D^2 data, blue dots as a function of time. Again D^2 vs t follows a linear trend for about 20 minutes. Therefore the experimental results show that the validity of the drying theoretical model and of the adopted approaches.

Conclusions

In this work, the results of an experimental study, where laser techniques are applied to acoustically levitated droplets of trehalose aqueous solutions are reported. The study of trehalose is particularly important for a better understanding of extremophiles, organisms that can survive in harsh conditions. It is shown how acoustic levitation, combined with spectroscopic techniques, allows to explore a wide disaccharide concentration range and to test the dependence of the diameter law as a function of lag time, i.e. D^2 versus time. The aim is to perform Raman and IR spectroscopic analyses and to test the validity of the diameter law. By these analyses, it emerges that the behaviour of D^2 vs t follows a linear trend for about 20 minutes, reaching then a plateau at longer time. The result is confirmed by the analysis of the intramolecular region, showing that trehalose is able to avoid total water evaporation, this property being essential forable to survive under extreme environmental conditions thanks to the synthesis of trehalose. These extreme environments are characterized, for example, by high aridity, very hot and low temperatures, high salinity and high values of pH. Just to report some example, the *Atacama Desert* (Chile, USA), is one the most arid place on Earth, with a year average rain fall lower than 0.08 mm, where it is possible to find *Selaginella lepidophylla*; such a plant adapts itself to a condition of prolonged òt in this environment.

Another extreme environment is furnished by the *Death Valley* (USA), this region holding the record for the World highest air temperature equal tqo 56.7 ° C in July 1913; concerning the extremophiles that are present in this extreme environment, it is possible to mention *Bacillus Vallismortis* which ferments trehalose.

The obtained results put into evidence that trehalose molecules are capable to strongly bond themselves to water molecules, so avoiding water evaporation under high temperature and drought conditions.

This research was supported by University of Messina Research & Mobility 2016 Project (project code RES_AND_MOB_2016_ TORRISI).

References

1. H.T. Mohamed, V. Untereiner, G.D. Sockalingum GD, et al. Glycoconj J. **34**, 309 (2017)
2. C. Branca, S. Magazù, G. Maisano, et al. J. Phys. Condens. Matter **10**, 10141 (1998)
3. M. I. Santos, E. Gerbino, E. Tymczyszyn et al. Foods, **4**, 283 (2015)
4. C. Branca, S. Magazu, G. Maisano et al. J. Phys. Cond. Matter **11**, 3823 (1999)
5. S.Y. Lin, M. J. Li and W.T. Cheng, Spect. **21**, 1 (2007)
6. L. Hennet, I. Pozdnyakova, V. Cristiglio et al. J. Phys. Condens. Matter **19**, 455210 (2007)
7. D. L. Hennet, L., Cristiglio, V., Kozaily, et al., Eur. Phys. J. Spec. Top. **196**, 151 (2011)
8. N. Acharya, S. A. Acharya, V. N. Kanika et al. Biochem. J. **405**, 503 (2007)
9. F. Aliotta, M.E. Fontanella, S. Magazù, U. Wanderlingh, Prog. Colloid Polym. Sci. **84**,483 (1991)
10. A. Apicella, B. Cappello, M. A. Del Nobile, et al. Biomaterials **14**, 83 (1993)
11. B. Arumugam, V., Akalya, A., Rajasekaran, J. F., Balakrishnan, J. Polym. Mater. **17**, 371 (2000)

12. G. Mathiak, J. Brillo, I. Egry, et al. Microgravity Sci. Technol. **18**, 67 (2006)
13. G. Mathiak, I. Egry, L. Hennet, et al. in Int. J. Thermophys. **26**,1151 (2005),
14. A. Ashkin, J. M. Dziedzic, J. E. Bjorkholm, and S. Chu, Opt. Lett. **11**, 288 (1986)
15. W. J. K. R. Benmore, C.J., Phys. Rev. X **1**, 01104 (2011)
16. J. K. R. DeVos, J. K., Hampton, D. S., Merkley et al., Int. J. Microgravity Sci. Appl. **9**, 146 (1992)
17. J. W. E. Drewitt, L. Hennet, A. Zeidler et al., Phys. Rev. Lett. **109**, 235501 (2012)
18. C. J. Benmore, J. K. R. Weber, A. N. Tailor et al., J. Pharm. Sci. **102**, 1290 (2013)
19. P. C. Biswas, A., Weber J.K.R., Nordine, J. Mater. Res. **10**, 1823 (1995)
20. M. Li, W., Kim, Y., Lia, J., Lee, Soft Matter **10**, 5231 (2014)
21. B. Jonsson, B. Lindman, and K. Holmberg, IEEE Electr. Insul. Mag. **14**, 42 (1998)
22. F. Kozaily, J., Hennet, L., Fischer et al. Phys. Status Solidi Curr. Top. Solid State Phys. **8**, 3155 (2011)
23. S. Cerveny, Á. Alegría, and J. Colmenero, Phys. Rev. E **77**, 031803 (2008)
24. S. Magazù, F. Migliardo, F. Affouard et al. J. Chem. Phys. **132**, 184512 (2010)
25. S. Magazù, F. Migliardo, and A. Benedetto, Rev. Sci. Instrum. **82**, 184512 (2011)
26. S. L. Bradbury, W. B. Jakoby, Proc. Natl. Acad. Sci. U. S. A. **69**, 2373 (1972)
27. P.-O. I. Cuello, G. J., Cristiglio, V., Hennet, L., in J. Phys. Conf. Ser. **549**, 012002 (2014)
28. I. Egry, a. Diefenbach, W. Dreier, and J. Piller, Int. J. Thermophys. **22**, 569 (2001)
29. A. Braun, P. C. Stenger, H. E. Warriner, J et al. Biophys. J. **93**, 123 (2007)
30. S. Magazù, E. Calabrò, M. T. Caccamo, and A. Cannuli, Curr. Chem. Biol. **10**, (2016)
31. R. A. Krishnan, S., Nordine, P. C., Weber, J. K. R., Schiffman, High Temp. Sci. **31**, 45 (1991)
32. S. L. Bradbury and W. B. Jakoby, Proc. Natl. Acad. Sci. U. S. A. **69**, 2373 (1972)
33. M. Li, W., Kim, Y., Lia, J., Lee, Soft Matter **10**, 5231 (2014)
34. S. Magazù, Phys. B **226**, 92 (1996)
35. S. Magazù, J. Mol. Struct. **523**, 47 (2000).
36. K. Venkatramanan, Int. Rev. Chem. Eng. **3**, 308 (2011)
37. L. C. P. Wang, T. G., Anilkumar, A.V., J. Fluid Mech. **308**, 1 (1996)
38. J. K. R. Weber, S. Krishnan, S. Ansell, A. A. . Hixson, and P. P. C. Nordine, Phys. Rev. Lett. **84**, 3622 (2000).
39. R. J. K. Weber, C. J. Benmore, S. K. Tumber et al. Eur. Biophys. J. **41**, 397 (2012)
40. J. K. R. Weber, C. A. Rey, J. Neuefeind, and C. J. Benmore, Rev. Sci. Instrum. **80**, (2009)
41. R. J. K. Weber, C. J. Benmore, S. K. Tumber et al. Eur. Biophys. J. **41**, 397 (2012)
42. M. T. Caccamo, A. Cannuli, E. Calabrò, and S. Magazù, IOP Conf. Ser.: Mater. Sci. Eng. **199** 012119 (2017)
43. S. Magazù, F. Migliardo, and M. T. Caccamo, Adv. Mater. Sci. Eng. **2013**, (2013)
44. F. Migliardo, M. T. Caccamo, and S. Magazù, J. Non. Cryst. Solids **378**, 144 (2013)
45. F. Migliardo, S. Magazù, and M. T. Caccamo, J. Mol. Struct. **1048**, 261 (2013)
46. F. Migliardo, S. Magazù, M.T. Caccamo, Food Biophys. **9**, 99 (2014)
47. S. Magazù, F. Migliardo, M. T. Caccamo, 2012, J. Phys. Chem. B, **116**, 9417 (2012).
48. W. K. Rhim, S. K. Chung, D. Barber et al. Rev. Sci. Instrum. **64**, 2961 (1993).
49. B. Cappello, M. A. Del Nobile, M. I. La Rotonda et al. Farmaco **49**, 809 (1994).
50. T. K. Rostovtseva, E. M. Nestorovich, S. M. Bezrukov, Biophys. J. **82**, 160 (2002).
51. O. Smith, G.D., Bedrov, D. Borodin, J. Am. Chem. Soc. 122, 9548 (2000).
52. J. J. Towey, K. Soper, L. Dougan, Phys. Chem. Chem. Phys. **13**, 9397 (2011a).
53. S. Condello, E. Calabrò, D. Caccamo et al. J., Amino Acids, 42: 775(2012)
54. Magazù, S., Migliardo, F., Benedetto, et al. in Dynamics of Biological Macromolecules by Neutron Scattering, 22 (2011)
55. Branca, C., Maccarrone, S., Magazù, et al., J. Chem. Phys. **122**, 174513 (2005)
56. Branca, C., Magazù, S., Migliardo et al., Physica A: Statistical Mechanics and its Applications, **304**, 314 (2002)
57. B. Varga, F. Migliardo, E. Takacs et al. Chem. Phys. **345**, 250 (2008).
58. R. Chelli, P. Procacci, G. Cardini et al. Phys. Chem. Chem. Phys. 1, 879 (1999)
59. J. Chen, S. K. Spear, J. G. Huddleston et al. Green Chem. **7**, 64 (2005)
60. P. J. Flory, Cornell Univ. Press. Ithaca, New York **399** (1953).
61. A. K. Gangopadhyay, G. W. Lee, K. F. Kelton et al., Rev. Sci. Instrum. **76**, (2005).
62. P. Gonzalez-Tello, F. Camacho, G. Blazquez, J. Chem. Eng. Data **39**, 611 (1994).
63. R. Grandori, I. Matecko, P. Mayr, N. Muller, J. Mass Spectrom. **36**, 918 (2001).
64. C. Granier, J. Potard, in 6th Eur. Symp. Microgravity. Bordeaux, Fr. ESA **SP-256** 421 (1987)
65. S. A. Hu, T., Manjula, B. N.; Li et al. Biochem. J. **402**, 143 (2007)
66. G. Jacobs, I. Egry, K. Maier et al. Rev. Sci. Instrum. **67**, 3683 (1996)
67. B. Jonsson, B. Lindman, and K. Holmberg, IEEE Electr. Insul. Mag. **14**, 42 (1998)
68. F. Kozaily, J., Hennet, L., Fischer, H.E. et al. Phys. Status Solidi Curr. Top. Solid State Phys. **8**, 3155 (2011)
69. Marchese, N., Cannuli, A., Caccamo, M.T. et al. Biochim. Biophys. Acta - General Subjects, **1861**, 3661(2017

70. S. Magazù, F. Migliardo, A. Benedetto, et al. J. Non. Cryst. Solids, **357**, 664 (2011)
71. D. Barreca, G. Laganà, S. Ficarra et al. Biophys. Chem. **147**, 146 (2010)
72. S. E. Pagnotta, M. A. Ricci, F. Bruni, S. McLain, and S. Magazù, Chem. Phys. **345**, 159 (2008)
73. P. Ballone, M. Marchi, C. Branca et al., J. Phys. Chem. B **104**, 6313 (2000)
74. S. Magazù, G. Maisano, H. D. Middendorf et al. J. Phys. Chem. B **102**, 2060 (1998).
75. S. Magazù, G. Maisano, P. Migliardo et al. J. Chem. Phys. **109**, 1170 (1998)
76. M. T. Caccamo, G. Castorina, F. Colombo, V., Insinga, E. Maiorana, S. Magazù, Atmosph. Res. **198**, 22 (2017)
77. M. T. Caccamo, S. Magazù, Spectrosc. Lett. **50**, 130 (2017).
78. R. Patil, Procedia Comp. Sc. **54**, 849 (2015)
79. M. Srivastava, C. Lindsay Anderson, Jack H. Freed IEEE Access. **4**, 3862 (2016)
80. M. T. Caccamo, S. Magazù, Vib. Spectrosc. **85**, 222 (2016).
81. M.T. Caccamo, S. Magazù, Appl. Spectrosc. **71**, 401 (2017).
82. M.T. Caccamo, S. Magazù, Polym. Testing, **62**, 311(2017)
83. A. N. Hayhurst, M. S. Parmar, Combustion and Flame **130**, 361 (2002)
84. Peiyong Ni, Appl. Mathem. Modell. **34**, 2370 (2010)

Physical investigations on the radiation damage of graphene oxide by IR pulsed laser

Letteria Silipigni[1*], *Lorenzo* Torrisi[1,2], *Mariapompea* Cutroneo[2,3]

[1]Dipartimento di Scienze Matematiche e Informatiche, Scienze Fisiche e Scienze della Terra, MIFT, Università di Messina, V.le F.Stagno d'Alcontres 31, 98166 S. Agata, Messina Italy
[2]Dottorato di Ricerca in Fisica, MIFT, Università di Messina, Italy
[3]Nuclear Physics Institute, CAS, 25068 Rez, Czech Republic

Abstract. Graphene oxide foils were irradiated by Nd:YAG laser at the moderated intensities of the order of 10^8 W/cm^2 in vacuum. Measurements of atomic emission during the laser irradiation were performed with ion collectors and mass spectrometry, demonstrating a strong emission of carbon, oxygen and hydrogen atoms. Further investigations of the irradiated graphene oxide foils were carried out on pristine and laser irradiated samples by using Rutherford backscattering spectrometry and X-ray photoelectron spectroscopy. Results indicate that graphene oxide losses oxygen and hydrogen during the irradiation, changing its carbon content and its chemical and physical properties.

1 Introduction

Carbon has different crystallographic structures (allotropes), of which the most common ones are graphite and diamond.

Graphite has a three-dimensional layered structure in which tetravalent atoms of carbon are connected between them by three covalent bonds forming a hexagonal network structure of which any layer is called *graphene*. Different graphene layers are parallel between them. The bonds between layers are weak, thus the layers can slide easily over each other. The distance between layers is 0.335 nm. Even if *graphene* only refers to an individual layer of graphite's planar honeycomb carbon lattice, the term *graphene* has also been adopted for up to 10 layers of carbon sheets by convention [1]. Graphene has peculiar physical and chemical properties, including excellent optical and mechanical properties, such as unprecedented highly versatile Young's modulus and tensile strength, high transmission over a wide wavelength range, high electrical and thermal conductivity, high biocompatibility, high surface effects with respect to volume ones [2].

Due to these properties the applications of *graphene* and its derivatives are wide range of; they can be employed as sensors, membranes, composite materials, biomaterials, in microelectronics, catalysis, etc.

Graphene oxide (GO) is defined as a monolayer of *graphene* with various oxygen-containing functional groups mainly in the form of hydroxyl (C-OH) and epoxy (C-O-C) groups attached on the basal plane with a very small contribution of carbonyl (C=O) and carboxyl (O-C=O) groups which predominantly decorate the edges of the graphene sheets [3]. Therefore *graphene oxide* (GO) is a compound of carbon, oxygen, and hydrogen in variable ratios. It is generally obtained by treating graphite with strong oxidizers. *Graphene oxide* layers generally are about 1.1 nm thick; they are hydrophilic and easily hydrated when exposed to water vapor or immersed in liquid water. *Graphene oxide* decomposes when it is rapidly heated at moderately high temperatures (~300 °C) or chemically treated with specific solvents, with formation of finely dispersed amorphous carbon and radicals [4]. The structure can be intercalated with many different atomic and molecular species conferring different properties to the GO layers. GO thin films, of the order of 100 nm in thickness, can be prepared to be near impermeable to gases, liquids or strong chemical agents.

Fig. 1 shows a typical graphene layer (a) and a graphene oxide layer (b) containing different functional groups.

a)
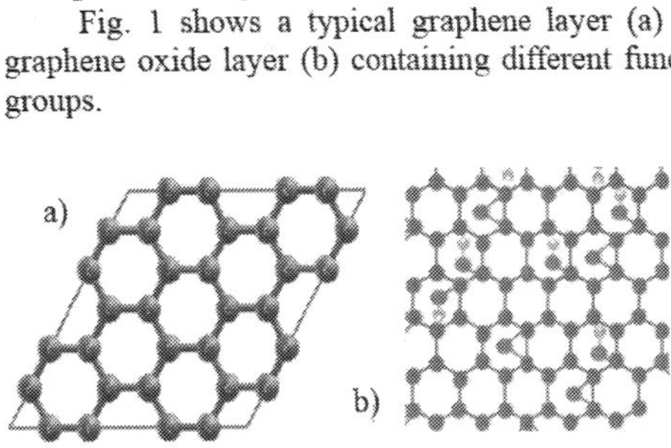
b)

Fig. 1. Typical graphene layer (a) and graphene oxide layer containing different functional groups (b).

* Corresponding author: lsilipigni@unime.it

The electrical conductivity of GO is strongly influenced by such functional groups and its band gap could be tuned from an insulator to a semimetal by the selective reduction of GO, where the term *reduction* is defined as *a loss of oxygen*. Generally GO has a low sheet resistance (~100-1000 Ω per square), and a high optical transmittance (~80% at 550 nm) [2].

Often structural defects are maintained low and their concentration depends on the synthesis process that is used to grow GO thin films [5].

In this paper the damage of GO due to fast and energetic laser pulses, of 3 ns and up to 500 mJ/cm^2, was investigated.

2 Materials and Methods

Graphene oxide foils were prepared by the graphite oxidation by using the Hummer's method as described in detail in the literature [6]. Thin GO foils, 7 µm in thickness and 4 cm^2 in surface, were employed in such investigation. Their synthesis was obtained at CANAM laboratory in Rez (Czech Republic).

An unfocalized Nd:YAG laser, at the 1064 nm wavelength, 3 ns pulse duration, 500 mJ maximum pulse energy, 1 cm^2 spot size, was employed in the single pulse mode, up to an intensity of about 1.7x10^8 W/cm^2. This laser was used to irradiate GO foils in a vacuum chamber by using different laser pulse energies. 50 mJ, 75 mJ, 100 mJ, and single and double laser pulses, up to 500 mJ onto a surface of 1 cm^2 were employed. The laser irradiation was monitored in-situ by using a Joule-meter for the laser pulse energy control. An ion collector (IC), connected in the time-of-flight (TOF) configuration to a fast oscilloscope for the measurements of the carbon and oxygen ion emission and a mass quadrupole spectrometer (MQS) for the gas degassing diagnostics were employed. A scheme of the used experimental set up is reported in Fig. 2.

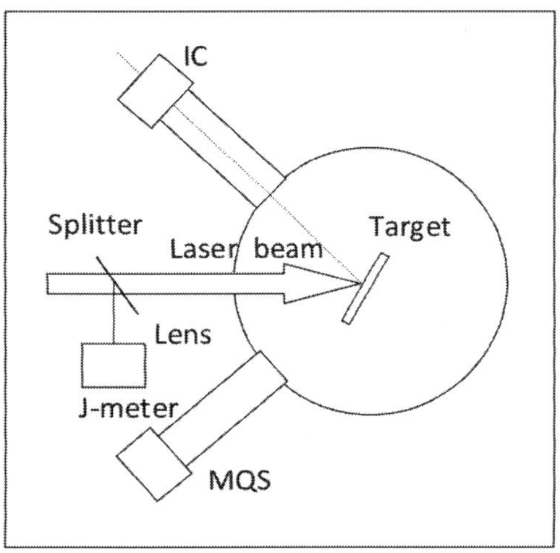

Fig. 2. Experimental set-up.

IC gives TOF spectra from which it is possible to evaluate the ion emission velocities and, as a first approximation, the charge states of the emitted ion species [7].

MQS provides spectra of the atomic and molecular species produced by the laser ablation of the sample surface; they were recorded from 1 up to 200 amu.

Rutherford backscattering spectrometry (RBS) analysis was performed by using a 2.0 MeV helium beam in order to measure the composition and the stoichiometry of the sample before and after the laser irradiation [8].

X-ray photoemission spectroscopy (XPS) was employed to analyze the structure of the graphene oxide foils before and after the laser irradiation. It is a surface-sensitive analytical technique that is useful to determine the chemical environment of atoms, in this case GO carbon atoms. XPS analysis was carried out on a VG Scientific system operating, under 3x10^{-8} mbar, to an anodic voltage of 10 kV in the constant pass-energy mode at 20 eV with an energy resolution [full width at half maximum (FWHM)] of 1.0 eV as detected by the Ag 3d$_{5/2}$ emission. The spectrometer used a non-monochromatic Al Kα X-ray source ($h\nu$ = 1486.6 eV) operating at 320 W. The channeltron voltage was fixed at 2.5 kV during each run. Owing to sample charging, the carbon C 1*s* line at 285.0 eV was used as reference for the determination of the binding energy position of all the investigated core levels. A Shirley background function was utilized to subtract the inelastic background at all the XPS spectra.

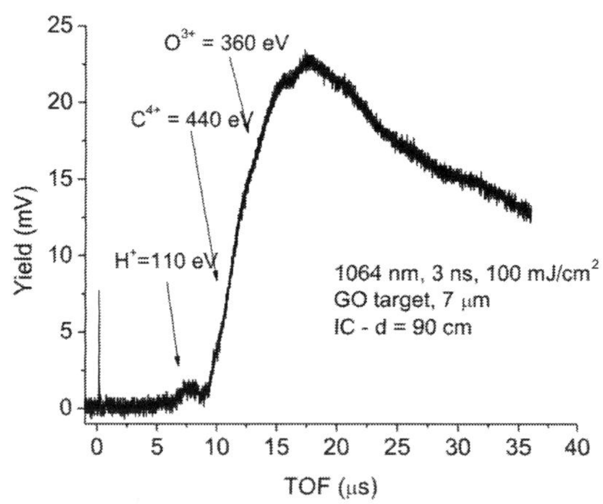

Fig. 3. IC-TOF spectrum of ions emitted during the GO laser irradiation.

3 Results

The laser irradiation of GO generates a plasma in front of the target emitting visible light, UV, soft X-rays, electrons and ions, as reported in previous investigations [9]. The different electron and ion mobility produces a non equilibrium plasma accelerating ions along the normal to the target surface.

Fig. 3 shows a typical IC-TOF spectrum recorded during a laser shot irradiation of GO placed at 90 cm distance from the IC detector. The spectrum shows the emission of protons, carbon and oxygen ions. The protons are detected at a kinetic energy of 110 eV: on the basis of the Coulomb Boltzmann Shifted (CBS) distribution, it means that the acceleration is of 110 eV per charge state. The carbon and oxygen ions are detected at a maximum kinetic energy of 440 eV and 360 eV, respectively. Thus, measurements indicate a maximum charge state of 4+ and 3+ for the carbon and oxygen ions, respectively. Such charge states indicate that plasma electrons have energy up to the potential ionization energy to obtain C^{4+} and O^{3+}, which correspond to 64 eV and 55 eV, respectively. Because the ionization potential of O^{4+} is 77 eV, the maximum electron energy should be lower than this value. Assuming the maximum electron energy to be about 65 eV and the mean energy of the Boltzmann distribution to be a third of this maximum value, the electron mean energy of the plasma is of about 21 eV. Thus the mean electron plasma temperature ($kT = 2E/3$) corresponds to about 14 eV.

During the plasma formation the MQS spectrometer detects the emitted atoms and molecular species. Our analysis was performed by fixing some masses of interest, such as 1, 12 and 16 amu, corresponding to the atomic presence of H, C and O, respectively.

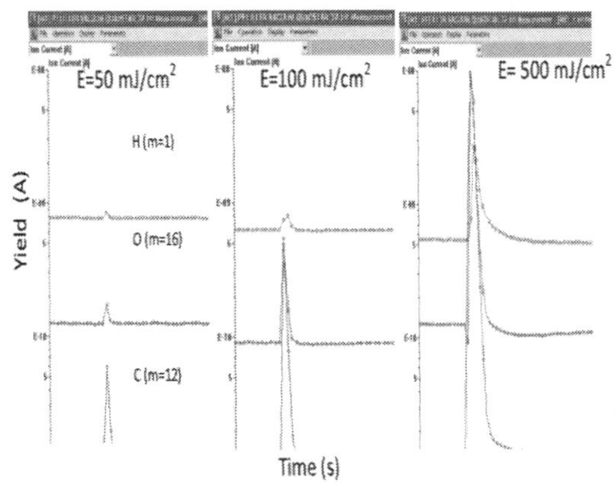

Fig. 4. Typical MQS spectra of emitted C, O and H as a function of the time and of the laser pulse energy density.

Fig. 4 reports three typical MQS spectra of the emitted species (H, O and C) as a function of different laser pulse energy density, from 50 mJ/cm^2 up to 500 mJ/cm^2. Such fixed masses were plotted as a function of the time in condition of laser switched off and laser switched on, in order to evince their net yield during the laser irradiation. Spectra indicate that the GO emits carbon, oxygen and hydrogen, and other molecular groups containing oxygen, proportionally to the laser released energy. Results show that the laser ablation produces a strong emission of hydrogen, carbon and oxygen, reducing the content of oxygen and increasing the residual carbon content of the sample.

MQS yields are proportional to the GO composition, which was controlled by using a 2.0 MeV helium RBS analysis. These spectra have evinced a composition in which oxygen decreases significantly with the laser irradiation, proportionally to the absorbed energy. Fig. 5 reports two typical RBS spectra relative to the analysis of the pristine graphene oxide (a) and the laser reduced graphene oxide (b).

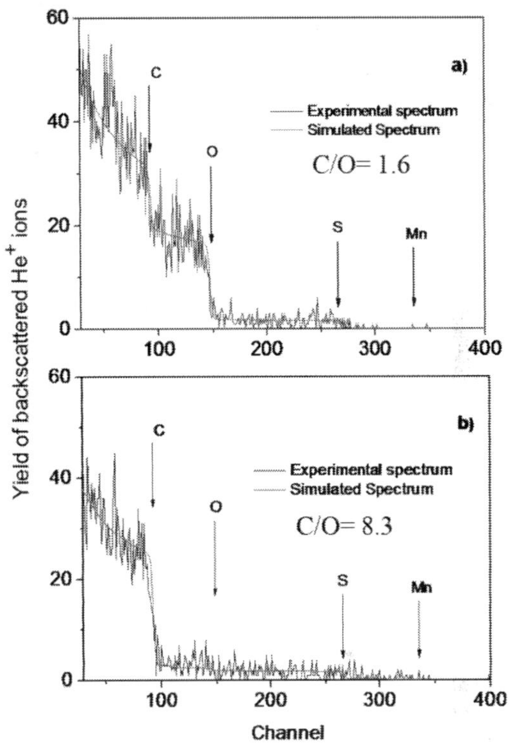

Fig. 5. RBS spectra relative to the pristine GO (a) and to the laser irradiated GO (b).

The first spectrum is rich in oxygen, while in the second one there is a strong oxygen reduction. The continuum lines represent the simulated spectra by using SIMNRA code [10]. Moreover, S and Mn elements are present as contaminants in the pristine sample. The quantitative analysis has demonstrated that the carbon to oxygen atomic ratio, C/O, is 1.6 in the pristine GO sample and it increases to 8.3 in the laser irradiated GO sample by using a laser pulse energy density of about 400 mJ/cm^2.

XPS analyses were performed in the pristine and laser irradiated GO foils, in order to investigate about the chemical bonds of carbon with the different functional groups present in the graphene oxide structure as a function of the laser energy density irradiation. Fig. 6 shows the comparison between the XPS spectra, performed in the C1s and O1s core level regions for the pristine (*a*) and laser irradiated GO (*b, c*) foils in order to evince the carbon and the oxygen content and their chemical groups, respectively. The laser treated GO foils were irradiated at 50 mJ/cm^2 (*b*) and at 200 mJ/cm^2 (*c*) energy densities respectively.

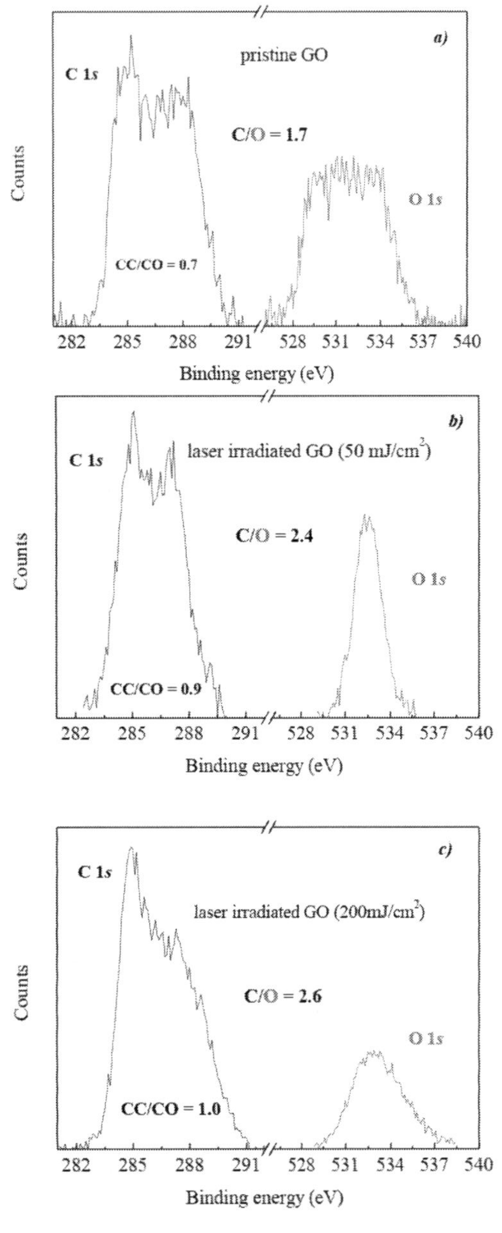

Fig. 6. Comparison of the C1s and O1s core level XPS spectra revealed for the pristine GO (*a*) and laser irradiated GO foils at 50 mJ/cm^2 (*b*) and 200 mJ/cm^2 (*c*).

The untreated GO foil high-resolution C1s XPS spectrum (see fig.6*a*) consists of a peak at about 285.0 eV, followed by a second large structure at about 288.0 eV that can be fit to peaks at about 286.8 eV and at 288.4 eV. These features can be attributed to C–C and C=C bonds (which cannot be detected separately due to the XPS spectrometer resolution), C–O (in epoxy and hydroxyl groups) and C=O (in carbonyl groups) links, respectively [11].

In the following CC refers to the sum of the C–C and C=C bonds, while CO designates all combinations of links between C and O atoms.

When GO is laser irradiated the ratio of the different peaks changes. In particular at 50 mJ/cm^2 (see Fig. 6*b*) the 285.0 eV peak, due to the C-C and C=C links, and the C-O peak at about 287.1 eV increase as well the C=O peak at about 289.2 eV decreases; at 200 mJ (see Fig. 6*c*) the 285.0 peak continues to grow while the C-O and C=O peaks at about 287.1 eV and 288.8 eV respectively decrease. Consequently, in going from the pristine GO to laser irradiated GO (200 mJ/cm^2), the ratio between the areas of the CC and CO peaks increases from 0.7 to 1.0, indicating a partial removal of oxygen-containing functional groups and, therefore, a partial reduction of GO in good agreement with RBS and MQS data.

As regards the O1s XPS spectra, initially the O1s spectrum indicates the presence of two large peaks at about 530.0 eV and 533.3 eV (see Fig. 6*a*), due to C=O and C-OH groups respectively, according to the literature [12]. After the 50 mJ laser irradiation, the intensity of the peak at lower binding energies decreases (see Fig. 6*b*) up to disappear at 200 mJ (see Fig. 6*c*), while at this laser energy density the presence of the only peak at about 533.1 eV indicates that C-OH groups are still present and are difficult to further reduce in good agreement with the literature [12]. The atomic ratio of carbon to oxygen (the area of the C1s peak corrected for its tabulated atomic sensitivity factor divided by the area of the O1s peak corrected for its tabulated atomic sensitivity factor) was calculated after each treatment and the obtained results are shown in Table I.

Table I. The values of the C1s/O1s atomic concentration ratio obtained by high resolution XPS spectra.

	C1*s*/O1*s* atomic ratio
Untreated GO foil	1.7
50 mJ laser irradiated GO foil	2.4
200 mJ laser irradiated GO foil	2.6

The XPS spectra deduced C/O atomic ratio value, reported in Table I for the pristine GO foils is in agreement with that derived by the RBS spectra. More detailed investigations will be carried out shortly to shed light on the matter.

4 Discussion and conclusions

As reported in the literature, a key topic in the research and applications of GO is its reduction, which partly restores the structure and properties of graphene.

In this work, we have demonstrated that the pulsed laser beam can be used to modify the chemical and physical properties of GO by reducing its content in oxygen and increasing the one in carbon.

A strong emission of carbon, oxygen, hydrogen atoms is produced by the laser irradiation as indicated by the IC-TOF and MQS spectra.

By means of the RBS and XPS techniques, we have observed that the laser treatment at different laser pulse energy densities generates GO with defined C/O atomic ratios and specific oxygen functionalities.

In particular, the analysis of the XPS spectra of the C1s and O1s core levels indicate that a loss of some oxygen functional groups occurs in GO upon laser irradiation. This loss increases as laser density energy grows and it has been confirmed by the RBS data.

Reported measurements are in agreement with the literature describing similar studies performed by using ion beam lithography of GO transforming the insulator pristine GO in the final conductive rich carbon graphene [13].

Our results indicate that pulsed laser irradiation is a good tool to reduce GO and it has important advantages over conventional chemical or thermal treatments since it is rapid and it can be applied directly to the GO film deposited, not requiring dangerous or toxic reagents.

However, the oxidation/reduction control by means of the laser irradiation must be further investigated in details to better understand how to tune the band gap in GO by varying the oxygen concentration as a function of the laser parameters (wavelength, pulse energy, pulse duration,…). For this purpose a more detailed XPS analysis will be undertaken together with an accurate Raman investigation. A deeper understanding of the oxidation and reduction processes will allow us to better control the attaching and elimination of some functional groups to some specific locations on the carbon plane in order to realize innovative devices in various fields ranging from electronics, energetics to optoelectronics.

Authors thank the CANAM (Center of Accelerators and Nuclear Analytical Methods) infrastructure (LM2015056). This work was supported by the "Research and Mobility" project of Messina University No. 74893496, scientifically coordinated by Professor L. Torrisi.

References

1. M. Aliofkhazraei , N. Ali, W.I. Milne, C.S. Ozkan, S. Mitura, J.L. Gervasoni (eds.), *Graphene science handbook-Mechanical and chemical properties*, CRC Press, T&F Group, Boca Raton (2016).
2. M. Aliofkhazraei , N. Ali, W.I. Milne, C.S. Ozkan, S. Mitura, J.L. Gervasoni (eds.), *Graphene science handbook- Electrical and Optical properties*, CRC Press, T&F Group, Boca Raton (2016).
3. D.R. Dreyer, S. Park, C.W. Bielawski and R.S. Ruoff, *Chem.Soc.Rev.* **39**, 228–240 (2010).
4. Z. Wang, D. Xu, Y. Huang, Z. Wu, L. Wang and X. Zhang, *Chem. Commun.* **48**, 976–978 (2012)
5. C. Gomez-Navarro, J.C. Meyer, R.S. Sundaram, A. Chuvilin, S. Kurasch, M. Burghard, K. Kern, U. Kaiser,*Nano Lett.* **10**, 1144–1148 (2010).
6. J. Chen, B. Yao, C. Li, G. Shi, *Carbon* **64**, 225–229 (2013)
7. L. Torrisi, G. Ceccio, N. Restuccia, E. Messina, P. G. Gucciardi, M. Cutroneo, *Laser and Particle Beams* **35**(2), 294-303 (2017)
8. W.K. Chu, J.W. Mayer and M.A. Nicolet (eds.), *Backscattering Spectrometry*, Academic Press, New York, (1978)
9. L Torrisi, *Nukleonika*, **60** (2), 207-212 (2015)
10. SIMNRA Code, actual website 2017: http://home.mpcdf.mpg.de/~mam/
11. G. Sobon, J. Sotor, J. Jagiello, R. Kozinski, M. Zdrojek, M. Holdynski, P. Paletko, J. Boguslawski, L. Lipinska, and K.M. Abramski, *Opt. Express* **20**, 19463- 19473 (2012).
12. D. Yang, A. Velamakanni, G.Bozoklu, S. Park, M. Stoller, R.D. Piner, S.Stankovich, I. Jung, D.A.Field, C.A.Ventrice Jr, R.S.Ruoff, *Carbon* **47**,145-152 (2009)
13. D. E. Lobo, J. Fu, T. Gengenbach, and M. Majumder, *Langmuir* **28**, 14815–14821 (2012)p

AUTHOR INDEX

AUTHOR INDEX